T0331651

Smart Systems Design, Applications, and Challenges

João M.F. Rodrigues
Universidade do Algarve, Portugal & LARSyS, Institute for Systems and Robotics, Lisbon, Portugal

Pedro J.S. Cardoso
Universidade do Algarve, Portugal & LARSyS, Institute for Systems and Robotics, Lisbon, Portugal

Jânio Monteiro
Universidade do Algarve, Portugal & INESC-ID, Lisbon, Portugal

Célia M.Q. Ramos
Universidade do Algarve, Portugal & CIEO, Portugal

A volume in the Advances in Computational Intelligence and Robotics (ACIR) Book Series

Published in the United States of America by
IGI Global
Engineering Science Reference (an imprint of IGI Global)
701 E. Chocolate Avenue
Hershey PA, USA 17033
Tel: 717-533-8845
Fax: 717-533-8661
E-mail: cust@igi-global.com
Web site: http://www.igi-global.com

Library of Congress Cataloging-in-Publication Data

Names: Rodrigues, João M. F., editor. | Cardoso, Pedro J. S., editor. | Monteiro, Janio, editor. | Ramos, Célia M. Q., editor.
Title: Smart systems design, applications, and challenges / João M.F. Rodrigues, Pedro J.S. Cardoso, Jânio Monteiro and Célia M.Q. Ramos, editors.
Description: Hershey, PA : Engineering Science Reference, 2020. | Includes bibliographical references and index. | Summary: "This book explores new trends and applications in smart systems"-- Provided by publisher.
Identifiers: LCCN 2019036991 (print) | LCCN 2019036992 (ebook) | ISBN 9781799821120 (hardcover) | ISBN 9781799821137 (paperback) | ISBN 9781799821144 (ebook)
Subjects: LCSH: System design. | Smart materials. | Conscious automata--Design and construction.
Classification: LCC TA168 .S556 2020 (print) | LCC TA168 (ebook) | DDC 004.2/1--dc23
LC record available at https://lccn.loc.gov/2019036991
LC ebook record available at https://lccn.loc.gov/2019036992

This book is published in the IGI Global book series Advances in Computational Intelligence and Robotics (ACIR) (ISSN: 2327-0411; eISSN: 2327-042X)

British Cataloguing in Publication Data
A Cataloguing in Publication record for this book is available from the British Library.

Advances in Computational Intelligence and Robotics (ACIR) Book Series

Ivan Giannoccaro
University of Salento, Italy

ISSN:2327-0411
EISSN:2327-042X

Mission

While intelligence is traditionally a term applied to humans and human cognition, technology has progressed in such a way to allow for the development of intelligent systems able to simulate many human traits. With this new era of simulated and artificial intelligence, much research is needed in order to continue to advance the field and also to evaluate the ethical and societal concerns of the existence of artificial life and machine learning.

The **Advances in Computational Intelligence and Robotics (ACIR) Book Series** encourages scholarly discourse on all topics pertaining to evolutionary computing, artificial life, computational intelligence, machine learning, and robotics. ACIR presents the latest research being conducted on diverse topics in intelligence technologies with the goal of advancing knowledge and applications in this rapidly evolving field.

Coverage

- Automated Reasoning
- Robotics
- Neural Networks
- Evolutionary Computing
- Pattern Recognition
- Natural Language Processing
- Artificial Intelligence
- Computer Vision
- Machine Learning
- Intelligent control

IGI Global is currently accepting manuscripts for publication within this series. To submit a proposal for a volume in this series, please contact our Acquisition Editors at Acquisitions@igi-global.com or visit: http://www.igi-global.com/publish/.

The Advances in Computational Intelligence and Robotics (ACIR) Book Series (ISSN 2327-0411) is published by IGI Global, 701 E. Chocolate Avenue, Hershey, PA 17033-1240, USA, www.igi-global.com. This series is composed of titles available for purchase individually; each title is edited to be contextually exclusive from any other title within the series. For pricing and ordering information please visit http://www.igi-global.com/book-series/advances-computational-intelligence-robotics/73674. Postmaster: Send all address changes to above address.

Titles in this Series

For a list of additional titles in this series, please visit: www.igi-global.com/book-series

Managerial Challenges and Social Impacts of Virtual and Augmented Reality
Sandra Maria Correia Loureiro (Business Research Unit (BRU-IUL), Instituto Universitário de Lisboa (ISCTE-IUL), Lisboa, Portugal)
Engineering Science Reference • copyright 2020 • 280pp • H/C (ISBN: 9781799828747) • US $195.00

Innovations, Algorithms, and Applications in Cognitive Informatics and Natural Intelligence
Kwok Tai Chui (The Open University of Hong Kong, Hong Kong) Miltiadis D. Lytras (The American College of Greece, Greece) Ryan Wen Liu (Wuhan University of Technology, China) and Mingbo Zhao (Donghua University, China)
Engineering Science Reference • copyright 2020 • 300pp • H/C (ISBN: 9781799830382) • US $235.00

Avatar-Based Control, Estimation, Communications, and Development of Neuron Multi-Functional Technology Platforms
Vardan Mkrttchian (HHH University, Australia) Ekaterina Aleshina (Penza State University, Russia) and Leyla Gamidullaeva (Penza State University, Russia)
Engineering Science Reference • copyright 2020 • 355pp • H/C (ISBN: 9781799815815) • US $245.00

Handbook of Research on Fireworks Algorithms and Swarm Intelligence
Ying Tan (Peking University, China)
Engineering Science Reference • copyright 2020 • 400pp • H/C (ISBN: 9781799816591) • US $295.00

Handbook of Research on Emerging Trends and Applications of Machine Learning
Arun Solanki (Gautam Buddha University, India) Sandeep Kumar (Amity University, Jaipur, India) and Anand Nayyar (Duy Tan University, Da Nang, Vietnam)
Engineering Science Reference • copyright 2020 • 674pp • H/C (ISBN: 9781522596431) • US $375.00

AI Techniques for Reliability Prediction for Electronic Components
Cherry Bhargava (Lovely Professional University, India)
Engineering Science Reference • copyright 2020 • 330pp • H/C (ISBN: 9781799814641) • US $225.00

Artificial Intelligence and Machine Learning Applications in Civil, Mechanical, and Industrial Engineering
Gebrail Bekdaş (Istanbul University-Cerrahpaşa, Turkey) Sinan Melih Nigdeli (Istanbul University-Cerrahpaşa, Turkey) and Melda Yücel (Istanbul University-Cerrahpaşa, Turkey)
Engineering Science Reference • copyright 2020 • 312pp • H/C (ISBN: 9781799803010) • US $225.00

701 East Chocolate Avenue, Hershey, PA 17033, USA
Tel: 717-533-8845 x100 • Fax: 717-533-8661
E-Mail: cust@igi-global.com • www.igi-global.com

Editorial Advisory Board

Table of Contents

Foreword xviii

Preface xx

Acknowledgment xxiii

Chapter 1
Augmented Intelligence: Leverage Smart Systems 1
João M.F. Rodrigues, Universidade do Algarve, Portugal & LARSyS, Institute for Systems and Robotics, Lisbon, Portugal
Pedro J.S. Cardoso, Universidade do Algarve, Portugal & LARSyS, Institute for Systems and Robotics, Lisbon, Portugal
Jânio Monteiro, Universidade do Algarve, Portugal & INESC-ID, Lisbon, Portugal
Célia M.Q. Ramos, Universidade do Algarve, Portugal & CIEO, Portugal

Chapter 2
Deep Learning on Edge: Challenges and Trends 23
Mário P. Véstias, INESC-ID, ISEL, Instituto Politécnico de Lisboa, Portugal

Chapter 3
Smart Blockchain: A Formidable Combination of Analytics and Blockchain – A Survey 43
Gutha Jaya Krishna, Institute for Development and Research in Banking Technology, India
Vadlamani Ravi, Institute for Development and Research in Banking Technology, India
Bheemidi Vikram Reddy, Institute for Development and Research in Banking Technology, India
Mohammad Zaheeruddin, Institute for Development and Research in Banking Technology, India

Chapter 4
The Significance of the Study of the Brain's Hippocampus for the Progress of Biologically-Inspired Computational Systems 63
Konstantinos Domdouzis, Sheffield Hallam University, UK

Chapter 5
Monitoring, Predicting, and Optimizing Energy Consumptions: A Goal Toward Global Sustainability 80
Pedro J. S. Cardoso, Universidade do Algarve, Portugal & LARSyS, Institute for Systems and Robotics, Lisbon, Portugal
Jânio Monteiro, Universidade do Algarve, Portugal & INESC-ID, Lisbon, Portugal
Cristiano Cabrita, Universidade do Algarve, Portugal & Centre of Intelligent Systems, IDMEC, Portugal
Jorge Semião, Universidade do Algarve, Portugal & INESC-ID, Lisbon, Portugal
Dario Medina Cruz, Universidade do Algarve, Portugal
Nelson Pinto, Universidade do Algarve, Portugal
Célia M.Q. Ramos, Universidade do Algarve, Portugal & CIEO, Portugal
Luís M. R. Oliveira, Universidade do Algarve, Portugal & CISE, Portugal
João M. F. Rodrigues, Universidade do Algarve, Portugal & LARSyS, Institute for Systems and Robotics, Lisbon, Portugal

Chapter 6
The Exploration of Autonomous Vehicles 108
Anthony J. Gephardt, Waynesburg University, USA
Elizabeth Baoying Wang, Waynesburg University, USA

Chapter 7
Average Speed of Public Transport Vehicles Based on Smartcard Data 123
Vera Costa, INESC TEC, Faculdade de Engenharia, Universidade do Porto, Portugal
José Luís Borges, INESC TEC, Faculdade de Engenharia, Universidade do Porto, Portugal
Teresa Galvão Dias, INESC TEC, Faculdade de Engenharia, Universidade do Porto, Portugal

Chapter 8
An In-Depth Study of Mobile Ticketing Services in Urban Passenger Transport: State of the Art and Future Perspectives 145
Marta Campos Ferreira, Faculty of Engineering, University of Porto, Portugal
Teresa Galvão Dias, Faculty of Engineering, University of Porto, Portugal
João Falcão e Cunha, Faculty of Engineering, University of Porto, Portugal

Chapter 9
Leveraging IoT Framework to Enhance Smart Mobility: The U-Bike IPBeja Project 166
Isabel Sofia Brito, Polytechnic Institute of Beja, Portugal & Group on Reconfigurable and Embedded Systems, UNINOVA, Portugal
Luís Murta, Polytechnic Institute of Beja, Portugal
Nuno Loureiro, Polytechnic Institute of Beja, Portugal
Pedro Rodrigo Duarte Pacheco, Easycicle, Portugal
Pedro Bento, Polytechnic Institute of Beja, Portugal

Chapter 10
Realizing IoE for Smart Service Delivery: Case of Museum Tour Guide.. 186
Umar Mahmud, Foundation University Islamabad, Pakistan
Shariq Hussain, Foundation University Islamabad, Pakistan
Arif Jamal Malik, Foundation University Islamabad, Pakistan
Sherjeel Farooqui, Foundation University Islamabad, Pakistan
Nazir Ahmed Malik, Department of Computer Science, Bahria University, Islamabad, Pakistan

Chapter 11
Lower Memory Consumption for Data Transmission in Smart Cloud Environments With CBEDE Methodology.. 216
Reinaldo Padilha França, State University of Campinas (UNICAMP), Brazil
Yuzo Iano, State University of Campinas (UNICAMP), Brazil
Ana Carolina Borges Monteiro, State University of Campinas (UNICAMP), Brazil
Rangel Arthur, State University of Campinas (UNICAMP), Brazil

Chapter 12
Do It Fluid: Innovation in Smart Conversational Services Through the Flow Design Approach 238
John Knight, Aalto University, Finland
Rachel Jones, Instrata Ltd, UK
Deniz Sayar, Izmir University of Economics, Turkey
Damian Copeland, ICR Speech Solutions, UK
Daniel Fitton, University of Central Lancashire, UK

Chapter 13
The Edgard Project: Towards a Discussion of Ethical Tech in Art and Design 259
Fernando Luiz Fogliano, Senac University Center, Brazil
André Gomes de Souza, Senac University Center, Brazil
Guilherme Henrique Fidélio de Freitas, Senac University Center, Brazil
Rachel Lerner Sarra, Senac University Center, Brazil
Juliana Pereira Machado, Senac University Center, Brazil

Chapter 14
Multisensory Experiences in Virtual Reality and Augmented Reality Interaction Paradigms........... 276
Inma García-Pereira, Universitat de València, Spain
Lucía Vera, Universitat de València, Spain
Manuel Pérez Aixendri, Universitat de València, Spain
Cristina Portalés, Universitat de València, Spain
Sergio Casas, Universitat de València, Spain

Detailed Table of Contents

Foreword xviii

Preface xx

Acknowledgment xxiii

Chapter 1
Augmented Intelligence: Leverage Smart Systems 1
João M.F. Rodrigues, Universidade do Algarve, Portugal & LARSyS, Institute for Systems and Robotics, Lisbon, Portugal
Pedro J.S. Cardoso, Universidade do Algarve, Portugal & LARSyS, Institute for Systems and Robotics, Lisbon, Portugal
Jânio Monteiro, Universidade do Algarve, Portugal & INESC-ID, Lisbon, Portugal
Célia M.Q. Ramos, Universidade do Algarve, Portugal & CIEO, Portugal

Smart systems make decisions incorporating data available from different sensing in a way to control and make smart actions. In this context, smart actions consist in augmenting user's actions and/or decisions by using devices or additional information. Those actions could and should be different from user to user, depending on its characteristics and needs. To obtain smart actions adapted to the users, it is necessary to detect the user's individualities on-the-fly. This chapter focuses on how augmented intelligence can leverage smart systems, addressing: (a) the definitions and relations of artificial intelligence, augmented intelligence, and smart systems, namely the state of the art on how to extract human features that can be used to develop augmented intelligent systems (using only computer vision methods); (b) a brief explanation of a "describing people integrated framework", a framework to extract user information automatically without any user intervention; and (c) a description of several implemented smart systems, including a future work perspective.

Chapter 2
Deep Learning on Edge: Challenges and Trends 23
Mário P. Véstias, INESC-ID, ISEL, Instituto Politécnico de Lisboa, Portugal

Deep learning on edge has been attracting the attention of researchers and companies looking to provide solutions for the deployment of machine learning computing at the edge. A clear understanding of the design challenges and the application requirements are fundamental to understand the requirements of the next generation of edge devices to run machine learning inference. This chapter reviews several aspects of deep learning: applications, deep learning models, and computing platforms. The way deep learning

is being applied to edge devices is described. A perspective of the models and computing devices being used for deep learning on edge are given, as well as what challenges face the hardware designers to guarantee the vast set of tight constraints like performance, power consumption, flexibility, etc. of edge computing platforms. Finally, a trends overview of deep learning models and architectures is discussed.

A blockchain is a digitized, decentralized, open system of records. Of late, there is a phenomenal spurt in the research and application activities of the catch-all phrase analytics, which subsumes machine learning, text mining, classical optimization, artificial intelligence, evolutionary computing, visual analytics, big data analytics, etc. in many a diverse field. Consequently, even new technology like blockchain is not left behind. This chapter presents a comprehensive survey of 33 papers that appeared between 2016 and 2018 under the theme, 'Analytics and Blockchain', which focuses on how analytical approaches and blockchain implementation are symbiotically related to improve their overall performance in solving various real-world problems. The core idea behind the survey is to facilitate the reader to appreciate the utility of analytical methods to the design, implementation, and application of blockchain and suggesting future directions for further research.

The field of Artificial Intelligence faces unprecedented progress. It is expected that the use of Artificial Intelligence to different sectors of science and economy will be increased. This is also shown by the fact that at the moment, Artificial Intelligence is characterised by popularity which is proven through the constant presentations on the news. This chapter shows how the study of the brain's hippocampus can further progress the field of Artificial Intelligence. The chapter presents indicative examples of the literature that show how the study of the hippocampus can lead to the development of specific applications. It also shows the impact to the development of biologically-inspired systems through the analysis of specific capabilities of the hippocampus. A number of conclusions are drawn in relation to the significance of the study of the brain's hippocampus for the development of new applications.

Energy consumption and, consequently, the associated costs (e.g., environmental and monetary) concern most individuals, companies, and institutions. Platforms for the monitoring, predicting, and optimizing energy consumption are an important asset that can contribute to the awareness about the ongoing usage levels, but also to an effective reduction of these levels. A solution is to leave the decisions to smart system, supported for instance in machine learning and optimization algorithms. This chapter involves those aspects and the related fields with emphasis in the prediction of energy consumption to optimize its usage policies.

This chapter explores the world of autonomous vehicles. Starting from the beginning, it covers the history of the automobile dating back to 1769. It explains how the first production automobile came about in 1885. The chapter dives into the history of auto safety, ranging from seatbelts to full-on autonomous features. One of the main focuses is the creation and implementation of artificial intelligent (AI), neural networks, intelligent agents, and deep Learning Processes. Combining the hardware on the vehicle with the intelligence of AI creates what we know as autonomous vehicles today.

In public transport, traveler dissatisfaction is widespread, due to long waits and travel time, or the low frequency of the service provided. Public transport providers are increasingly concerned about improving the service provided. To improve public transport, detailed knowledge of the network and its weaknesses is necessary. An easy and cheap way to achieve this information is to extract knowledge from the data

Widespread use of numerous hand-held smart devices has opened new avenues in computing. Internet of things (IoT) is the next big thing resulting in the 4th industrial revolution. Coupling IoT with data collection, storage, and processing leads to Internet of everything (IoE). This work outlines the concept of smart device and presents an IoE ecosystem. Characteristics of IoE ecosystem with a review of contemporary research is also presented. A comparison table contains the research finding. To realize IoE, an object-oriented context aware model is presented. This model is based on Unified Modelling Language (UML). A case study of a museum guide system is outlined that discusses how IoE can be implemented. The contribution of this chapter includes review of contemporary IoE systems, a detailed comparison, a context aware IoE model, and a case study to review the concepts.

Smart telecoms will deliver lasting improvements to business productivity and enduring consumer benefits that raise the quality of life by enabling telecommuting, telemedicine, entertainment, access to e-government, and a wealth of other online services. And we'll need next-generation digital platforms on which telecom providers can create and deliver all kinds of services. Therefore, this chapter develops a method of data transmission based on discrete event concepts. This methodology was named CBEDE. Using the MATLAB software, the memory consumption of the proposed methodology was evaluated, presenting the great potential to intermediate users and computer systems, ensuring speed, low memory consumption, and reliability. With the differential of this research, the use of discrete events applied in the physical layer of a transmission medium, the bit itself, being this to low-level of abstraction, the results show better computational performance related to memory utilization, showing an improvement of up to 79.89%.

This chapter draws on practical experience in designing, delivering, operating, and innovating conversational services. The article summarises the current context for these distinctively new kinds of services and provides an overview of the relevant technologies and common platforms used in commercial service production. The chapter explores the broader commercial context for smart voice-oriented services and provides an applied framework to aid service innovation. The two concluding parts move into service production, outline a grounded design approach (FLOW) for maximising service flow, and discuss future research directions, specifically how design anthropology can help in radical service innovation.

Chapter 13
The Edgard Project: Towards a Discussion of Ethical Tech in Art and Design 259
Fernando Luiz Fogliano, Senac University Center, Brazil
André Gomes de Souza, Senac University Center, Brazil
Guilherme Henrique Fidélio de Freitas, Senac University Center, Brazil
Rachel Lerner Sarra, Senac University Center, Brazil
Juliana Pereira Machado, Senac University Center, Brazil

As Artificial Intelligence technologies advance, their use becomes increasingly widespread, and what was once a fantasy of being able to communicate with a virtual being is now part of our everyday lives. New issues arise with every new technology, and discussions are needed to avoid significant problems. By looking at what is happening now and at the impact that AI has and will continue to have, one needs to remember history and its contradictions with a critical mind to have a more ethical approach to technology innovations. This chapter focuses on the Edgard Project, an intersection between Contemporary Art and Conversational Interface Design. Consisting of a chatbot named Edgard, the project emerges as an artistic approach to encourage critical thinking about the technology that gives him "life" and highlights the contradictions of AI through its ironic discourse and outdated interface.

Chapter 14
Multisensory Experiences in Virtual Reality and Augmented Reality Interaction Paradigms........... 276
Inma García-Pereira, Universitat de València, Spain
Lucía Vera, Universitat de València, Spain
Manuel Pérez Aixendri, Universitat de València, Spain
Cristina Portalés, Universitat de València, Spain
Sergio Casas, Universitat de València, Spain

Multisensory stimuli can be integrated in systems that make use of different paradigms, such as Virtual Reality (VR), Augmented Reality (AR) or, in a wider sense, Mixed Reality (MR), enhancing user experiences within the virtual content. However, despite the many technological solutions that exist (both hardware and software), only visual and sonic stimuli can be considered as highly integrated in consumer-grade applications. This chapter addresses the current state of the art in multisensory experiences, taking also in consideration the aforementioned interaction paradigms, and brings the benefits and challenges. As an example, authors introduce ROMOT, a RObotic 3D-MOvie Theatre, that supports and integrates various types of displays and interactive applications, providing users with multisensory experiences.

Chapter 15
Virtual and Augmented Reality for the Visualization of Summarized Information in Smart Cities: A Use Case for the City of Dubai 299
Sergio Casas, Universitat de València, Spain
Jesús Gimeno, Universitat de València, Spain
Pablo Casanova-Salas, Universitat de València, Spain
José V. Riera, Universitat de València, Spain
Cristina Portalés, Universitat de València, Spain

In this chapter, authors deal with the problem of visualizing summarized information in a complex system like a smart city. They introduce the topic of smart city in the context of the information revolution that is taking place in the world. Next, they review how this information can be visualized, highlighting

immersive 3D methods such as Virtual Reality (VR) and Augmented Reality (AR), which are particularly suitable for these applications, since 2D information does not usually induce a focused and sustained attention. The chapter describes and shows a use case in which VR and Spatial AR (SAR) are used in a smart city system to visualize summarized information about the state and management of the city. The SAR system relies on a multi-projector mapping procedure, and therefore authors also explain the technical details that the calibration and implementation of this type of AR application requires.

Reluctance and even resistance for the city transformation into a smart tourism destination (STD) may occur on the edge field of smart systems, even if over time, often with political changes, this resistance breaks down. This situation is occurring in Cordoba (Spain) which held the record of four inscriptions in the World Heritage Lists granted by UNESCO. The concept of smart tourism is recent. As a research domain, it is only emerging. So, few analyses and case studies were elaborated on it. This chapter shows how this reluctance is coming from stakeholders and how the transformation of Cordoba into a smart destination can be made harmoniously. Three scenarios are possible for this transformation either following actual tendencies or looking for two other paths, the green one or the intercultural one; how it could be done and its impacts on the City and the Province of Cordoba's economy, society, culture, and environment.

This chapter analyzes a thematic tourist park using a logarithmic model. The study shows that the application of the model allows the management of different routes to optimize waiting times and take advantage of the time allocated to the fun. The theoretical concepts were Dijkstra Algorithms. The investigation is exploratory in a single place. The data were obtained during the stay in the park of the year 2018 and was concentrated on a database. The research concludes that the application using the Dijkstra Algorithm on a mobile dispositive can determine the best places to visit and improve the experience in a thematic park.

New technologies have helped to improve the tourism sector and to develop strategies that resulted in the so-called smart destinations, underpinned and transformed by modern information and communication

Foreword

We are living a second wave in which technologies related to artificial intelligence and robotics have found a resurgence. I have been devoting my time to these areas of research since 1990, and I confess that this revival was a thing that was not in my mind. I was used to work on these lines that nobody seemed to pay attention, except for the scientific community specialized in the fields mentioned above. Neither the companies nor society knew or understood these science fields, and people relate these technologies with science fiction movies instead with actual developments.

I have asked myself several times about the reason for this new interest, which is also accompanied by a greater knowledge of the aforementioned areas than several years ago. As by using an engineering analysis, when something happens in a system, it is not due to a single cause, it usually because of several combined factors. I will start analyzing them. The first one focuses on education. Practically, twenty years ago there was neither degrees nor masters fully specialized in artificial intelligence and robotics. These disciplines were previously spread along different engineering curricula and following a unique point of view adapted to the contents of a particular degree. There were no studies with an integrated and multifaceted vision of these disciplines. On the other hand, artificial intelligence and robotics are integrative scientific fields dependent on other disciplines. Tomás Lozano Pérez defined them as the implementation of the perception-planning-action cycle. In this way, it has been required the adequate maturation of scientific-technical developments in order to carry out prototypes of application which integrates in a proper way the mentioned cycle. In this sense, it has been moved from an atomized research scheme with highly specialized research teams to other approaches with integrative research based on interdisciplinary teams.

Complementing the two factors mentioned above (education and research) related to the classic functions of universities, external causes have also converged. The rise of the third university mission, the transfer of technology to the productive sector, has been based on a change in the mentality of researchers: now the application of the research results is a main goal and it was stablished alliances with the productive sector in order to develop market based applications. These alliances, now, are strategic in such a way that the companies feel the universities as a source of knowledge to develop new applications with innovative approaches, and no there are doubt about artificial intelligence and robotics are very promising sources for this. On the other hand, the lower costs of the devices that support the applications of these two disruptive technologies as well as their increasing processing power makes it more attractive for companies to develop artificial intelligence-based applications. Moreover, the European Government promotes the development of technologies related with the society challenges. The concept of Smart City covers several of these challenges, and its main objective is defined as provide more services to citizens based on new technologies and at the same time reaching more sustainable cities in their broader concept. Energy management, public transport or smart tourism are fields of application that derive directly from the implementation of this smart city concept. All of them need artificial intelligence and robotics.

This book, which I have had the pleasure of writing this foreword, and that you dear reader, you will start to enjoy, it combines in a single publication everything that I have related in the previous paragraphs. It includes chapters with specialized areas such as data science, virtual reality or Deep-Learning; as well as other more integrative ones such as smart tourism, autonomous vehicles, bio-inspected computing, or the inclusion of the ethics. On the other hand, it addresses specific applications in the field of smart cities, focusing on transport, new services or tourism management. Thus, it is a very complete work that is like a manual useful to know the state of the art of artificial intelligence in both its scientific aspect and its applications, especially focused on the field of smart cities.

Sincerely, I hope that you enjoy reading these pages, either the complete book or the chapters that are related with your field of expertise. And, of course, be curious to continue moving forward in the proposed topics. Over the shoulders of giants, dear reader.

Víctor F. Muñoz-Martínez
University of Malaga, Spain

Preface

In his classic paper on computing machinery and intelligence, from 1950, Turing states the difficulty of writing an intelligent program: "*At my present rate of working I produce about a thousand digits of programme a day, so that about sixty workers, working steadily through the fifty years might accomplish the job, if nothing went into the wastepaper basket. Some more expeditious method seems desirable*". In the same paper, Turing also proposes a method to achieve it, i.e., how to develop an intelligent program, stating that it could be done by developing a child's mind ("simple program") and then educate it ("to develop a more complex program"): "*Presumably the child brain is something like a notebook as one buys it from the stationer's. Rather little mechanism, and lots of blank sheets. (Mechanism and writing are from our point of view almost synonymous). Our hope is that there is so little mechanism in the child brain that something like it can be easily programmed*". More recently, Dave Ferrucci, creator of IBM's Watson, founder and CEO of Elemental Cognition, said: "*The next challenge for the Artificial Intelligence industry is for machines to go beyond simple pattern matching and to autonomously learn to understand and reason about the world the way people do. (...) The demand for machines to explain and justify answers in terms humans can understand will grow with our desire to leverage computers in the process of human decision making, especially in areas where the reasons matter.*" Clearly, the "intelligence" in what is considered a smart system changed completely in 70 years but, even after all these years, the hot topics are still the same: how to write an "intelligent program"? How to implement a "smart system"?

A recent Gartner's article presents the top 10 strategic technologies trends for the present and the following years. One of these trends involve autonomous things, which include cars, robots and drones, but also other devices like appliances. Autonomous things use artificial intelligence (AI) to perform tasks, traditionally made by humans. Another trend is augmented analytics, where the huge amount of data generated nowadays, in fields such as businesses or internet of things (IoT), can be automatically analyzed by algorithms that try to evaluate many hypothesis and hidden patterns, instead of only testing a minor set of possibilities, conceivably biased by personal views. Other trends comprise empowered edge, smart spaces, and digital ethics. All these trends include smart systems and artificial intelligence, impacting in our day to day life, in our cities, and even in our free time. However, smart systems when connected to AI are still closely associated with some popular misconceptions, that cause the general public to either have unrealistic fears about it, or to expect too much from it, namely on how it will change our workplace and life in general. It is important to show that such fears are unfounded, and that new trends, technologies, and smart systems will be able to improve the way we live, benefiting the society, without replacing humans in their core activities.

Preface

The main book's focus are smart systems and artificial intelligence, presenting the state-of-the-art technologies and available systems in this domain. The publication tries to explain such solutions from an augmented intelligence perspective, showing that these technologies can be used to benefit humans, instead of replacing them, by augmenting the information and actions of our daily life. At the same time, a manual/book that explains with correctness, high quality, and accuracy some topics of AI to professionals and to young students which are entering this area was developed. The book addresses smart systems that incorporate functions of sensing, actuation, and control in order to describe and analyze a situation and make decisions based on the available data in a predictive or adaptive manner. Subjects related to smart tourism, destinations, mobility and cities are also addressed.

The target audience are professionals, researchers, students, or simple curious that are interested in the new trends in smart systems and its applications. This book can help teachers, engineers, investigators, and professionals in the smart systems engineering, to improve their knowledge about the contemporary theories, technologies, and tools available, and in this way be more demanding and capable of improving their products or teaching. The objective was to show and reinforce the interconnections between the five areas of artificial intelligence, machine learning, computer vision, sensors network, and smart tourism, and between these and the main book area, smart systems.

The first chapters reflect generics concepts of this area. Chapter 1 addresses what is a smart system, smart actions augmented intelligence and artificial intelligence, it also presents several smart systems. Chapter 2 presents the challenges and trends of deep learning on edge, one issue that has been attracting the attention of researchers and companies looking to provide solutions for the deployment of machine learning computing at the edge.

In Chapter 3 an emergent topic is addressed, smart blockchain, presenting a detailed survey over this topic. This introduction to the smart systems field is finished by a very interesting chapter (number 4) that addresses biologically-inspired computational systems, and how brain's Hippocampus can influence this.

After this initial 4 chapters dedicated to the introduction of the subject, the next chapters are dedicated to our cities, our smart systems, and sustainability issues.

Chapter 5 addresses the monitoring, predicting, and optimization of energy consumptions in our homes and companies. The following chapters explore our mobility in the cities. Chapter 6 explores the world of autonomous vehicles, and in Chapter 7 concentrates on the average speed of public transport vehicles based on smartcard data.Chapter 8 focus in the mobile ticketing services in urban passenger transport, also related with mobility, Chapter 9 presents the case of U-Bike, an IoT framework to enhance smart mobility using bikes.

The next chapters present some models that can be used in the city, in different smart systems. Chapter 10 relates IoT with internet of everything (IoE) where the person, the data and the processes are interconnected. In Chapter 11 the authors focus on a smart cloud environment by CBEDE methodology. Chapter 12 focuses on the novelty in smart conversational services through the flow design approach. Also focused in the design, but now relating with a discussion of ethical in arts and smart technology is Chapter 13.

The next two chapters focus on smart systems in virtual and augmented reality. Being, Chapter 14 a multisensory experience in virtual reality and augmented reality interaction paradigms, and Chapter 15 a system for visualization of smart cities, in the case the city of Dubai.

The final chapters are related with smart tourism namely, Chapter 16 presents the case of the Cordoba City in Spain, being Chapter 17 related with logistics model to optimize times at attractions in a thematic amusement park. The final chapter, Chapter 18 presents a gamification tool for smart tourism.

As a conclusion, the book brings together a comprehensive collection of research trends on the edge field of smart systems, connecting the fields of artificial intelligence, machine learning, computer vision, sensors network, and smart tourism from a set of international experts on the theoretical, design, evaluation, implementation and use of innovative technologies on these fields. It integrates in a single book many point of views that usually are not integrated in a place, as these subjects are scattered in many articles (e.g., proceedings, journals, and internet). The book focuses on the applications, probably making it more compelling for a majority of potential users. Finally, with this book we hope to have contributed to the construction and promotion of knowledge that we believe will foster the discovery of new scientific paths.

Acknowledgment

We would like to acknowledge all the authors that contribute to the success of the book, all anonymous reviewers and our research centers, that are supported by the Portuguese Foundation for Science and Technology (FCT), projects LARSyS (UID/EEA/50009/2019), CIAC (UID/Multi/04019/2019), CinTurs (UID/SOC/04020/2019), CEFAGE (UID/ECO/04007/2019), and INESC-ID (UID/CEC/50021/2019).

Chapter 1
Augmented Intelligence:
Leverage Smart Systems

João M.F. Rodrigues
https://orcid.org/0000-0002-3562-6025
Universidade do Algarve, Portugal & LARSyS, Institute for Systems and Robotics, Lisbon, Portugal

Pedro J.S. Cardoso
https://orcid.org/0000-0003-4803-7964
Universidade do Algarve, Portugal & LARSyS, Institute for Systems and Robotics, Lisbon, Portugal

Jânio Monteiro
https://orcid.org/0000-0002-4203-1679
Universidade do Algarve, Portugal & INESC-ID, Lisbon, Portugal

Célia M.Q. Ramos
https://orcid.org/0000-0002-3413-4897
Universidade do Algarve, Portugal & CIEO, Portugal

ABSTRACT

Smart systems make decisions incorporating data available from different sensing in a way to control and make smart actions. In this context, smart actions consist in augmenting user's actions and/or decisions by using devices or additional information. Those actions could and should be different from user to user, depending on its characteristics and needs. To obtain smart actions adapted to the users, it is necessary to detect the user's individualities on-the-fly. This chapter focuses on how augmented intelligence can leverage smart systems, addressing: (a) the definitions and relations of artificial intelligence, augmented intelligence, and smart systems, namely the state of the art on how to extract human features that can be used to develop augmented intelligent systems (using only computer vision methods); (b) a brief explanation of a "describing people integrated framework", a framework to extract user information automatically without any user intervention; and (c) a description of several implemented smart systems, including a future work perspective.

DOI: 10.4018/978-1-7998-2112-0.ch001

INTRODUCTION

Dave Ferrucci, creator of IBM's Watson supercomputer, founder and CEO of Elemental Cognition, recently said: "The next challenge for the artificial intelligence industry is for machines to go beyond simple pattern matching and to autonomously learn to understand and reason about the world the way people do. [...] The demand for machines to explain and justify answers in terms humans can understand will grow with our desire to leverage computers in the process of human decision making, especially in areas where the reasons matter." (Hastreiter, 2017).

A Gartner's article, published in October 2018, presents the top 10 strategic technology trends for 2019 and following years (Panetta, 2018). One of these trends involves autonomous things, which includes cars, robots and drones, and other devices like appliances. Autonomous things use artificial intelligence (AI) to perform tasks, traditionally made by humans. Another trend is augmented analytics, where the huge amount of data generated nowadays, in fields such as businesses or internet of things (IoT), can be automatically analysed by algorithms that try to evaluate many hypothesis and hidden patterns, instead of only testing a minor set of possibilities, peradventure biased by personal views. Other trends comprise edge computing, a topology where the information processing and content collection and delivery are placed closer to the information sources. Smart spaces and digital ethics are examples of other trends.

All those trends include smart systems and AI, impacting in our day to day life, in our cities, and even in our free time. In this context, smart systems are stand-alone systems or systems that help users to make decisions by incorporating available data from different sensing, in a way to control and make smart actions, e.g., augmenting user's actions and/or decisions, by using devices or additional "smart" information. The interfaces that allow those smart actions (i.e., smart interactions with users) could and should be different for each user, depending on the user's characteristics and needs. Several examples of different interfaces for the same purpose can be presented: a kid that does not know how to read, cannot interact with a device using text; a senior with locomotion difficulties cannot use an application that sends him to climb stairs to go from point A to B; a person in a wheelchair cannot reach information or a touch interface placed at the same height as for a 2 meters "basketball player"; at home, or at a hotel, different persons have different needs for the rooms in term of luminosity or temperature; or blind or deaf persons need interfaces adapted to themselves. In short, there is an endlessness number of examples, including examples of robots or appliances that in the present and in the future will (intelligently) communicate with the users (humans).

Depending on context and authors, there are several definitions for augmented intelligence (Pasquinelli, 2015; Araya, 2018, 2019). This chapter refers to the effective use of information technology in augmenting human intelligence, and/or helping humans to augment their actions, so called smart actions. The fundamental idea is the use of AI to augment the user's actions, i.e., to assist the human (worker), not replace him. A more detailed explanation can be found in the following section.

In summary, interfaces or interactions with users where *one-size-fits-all* is not and cannot be a solution in a near future. However, two problems in building adaptive interfaces can easily be identified; (a) how to harvest user's information without his/her intervention (but with his/her consent), in a way to adapt the user interface, the room, the house, the "city", to the needs and preferences of each individual user - put the augmented intelligence to the service of humans; and (b) how to minimize or even eradicate the fear that some users have of AI (and related fields).

Relatively to the latter, (b), artificial intelligence is still closely associated with some popular misconceptions, which cause the general public to either have unrealistic fears about AI, or to expect too much from it, especially in the way it will change our life in general. Latter discussed in this chapter; it is important to show that such fears are unfounded. On the opposite, new trends, technologies, and smart systems will be able to improve the way we live, benefiting the society, without replacing humans in their core activities.

This chapter focuses is the blend between augmented intelligence, smart systems, adaptive user interfaces, and a computer vision framework for describing people, addressing three main topics: (i) the contextualization, definitions, and relations of AI, augmented intelligence, and smart systems, including the state of the art in how to extract human features that can be used for augmented (intelligent) systems using only traditional RGB cameras – computer vision methods; (ii) a brief explanation of a describing people integrated framework (DPIf), a framework to extract user's information automatically without any user intervention; and (iii) a description of several implemented systems, from augmented reality to tourism, energy, and routing (user navigation) where the augmented intelligence is present and also where the framework (DPIf) is applied or can be applied. The chapter will also address the associated perspectives for future work.

In the following section the backgrounds will be presented, together with the computer vision state of the art tools for person's attributes detection, followed by a section that presents the DPIf. Examples of smart systems and their relation with augmented intelligence, some of those already integrating (the principles of) DPIf, will be presented next. The final sections present a conclusion, future work, and tendencies.

Background

Smart systems have the unique ability to optimize their performance under a variety of inputs and recover quickly from a wide variety of disturbances. This ability depends on both the cognitive and physical capabilities of such systems (Jones *et al.*, 2015). In current human-computer interaction (HCI), interfaces are proposed to enable easy access to the proliferating functions and services. However, the increasing intelligence of machines leads to a shift from HCI to human-machine cooperation (HMC) (van Maanen *et al.*, 2005). As such, machines should either be designed to cooperate, or to learn how to cooperate, with humans, so they will be able to assess and adapt to human's goals. There is a growing need for humans and machines to understand each other's reasoning and behavior (van Maanen *et al.*, 2005; Crandall *et at.*, 2018). For HMC the aim is to support individual user characteristics, tasks, and contexts in order to establish a HMC in which the computer provides the "right" information and functionality, at the "right" time, and in the "right" way (Fischer, 2001; van Maanen *et al.*, 2005).

In this context, two important definitions much be stressed: (a) augmented intelligence and (b) adaptive user interfaces (AUI). The first one, augmented intelligence, can be considered as an alternative conceptualization of AI that focuses on AI's assistive role in advancing human capabilities. The usual use of the term AI grossly discounts the importance of human creativity. Augmented intelligence reflects the ongoing impact of AI in amplifying human innovation (Araya, 2019), it emphasizes the fact that technology is designed to enhance human intelligence rather than replace it. In reality, it is what AI is doing in most cases nowadays. The augmented intelligence term also helps to remove some unrealistic fears and improbable expectations from the general public, especially on how about autonomous robots and other intelligent systems will change the workplace and life in general. The choice of the word augmented, which means "to improve", reinforces the human intelligence and the creativity role, which plays an enormous part when using machine learning algorithms to discover relationships and solve problems.

The second definition, AUI, consists on interfaces that adapt to each user, i.e., they should differentiate the needs and goals of each individual, leading to the need to address several challenges, as already mentioned in the Introduction section. Furthermore, user interfaces are a designation that usually is more related with software applications but, here, a broader definition is considered, amplifying it to the space where interactions between humans and machines occur. One example is the design of autonomous robots that can cooperate with people in their daily tasks in a human-like way.

As also mentioned in the Introduction, one of the problems to build AUI is how to harvest user's information without the user intervention (possibly with consent or not associating the information with a user – forgetting the personal information after the interaction), in a way to adapt the user interface, the software application (Rodrigues *et al.*, 2017), the robot, the room, the house, etc. to his/her needs and preferences. This non-intrusive mode of working is the way to have (augmented) intelligence leveraged with (smart) systems.

Nevertheless, it is important to stress that humans communicate mainly through speech, though they are endowed with an essential non-verbal communication capacity in social interactions. The way they behave, their body expression, as well as their facial expression, reflect their emotional state which are key elements in communication. One of the topics of this chapter is on the visual description of the human user in bio-characteristics terms (e.g., height, age, gender, and facial expression) and physical disability, by characterizing the objects that are in use by the person. In future, pose information will also be addressed, allowing the development of better AUI. Also, out of the chapter's focus, all available complementary information can be integrated such as sound, social networks data, or historical information stored in the cloud or in the interacting devices.

As mentioned, this chapter discusses an integrated framework that describes persons in a non-intrusive way, using a single (RGB) monocular camera, computer vision, and machine learning techniques. The framework detects single human silhouettes, pose, and the user's face. From the collected data, the framework recognizes and classifies humans in a number of ways, such as height, age, gender, emotions, also clothe classification, trousers, skirt, T-shirt, glasses, phones, etc. The framework's main idea (in the future) is to use minimum computer power to on-the-fly characterize users in a non-intrusive way, creating a temporary identification that is forgotten as soon as the user ends the interaction (preserving this way the user's data). This information will allow to build AUI, i.e., the customization of a different UI for each person (Rodrigues *et al.*, 2017).

There is a set of basic emotions universally recognizable through facial expressions (e.g., sadness, joy, disgust, anger, fear, and surprise) that ensures a level of communication and cross-cultural interpretation (Matsumoto & Ekman, 2008). Nowadays there are several application programming interfaces (API) and software development kits (SDK) available for the recognition of facial expressions (recognize moods), such as: Emotient, Affectiva, EmoVu, Nviso, Kairos, Project Oxford by Microsoft, Face Reader by Noldus, Sightcorp, SkyBiometry, Face++, Imotions, CrowdEmotion, or FacioMetrics (Doerrfeld, 2016). Several examples are based on OpenCV (OpenCV, 2019), an open source computer vision library, including the ones proposed in (Beyeler, 2015; Silva, 2016; van Gent, 2016; Shaikh, 2016), the tutorials connecting Python and OpenCV (Toscano, 2017), or the applications connecting OpenCV with convolutional neural networks (CNN) (RTFE, 2019), implemented by Keras (2019). Other examples using Keras can be found in the works of Arriaga *et al.* (2017) or Kumar (2018).

A recent publication by Baltrusaitis *et al.* (2018) defines facial behavior as a set of facial landmark locations, head pose, eye gaze, and facial expressions. Nevertheless, not all these features are mandatory for estimating facial emotions. The same authors also show a comparison of methods, including their method - OpenFace 2.0. They compare methods in their availability to be freely used for research purposes, their availability for model training source code, their availability of model fitting/testing/runtime source code, and their availability of model fitting/testing/runtime executables. OpenFace 2.0 uses a CNN based face detector and an optimized facial landmark detection algorithm, allowing them to define facial behavior based on the above-mentioned facial landmark location, head pose, eye gaze, and facial expressions.

Face++ (2019) is a face detection, recognition, and comparing tool, as well as a face related attributes estimator including age, gender, emotion, head pose, eye status, and ethnicity. The tool also includes body recognition and body attributes estimation, including gender and upper and lower body clothing colors. Another tool is the (Microsoft) Azure Cognitive Services Face API which provides algorithms to be used in face verification, face detection, and emotion recognition (AzureFace, 2019).

One important focus, not yet fully explored, are facial expressions in elderly, although there is some literature on this subject (Hsiao & Hsieh, 2014; Yu *et al.*, 2015; Rukavina *et al.*, 2016; Yaddaden *et al.*, 2016; Castillo *et al.*, 2016; Muhammad *et al.*, 2017). Another important focus is gender and age estimation for which, once again, there are several libraries that allow us to detect those features (some already mentioned). For person's gender, some methods use the OpenCV library (Datta, 2016; Nourani, 2017; OpenCVgender, 2019). The Intel's Computer Vision Library (Bradski & Pisarevsky, 2000) and the OpenVINO toolkit (IntelDev, 2019), also from Intel, can be used for gender and for objects detection, among other features, including other AI solutions.

Levi and Hassner (2015) presented age and gender classification using deep neural networks (DNN). Annalakshmi *et al.* (2018) use spatially enhanced local binary patterns and histogram of oriented gradients features for gender classification. Ponnusamy (2019) presents CVlib which can be used for face and gender detection as well as object detection, it uses Keras. Zhang (2018) also uses Keras to achieve face and gender detection.

OpenPose is a solution for human body detection (Cao *et al.*, 2017, 2018; Hidalgo *et al.*, 2019), one of many viable solutions that allow us to detect the human skeleton, human region of interest (RoI), as well as faces (facial RoI). For height estimation, one solution was proposed by Momeni-K *et al.* (2012) which is applied over the human RoI estimation. This method requires prior knowledge of the camera's height, angle, and vanishing point of its view.

Most recent methods use AI techniques, with a large percentage of them focusing in CNN or in DNN. For that there are a few machine learning platforms such as, the already mentioned, Keras (Keras, 2019) or TensorFlow (Girija, 2016; Abadi *et al.*, 2016; TensorFlow, 2019).

In the specific case of AR, there are several approaches for object recognition when utilizing image base AR, usually using "marker recognition" (Rodrigues *et al.*, 2018). Markers are objects – images of objects – that exist in the environment and are used as templates. Every time the system detects a marker, the AR information is overlapped in the screen or projected into the environment (depending of the AR type). The marker recognition for AR is usually achieved using descriptors. To create the descriptors, it is necessary to detect keypoints, i.e., interest points. They are spatial locations, or points in the image that define what is interesting or what stands out in the image. Typically, this is a computationally expensive task (Terzic *et al.*, 2015). However, although keypoints indicate the location of specific events in an image, they do not contain information about the regions where they are located: the local image structure. For keypoint matching across images it is necessary to take the local image information around a keypoint and to code this, building a local image descriptor for every single keypoint. These descriptors must be robust to image variations, like changes in illumination, rotations, translations, etc.

Descriptors characterize local image regions by means of a compact numerical representation which should be consistent under a wide range of image transformations. There are very robust and reliable histogram-based descriptors such as SIFT (Lowe, 2004) and SURF (Bay *et al.*, 2008), which are floating-point descriptors, where image regions are often described in terms of gradient histograms collected in predefined sampling regions (Bay et al., 2008; Lowe, 2004). Although such histograms have been considered the state of the art for a long time, they have a large dimensionality, require a lot of storage capacity, and this results in comparably slow matching. There has been a more recent trend towards binary descriptors such as ORB (Rublee *et al.*, 2011) and BRISK (Leutenegger *et al.*, 2011), which have reduced memory and bandwidth requirements, and faster extraction and matching. This makes them ideal for a growing number of real-time applications (including AR) and embedded devices with low computational power. Some recent binary descriptors have been shown to be on pair with histogram-based methods, or even excelled them in some test scenarios (Trzcinski *et al.*, 2013). Recently, focus has shifted towards learning efficient descriptors, which was aided by the availability of large datasets of registered image patches (Brown *et al.*, 2011). More details about descriptors and a complete state of the art can be found in (Saleiro *et al.*, 2017).

There is a huge number of libraries and algorithms that can/do generic object detection and classification, with some of those already having several objects classified, but allowing the easily teaching of more. Many of them can do object classification and extract some of the mentioned human attributes. The API that integrate more features are: the Azure Cognitive Services Directory (AzureVision, 2019), the Amazon Rekognition (Rekognition, 2019), the Google Cloud Vision (GoogleVision, 2019), the OpenVINO toolkit from Intel (IntelDev, 2019), and, of course, the OpenCV library (OpenCV, 2019). More exist, like the already mentioned CVlib (Ponnusamy, 2019). Nevertheless, collecting and annotating images from the real world is too demanding in terms of labour and money investments and is usually inflexible to build datasets with specific characteristics. There are several ways to try to minimize this: one is to use small datasets instead of using large datasets for training, such as in (Raj *et al.*, 2019). In this case, the authors present a method of unsupervised pre-training to effectively train the network and recognize objects using a CNN which is pre-trained using a sparse auto-encoder. The region proposals for the objects are forwarded to a CNN for feature extraction and finally into a fully connected layer for classification. A different solution is to train and test object recognition with virtual images (Tian *et al.* 2018).

As a brief conclusion, it is now possible to characterize any user in real time, even for constrained devices and using commodity cameras, retrieving information about the gender, age, height of the person, as well as some basic emotions. In addition, it is possible to characterize the user with objects that he/she is caring/using, for instance if a user in a wheelchair, caring bags, or walking with a dog. All this information allows us to create augmented intelligence, to develop smart systems to adapt different UI. In the next section, the architecture for the describing people integrated framework will be briefly presented.

Describing People Integrated Framework

Describing people integrated framework (DPIf) is a vision-based library for acquiring human attributes and classify objects. Figure 1 shows the generic architecture of the framework for which an initial prove-of-concept was implemented by Turner *et al.* (2019). In this scheme, the algorithm starts by detecting a person and, in particular, it focusses on the person's face. After the face being detected, the distance to the camera is estimated by evaluating the size of the face RoI (*df*). If the face is estimated to be at least

Figure 1. Describing people integrated framework for acquiring human attributes

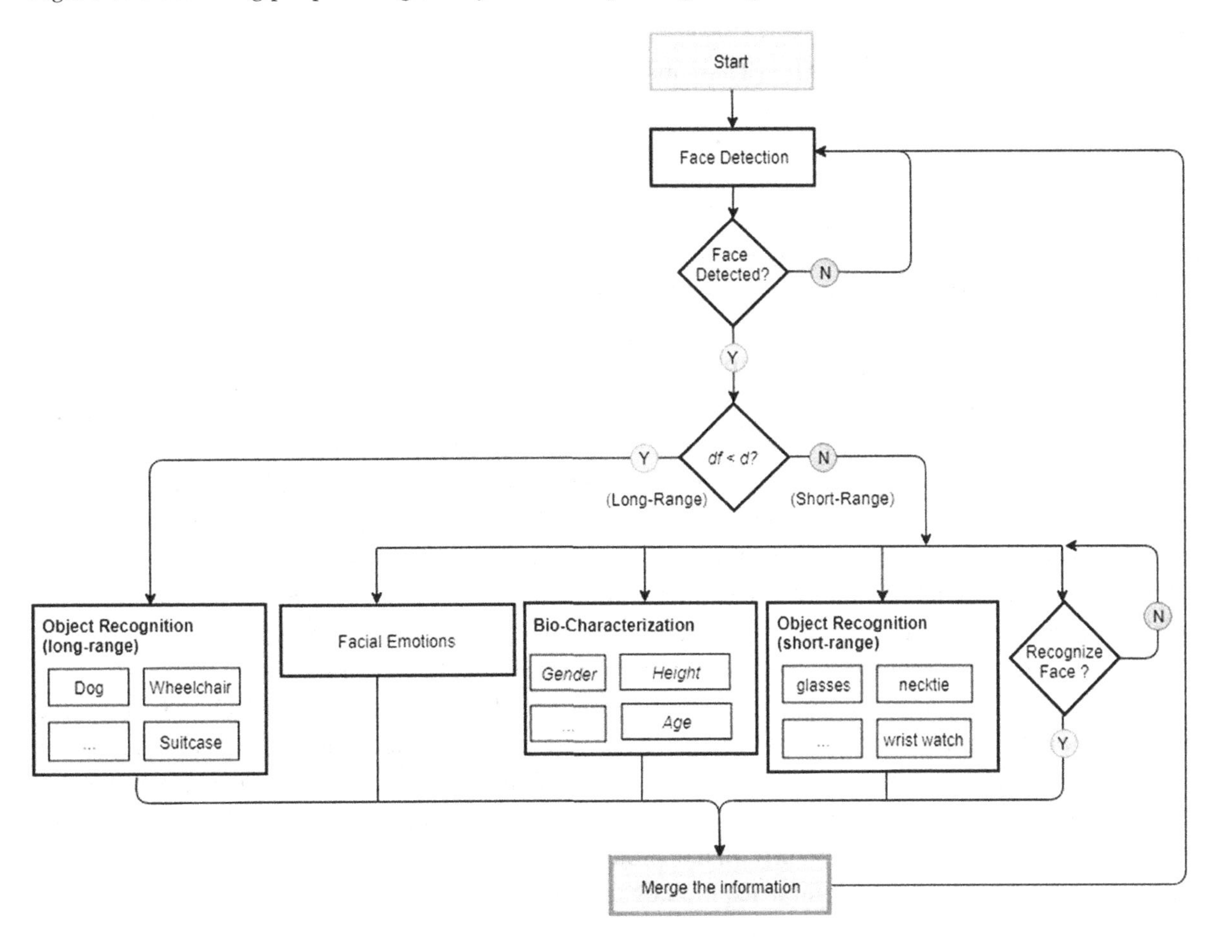

at a specific distance (*d*), i.e., some predefined height RoI size, then it is considered as a *long-range* detection, and "bigger" / "less detailed" objects are computed. When the person gets closer to the camera, then the remaining attributes are detected, from facial emotions, gender, and age, to close smaller objects, like glasses. It important to stress that information is acquired in parallel, not being mandatory the acquisition of all features for all applications. The methods used to recognise the objects and the human characteristics are detailed in Turner *et al.* (2019).

Next section shows some diversified examples of smart systems, developed or in development coordinated by the authors. One of the key line transversal to all systems was the smart systems and the augmented intelligence.

Applied Smart Systems

In this section some examples of smart systems that include augmented intelligence are described.

Smart Bus STOP

ACCES4ALL - Accessibility for All in Tourism (Rodrigues *et al.*, 2018b) is a project that aims to study the "accessibility for all" and it works towards improving sustainable mobility, senior tourism, social inclusion, and territorial marketing, while taking the needs of the elderly and other people with reduced mobility into consideration.

One of the ACCES4ALL modules is an eco-friendly Smart Bus Stop (SBS) (Rodrigues *et al.*, 2018b). The SBS system has several cameras, both inside and outside the bus stop shelter. The vision-based system architecture of the SBS is divided in 5 main blocks, namely: (a) face detection; (b) object recognition; (c) face recognition; (d) human attributes compilation; and (e) adaptation of the AUI according with the human attributes compilation. The SBS uses the DPIf (Turner *et al.*, 2019) and more details about it can be found in (Rodrigues *et al.*, 2018b). It is important to stress that, in a way to protect the user's personal identification, no prior information or identification is required as input to the system. So, for the first-time that the user appears in from of the camera, he/she is not recognized, and the system has to acquire enough information to create an identification for that user. The facial recognition does the trick. In this work, one identification (ID) is assigned to a person, so the computed attributes can be assigned to that specific ID (person). In other words, if the person is not recognized, the system will treat he/she as a new person and proceed to start calculating their attributes. The real identity of the person is not relevant for the system, only the preferences that the system learns on-the-fly.

Figure 2 shows a sequence of images from the working DPIf, from left to right, top to bottom: the face detection, the facial emotions, the training of the system with the face for latter identification, and finally, gender, height (in meters), age, and objects classification (in this case the detection of wheelchairs, white canes, bags, and dogs is active). After the detection of this information, the user interface of the bus station is adapted accordantly. In this project, the augmented intelligence is applied in the UI, which is adapted with the information acquired by the DPIf.

Five Senses AR

The second example is the Mobile Five Senses Augmented Reality System for Museums (M5SAR) project which aims at the development of an enhanced augmented reality system, to be a guide in cultural, historical and museum events, complementing or replacing the traditional orientation given by tour guides, directional signs, or maps (Rodrigues *et al.*, 2018c, 2018d).

The M5SAR project novelty is to extend the AR, exploring the human's five senses, by presenting for each museum object a narrative explaining or representing its story (Rodrigues *et al.*, 2017). The complete system consists of a smartphone application (App) and a physical device here referred as "portable device". The portable device is integrated with the smartphone to complete the exploration of the 5 human senses: sight, hearing, touch, smell, and taste. The device is portable and small (see Fig. 3), but it adds touch, smell and taste experiences to complete the augmented system (Sardo *et al.*, 2017). In summary, the system has the following main features: (i) an App that can be installed in any mobile device (independent of the operating system) which (ii) detects the museum objects using computation and images acquired with the mobile device's rear camera, and (iii), for each object of the museum (after being detected), return contents (text, audio, and video) to the users. The App has (iv) an adaptive user interface, that can adapt to the user on-the-fly and includes (v) a smart route navigation and location

Figure 2. A sequence of the vision-system detecting the human attributes

module. For the most important objects (masterpieces), (vi) the system gives 5 senses contents, i.e., text, audio, video, touch (cold, hot, and/or vibration), smell, and taste.

In term of architecture, the following main modules were included:

(a) A local and a *cloud* server: the local server (located in the museum) stores the necessary contents (i.e., audio and video contents, actions to activate the hardware device, and museum AR object's markers – image descriptors) and coordinates the communications between the *cloud* server and mobile devices. The *cloud* server is used for the installation of the App, generic information about museums storage (e.g., localization of the museum, open and closing hours, etc.). The contents in the App, about the museum's objects, are activated only if the person is inside the museum. The local server enables a minimization of the costs, as well as latency, in network terms.

(b) The mobile App is divided in three main sub-modules: (b.i) Adaptive user interfaces, where the contents are presented accordingly with the user's type, being pre-defined 5 types of users: expert, casual, senior, child, and family; (b.ii) The object recognition sub-module, where the museum objects are recognized, from 2D paintings to 3D statues; And the (b.iii) location and adaptive navigation which can create navigation routes that adapts to each user, including parameters such as the time he/she wants to spend in the museum and the objects that he/she wants to see. The App

Figure 3. M5SAR system working in the Faro Municipal Museum (Faro, Portugal)

also allows a "family" mode to know where the family members are in the museum (Cardoso *et al.*, 2018, Rodrigues *et al.*, 2018, 2018c, 2018d).

(c) Two types of devices connected to the mobile device via Bluetooth were considered: (c.i) Beacons, which are employed for the users localization in a museum section, and also works as a trigger to the deploy of contents in the mobile device (the goal was to always occupy the least amount of mobile memory, and do the least amount of communications between local server and mobile device) and the (c.ii) portable device for touch, taste, and smell sensations used to enhance the five senses (Sardo *et al.*, 2017).

In this project, a small component of DPIf for the AUI is used. For more details see (Rodrigues *et al.*, 2017; Rodrigues *et al.*, 2018, 2018c, 2018d). Example of the M5SAR system working in Faro museum can be found in Fig. 3.

Holographic Public Relations

The project PRHOLO: The realistic holographic public relations (Rodrigues *et al.*, 2016) is a real size human public relation that can be digitally represented using avatars or videos of a prerecorded person. The system uses a 360° holographic representation, based on Pepper's Ghost. The main drawback of this technique is the requirement of a large space while, as advantage, it allows capturing a client's attention due to its attractiveness.

The system uses natural interaction (Alves *et al.*, 2018). The installation consists in a 360° holographic representation of volumetric avatars or objects, and optionality also a screen where a set of menus with videos, images, and textual contents are presented. The hologram can follow the user's movements, showing advertising of a product of the company or a face. It uses several (typically between 4 to 8) Microsoft Kinects for enabling natural interaction through gestures and speech, while building several statistics of the visualized content, including the number of people around the installation, how much time each person looks to the installation, and how much time a person interacts with the installation using gestures or voice.

The DPIf was not used in the present architecture, human attributes were extracted using the Kinect ("3D sensor") and those attributes were used to adapt the hologram to each user. The augmented intelligence is achieved by the voice communication between the installation and the user, as well as the

Figure 4. The PRHOLO prototype version

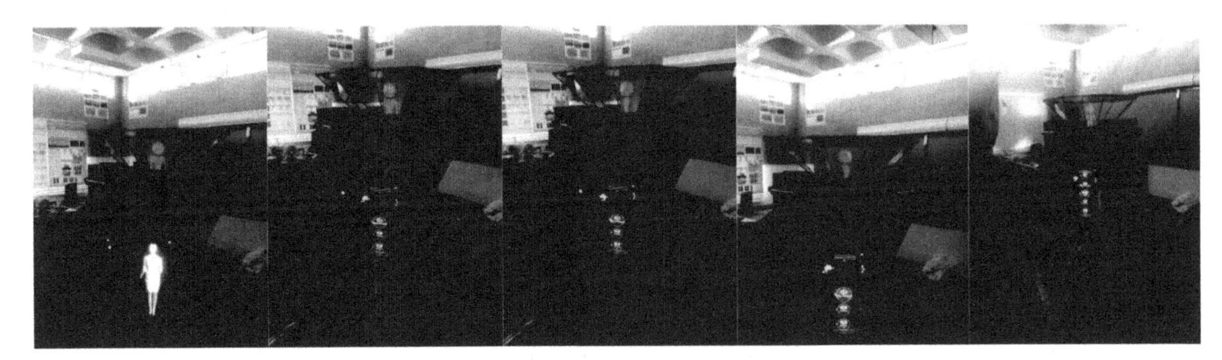

interaction with the 3D avatar/object with the user. Figure 4 shows an example of the prototype version of the system. More details about the implementation can be found in (Rodrigues *et al.*, 2016, Alves *et al.*, 2018). The commercial version available at http://prholo.com/.

Pool Live Aid

PoolLiveAid project (Sousa *et al.*, 2016) is and AR projected system intended to help and assist amateur players by using the pool table as an interface, showing on-the-fly a prediction of what will happen when the player hits the white ball in its current position; menus or other visual effects can also be projected and accessed over the table (or in an outside monitor).

The system is also based on a Kinect sensor and a projector placed above the table. The Kinect sensor is responsible for capturing the game, which is then processed by a standard computer, enabling detection of game elements such as the table's rails (borders), cue direction, and balls' positions, which are all used to predict a trajectory. The output result is then forwarded in real time to a projector, showing what might be the result of that shot on the pool table (Sousa *et al.*, 2016).

In this case, the augmented intelligence allows the prevision of the expected trajectories when the user moves the cue near the white ball. In addition (not implemented at this time), information about the user could be used to configure advertisements to be projected in the table. Figure 5 shows the system working in a demonstration event.

Figure 5. PoolLiveAid system working in a demonstration event

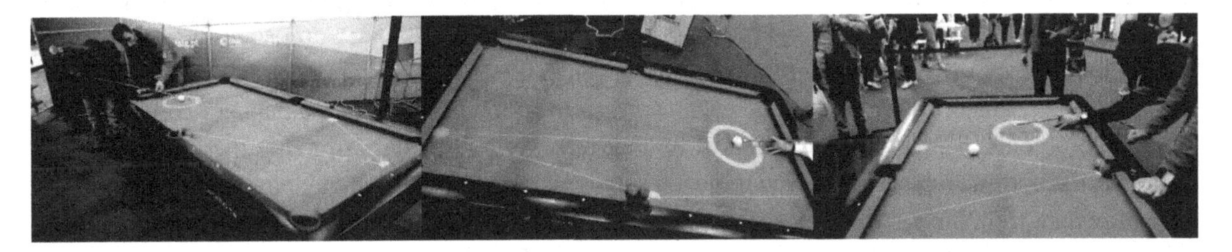

Religious Tourism Experience

Religious Tourism Experience Model (RTEM) project aims to build a localization based (GPS) AR App to develop new cultural experiences, promoting tourism while enhancing the potential of cultural religious heritage sites (Ramos *et al.*, 2016). By taking in consideration the location, time, temperature, and other characteristics of the context where the tourist is located, the system uses the navigation module to do (smart) recommendations. In this App, tourists can interact with point-of-interest (PoI) and, at the same time, receive clearer information, which makes the interpretation of the sites easier. When a tourist/resident visits a specific heritage/religious element, the system contributes in three different dimensions to the associated experience (Ramos *et al.*, 2016): (a) enjoying the heritage, adding value to the trip (touristic experience); (b) acquiring knowledge about a patrimonial element and increase the cultural knowledge (cultural experience); and (c) interacting with the technological application while having fun, and sharing their opinion and comments on social media (technological experience).

As mentioned previously, the RTEM App (Ramos *et al.*, 2018) has (i) a smart navigation module (smart itinerary), (ii) business intelligence tools combined with a machine learning approach to analyse the information collected in social media, to potentiate the discovery of new consumption patterns of tourists and residents, and (iii) intelligent interfaces with accessibility concerns that permit to develop an interface with personalized information according to each tourist profile (e.g., age and limitations) (Ramos & Garcia, 2019). In addition, (iv) the applications present stories associated with the region or destination, by creating gamified thematic itineraries connected to religious monuments that tourists have to follow and visit. If they complete the itinerary they are presented with rewards, such as a "free coffee" or a "ticket to a local event" (Ramos *et al.*, 2016).

From left to right, Fig. 6 starts by showing a map of the Algarve where PoI, represented by red circles, correspond to cities. For each of those PoIs (cities) there is a secondary level of new PoIs, being an example showed in the figure's second map. In the latter case, each of those PoI corresponds to a religious monument in Faro, and each has many challenges, as shown in the third and fourth image, the Faro's cathedral. Entering the cathedral, fourth image, new PoIs (third level level) are available. In the present case, the user must see (point the mobile to) all marked objects inside the cathedral. If the visit is completed, he/she is offered with a coffee. The application works as a game, where the user must finish each level, i.e., visit each PoIs.

In this project, it is projected to apply DPIf in the future, in a way to adapt the UI to each user, and by selecting which "buildings" and objects to see, in which order, enhancing the user's experience.

Intelligent Fresh Food Fleet

Intelligent Fresh Food Fleet Router (i3FR) was a project aiming to, among other things, do real-time distribution planning of dry, fresh, and frozen goods (Cardoso *et al.,* 2015a; Cardoso *et al.,* 2015b). The proposed optimization system integrates with an existing enterprise resource planning (ERP) software without causing major disruption to the current distribution process of companies and can deal with dynamic problems, namely, the inclusion of new orders while already optimizing the routes without restarting the optimization process. The project had two main modules developed: (a) a hardware system to acquire data from the distribution vehicles, that will be used to validate the smart routes and provide information about other parameters, such as the vehicles chambers' temperatures, vehicles' fuel cost, travelled kilometres, and the vehicles' GPS data; and (b) a routing optimization platform which computes

Figure 6. Religious tourism experience (see text)

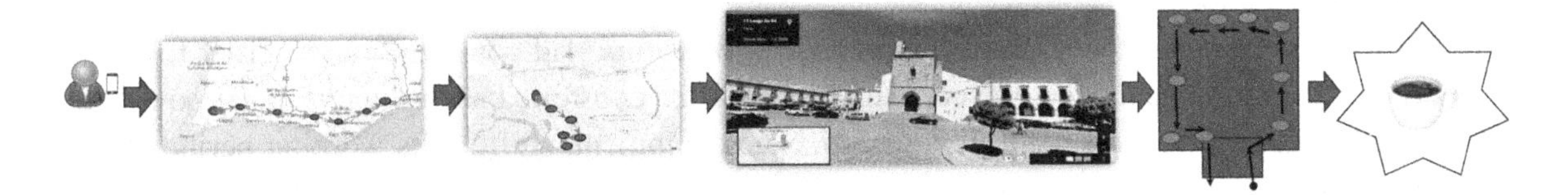

routes from the depots to the delivery points using cartographic information and takes into consideration multiple objectives.

The i3FR smart routing system is divided in four main components (see Fig. 7 left), namely: (i) i3FR-optimizer which is where the actual route optimization algorithm is implemented, taking into account a multiple objectives variant of a stochastic capacitated vehicle routing problem with time windows; (ii) i3FR-hub which implements all the communications occurring inside and to the exterior of the i3FR routing system; (iii) i3FR-database which operates on top of a non-relational database to provide local storage of information relevant to the optimization procedure (e.g., data retrieved from the ERP, cartographic information, and computed routes); and (iv) i3FR-Maps which is a cartography subsystem to retrieve and store routing information from other systems (e.g., from Google Maps or from an Open Source Routing Machine server (Luxen & Vetter, 2011)).

The implemented solution presents augmented intelligence in terms of navigation, creating "smart" on-the-fly routes in accordance with the user's constraints, i.e., lets the users deal with late orders and different states for the routes on-the-fly, which allows to do a phased picking and loading of the vehicles. As mere examples, some results for the Algarve's region are presented in Fig. 7 right.

Smart Systems for Energy Management and Optimization

Smart systems are currently being used in energy management, to increase the energy efficiency of buildings, reduce energy costs, and promote a better usage of the energy that is generated from renewable sources. In fact, the global trend is to increase the amount of energy generated from renewable energy sources. Moreover, different from traditional electrical grids, where the generation was centralized, cur-

Figure 7. Left: i3FR system's global diagram. Right: example of a vehicle route represented over the OpenStreetMap API.

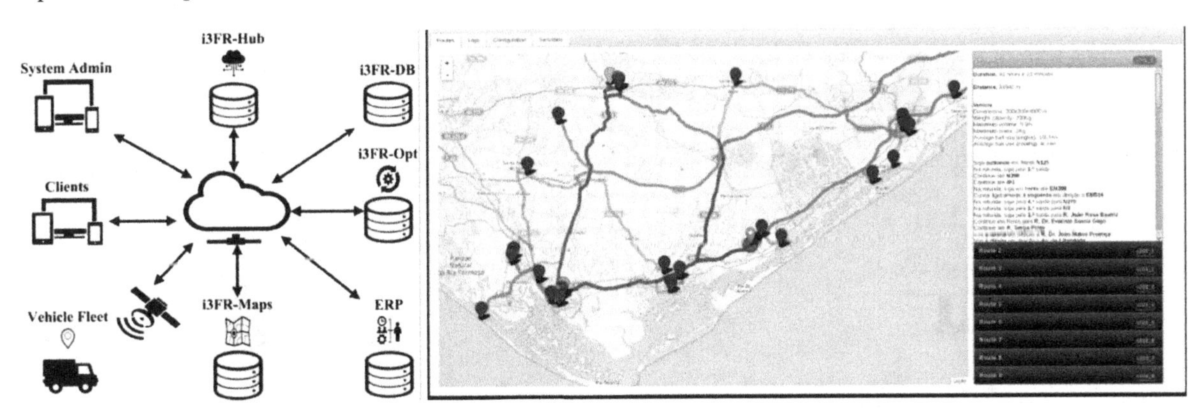

rently, renewable energy generation is being introduced, in a distributed way, in the periphery of electric grids, at the distribution level.

As there must be a constant equilibrium between generation and consumption, in order to use all the energy that is being generated, the consumption must be controlled according with the power that is being produced. This means controlling the electric equipment, shifting its working periods to the moments when generation is available. This may also mean storing the energy when there is surplus of energy, and to consume it when there is lack of generation.

In any of these cases storage and consumption must be controlled according with the current and yet to come generation levels. This requires bidirectional communication with the equipment that consume energy, but also optimization algorithms to schedule these loads, reducing costs and maximizing the energy that results from renewable energy sources. The devices that perform these tasks are called Energy Management Systems (EMSs) (Tie & Tan, 2013).

In this context, the AGERAR project (Storage and Management of Renewable Energy in Commercial and Residential Applications) aims to promote energy efficiency and sustainability in commercial and residential microgrids, increasing the use and improving the management of renewable energies thanks to energy storage systems and the use of information and communication technologies (Agerar, 2019).

In order to achieve these objectives, the project includes the design, development and evaluation of sensors, communications equipment, and web-based platforms for real-time monitoring of microgrids with renewable energy. Figure 8 presents the network architecture of the LoraWAN wireless sensor network developed for the monitoring of the electrical grid and the protocols used in each of the interfaces.

Figure 8. Network architecture of the LoraWAN wireless sensor network used in the monitoring of the electrical grid

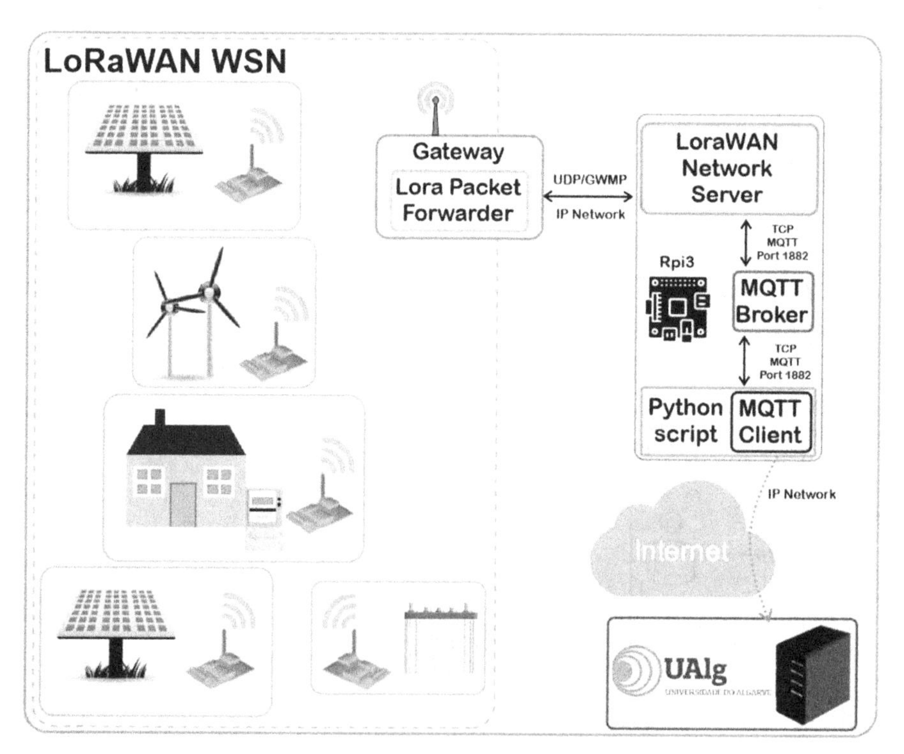

The developed equipment was installed and validated in the experimental facility of the Energy Laboratory of the Instituto Nacional de Técnica Aeroespacial (INTA), in Spain.

Based on these sensor network, specific activities were also defined to enable the implementation of algorithms for load scheduling, building a predictive control type of EMS. The project has also included algorithms for load scheduling of electric vehicles.

CONCLUSION

This chapter illustrates the relation and application of a computer vision framework for describing humans within augmented intelligence systems. The focus of the chapter is to illustrate that augmented intelligence can be applied in many ("all") recent applications, from augmented reality, to navigation, to tourism, to energy.

It is important to stress, that none of these smart systems has the goal to replace humans, but instead to give a better quality of life to them. Nevertheless, it is expected that some of the jobs that exists in the past will be changed to new jobs that already are starting to exist or will appear soon.

As a final conclusion, we are in the middle of an industrial revolution, Industry 4.0 (Lasi *et al.*, 2014), which is the name given to automation and data exchange in manufacturing technologies, industry with smart systems. It includes cyber-physical systems, the internet of things, cloud computing and cognitive computing, and basically augmented intelligence. The same happens also with the Tourism 4.0 (Tourism4.0., 2018), where the digital e the augmented intelligence has a big component.

FUTURE RESEARCH DIRECTIONS

The future research will focus on human behaviour, by the connection with body expression as well as facial expression to reflect the emotional state as key element in communication, in a way to achieve human-machine cooperation. In the future, it is fundamental to allow our systems to interact with humans like the communication human-human occurs, interpreting small behaviour signals. The focal point will be the study of strategies to implement systems that allow machines to cooperate, or to learn how to cooperate, with humans, in a way they will be able to assess and adapt to human goals.

Establish HMC in which the machine/computer provides the "right" information and functionality at the "right" time and in the "right" way to humans will be the holy grail.

ACKNOWLEDGMENT

This work was supported by the Portuguese Foundation for Science and Technology (FCT), project LARSyS (UID/EEA/50009/2019), CIAC (UID/Multi/04019/2019), CinTurs (UID/SOC/04020/2019), CEFAGE (UID/ECO/04007/2019), ACCES4ALL: Accessibility for All in Tourism (SAICT-POL/23700/2016), Portugal2020, CRESC2020, PO Norte 2020, FEDER (Fundo Europeu de Desenvolvimento Regional) and by the European Union, under the FEDER and INTERREG programs, in the scope of the AGERAR (0076_AGERAR_6_E) project.

REFERENCES

Abadi, M., Barham, P., Chen, J., Chen, Z., Davis, A., Dean, J., ... Kudlur, M. (2016) Tensorflow: A system for large-scale machine learning. In *12th Symposium on Operating Systems Design and Implementation*, pp. 265-283.

Agerar. (2019). Agerar Project, Retrieved from http://institucional.us.es/agerar/, accessed 2017/10/11

Alves, R., Sousa, L., Negrier, A., Rodrigues, J. M. F., Monteiro, J., Cardoso, P., ... Bica, P. (2018). Interactive 360 Degree Holographic Installation. *International Journal of Creative Interfaces and Computer Graphics*, *8*(1), 20–38. doi:10.4018/IJCICG.2017010102

Annalakshmi, M., Roomi, S. M. M., & Naveedh, A. S. (2018). A hybrid technique for gender classification with SLBP and HOG features. *Cluster Computing*, 1–10. doi:10.100710586-017-1585-x

Araya, D. (2018). *Augmented Intelligence: Smart Systems and the Future of Work and Learning*. Peter Lang International Academic Publishers. doi:10.3726/b11342

Araya, D. (2019). Forges: 3 Things You Need to Know About Augmented Intelligence, Retrieved from https://goo.gl/oVUsHf, in 2019/03/25

Arriaga, O., Valdenegro-Toro, M., & Plöger, P. (2017) Real-time convolutional neural networks for emotion and gender classification. *arXiv preprint arXiv:1710.07557*

AzureFace. (2019). Azure Face API, Retrieved from https://azure.microsoft.com/en-us/services/cognitive-services/face/, last accessed 2019/04/09

AzureVision. (2019). Cognitive Services Directory – Vision (2019). Retrieved from https://azure.microsoft.com/en-us/services/cognitive-services/directory/vision/

Baltrusaitis, T., Zadeh, A., Lim, Y. C., & Morency, L. P. (2018) Openface 2.0: Facial behavior analysis toolkit. In *Procs 13th IEEE International Conference on Automatic Face & Gesture Recognition (FG 2018),* pp. 59-66, IEEE.

Bay, H., Tuytelaars, T., & van Gool, L. (2008). SURF: Speeded-up robust features. *Computer Vision and Image Understanding*, *110*(3), 346–359. doi:10.1016/j.cviu.2007.09.014

Beyeler, M. (2015). OpenCV with Python Blueprints. Learning to Recognize Emotions on Faces; Retrieved from https://www.packtpub.com/mapt/book/application_development/9781785282690/7/ch07lvl1sec53/facial-expression-recognition, accessed 2017/10/11.

Bradski, G. R., & Pisarevsky, V. (2000) Intel's Computer Vision Library: applications in calibration, stereo segmentation, tracking, gesture, face and object recognition. In *Procs IEEE Conference on Computer Vision and Pattern Recognition (CVPR 2000)*, vol. 2, pp. 796-797 10.1109/CVPR.2000.854964

Brown, M. 2011. Learning Local Image Descriptors Data. Retrieved from http://www.cs. ubc.ca/mbrown/patchdata/patchdata.html, accessed 2017/10/11

Cao, Z., Hidalgo, G., Simon, T., Wei, S. E., & Sheikh, Y. (2018). OpenPose: realtime multi-person 2D pose estimation using Part Affinity Fields. *arXiv preprint arXiv:1812.08008.*

Cao, Z., Simon, T., Wei, S. E., & Sheikh, Y. (2017). Realtime multi-person 2D pose estimation using part affinity fields. In *Proceedings of the IEEE Conference on Computer Vision and Pattern Recognition* (pp. 7291-7299). 10.1109/CVPR.2017.143

Cardoso, P. J., Schütz, G., Mazayev, A., Ey, E., & Corrêa, T. (2015b). A solution for a real-time stochastic capacitated vehicle routing problem with time windows. *Procedia Computer Science*, *51*, 2227–2236. doi:10.1016/j.procs.2015.05.501

Cardoso, P. J. S., Guerreiro, P., Monteiro, J., & Rodrigues, J. M. F. (2018). Applying an Implicit Recommender System in the Preparation of Visits to Cultural Heritage Places. In M. Antona, & C. Stephanidis (Eds.), *Universal Access in Human-Computer Interaction 2018, LNCS 10908* (pp. 421–436)., doi:10.1007/978-3-319-92052-8_33

Cardoso, P. J. S., Schütz, G., Mazayev, A., & Ey, E. (2015a). Solutions in Under 10 Seconds for Vehicle Routing Problems with Time Windows using Commodity Computers. In *Proceedings 8th International Conference on Evolutionary Multi-Criterion Optimization (EMO 15),* pp. 418-432, March 29-April 1. Cham, Switzerland: Springer.

Castillo, J. C., Castro-González, Á., Fernández-Caballero, A., Latorre, J. M., Pastor, J. M., Fernández-Sotos, A., & Salichs, M. A. (2016). Software architecture for smart emotion recognition and regulation of the ageing adult. *Cognitive Computation*, *8*(2), 357–367. doi:10.100712559-016-9383-y

Crandall, J. W., Oudah, M., Ishowo-Oloko, F., Abdallah, S., Bonnefon, J. F., Cebrian, M., ... Rahwan, I. (2018). Cooperating with machines. *Nature Communications*, *9*(1), 233. doi:10.103841467-017-02597-8 PMID:29339817

Datta, S. (2016). Learning OpenCV 3 Application Development. Chapter 6. Face Detection Using OpenCV. Retrieved from https://www.packtpub.com/mapt/book/application_development/9781784391454/6/ch06lvl1sec64/gender-classification, accessed 2019/06/05

Doerrfeld, B. (2016). 20+ Emotion Recognition APIs That Will Leave You Impressed, and Concerned. Retrieved from https://nordicapis.com/20-emotion-recognition-apis-that-will-leave-you-impressed-and-concerned/, accessed 2017/10/11.

Face++. (2019). Face++ Cognitive Services. Retrieved from https://www.faceplusplus.com, accessed 2017/10/11

Fischer, G. (2001). User modeling in human–computer interaction. *User Modeling and User-Adapted Interaction*, *11*(1-2), 65–86. doi:10.1023/A:1011145532042

Girija, S. S. (2016). Tensorflow: Large-scale machine learning on heterogeneous distributed systems. Software available from tensorflow.org.

GoogleVision. (2019). Google Cloud Vision. Retrieved from https://cloud.google.com/vision/, last accessed 2019/04/09

Hastreiter, N. (2017). Huffpost: What Impact Will AI Have In The Next 20 Years? Retrieved from https://goo.gl/SdtUF3, in 2019/03/25

Hidalgo, G., Cao, Z., Simon, T., Wei, S., Joo, H., Sheikh, Y., . . . Raaj, Y. (2019). OpenPose. Retrieved from https://github.com/CMU-Perceptual-Computing-Lab/openpose, accessed 2019/06/05

Hsiao, K. A., & Hsieh, P. L. (2014). Age difference in recognition of emoticons. In *Proceedings International Conference on Human Interface and the Management of Information* (pp. 394-403). Cham, Switzerland: Springer.

IntelDev. (2019) OpenVINO: Develop Multiplatform Computer Vision Solutions. Retrieved from https://software.intel.com/en-us/openvino-toolkit, accessed 2019/06/05

Jones, A., Subrahmanian, E., Hamins, A., & Grant, C. (2015). Humans' critical role in smart systems: A smart firefighting example. *IEEE Internet Computing*, *19*(3), 28–31. doi:10.1109/MIC.2015.54

Keras. (2019). Keras: The Python Deep Learning library. Retrieved from https://keras.io/, last accessed 2019/04/09

Kumar, A. (2018) Demonstration of Facial Emotion Recognition on Real Time Video Using CNN: Python & Keras. Retrieved from https://appliedmachinelearning.blog/2018/11/28/demonstration-of-facial-emotion-recognition-on-real-time-video-using-cnn-python-keras/, last accessed 2019/04/09.

Lasi, H., Fettke, P., Kemper, H. G., Feld, T., & Hoffmann, M. (2014). Industry 4.0. *Business & Information Systems Engineering*, *6*(4), 239–242. doi:10.100712599-014-0334-4

Leutenegger, S., Chli, M., & Siegwart, R. Y. (2011) BRISK: binary robust invariant scalable keypoints. *Proc. Int. Conf. Computer Vision (ICCV)*, 2548–2555

Levi, G., & Hassner, T. (2015). Age and gender classification using convolutional neural networks. In *Proc. of the IEEE conference on computer vision and pattern recognition workshops* (pp. 34-42). Retrieved from https://talhassner.github.io/home/publication/2015_CVPR, last accessed 2019/04/09

Lowe, D. G. (2004). Distinctive image features from scale-invariant keypoints. *International Journal of Computer Vision*, *60*(2), 91–110. doi:10.1023/B:VISI.0000029664.99615.94

Luxen, D., & Vetter, C. (2011). Real-time routing with OpenStreetMap data. In *Proc. of the 19th ACM SIGSPATIAL international conference on advances in geographic information systems* (pp. 513-516). New York, NY: ACM.

Matsumoto, D., & Ekman, P. (2008). Facial expression analysis. *Scholarpedia*, *3*(5), 4237. doi:10.4249cholarpedia.4237

Momeni-K, M., Diamantas, S. C., Ruggiero, F., & Siciliano, B. (2012). Height Estimation from a Single Camera View. In *Procs Int. Conf. on Computer Vision Theory and Applications*, pp. 358-364.

Muhammad, G., Alsulaiman, M., Amin, S. U., Ghoneim, A., & Alhamid, M. F. (2017). A Facial-Expression Monitoring System for Improved Healthcare in Smart Cities. *IEEE Access: Practical Innovations, Open Solutions*, *5*, 10871–10881. doi:10.1109/ACCESS.2017.2712788

Nourani. (2017). Gender classification by LBP. Retrieved from https://github.com/nourani/LBP, accessed 2019/06/05.

OpenCV. (2019). Open Source Computer Vision Library. Retrieved from https://opencv.org/, last accessed 2019/04/09.

OpenCV. (2019). Gender Classification with OpenCV. Retrieved from https://docs.opencv.org/3.0-last-rst/modules/face/doc/tutorial/facerec_gender_classification.html, last accessed 2019/04/09

Panetta, K. (2018) Gartner Top 10 Strategic Technology Trends for 2019. Retrieved from https://goo.gl/zqgoXP, in 2019/03/25.

Pasquinelli, M. (2015). Alleys of your mind: augmented intelligence and its traumas (p. 212). Meson Press.

Ponnusamy, A. (2019) cvlib: A high level easy-to-use open source Computer Vision library for Python. Retrieved from https://www.cvlib.net/, accessed 2019/06/05

Raj, A., Gandhi, K., Nalla, B. T., & Verma, N. K. (2019). Object Detection and Recognition Using Small Labeled Datasets. In *Computational Intelligence: Theories, Applications, and Future Directions-Volume II* (pp. 407–419). Singapore: Springer.

Ramos, C. M. Q., & Garcia, A. M. (2019). Analysis of the Contribution of ICT to Cultural and Religious Tourism: In Communicating Religious Heritage to Visitors and Tourists. In J. Álvarez-García, M.C. Río Rama, & M. Gómez-Ullate (Eds.), Handbook of Research on Socio-Economic Impacts of Religious Tourism and Pilgrimage (pp. 167-194). Hershey, PA: IGI Global.

Ramos, C. M. Q., Henriques, C., & Lanquar, R. (2016). Augmented Reality for Smart Tourism in Religious Heritage Itineraries: Tourism Experiences in the Technological Age. In J. M. F. Rodrigues, P. Cardoso, J. Monteiro, & M. Figueiredo (Eds.), Handbook of Research on Human-Computer Interfaces, Developments, and Applications (pp. 250-278). Hershey, PA: IGI Global.

Ramos, C. M. Q., Henriques, C., & Rodrigues, J. M. F. (2018) Religious Tourism Experience Model (RTEM): A Recommendation Model for Dissemination of Cultural and Religious Heritage. In J. M. F. Rodrigues, C. M. Q. Ramos, P. Cardoso, & C. Henriques (Eds.), Technological Developments for Cultural Heritage and eTourism Applications (pp. 1-29). Hershey, PA: IGI Global.

Rekognition. (2019) Amazon Rekognition. Retrieved from https://docs.aws.amazon.com/rekognition/, last accessed 2019/04/09

Rodrigues, J. M. F., Alves, R., Sousa, L., Negrier, A., Cardoso, P. J. S., Monteiro, J., . . . Bica, P. (2016). PRHOLO - 360° Interactive Public Relations, Chapter 7 of the Handbook of Research on Human-Computer Interfaces, Developments, and Applications (pp. 166 – 191). Hershey, PA: IGI Global. doi:10.4018/978-1-5225-0435-1.ch007

Rodrigues, J. M. F., Cardoso, P. J. S., Lessa, J., Pereira, J. A. R., Sardo, J. D. P., Freitas, M., . . . Bica, P. (2018). An Initial Framework to Develop a Mobile 5 Sense Museum System, Chapter 5 of Technological Developments for Cultural Heritage and eTourism Applications (pp. 97- 119). Hershey, PA: IGI Global. Doi:10.4018/978-1-5225-2927-9.ch005

Rodrigues, J. M. F., Cardoso, P. J. S., Lessa, J., Pereira, J. A. R., Sardo, J. D. P., Freitas, M., . . . Bica, P. (2018d). An Initial Framework to Develop a Mobile 5 Sense Museum System, Chapter 5 of Technological Developments for Cultural Heritage and eTourism Applications (pp. 97- 119). Hershey, PA: IGI Global. Doi:10.4018/978-1-5225-2927-9.ch005

Rodrigues, J. M. F., Martins, M., Sousa, N., & Rosa, M. (2018b) IoE Accessible Bus Stop: an initial concept. In *Proc. 8th International Conference on Software Development and Technologies for Enhancing Accessibility and Fighting Info-exclusion* (DSAI 2018), pp. 137-143. New York, NY: ACM. DOI: 10.1145/3218585.3218659

Rodrigues, J. M. F., Pereira, J. A. R., Sardo, J. D. P., Freitas, M. A. G., Cardoso, P. J. S., Gomes, M., & Bica, P. (2017). Adaptive Card Design UI Implementation for an Augmented Reality Museum Application. In M. Antona, & C. Stephanidis (Eds.), *Universal Access in Human-Computer Interaction 2017, Part I, LNCS 10277* (pp. 433–443). doi:10.1007/978-3-319-58706-6_35

Rodrigues, J. M. F., Veiga, R. J. M., Bajireanu, R., Lam, R., Pereira, J. A. R., Sardo, J. D. P., ... Bica, P. (2018c). Mobile Augmented Reality Framework – MIRAR. In M. Antona, & C. Stephanidis (Eds.), *Universal Access in Human-Computer Interaction 2018, LNCS 10908* (pp. 102–121). doi:10.1007/978-3-319-92052-8_9

RTFE. Real Time Facial Expression Recognition with Deep Learning. Retrieved from https://github.com/a514514772/Real-Time-Facial-Expression-Recognition-with-DeepLearning, last accessed 2019/04/09

Rublee, E., Rabaud, V., Konolige, K., & Bradski, G. R. (2011). Orb: an efficient alternative to SIFT or SURF. *Proc. Int. Conf. Computer Vision (ICCV)*, 2564–2571 10.1109/ICCV.2011.6126544

Rukavina, S., Gruss, S., Hoffmann, H., & Traue, H. C. (2016). Elderly People Benefit More from Positive Feedback Based on Their Reactions in the Form of Facial Expressions during Human-Computer Interaction. *Psychology (Irvine, Calif.)*, *7*(09), 1225–1230. doi:10.4236/psych.2016.79124

Saleiro, M., Terzic, K., Rodrigues, J. M. F., & du Buf, J. M. H. (2017). BINK: Biological Binary Keypoint Descriptor. *Bio Systems*, *162*, 147–156. doi:10.1016/j.biosystems.2017.10.007 PMID:29031966

Sardo, J. D. P., Semião, J., Monteiro, J. M., Esteves, E., Pereira, J., Freitas, M., Rodrigues, J. M. F. (2017). Portable Device for Touch, Taste and Smell Sensations in Augmented Reality Experiences, In Proceedings INCREaSE, SPRINGER LNCS XXX, pp. 307–322. doi:10.1007/978-3-319-70272-8_26

Shaikh, I. (2016). Emotion Recognition Using Scikit-learn & OpenCV. Retrieved from https://github.com/its-izhar/Emotion-Recognition-Using-SVMs, accessed 2017/10/11

Silva, A. (2016). Facial Expression Recognition with OpenCV 3+. Retrieved from http://andrew-silva.com/2016/11/09/facial-expression-recognition-with.html, accessed 2017/10/11

Sousa, L., Alves, A., & Rodrigues, J. M. F. (2016). Augmented reality system to assist inexperienced pool players. *Computacional Visual Media*, *2*(2), 183–193. doi:10.100741095-016-0047-3

TensorFlow. (2019). TensorFlow: An end-to-end open source machine learning platform. Retrieved from https://www.tensorflow.org/, last accessed 2019/04/09

Terzic, K., Rodrigues, J. M. F., & du Buf, J. M. H. (2015). BIMP: A real-time biological model of multi-scale keypoint detection in V1. *Neurocomputing*, *150*, 227–237. doi:10.1016/j.neucom.2014.09.054

Tian, Y., Li, X., Wang, K., & Wang, F. Y. (2018). Training and testing object detectors with virtual images. *IEEE/CAA Journal of Automatica Sinica, 5*(2), 539-546.

Tie, S. F., & Tan, C. W. (2013). A review of energy sources and energy management system in electric vehicles. *Renewable & Sustainable Energy Reviews*, *20*, 82–102. doi:10.1016/j.rser.2012.11.077

Toscano, L. (2017) Emotions Detection Via Facial Expressions with Python & OpenCV. Retrieved from https://www.apprendimentoautomatico.it/articles/17/emotions-detection-via-facial-expressions-with-python-opencv, accessed 2017/10/11

Tourism4.0. (2018). What is Tourism 4.0? Available at https://www.tourism4-0.org/ last accessed 2019/07/08.

Trzcinski, T., Christoudias, C. M., Fua, P., & Lepetit, V. (2013) Boosting binary keypoint descriptors. *Proc. Int. Conf. Computer Vision and Pattern Recognition (CVPR)*, 2874–2881.

Turner, D., Rodrigues, J. M. F., & Rosa, M. (2019). Describing People: An Integrated Framework for Human Attributes Classification, accepted INternational CongRess on Engineering and Sustainability in the XXI cEntury – INCREaSE 2019.

van Gent, P. (2016). Emotion Recognition With Python, OpenCV and a Face Dataset. A tech blog about fun things with Python and embedded electronics. Retrieved from http://www.paulvangent.com/2016/04/01/emotion-recognition-with-python-opencv-and-a-face-dataset/, accessed 2017/10/11

van Maanen, P. P., Lindenberg, J., & Neerincx, M. A. (2005). Integrating human factors and artificial intelligence in the development of human-machine cooperation. In *Proc. of the 2005 international conference on artificial intelligence (ICAI'05)*

Yaddaden, Y., Bouzouane, A., Adda, M., & Bouchard, B. (2016). A New Approach of Facial Expression Recognition for Ambient Assisted Living. In *Proceedings of the 9th ACM International Conference on PErvasive Technologies Related to Assistive Environments* (p. 14). New York, NY: ACM. 10.1145/2910674.2910703

Yu, X., Salmon, C. T., & Leung, C. (2015). Emotional interactions between artificial companion agents and the elderly. In *Proceedings of the 2015 International Conference on Autonomous Agents and Multiagent Systems* (pp. 1991-1992). International Foundation for Autonomous Agents and Multiagent Systems.

Zhang, C. (2018). Easy Real time gender age prediction from webcam video with Keras. Retrieved from https://www.dlology.com/blog/easy-real-time-gender-age-prediction-from-webcam-video-with-keras/, last accessed 2019/04/09

ADDITIONAL READING

Amos, B., Ludwiczuk, B., & Satyanarayanan, M. Openface (2016) A general-purpose face recognition library with mobile applications. CMU School of Computer Science Cardoso, P. J., Monteiro, J., Semião, J. & Rodrigues, J. M. F. (2019) Handbook of Research on Harnessing the Internet of Everything (IoE) for Accelerated Innovation Opportunities, (pp. 1-300). Hershey, PA: IGI Global, ISBN13: 9781522573326, ISBN10: 1522573321, EISBN13: 9781522573333. Doi:10.4018/978-1-5225-7332-6

Ekman, P., & Friesen, W. (1978). *Facial action coding system (FACS): Manual*. Palo Alto: Consulting Psychologists Press.

FDHC. Face Detection using Haar Cascades, https://docs.opencv.org/4.1.0/d7/d8b/tutorial_py_face_detection.html, last accessed 2019/04/09

Rodrigues, J. M. F., Cardoso, P., Monteiro, J., & Figueiredo, M. (Eds.). (2016) Handbook of Research on Human-Computer Interfaces, Developments, and Applications, IGI Global, 643pp. ISBN13: 9781522504351; ISBN10: 1522504354; EISBN13: 9781522504368. Doi:10.4018/978-1-5225-0435-1

Rodrigues, J. M. F., Ramos, C. M., Cardoso, P. J., & Henriques, C. (2018). Handbook of Research on Technological Developments for Cultural Heritage and eTourism Applications (pp. 1-506). Hershey, PA: IGI Global. doi: , ISBN13: 9781522529279, ISBN10: 1522529276, EISBN13: 9781522529286 doi:10.4018/978-1-5225-2927-9

Viola, P., & Jones, M. (2001). Rapid object detection using a boosted cascade of simple features. *CVPR*, *1*(1), 511–518.

KEY TERMS AND DEFINITIONS

Augmented Intelligence: can be considered as an alternative conceptualization of artificial intelligence that focuses on AI's assistive role in advancing human capabilities.

Computer Vision: Computer vision is an interdisciplinary scientific field that deals with how computers can be made to gain high-level understanding from digital images or videos.

Smart Actions: In the context of this chapter, consist in augmenting user's actions and/or decisions, by using devices or additional information.

Smart Grid: An electrical grid that incorporates ICT to enable both machine-to-machine and human-to-machine communications and the optimization of the managed resources.

Smart Routing: Intelligent system for navigation of vehicles, based on ICT technologies.

Smart Systems: Are stand-alone systems or systems that help users to make decisions incorporating available data from different sensing, in a way to control and make smart actions.

Smart Tourism: Refers to the application of information and communication technology (ICT), mobile communication, cloud computing, artificial intelligence, and virtual reality, for developing innovative tools and approaches to improve tourism.

Chapter 2
Deep Learning on Edge: Challenges and Trends

Mário P. Véstias
https://orcid.org/0000-0001-8556-4507
INESC-ID, ISEL, Instituto Politécnico de Lisboa, Portugal

ABSTRACT

Deep learning on edge has been attracting the attention of researchers and companies looking to provide solutions for the deployment of machine learning computing at the edge. A clear understanding of the design challenges and the application requirements are fundamental to understand the requirements of the next generation of edge devices to run machine learning inference. This chapter reviews several aspects of deep learning: applications, deep learning models, and computing platforms. The way deep learning is being applied to edge devices is described. A perspective of the models and computing devices being used for deep learning on edge are given, as well as what challenges face the hardware designers to guarantee the vast set of tight constraints like performance, power consumption, flexibility, etc. of edge computing platforms. Finally, a trends overview of deep learning models and architectures is discussed.

INTRODUCTION

Machine learning algorithms and, in particular, deep learning brought Artificial Intelligence to many application domains. In a deep learning workflow data is gathered and prepared for training the machine learning model. In the training step, deep learning models are trained with a large set of known instances so that they can classify new inputs not used during the training step. These trained models are then deployed for inference. Inference is when the trained model is used to classify new and unknown data instances.

Training is computationally heavy and requires high-performance computing platforms that still take hours or even days to train large deep learning models. Inference is orders of magnitude less demanding in terms of computation and can also be deployed in the same computing platform used for training. The common process is to use the high-performance computing platform for both training and inference. In many cases, data to be processed by the deep neural model is received from an edge device (any

DOI: 10.4018/978-1-7998-2112-0.ch002

hardware device that serves as an entry point of data and may store, process and/or send the data to a central server) and the inference result is sent back to the edge device. However, in a vast set of applications (security, surveillance, facial recognition, autonomous car driving, industrial, etc.) this round-trip method of doing inference is inefficient or unfeasible. Running the inference near the source of data is advantageous and in some cases necessary so that important information can be extracted in site and if necessary at real-time instead of sending data to the cloud and wait for the inference classification. Whenever the communication latency and data security violations are undesirable, like autonomous vehicles, local processing at the sensor is a requirement. In these cases, inference is done at the edge avoiding long data communications and high computing latencies. For these reasons, many deep learning tasks are migrating from the cloud of high-performance computing platforms to the low cost, low density embedded devices at the edge.

This brings new problems and open issues to the design of machine learning models at edge devices since running deep learning on edge is subject to different performance, memory and cost requirements then those considered by cloud computing design processes. Cloud inference is focused on delivering high performance inference with the highest model accuracy. Edge inference benefits from high accuracy models but achieving the highest accuracy is not the main metric. Cost, performance, energy, real-time, size are some of the most important design parameters considered when implementing computing platforms for edge inference on edge.

Deep learning on edge has been attracting the attention of researchers and companies looking to provide solutions for the deployment of machine learning computing at the edge. A clear understanding of the design challenges and the application requirements are fundamental to understand the requirements of the next generation of edge devices to run machine learning inference.

In this chapter several aspects of deep learning: applications, deep learning models and computing platforms, will be reviewed. Then the way deep learning is being applied to edge devices will be describes. A perspective of the models and computing devices that are being used for deep learning on edge will be given, what challenges are facing the hardware designers to guarantee the vast set of tight constraints like performance, power consumption, flexibility, etc. of edge computing platforms. Finally, a trends overview of deep learning models and architectures will be discussed.

BACKGROUND

Machine learning is a subfield of artificial intelligence whose objective is to give systems the capacity to learn and improve by its own without being explicitly programmed to do it. Machine learning algorithms extract features from data and build models from it so that new decisions and new outcomes are produced without being programmed a priori with these models and rules.

There are many types of machine learning algorithms with different approaches and application targets: Bayesian (Barber, 2012), clustering (Bouveyron et al., 2019), instance-based (Keogh, 2011), ensemble (Zhang, 2012), artificial neural network (Haykin, 2008), deep learning network (Patterson & Gibson, 2017), decision tree (Quinlan, 1992), association rule learning (Zhang & Zhang, 2002), regularization (Goodfellow et al., 2016), regression (Matloff, 2017), support-vector machine (Christmann & Steinwart, 2008) and others.

Different problems require different models and algorithms and so each of these algorithms apply to different types of data sets and applications. All these algorithms can be broadly classified according to the learning style: supervised, unsupervised and semi-supervised. Supervised machine learning algorithms (e.g., regression, artificial neural network and deep learning network) train the model of the algorithm with training data. Each instance of the training data has an associated label that identifies the expected result for each particular input. In the training process, the model is corrected and adjusted according to the expected outcomes. In the unsupervised class of algorithms (e.g., dimensionality reduction, k-means clustering, etc.), the training data do not have an expected outcome. The algorithms in this class try to extract features from the input data and cluster input data according to these features without any previous knowledge of the input data characteristics. The semi-supervised algorithms mixes both previous classes, that is, there is a desired outcome but the model must learn features to classify data.

Among the many machine algorithms this chapter is concerned with deep learning algorithms whose ground are artificial neural networks (ANN). ANNs are inspired by the structure of the human brain consisting of interconnected neurons. Theoretically, an ANN is a universal model capable to learn any function (Hornik et al., 1989). Deep learning is basically deep artificial neural networks with several and more complex layers designated deep neural networks. Since the introduction of deep learning that several models have been proposed, like convolutional neural network, recurrent neural network, deep belief network, deep Boltzmann machine, Kohonen self-organizing neural network, modular neural network and stacked auto-encoder.

Deep Learning

The grounds of deep learning models are artificial neural networks. Before proceeding with a description of deep neural networks the following section describes the fundamentals of ANNs

Figure 1. Artificial neural network

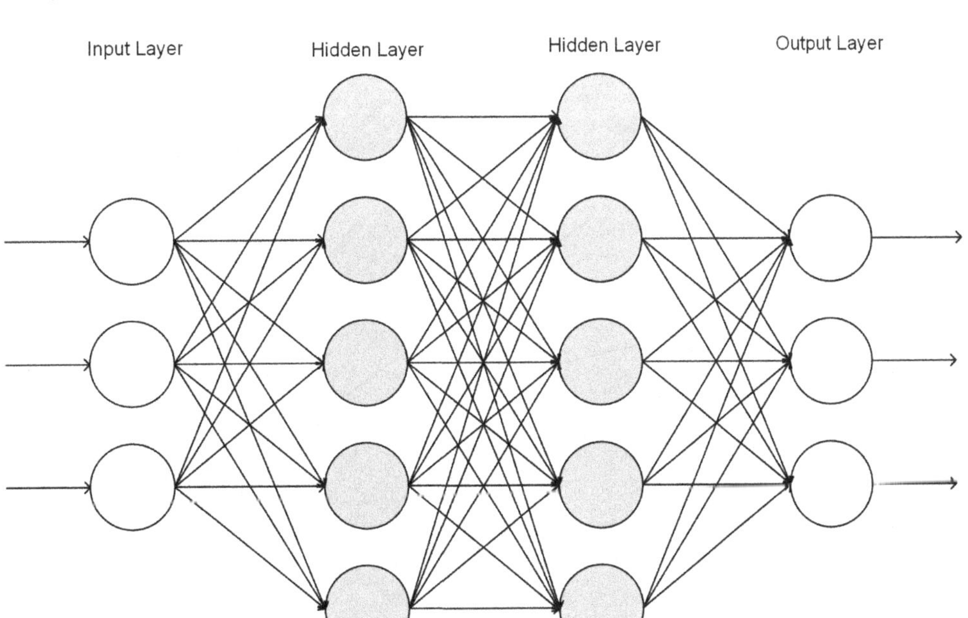

Artificial Neural Network

An artificial neural network (Haykin, 2008) consists of a basic structure known as perceptron or neuron organized in a series of layers. The first layer is the input layer, the last one is the output layer and all the other layers between the input and the output layer are known as hidden layers (see figure 1).

Neurons in the input layer receive input data and generate outcomes for the neurons of the first hidden layer, while the output layer receives the outcomes from the last hidden layer and produces the classification associated with the input data. These are feedforward networks which the underlying graph contains no feedback connections or cycles and are the focus of this chapter. Each neuron of an artificial neural network generate an output which is a function of all its inputs and sends it to all nodes of the next layer, except the output layer that does not have a next layer and so the outcomes of the neuron are the output results. The first artificial neural networks had only one hidden layer. Recently, this number has increased considerably and according to (Bengio, 2009) when a neural network has more than three layers is referred to as deep neural network.

A perceptron encodes n inputs, $\{x_1, x_2, x_3,\ldots, x_n\}$, from neurons of the previous layer using a vector of weights or parameters $\{w_1, w_2, w_3,\ldots, w_n\}$ associated with the connections between previous perceptrons and the target perceptron that determines the importance of the corresponding input to the perceptron being calculated. Each perceptron still has an additional bias value that is used to shift the output to better fit the data (see figure 2).

The output of a perceptron, y, its prediction value, is computed as a function of the weights and the bias:

$$y = f\left(b + \sum_{k=0}^{n-1} w_k x_k\right)$$

Figure 2. Perceptron or neuron

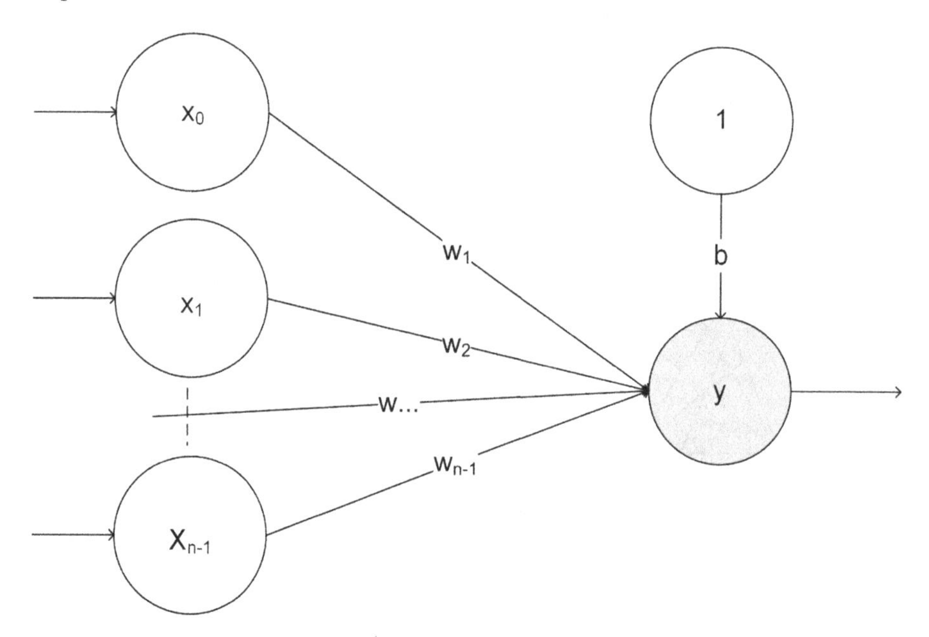

Function *f(.)* referred to as activation function determines the output of the neuron. In its simplest form the activation function is binary, that is, the output is inactive, '0', or active, '1'. While simple, requires that many neurons are used for a non-linear separation of classes. A linear function can be used instead that linearly rates the output between two values. These functions exhibit similar problems of a binary function and since the output is unbounded, it leads to an unstable convergence. Instead, normalized functions are used with better properties for classification and learning stability (Nwankpa et al, 2018).

One of the first activation functions was the sigmoid given by $\frac{1}{1+e^{-z}}$ that predicts the probability of the output with values between 0 and 1, and the hyperbolic tangent, $(e^x - e^{-x})/(e^x + e^{-x})$ that increases the output range to]-1, 1[. Many other activation functions were proposed during the last decades but one of the most recently used is the ReLU (Rectified Linear Unit) that is 0 if the output is less or equal than 0, and 1 otherwise. In (Glorot et al., 2011) it was shown that ReLU leads to better training of deep neural networks. Previous activation functions apply to a single set. In many models, the output layer provides outputs for multiple classes. To associate values to multiple classes the softmax function is used. This function takes as input a vector of *k* values and normalizes it into a probability distribution of *k* probabilities as follows:

$$f_i(x) = \frac{e^{x_i}}{\sum_{j=1}^{K} e^{x_j}}$$

The weights of a neural network model must be adjusted for each specific problem and for the best network accuracy. Determining all weights of the network for best accuracy is known as the training step. Training can be supervised, unsupervised or semi-supervised. In the supervised training the set of weights, *W*, is found with the help of an objective function that quantifies the error, *E*, between the measured outputs for a particular set of weights, y_m, and the expected outputs, t_m, through all *N* data inputs, x_n. The error is calculated as the sum of the squared error:

$$E(W) = \sum_{n=1}^{N}\sum_{m=1}^{M}\left(y_m(x_n, W) - t_m\right)^2$$

The training algorithm iteratively runs the forward propagation and the backpropagation steps. Starting with an initial set of weights, an input is applied at the input layer of the network and propagated until the output layer to find an output classification. The loss function, E(W), is then applied to determine the mean squared error between the obtained output and the required output. This finishes the forward propagation step.

Then the backpropagation step starts with the objective to adjust the weights to minimize the error function. This is achieved by propagating backwards the value of the loss function to all neurons that contribute to each output neuron. Neurons with a higher contribution to the loss function value of an output neuron receive a higher fraction of this value. When all neurons have received its loss fraction, weights are adjusted to reduce the loss. The adjustment of weights is done with the *gradient descent* (Ruder, 2016) technique as follows:

$$\Delta W[i] = -^{3}\left(\frac{\partial E_n}{\partial w[i]}\right)$$

In gradient descendent, weights are incrementally changed based on the derivative of the loss function and a learning rate, γ. It means that the loss function must be differentiable.

Unless the network is very simple, the model do not achieve 100% accuracy. Therefore, a criteria must be used to stop the training process. Normally the process stops when the accuracy improvement between two training iterations is below a certain threshold.

Gradient descent is an heuristic method and so it does not guarantee the global minimum. Finding a good local minimum depends on weight initialization. Some works have shown that better results can be achieved if the initial weights are randomly chosen within specific ranges (Glorot & Bengio, 2010).

Unsupervised training follows a different process since there are no expected outputs. The model is trained using extracted features from the input data. Semi-supervised training is a mix of both techniques where training is done with some labeled data and some unlabeled data.

Deep Neural Network

Deep neural networks (DNN) are an extension of the traditional artificial neural networks with more layers and different types of layers. Therefore in DNN each layer is trained with the output features extracted by the previous layer. So as the data progress through the network model more features are aggregated that represent more complex representations, like a hierarchy of features.

Unlike most traditional machine learning algorithms, deep learning is able to extract features from the input data without being explicitly programmed to do it. An important fact about deep learning is that the more data is used to train the network the better the accuracy, contrary to other machine learning algorithms. Since the size of the DNN can be freely increased it means that it can be applied to any complex classification problem with a high dimensionality of features.

The concept of a DNN with multiple layers is not new but its feasibility is recent. A neural network with many layers requires intensive computations to be trained. The required computational power is now possible with the recent high-performance computing platforms. To achieve high accuracy DNN also need a large set of data instances to train the network and it is a fact that today designers and developers have access to huge amounts of data to do it.

So, the accuracy of deep neural networks gets better with larger models and when trained with more data. The consequence is that both training and inference of DNN requires high memory size to store weights and feature maps and computational capacity to train it in reasonable times.

Since its introduction as a machine learning model, several types of deep neural networks were proposed differing mainly on how neurons are interconnected. In the following the most important types of DNNs will be briefly explained.

Types of Deep Neural Networks

Feed-Forward Neural Network is the traditional neural network model as explained in the previous section. All layers are dense and have the same structure. Dense layers means that all neurons of a layer connect to all neurons of the previous layer, except the first layer that receives inputs. Theoretically these networks can model any relationship between inputs and outputs with enough hidden neurons but this may lead to impractical implementations and so different neural networks models are adopted.

Convolutional neural networks (CNN) were introduced in (LeCun et al., 1989) to image classification. A feed-forward neural network can be naturally applied to classify image. The problem is that input pixels are modeled with input neurons and so, for reasonably sized images the number of neurons of the input layer is large which requires many parameters from the input to the first hidden layer. Considering images of size 128 × 128, the first layer would have 2^{14} neurons. Assuming the next layer with the same number of nodes, the first hidden layer would require 2^{28} weights. Since a deep neural networks is being used, this number easily increases to an order that turns the training process too hard.

Instead of using a neural network with dense layers, the interconnections between layers take into consideration the type of input data (LeCun, 1989). CNNs takes advantage of the spatial correlation between neighbor pixels to establish dependencies between neurons of different layers, that is, the output of a neuron is the result of the convolution between a small window of weights and the respective output of neurons of the input map. These layers are therefore designated convolutional layers. A CNN contains one or more convolutional layers. Each convolutional layer identify features of the image which are then correlated by the next convolutional layer to learn complex features.

The set of convolutional layers may be followed by one or more fully connected layers which are dense layers identical to those used in feed-forward layers. Since these layers interconnect all neurons of previous layers the complex features extracted so far are globally correlated.

Recurrent neural networks (RNN) were introduced in (Elman, 1990) and are basically dense networks with state. RNNs have a hidden state distributed through all neurons that allows them to store information of previous data. State information is updated in a non-linear way which permits the model to follow complex state sequences. This type of networks is very powerful but must be carefully designed to avoid the vanishing problem where weights converge to extreme values loosing previous information. RNNs have a vast set of applications including also those without an explicit association with a sequence of events. A picture, for example, can be processed as a sequence of pixels. A common application of RNNs is autocompletion where the information of a sequence is automatically determined.

The vanishing problem of RNNs happens because the state of the network is hard to keep for a long time. The **long-short term memory network** – LSTM - (Hochreiter, 1998; Hochreiter & Schmidhuber, 1997) reduces the vanishing problem of RNNs with the introduction of gates and an explicit memory to store states. The memory stores the state until a gate cell tells the memory to forget a state. LSTMs add a cell layer to remember information from a previous iteration of the model. With these improvements LSTM networks were able to execute complex tasks, like music composition.

RNN and LSTM networks may have an unpredictable behavior by following a non-deterministic path or oscillating. To overcome this instability problems the **Hopfield network** (HN) introduced in (Hopfield, 1982) can be used. The HN is the densest neural network since all neurons connect to all other neurons. The network is trained for a set of patterns and only these can be identified by the network, that is, for a particular input the network will converge to one of the stable patterns learned during training. The HN has been shown to be very limited in the number of patterns (15% of the number of neurons) it can learn because of the spurious minima. This limitation is associated with the fact that if two local minima correspondent to two training patterns are too close it may create a single local minima for both and therefore none of the two patterns will be memorized.

The **Boltzmann Machine** (BM) network is an unsupervised model that was introduced in (Hinton & Sejnowski, 1986). His model is similar to the Hopfield network but only considers input and hidden neurons. After a network update during training the input neurons become output neurons. During the learning process, the BM maximizes the product of probabilities assigned to the elements of the training set. BM are used for dimensionality reduction, classification, feature learning, among others.

Deep Belief Network (DBN) introduced in (Bengio et al., 2006) are probabilistic generative models with multiple layers of the so called latent variables. The first two layers have undirected connections while the next layers have directed connections between layers. This type of networks is used recognize and generate images and videos.

The **autoencoder** (AE) network model was designed for unsupervised learning and is used to encode an input with a representation with less number of dimensions (Bourlard & Kamp, 1988). A decoder is then used to decode the data and obtain the original data. So they are basically used to reduce the size of the inputs into a compact representation.

Deep Neural Networks in Practice

From among the many different neural network models, CNNs have gained most of the attention due to its good image classification results better than other deep neural networks and because there has been an exponential increase of applications requiring image classification. In the following, some of the most known CNN for its results and novelty will be described.

LeNet (LeCun et al., 1995) proposed one of the first convolutional neural network for hand digit classification with high accuracy. The model accepts and classifies grayscale images of size 32×32 according to ten different classes representing the ten possible digits. The network has a total of 60K weights and an overall accuracy above 99%.

AlexNet was the first large CNN (Krizhevsky, 2012) with very good results for image classification. Compared to LeNet, AlexNet is deeper and has 1000× more weights than LeNet. The input images are also larger (227×227×3). AlexNet has top-5 error rates around 17.0% and a top-1 error rate of 37.5% when used to classify images from ImageNet.

In 2013 a multiplayer deconvolutional neural network – ZefNet - was proposed in (Zeiler & Fergus, 2013). The authors propose a methodology to help in the design of the network based on a process to observe the network activity of neurons (Erhan, 2009; Le, 2013). Following the methodology, they were able to improve AlexNet to a top-5 error rate of 11.2%.

One year later the VGG neural network (Simonyan & Zisserman, 2014) increased the size of CNNs published so far with 138 million parameters. VGG improved previous year top-5 error rate to 7.3%. In spite of several filter size improvements to reduce the number of parameters, VGG still have a huge number of parameters requiring long training times and high inference times.

In the same year, a new CNN - GoogleNet (Szegedy et al., 2014; Szegedy, 2016) - introduced a new layer that has groups of convolutions running in parallel within a module designated Inception. In the inception module, several convolutional layers run in parallel like a small neural network inside a larger model. With 6.8 million parameters GoogleNet achieved a top-5 error rate of 6.7% for ImageNet.

A very deep neural network - ResNet (He et al., 2015) - has increased the number of layers to 152 and achieved a top-5 error rate of 3.6% in the ImageNet contest. Similar to GoogleNet, ResNet includes a new block named *Residual Block* where the output map of a series of two convolutional layers are added to the input of the block. An optimization of ResNet was proposed in (Xie et al., 2017) with the ResNeXt CNN with a top-5 error of 3.03%. Another improvement of ResNet was proposed in (Huang,

2018) - DenseNet - with a similar accuracy but with about half of the parameters. This was possible with a modification of the residual block where a layer has dependencies with all previous layers.

Recently, SENet (Hu, Shen & Sun, 2018) introduced a new network block based on the residual block – squeeze-and-excitation that emphasizes important features and cancel less useful ones. SENet won the ILSVRC competition in 2017 with a top-5 error rate of 2.25%.

Deep Learning on Edge

Deep learning algorithms are very demanding in terms of memory resources to store weights and computing power for training and inference. For these reasons, deep learning networks run of high-performance computing platforms. Data collected from edge sensors are sent to high-performance computing centers to be processed and when required the result is sent back to the edge to be presented or to help taking decisions.

Today edge devices are almost everywhere in a large set of applications in industrial environments, automotive, surveillance and security cameras, drones, satellites, medical equipment and new applications appear every day. All these devices collect an enormous amount of data to be processed by the central high-performance computing platform. This reduces the complexity and the energy required for edge devices. However, for an increasing set of applications, the processing of collected data to take decisions has to be done near the sensor for several reasons: unreliable channel transmission, large communication round-trip delay, real-time processing, security and privacy of data.

Hence, data processing algorithms and decision taking is migrating from the cloud to the edge, in particular deep learning models. The problem is that to run an inference of a deep learning network requires high computing and memory resources, scarce resources on an edge computing device. Even with the increasing capacity of edge and mobile devices, it is still a challenge to run deep neural networks within the embedded constraints of the device.

Until recently, the main concern of DNN designers was to achieve the best accuracy. With the advent of deep learning on edge, in the design and development of DNN accuracy is traded-off by energy, cost, computing resources and several other metrics associated with edge computing devices.

Two lines of research are being followed for the optimization of deep neural networks on edge: model optimization and optimized computing platforms for edge deep learning.

Model Optimization

As became evident from the neural networks describe above, the trend is to consider more layers and more weights. Sometimes, a new neural network is proposed that optimizes a previous one with the reduction of parameters but without ever reducing accuracy. DNN for edge and mobile processing reduces the number of computations and weights with eventually a slight reduction in accuracy. MobileNet (Howard et al, 2017) is a CNN for mobile devices that reduces the number of parameters by manipulating the kernels. It applies a single kernel to each input map and then combines the convolution outputs with a pointwise convolution. This leads to a reduction in the number of parameters and consequently the number of computations. Other optimizations were considered, like reducing the number of input and output maps and the image resolution. Different networks with different trade-offs were implemented achieving accuracies from 50% to 70%, a number of parameters from 0.5 to 4.2 millions and a number of multiply-accumulate (MAC) operations ranging from 41 to 559 millions. MobileNetv2 (Sandler et al., 2018) is an optimization of MobileNet that introduced some optimizations that reduce the number of parameters about 30% and the number of operations about 50% and still achieving higher accuracy.

ShuffleNet (Zhang et al., 2018) is another CNN proposed for mobile devices. Different convolutions are applied to separate parts of the input maps to reduce the number of operations. The output of convolutions are then shuffled so that the information from different groups can be mixed. The model has a complexity similar to MobileNetV2 but with better accuracy.

The optimizations proposed in the previous neural network models somehow changes the deep learning algorithm by considering different number and types of layers trying to reduce the number of parameters while keeping the accuracy. A different approach consists of reducing the complexity of the model at a lower level. In this approach, two types of techniques have been considered: data quantization and data reduction. Data quantization consists of techniques to reduce the arithmetic complexity and the number of bits used to represent parameters and activations. Data reduction is the set of techniques used to reduce the number of parameters or the volume of data transferred to the computing platform.

A commonly used data quantization technique is the conversion from single-precision floating-point representations to half-precision floating-point (Micikevicius et al., 2017) or 8-bit floating-point (Wang et al., 2018), fixed-point or integer representation. One advantage of this conversion is that the new representations are easier to implement and calculate than single-precision floating-point arithmetic. Also, with data represented with less bits the arithmetic operators are also less complex and so many operators can be implemented with the same silicon area.

Several works have shown that neural network models can use weights and activations represented with only 16 or even 8 bits and still keep accuracies close to the accuracy obtained with data represented with single-precision floating-point (Gysel et al., 2016), (Gupta et al., 2015), (Anwar et al., 2015), (Lin et al., 2016). Neural network models with weights and activations represented with a single bit have also been proposed - BNN (Binary Neural Networks) (Courbariaux et al., 2016; Umuroglu et al., 2016). BNN reduce considerably the bitwidth of data at the cost of some accuracy degradation. To reduce the impact of binarization over the network accuracy the number of weights must be considerably increased. Also, instead of using a binary representation, some works consider 2-bits to reduce the impact of the representation over the accuracy (Ubara et al., 2016).

A different approach to reduce the data volume of the network is to remove and compress data. In (Han et al., 2015) a DNN is compressed with punning and Huffman coding. Pruning is a process that removes some connections between neurons. For example, a reduction of about 90% of the weights belonging to the dense layers have a very small impact over accuracy. The disadvantage of pruning is that it introduces sparsity in the matrix of weights which complicates its implementation in hardware. When applied to dense layers, pruning is more efficient then when applied to convolutional layers, because the number of weights in dense layers is generally much higher than the set of all remaining weights.

Another technique to reduce the effects of large transfers of weights in fully connected layers is batching (Zhang et al., 2016). The batching technique stores several output feature maps of the last non-dense layers before being executed. It permits to reuse the same kernel for different input images.

Computing Platforms for Edge Deep Learning

Different technologies are available to deploy deep learning algorithms on edge devices. The right device depends on the design requirements, including delay, latency, area, energy, cost, flexibility, etc. Chips or devices for artificial intelligence at the edge try to optimize energy and performance efficiencies, that is, get the lowest energy consumption and enough performance to run a DNN model within design constraints.

General-purpose processors (CPU) can run any deep learning model and their programmability permits them to run any new DNN model without any modifications to the computing platform. The problem with CPUs is that they have a low energy and performance efficiency. GPUs (Graphics Processing Unit) are one of the most used platforms for training DNN because they are a many-core architecture with massive computing parallelism offering high-performance computing and at the same time offer a high level of programmability. They are energy and performance efficient but have high energy consumption which is infeasible for most edge platforms due to their restrictions on available energy. The highest performance and best energy efficiency is achieved with ASICs (Applications Specific Integrated Circuit). ASICs have limited programmability because the algorithm implementation is hardwired in silicon. Some hardware programmability can be considered at the cost of extra silicon to implement extra computing modules that are chosen according to the target algorithm to be implemented. FPGAs (Field-Programmable Gate Array) are more flexible than ASICs since the hardware can be reconfigured for new and different functions but are harder to reprogram than CPUs or GPUs. SoCs (Sytem-on-Chip) FPGAs are an attractive option to run deep learning models since they contain a general-purpose processor tightly connected to reprogrammable logic. The reprogrammable logic is used to design and implement the most time-consuming operations of DNNs, while the CPU is used to control the system, to run the less frequent operations and allowing the implementation of new functions whose hardware implementation is inefficient.

Most commercial solutions are based on ASICs since they provide the best solutions in terms of performance and energy consumption. Some companies provide IP (Intellectual Property) cores as DNN accelerators to be integrated in a computing system, while others provide full SoC solutions implemented on chip.

High-performance IP processors are common approaches to run machine learning algorithms. DesignWare EV6x (Synopsis, 2017) is an IP processor for vision processing on embedded devices. It consists of a 32-bit processor, a vector DSP and a dedicated accelerator for CNNs. The accelerator supports many CNN models, including regular and irregular CNNs, like GoogleNet, and supports 8 and 12 bits data quantization. The whole core has a peak performance of 4.5 TMACs with 2 TMACs/W.

DNA (Cadence, 2017) is another IP SoC processor from Cadence designed for the acceleration of deep neural networks on edge devices. The core integrates a Tensilica DSP, and the DNN accelerator. The architectures optimize the execution of the algorithm using techniques like zero-skipping (multiplications with zero are not computed), pruning, data compression and decompression. The core can be configured with different number of MACs and each MAC can be configured with different data representations (8 or 16 bits integer or 16-bits floating-point). The configuration with the highest performance has a peak performance of 12 TMACs with 3.7 TMACs/W.

NeuPro (Linley Group., 2018) is also an IP core for machine learning to be deployed in embedded devices for advanced driver-assistance systems, surveillance systems, among others. The core is a SoC with an accelerator that can execute any layer of a CNN and a vector processing unit and to control the accelerator and to run other functions not supported by the accelerator. The IP core is configurable in terms of number of MACs. MACs are configurable for the execution of MACs with different data sizes (8 or 16). The smallest configuration of the IP has a performance of 2 TOPs and the larger one has a performance of 12.5 TOPs.

In (Gyrfalcon Technology - 2018) a many core architecture with 168 processing units, each with local memory and a MAC unit, was proposed for audio and video processing, including deep learning networks. The core element of the architecture is an engine to speed-up matrix processing. With an operation frequency of 300 MHz the chip delivers 16.8 TOPs with a consumption of 700 mW corresponding to a power efficiency of 24 TOPs/W.

Movidius Myriad X processor (Intel, 2017) is a SoC vector processor with an accelerator for DNN inference at the edge. The MAC units support 16-bit fixed- and floating-point operations and 8-bit fixed-point. The accelerator has a peak performance of 1 TOPs and the whole processor has a total peak performance of 4 TOPs.

ASIC offer the best solutions but with a reduced flexibility. Deep learning networks are still in its infancy and therefore are constantly being modified and improved. Therefore, deploying an ASIC solution for deep learning is always a risk. Turning the ASIC architecture more flexibility reduces its silicon efficiency which reduces performance and increases energy consumption. These aspects open the set of available platforms for edge computing to reconfigurable devices. Coarse and fine-grained solutions were already proposed to run inference in low cost devices.

Eyeriss (Chen et al., 2017) is a coarse-grained reconfigurable accelerator for CNNs. It contains 168 processing elements connected with a network-on-chip (NoC). The NoC is configurable to adapt the dataflow of the architecture to the dataflow of the model to run. The architectures uses compression and decompression to reduce the data volume between the chip and external memory. With an operating frequency, the accelerator was tested with the inference of AlexNet with data quantized to 16-bits fixed-point has an energy efficiency of 166 GOPs/W with a measured average power of 278 mW.

DNPU is another coarse-grained reconfigurable processor (Shin et al., 2017) for CNNs and RNNs. The chip has dedicated units to the execution of convolutional layers. Pooling and activation function are executed by a centralized module shared by all convolution modules. Dense layers are implemented with a dedicated unit for matrix multiplications and multipliers can be configured (4, 8 or 16 bits fixed-point). The architecture with 4-bit multipliers has a peak performance of 1.2 TOPs with an energy efficiency of 3.9 TOPs/W.

DRP is a dynamically coarse-grained reconfigurable core to accelerate embedded machine learning algorithms (Fujii et al., 2018). The core has an array of dynamically reconfigurable processing elements. Both 16-bit fixed- and floating-point and binary precisions are supported. Dynamic reconfigurability is used to support large networks by reconfiguring the architecture for different layers at execution time. The chip achieved a performance near 1 TOPs.

Fine-grained reconfigurable devices, FPGAs, permit to optimize the hardware architecture for each particular deep learning model (Sze et al., 2017). The first FPGA implementations had the sole objective of improving performance and therefore considered high density FPGA devices (Shawahna et al., 2019). Now, with the necessity to deploy DNNs on edge devices, small to medium density FPGAs are also considered (Guo et al., 2018; Venieris et al., 2018). Recently, a solution was proposed to execute large CNNs in low density FPGAs (ZYNQ XC7Z020) with a peak performance of 400 GOPs (Véstias et al., 2018). With 8-bit fixed-point quantization, the architecture explores several levels of parallelism and proposes a method to run convolutions independently of the size of the convolution window. With all these optimizations, the architecture has a peak performance around 400 GOPs and an energy efficiency near 50 GOPs/W.

FUTURE RESEARCH DIRECTIONS

Deep learning models have improved during the last years. Typically, good accuracies are only achieved with large models. However, the evolution of DNN models have shown that with appropriate techniques it is possible to reduce the complexity of the models with a negligible accuracy loss. New models are needed that emphasize performance and energy efficiency. Binary neural networks are promising solutions with a great impact over hardware complexity and memory storage but still requires a lot of improvements to avoid large accuracy degradations.

Training and inference are still two separated steps. Training is done in high-performance computing platforms and the results are used by the same platform or by an edge device. Considering that an edge device is constantly receiving new data, these could be used to dynamically train the network to keep improving accuracy. Incremental training is executed on high-performance platforms, but the process could be also implemented in the edge device for the same reasons enumerated before.

Designing neural network models for specific problems is still an empirical process that leads to oversized networks with redundant parameters. It is important to better understand how particular layers and neurons influence the final accuracy and how to redesign the model so that the best accuracy is achieved. This will help to tailor models for specific applications, which is particularly important for edge devices.

Concerning the computing platforms, its design is somehow influenced by the fact that DNNs are still under constant research and evolution. Several ASIC solutions already exist but the risk is high since the architectures are optimized for particular neural networks. Any improvements or changes to the original network model reduces the efficiency of the ASIC solution since new functions or modules have to be executed by general purpose processors.

Reconfigurable architectures help us to overcome some of the limitations of ASICs since the hardware architecture can be upgraded on-board with new modules and/or operations. It is the only platform whose hardware can follow the constant evolution of DNNs. FPGAs allow optimized implementations of binary neural networks contrary to architectural solutions based only on CPUs or GPUs. A major problem of FPGA devices is that it is difficult to design them compared to implementations based on software only. Specific frameworks to automatically map neural networks on FPGA already exist but the results suffer some degradation compared to a hand-made design. The proliferation of reconfigurable devices as solutions for deep learning on edge depends on the availability of tools to automatically map neural network models on FPGA.

Given the heterogeneity of layers in high accuracy neural network models, it is important to consider flexible architectures with dedicated accelerators for the common operations requiring massive parallelism integrated with a high-performance processor that can execute the remaining operations or functions whose execution cannot be done by the accelerator. SoCs with a processor and dedicated hardware are the most appropriate solutions for these cases. Many of the ASIC architectures for deep learning proposed so far consider a SoC architecture. An example of this trend is the recently announced FPGA for deep learning (Xilinx, 2018) with software programmable processors, fine-grained reconfigurable hardware and an intelligent device for tasks associated with deep learning. The new FPGA upgrades previous devices with a new engine for deep learning inference.

Inference is still the only operation executed in deep learning platforms in the edge. However, the possibility to train or retrain a network in the edge opens the possibility of constant learning whenever new data is collected. Training still requires full precision and its computational complexity is much higher than that of inference. This mixed precision requires new architectures that can perform both training and inference with different data representations.

CONCLUSION

Deep learning algorithms have successfully improved the accuracy results of many machine learning algorithms. The set of applications that take advantage of these algorithmic improvements is increasing. Many of these applications are associated with edge devices and therefore running deep learning models on edge devices is now a major challenge.

This article describes the fundamentals of deep learning and known deep learning models proposed in the literature. Most of these models are only concerned with accuracy and only a few are optimized for mobile and edge computing. Neural network models for edge devices must be optimized even if this implies some accuracy degradation traded-off by lower energy consumption and improved execution times. Two main classes of optimizations have been applied so far: data quantization and data reduction. These reduce memory and computing requirements and in some cases without accuracy degradation.

ASICs and FPGAs are the most appropriate technologies for edge inference since they offer good energy and performance efficiencies. These metrics are better with ASICs, but FPGAs offer hardware flexibility to optimize the implementation of new neural network models. A brief description of recent commercial chips and published FPGA implementations were also given in this chapter.

REFERENCES

Anwar, S., Hwang, K., & Sung, W. (2015). Fixed point optimization of deep convolutional neural networks for object recognition. In *IEEE International Conference on Acoustics, Speech, and Signal Processing* (pp. 1131–1135). 10.1109/ICASSP.2015.7178146

Barber, D. (2012). *Bayesian Reasoning and Machine Learning*. Cambridge University Press.

Bengio, Y. (2009). Learning deep architectures for AI. *Foundations and Trends in Machine Learning*, *2*(1), 1–127. doi:10.1561/2200000006

Bengio, Y., Lamblin, P., Popovici, D., & Larochelle, H. (2006). Greedy layer-wise training of deep networks. In B. Schölkopf, J. C. Platt, & T. Hoffman (Eds.). In *Proceedings of the 19th International Conference on Neural Information Processing Systems* (153-160). Cambridge, MA.

Bourlard, H., & Kamp, Y. (1988). Auto-association by multilayer perceptrons and singular value decomposition. *Biological Cybernetics*, *59*(4-5), 291–294. doi:10.1007/BF00332918 PMID:3196773

Bouveyron, C., Celeux, G., Murphy, T., & Raftery, A. (2019). *Model-Based Clustering and Classification for Data Science: With Applications in R (Cambridge Series in Statistical and Probabilistic Mathematics)*. Cambridge, UK: Cambridge University Press. doi:10.1017/9781108644181

Cadence: Tensilica. (2017). DNA Processor IP For AI Inference.

Chen, Y., Krishna, T., Emer, J. S., & Sze, V. (2016). Eyeriss: An Energy-Efficient Reconfigurable Accelerator for Deep Convolutional Neural Networks. *IEEE Journal of Solid-State Circuits*, *52*(1), 127–138. doi:10.1109/JSSC.2016.2616357

Christmann, A., & Steinwart, I. (2008). *Support Vector Machines*. Springer-Verlag.

Courbariaux, M., & Bengio, Y. (2016) BinaryNet: Training Deep Neural Networks with Weights and Activations Constrained to +1 or -1. In CoRR, abs/1602.02830.

Elman, J. L. (1990). Finding Structure in Time. *Cognitive Science*, *14*(2), 179–211. doi:10.120715516709cog1402_1

Erhan, D., Bengio, Y., Courville, A., & Vincent, P. (2009). Visualizing higher-layer features of a deep network. *Univ. Montr.*, *1341*, 1.

Fujii, T., Toi, T., Tanaka, T., Togawa, K., Kitaoka, T., Nishino, K., ... Motomura, M. (2018). New generation dynamically reconfigurable processor technology for accelerating embedded AI applications. In *Symposium on VLSI Circuits* (41-42). 10.1109/VLSIC.2018.8502438

Glorot, X., & Bengio, Y. (2010). Understanding the difficulty of training deep feedforward neural networks. In *International Conference on Artificial Intelligence and Statistics* (249–256).

Glorot, X., Bordes, A., & Bengio, Y. (2011). Deep Sparse Rectifier Neural Networks. In *Proceedings of the Fourteenth International Conference on Artificial Intelligence and Statistics* (315-323).

Goodfellow, I., Bengio, Y., & Courville, A. (2016). *Deep Learning*. MIT Press.

Guo, K., Sui, L., Qiu, J., Yu, J., Wang, J., Yao, S., ... Yang, H. (2018). Angel-Eye: A Complete Design Flow for Mapping CNN Onto Embedded FPGA. *IEEE Transactions on Computer-Aided Design of Integrated Circuits and Systems*, *37*(1), 35–47. doi:10.1109/TCAD.2017.2705069

Gupta, S., Agrawal, A., Gopalakrishnan, K., & Narayanan, P. (2015) Deep Learning with Limited Numerical Precision. In *Proceedings of the 32nd International Conference on International Conference on Machine Learning:* Vol. 37. (1737–1746).

Gysel, P., Motamedi, M., & Ghiasi, S. (2016). Hardware-oriented Approximation of Convolutional Neural Networks. In *Proceedings of the 4th International Conference on Learning Representations*.

Han, S., Mao, H., & Dally, W. J. (2015). "Deep Compression: Compressing Deep Neural Network with Pruning, Trained Quantization and Huffman Coding". *CoRR*, abs/1510.00149.

Haykin, S. (2008). *Neural Networks and Learning Machines* (3rd ed.). Pearson.

He, K., Zhang, X., Ren, S., & Sun, J. (2015). Deep Residual Learning for Image Recognition. *Multimedia Tools and Applications*, *77*, 10437–10453.

Hinton, G. E., & Sejnowski, T. J. (1986). Learning and relearning in Boltzmann machines. In D. E. Rumelhart, J. L. McClelland, & CORPORATE PDP Research Group (Eds.), Parallel distributed processing: explorations in the microstructure of cognition. Vol. 1, MIT Press (282-317).

Hochreiter, S. (1998). The vanishing gradient problem during learning recurrent neural nets and problem solutions. *International Journal of Uncertainty, Fuzziness and Knowledge-based Systems*, *6*(02), 107–116. doi:10.1142/S0218488598000094

Hochreiter, S., & Schmidhuber, J. (1997). Long Short-Term Memory. *Neural Computation*, *8*(8), 1735–1780. doi:10.1162/neco.1997.9.8.1735 PMID:9377276

Hopfield, J. (1982). Neural networks and physical systems with emergent collective computational abilities. *Proceedings of the National Academy of Sciences of the United States of America*, *79*(8), 2554–2558. doi:10.1073/pnas.79.8.2554 PMID:6953413

Hornik, K., Stinchcombe, M., & White, H. (1989). Multilayer feedforward networks are universal approximators. *Neural Networks*, *2*(5), 359–366. doi:10.1016/0893-6080(89)90020-8

Howard, A. G., Zhu, M., Chen, B., Kalenichenko, D., Wang, W., Weyand, T., Andreetto, M., & Adam, H. (2017). MobileNets: Efficient Convolutional Neural Networks for Mobile Vision Applications. *CoRR*, abs/1704.04861.

Hu, J., Shen, L., & Sun, G. (2018). Squeeze-and-Excitation Networks, In *Proceedings IEEE Conference on Computer Vision and Pattern Recognition* (7132-7141). IEEE.

Huang, G., Liu, Z., Maaten, L., & Weinberger, K. (2018). Densely Connected Convolutional Networks. In *IEEE Conference on Computer Vision and Pattern Recognition.*

Hubara, I., Courbariaux, M., Soudry, D., El-Yaniv, R., & Bengio, Y. (2016). Binarized Neural Networks. In D. D. Lee, M. Sugiyama, I. Guyon, & R. Garnett (Ed.), Advances in Neural Information Processing Systems: Vol. 4107–4115. *Curran Associates, Inc.*

Intel. (2017). Movidius Myriad X VPU.

Keogh, E. (2011). Instance-Based Learning. In C. Sammut, & G. I. Webb (Eds.), *Encyclopedia of Machine Learning*. Boston, MA: Springer.

Krizhevsky, A., Sutskever, I., & Hinton, G. E. (2012). ImageNet Classification with Deep Convolutional Neural Networks. In Adv. Neural Inf. Process. Syst. 1–9.

Le, Q. V. (2013). Building high-level features using large scale unsupervised learning. In *IEEE International Conference on Acoustics, Speech and Signal Processing* (8595–8598). 10.1109/ICASSP.2013.6639343

LeCun, Y. (1989). Generalization and network design strategies. In Connectionism in Perspective.

LeCun, Y., Boser, B., Denker, J. S., Henderson, D., Howard, R. E., Hubbard, W., & Jackel, L. D. (1989). Backpropagation applied to handwritten zip code recognition. *Neural Computation*, *1*(4), 541–551. doi:10.1162/neco.1989.1.4.541

LeCun, Y., Jackel, L. D., Bottou, L., Cortes, C., Denker, J. S., Drucker, H., ... & Vapnik, V. (1995). Learning algorithms for classification: A comparison on handwritten digit recognition. In Neural networks: the statistical mechanics perspective, 261-276. Mech. Perspect.

Lin, D. D., Talathi, S. S., & Annapureddy, V. S. (2016). Fixed Point Quantization of Deep Convolutional Networks. In *Proceedings of the 33rd International Conference on International Conference on Machine Learning*. Vol. 48. (pp. 2849–2858).

Linley Group. (2018). Ceva NeuPro Accelerates Neural Nets.

Matloff, N. (2017). *Statistical Regression and Classification: from Linear Models to Regression* (1st ed.). Chapman and Hall. doi:10.1201/9781315119588

Micikevicius, P., Narang, S., Alben, J., Diamos, G. F., Elsen, E., García, D., … Wu, H. (2017). Mixed Precision Training. *CoRR*, abs/1710.03740.

Nwankpa, C., Ijomah, W., Gachagan, A. & Marshall, S. (2018). Activation Functions: Comparison of trends in Practice and Research for Deep Learning. *Corr*. abs/1811.03378.

Patterson, J., & Gibson, A. (2017). Deep Learning: A Practitioner's Approach. O'Reilley Media, 1st ed.

Quinlan, R. (1992). *C4.5: Programs for Machine Learning* (1st ed.). Morgan Kaufmann.

Ruder, S. (2016). An overview of gradient descent optimization algorithms. In CoRR.

Sandler, M. B., Howard, A. G., Zhu, M., Zhmoginov, A., & Chen, L. (2018). MobileNetV2: Inverted Residuals and Linear Bottlenecks. In *IEEE/CVF Conference on Computer Vision and Pattern Recognition* (4510-4520). 10.1109/CVPR.2018.00474

Shawahna, A., Sait, S. M., & El-Maleh, A. H. (2018). FPGA-Based Accelerators of Deep Learning Networks for Learning and Classification: A Review. *IEEE Access: Practical Innovations, Open Solutions*, *7*, 7823–7859. doi:10.1109/ACCESS.2018.2890150

Shin, D., Lee, J., Lee, J., & Yoo, H. (2017). 14.2 DNPU: An 8.1TOPS/W reconfigurable CNN-RNN processor for general-purpose deep neural networks. In *IEEE International Solid-State Circuits Conference* (240-241). 10.1109/ISSCC.2017.7870350

Simonyan, K., & Zisserman, A. (2014). Very deep convolutional networks for large-scale image recognition. In *arXiv preprint arXiv:1409.1556*.

Synopsys DesignWare. (2017). EV6x Vision Processors.

Sze, V., Chen, Y., Yang, T., & Emer, J. S. (2017). Efficient processing of deep neural networks: A tutorial and survey. *Proceedings of the IEEE*, *105*(12), 2295–2329. doi:10.1109/JPROC.2017.2761740

Szegedy, C., Vanhoucke, V., Ioffe, S., Shlens, J., & Wojna, Z. (2016). Rethinking the Inception Architecture for Computer Vision. In *IEEE Conference on Computer Vision and Pattern Recognition*, (2818-2826). 10.1109/CVPR.2016.308

Szegedy, C., Liu, W., Jia, Y., Sermanet, P., Reed, S., Anguelov, D., ... & Rabinovich, A. (2014). Going Deeper with Convolutions. *arXiv:1409.4842*.

Gyrfalcon Technology. (2018). Lightspeeur 2803S Neural Accelerator.

Umuroglu, Y., Fraser, N. J., Gambardella, G., Blott, M., Leong, P. H. W., Jahre, M., & Vissers, K. A. (2016). FINN: A Framework for Fast, Scalable Binarized Neural Network Inference. In CoRR, abs/1612.07119.

Venieris, S. I., & Bouganis, C. (2018). fpgaConvNet: Mapping Regular and Irregular Convolutional Neural Networks on FPGAs. *IEEE Transactions on Neural Networks and Learning Systems*, *30*(2), 326–342. doi:10.1109/TNNLS.2018.2844093 PMID:29994725

Véstias, M. P., Duarte, R. P., Sousa, J. T., & Neto, H. C. (2018). Lite-CNN: A High-Performance Architecture to Execute CNNs in Low Density FPGAs. In *28th International Conference on Field Programmable Logic and Applications* (pp. 393-399). 10.1109/FPL.2018.00075

Wang, N., Choi, J., Brand, D., Chen, C., & Gopalakrishnan, K. (2018). Training Deep Neural Networks with 8-bit Floating Point Numbers. *CoRR* abs/1812.08011.

Xilinx, V. (2018). The first adaptive compute acceleration platform (acap).

Zeiler, M. D., & Fergus, R. (2013). Visualizing and Understanding Convolutional Networks. arXiv. vol. 30 (pp. 225–231).

Zhang, C., & Ma, Y. (2012). *Ensemble Machine Learning*. New York: Springer-Verlag. doi:10.1007/978-1-4419-9326-7

Zhang, C., Wu, D., Sun, J., Sun, G., Luo, G., & Cong, J. (2016). Energy-Efficient CNN Implementation on a Deeply Pipelined FPGA Cluster. In *Proceedings of the International Symposium on Low Power Electronics and Design* (pp. 326–331). 10.1145/2934583.2934644

Zhang, C., & Zhang, S. (2002). Association Rule Mining. In *Lecture Notes in Artificial Intelligence*. Springer-Verlag.

Zhang, X., Zhou, X., Lin, M., & Sun, J. (2018) ShuffleNet: An Extremely Efficient Convolutional Neural Network for Mobile Devices. In *IEEE/CVF Conference on Computer Vision and Pattern Recognition* (pp. 6848–6856). 10.1109/CVPR.2018.00716

ADDITIONAL READINGS

Bishop, C. (2006). *Pattern Recognition and Machine Learning*. New York: Springer Verlag.

Erhan, D., Bengio, Y., Courville, A., Manzagol, P.-A., Vincent, P., & Bengio, S. (2010). Why does unsupervised pre-training help deep learning? *Journal of Machine Learning Research*, *11*, 625–660.

Erhan, D., Manzagol, P.-A., Bengio, Y., Bengio, S., & Vincent, P. (2009). The difficulty of training deep architectures and the effect of unsupervised pre-training. *In International Conference on Artificial Intelligence and Statistics* (153–160).

He, K., Gkioxari, G., Dollár, P., & Girshick, R. (2017). Mask R-CNN. In *International Conference on Computer Vision.*

Hinton, G. E., Osindero, S., & Teh, Y.-W. (2006). A fast learning algorithm for deep belief nets. *Neural Computation*, *18*(7), 1527–1554. doi:10.1162/neco.2006.18.7.1527 PMID:16764513

Jarrett, K., Kavukcuogl, K., Ranzato, M., & LeCun, Y. (2009). What is the best multi-stage architecture for object recognition? In *International Conference on Computer Vision* (2146–2153). 10.1109/ICCV.2009.5459469

Kalinowski, I., & Spitsyn, V. (2015). Compact Convolutional Neural Network Cascade for Face Detection. *CoRR*, abs/1508.01292.

Lawrence, S., & Giles, C. Lee, Tsoi, Ah C. & Back, A. (1997). Face Recognition: A Convolutional Neural Network Approach. In IEEE Transactions on Neural Networks. 8 (1): 98–113.

LeCun, Y., Kavukvuoglu, K., & Farabet, C. (2010). Convolutional networks and applications in vision. In *International Symposium on Circuits and Systems* (253–256).

Lei, T., Barziley, R., & Jaakkola, T. (2016). Rationalizing Neural Predictions. In *Proceedings of the Conference on Empirical Methods in Natural Language Processing* (107-117). 10.18653/v1/D16-1011

Matsugu, M., Mori, K., Mitari, Y., & Kaneda, Y. (2003). Subject independent facial expression recognition with robust face detection using a convolutional neural network. *Neural Networks*, *16*(5–6), 555–559. doi:10.1016/S0893-6080(03)00115-1 PMID:12850007

Pengcheng, Y., & Neubig, G. (2017). A Syntactic Neural Model for General-Purpose Code Generation, In *Proceedings of the 55th Annual Meeting of the Association for Computational Linguistics*, Vol. 1 (440-450).

Redmon, J., & Farhadi, A. (2017). YOLO9000: Better, Faster, Stronger. In *IEEE Conference on Computer Vision and Pattern Recognition* (6517-6525).

Scherer, D., Müller, A., & Behnke, S. (2010). Evaluation of pooling operations in convolutional architectures for object recognition". In *International Conference on Artificial Neural Networks* (92–101). 10.1007/978-3-642-15825-4_10

Xie, S., Girshick, R., Dollár, P., Tu, Z., & He, K. (2017). Aggregated Residual Transformations for Deep Neural Networks. In *IEEE Conference on Computer Vision and Pattern Recognition*. 10.1109/CVPR.2017.634

Zeiler, M. D. & Fergus, R.. (2013). Visualizing and understanding convolutional networks. *Computing Research Repository*, abs/1311.2901.

KEY TERMS AND DEFINITIONS

Activation Function: The activation function defines the output of a neuron given a set of inputs from the previous layer or data input.

Artificial Neural Network (ANN): It is a computing model based on the structure of the human brain with many interconnected processing nodes that model input-output relationships. The model is organized in layers of nodes that interconnect to each other.

Autoencoder: An unsupervised learning network and is used to encode an input with a representation with fewer dimensions.

Boltzman Machine: An unsupervised network that maximizes the product of probabilities assigned to the elements of the training set.

Convolutional Layer: A network layer that applies a series of convolutions to a block of input feature maps.

Convolutional Neural Network (CNN): A class of deep neural networks applied to image processing where some of the layers apply convolutions to input data.

Deep Belief Network: A probabilistic generative model with multiple layers of the so called latent variables tha keep the state of the network.

Deep Learning (DL): A class of machine learning algorithms for automation of predictive analytics.

Deep Neural Network (DNN): An artificial neural network with multiple hidden layers.

Edge Device: any hardware device that serves as an entry point of data and may store, process and/or send the data to a central server.

Feature Map: A feature map is a 2D matrix of neurons. A convolutional layer receives a block of input feature maps and generates a block of output feature maps.

Fully Connected Layer: A network layer where all neurons of the layer are connected to all neurons of the previous layer.

Hopfield Network: A dense neural network where all neurons connect to all other neurons.

Long-short Term Memory Network: A variation of recurrent neural networks to reduce the vanishing problem.

Machine Learning: A subfield of artificial intelligence whose objective is to give systems the ability to learn and improve by its own without being explicitly programmed to do it.

Network Layer: A set of neurons that define the network of a CNN. Neurons in a network layer are connected to the previous and to the next layer.

Perceptron: The basic unit of a neural network that encodes inputs from neurons of the previous layer using a vector of weights or parameters associated with the connections between perceptrons.

Pooling Layer: A network layer that determines the average pooling or max pooling of a window of neurons. The pooling layer subsamples the input feature maps to achieve translation invariance and reduce over-fitting.

Pruning: An optimization technique for deep neural networks that removes some connections between neurons to reduce the complexity of the network.

Recurrent Neural Network (RNN): A class of deep neural networks consisting of dense networks with state.

Semi-Supervised: A training process of neural networks that mixes supervised and unsupervised training.

Softmax Function: A function that takes as input a vector of k values and normalizes it into a probability distribution of k probabilities.

Supervised Training: A training process of neural networks where the outcome for each input is known.

Unsupervised Training: A training process of neural networks where the training set does not have the associated outputs.

Chapter 3
Smart Blockchain:
A Formidable Combination of Analytics and Blockchain – A Survey

Gutha Jaya Krishna
Institute for Development and Research in Banking Technology, India

Vadlamani Ravi
Institute for Development and Research in Banking Technology, India

Bheemidi Vikram Reddy
Institute for Development and Research in Banking Technology, India

Mohammad Zaheeruddin
Institute for Development and Research in Banking Technology, India

ABSTRACT

A blockchain is a digitized, decentralized, open system of records. Of late, there is a phenomenal spurt in the research and application activities of the catch-all phrase analytics, which subsumes machine learning, text mining, classical optimization, artificial intelligence, evolutionary computing, visual analytics, big data analytics, etc. in many a diverse field. Consequently, even new technology like blockchain is not left behind. This chapter presents a comprehensive survey of 33 papers that appeared between 2016 and 2018 under the theme, 'Analytics and Blockchain', which focuses on how analytical approaches and blockchain implementation are symbiotically related to improve their overall performance in solving various real-world problems. The core idea behind the survey is to facilitate the reader to appreciate the utility of analytical methods to the design, implementation, and application of blockchain and suggesting future directions for further research.

DOI: 10.4018/978-1-7998-2112-0.ch003

INTRODUCTION

Of late, blockchain technology with its phenomenal success has caught everyone's attention be it in academia or industry. Initially, blockchain technology, though not called so, was introduced in 1979 by Merkle as Merkle Tree (Merkle, 1979). But, the Bitcoin blockchain was conceptualized by Nakamoto in 2008 (Nakamoto, 2008). Then it was executed in the following year by Nakamoto as a central part of the computerized and secure money extensively progressed as bitcoin (Nakamoto, 2009). The core idea is a distributed ledger that works on consensus for all the trades on the network. Though with the usage of a blockchain, the bitcoin transformed into the essential modernized money to deal with the twofold spending issue without requiring a trusted expert or middleman and has been the inspiration for some of the modern applications. The primary function of blockchain is to verify and to handle data in a peer-to-peer network without any corruption or falsification.

We present the motivation and contribution of the survey in Section 2. In Sections 3 and 4, we describe the working of blockchain, its applications and analytical solutions respectively. The theme of the survey and the methodology employed to collect articles for the survey are described in Sections 5 and 6 respectively. The critique and analysis of the surveyed articles are presented respectively in Sections 7 and 8. Discussion, as well as future directions for further research, are presented in Section 9. Finally, we conclude the survey in Section 10.

MOTIVATION AND CONTRIBUTION

Motivation for the Survey

The motivation for the current survey is as follows:

- Lack of a single survey article, where the bidirectional relationship between and blockchain and analytics is reviewed and discussed
- To highlight possible issues and solutions with analytical techniques applied for/with blockchain.
- To showcase the phenomenal role played by analytics in solving the implementation and operational issues of blockchain
- To highlight the importance of blockchain in carrying out analytics projects in some application areas.

Contribution of the Survey

The contribution of the current survey is as follows:

- This is the first-of-its-kind survey involving blockchain and analytics.
- Providing readers, a critical analysis of the research studies reported at the intersection of analytics and blockchain highlighting their mutual importance.
- Presenting the issues and possible solutions in the relationship that analytics and blockchain seem to exhibit. By employing analytics, blockchain indeed becomes smart.

- Presenting a few future research directions in this area, which will be useful to the novice researchers and practitioners alike.

OVERVIEW OF BLOCKCHAIN AND APPLICATIONS

The working principle of the blockchain includes 1) keeping record of all data or transactions that are to be included in the ledger, 2) utilize a distributed system to verify the data or transactions, and 3) once verified, the data or transaction are added to the blockchain as blocks and can never be deleted or altered (Pierro, 2017).

When you hear of blockchain the terminology of "key" comes into picture for any data or transaction you feed into the blockchain. The data or transaction receives two keys. One is a public key, and another is a private key. These two together form a digital signature. These two keys are required to unlock the data held in a block of the blockchain. Apart from the public and private keys (i.e. digital signature), a timestamp and a unique coded identification number are also stored.

An important jargon introduced to the blockchain technology is that it is a decentralized or peer-to-peer architecture. This concept of decentralization adds to the security of the blockchain technology and this decentralized technology also does not suffer from a single point of failure as in the case of centralized technologies. Coming to the security aspect of the decentralization is when the transaction or data is added to the ledger, each of the copy is sent to all the nodes in the network. By doing so, an attempt to modify a data or transaction record will not be in sync with the other nodes in the network. But, even though with the advantages of decentralized architecture, it has the drawback of increased validation times which makes it slow compared to conventional validation approaches with a middleman or a centralized architecture. Decentralized architectures can be affected by network, communication, cost or scalability issues.

Each block in a blockchain is a permanent record of data or transactions. Or else it is a group of transactions or data records at a fixed timestamp. These blocks are added by miners. The miners solve a complex mathematical puzzle along with other miners to add the block to the chain, for which they use some amount of computational resources and are also given incentives for solving the puzzle. Solving these puzzles take quite some time (for example in bitcoin 10 minutes) which is an overhead. Even if one of the transaction validations fails the miner is affected to do the computations all over again. The blockchain is a single chain of all the solved and linked blocks shared with all factions in the decentralized system.

Even with these above-mentioned addendums, hackers and cryptanalysts are coming out with new ways to compromise the blockchain technology. Therefore, security and privacy are primary concerns.

Applications of Blockchain

Blockchain technology is successfully applied in some of the real-world application areas which include energy sector, cryptocurrency sector, big data sector, cloud computing, healthcare, Internet of Things (IoT), cybersecurity, etc. The following are the issues successfully addressed within these real-world areas which include: 1) security, 2) middleman free transactions, 3) ease or cost of implementation of the real-world solutions, and 4) managing data, etc.

Figure 1. Theme of the Survey

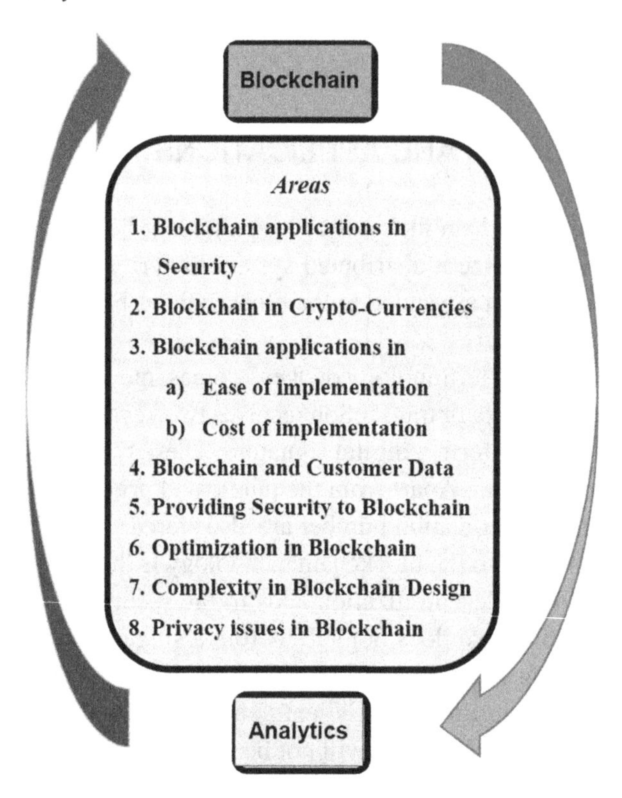

Analytical Solutions

In recent years, analytical methods, especially of predictive and prescriptive types, witnessed tremendous acceptance and success in numerous fields of science, engineering, medicine, finance, management, agriculture etc. Their effectiveness is demonstrated in various degrees in the above-mentioned diverse fields. These analytical methods include machine learning (subsuming fuzzy systems, rough sets, probabilistic machine learning), text mining (subsuming natural language processing), evolutionary computing (EC), big data analytics, artificial intelligence subsuming deep learning, etc. These include data-driven as well as non-data-driven solutions (for instance optimization where only objective functions are utilized). It is noteworthy that predictive analytics methods are all data-driven methods without exception. Thus, the data residing in a blockchain and the data entailed by blockchain during its operation make a perfect case for the application of predictive analytics.

The Theme of the Survey

The major issues in real-world applications of information technology are:

- Security: Real-world digital applications tend to have security as an issue. Possible breaches in network and data are common in any real-world applications. Therefore, these breaches must be controlled and reduced to the maximum extent possible for any of the business or industry to progress.

- Ease and Cost of Implementation: Most of the real-world implementations are moving towards easy and cost-effective solutions which make their dealings easy and profitable.
- Customer Data: Data related issues include privacy, security, volume, easy flow, etc. In some of the industries, data is of paramount importance and the above-mentioned issues need to be addressed effectively.
- Apart from the above, quite a few countries are moving towards a secure, paper-less and middle-man free currency where IT has tremendous application potential. Crypto-Currencies are solutions for this requirement. This trend is exacerbated with the advent and popularization of Bitcoin.

Blockchain intertwined with analytics is the solution to the issues mentioned above.
The major issues that arise while in implementing blockchain are:

- Security: Blockchain is decentralised and works majorly on trust within parties. Therefore, providing security to blockchain assumes a lot of significance.
- Optimization: There are many pockets such as cost, network, where blockchain technology needs to employ optimization techniques to solve the problem not only scientifically but also effectively,
- Complexity: The modern blockchains are becoming very complex in terms of structure and computations. Therefore, it needs complexity reduction on an urgent basis, if its proponents want its popularity and success zoom further.
- Privacy: Blockchains, if managed by third parties, tend to pose privacy issues. Fortunately, analytics can play stellar role in solve each of these issues.

The interplay between the blockchain and analytics is best manifested by their intimate relationship (see Figure 1). One must recognize the fact that while the blockchain solves some problems, it also entails new ones. In the first case, with the help of analytics, it makes more significant impact, while in the second case, analytics helps blockchain implementers to solve its operational issues. Thus, this relationship has two distinct strands or flavours: (i) Analytics with blockchain and (ii) analytics for blockchain.

Flow Diagram of the Survey Process

In the flow diagram (see Figure 2), there are two flows towards considering a paper for the survey process 1) ***"Analytics with Blockchain"*** and 2) ***"Analytics for Blockchain".*** In the survey process, firstly, standard online research article databases are searched for journal articles, conference proceedings and book chapters. In blockchain with analytics part of the flow, research articles that focus on how analytical approaches are applied in fusion with blockchain technology for solving various real-world problems are searched and if the above criteria are met, then they are accepted for the inclusion in the survey. In Analytics for blockchain part of the flow, research articles that focus on how analytical approaches are used within blockchain implementation to improve its overall performance are searched and if the above criteria are met, then they are accepted for the inclusion in the survey. The articles where either blockchain or analytics is not present are excluded i.e. for instance analytics related to bitcoin prediction where blockchain technology is not the primary concern is excluded.

Figure 2. Flow of the survey process

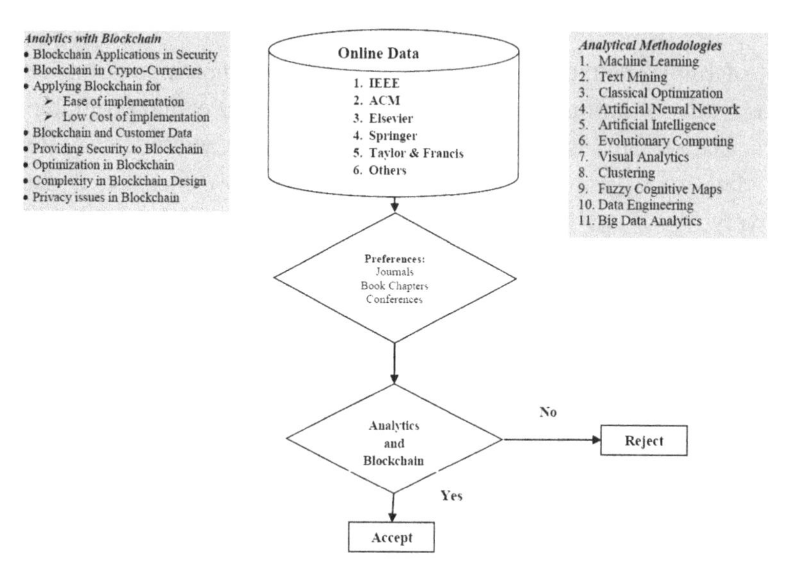

Earlier Surveys on Blockchain

There are around six (6) surveys on blockchain not on the current theme. But, on various blockchain application and blockchain technology related fields. The application related blockchain surveys include Big Data (Karafiloski & Mishev, 2017), Healthcare (Hölbl, Kompara, Kamišalić, & Nemec Zlatolas, 2018), Internet-Of-Things (Fernandez-Carames & Fraga-Lamas, 2018), AI (Salah, Rehman, Nizamuddin, & Al-Fuqaha, 2019). Blockchain technology related surveys include security (Li, Jiang, Chen, Luo, & Wen, 2017), current research in blockchain (Yli-Huumo, Ko, Choi, Park, & Smolander, 2016).

"SMART BLOCKCHAIN": COLLABORATION BETWEEN ANALYTICS AND BLOCKCHAIN

Analytics with Blockchain

In what follows, we review works where analytics and blockchain are intertwined together for better results.

a) Ease and Cost of Implementation

Blockchain technology has enabled banks in the financial empowerment by including financially left out people. The process is often called financial inclusion. In (Larios-Hernández, 2017), fuzzy-set Qualitative Comparative Analysis (fsQCA) is employed to decide the financial practice and reasons for the lack of a bank account, which can be utilized by the blockchain entrepreneurs to target the concerned segments. The proposed technology in future helps in the inclusion of left out persons and boost the financial growth of the country.

The energy sector is where both analytical solutions (i.e. optimization) and blockchain technology are getting seamlessly fused. In (T. Zhang, Pota, Chu, & Gadh, 2018), optimization with a Monte Carlo method and a blockchain based cryptocurrency are both employed in a real-time system for a monetary incentive of electric vehicle users for charging with renewable energy. In future, this application can be extended to areas like electric vehicle charging of self-driving cars and renewable energy sector, where incentives with blockchain technology as well as analytical solutions can be combined.

In (Xu, Wang, & Guo, 2017), blockchain-based decentralized resource management developed to avoid the energy cost of the scheduler in data centres. One central data centre fault will not affect the functioning of the resource management, thereby providing robustness. To lessen the total cost of request migration in data centres, a Reinforcement Learning-based request migration framework is developed. Numerical results show the better functioning of the model than that of the benchmark approaches.

In (Shih, Shih, & Wu, 2018), an infrastructure for multi-pollutant air quality deterioration early warning system was developed in big data analytics using Spark platform for the generation of various reports. This system is integrated with blockchain taking the advantage of the immutability feature of blockchain, where transaction records if added to the blockchain cannot be altered or changed thereby providing a consistent method to conduct continuous evaluation of the system.

Ability to provide a high level of information security by the blockchain technology lets the developers build a reliable and transparent system. Therefore in (Pokrovskaia, 2017), blockchain technology is combined with artificial neural network techniques to build an effective taxation, financial and regulatory mechanism. Multicriteria assessment of the assets or actions of effective and socially acceptable evaluations based on cognitive judgements using neural networks is performed. The fusion of both neural networks for assessment and the blockchain for a reliable and transparent system provides us with a suitable, secure and easy to implement applications in other areas too.

In peer-to-peer energy markets, operative restrictions should be maintained, and payments should be rendered justly without a centralised control. This turns out to best fit for applying blockchain technology. Therefore in (Munsing, Mather, & Moura, 2017), a distributed optimization and control problem of electricity distribution network is chosen and blockchain technology is employed for rendering the payments of the micro-grid network securely as well as transparently.

In (C. Zhang, Romagnoli, Zhou, & Kraft, 2017), short review highlighting three prospects of the next generation urban energy system was presented. These three prospects include 1) ontology-driven data framework development, 2) blockchain enabled peer-to-peer network, and 3) interdisciplinary partnership between data scientist, urban planner as well as an energy engineer for the digital transition of the urban energy system.

b) Crypto-Currency

In (Laskowski & Kim, 2016), a scalable proof-of-concept pipeline is established for the analysis of semi-structured data posted on online social platforms using Nature Language Processing. This data is fused into the crypto-currency marketplace for getting a complete picture of a complex blockchain ecosystem. In future, scaling this system to the cloud, performing real-time analysis of multiple data streams with sentiment analysis as well as machine learning capabilities can be encompassed with the system.

c) Customer Data

In (Roman-Belmonte, De la Corte-Rodriguez, & Rodriguez-Merchan, 2018), it is stated that blockchain technology can be fused with big data technology which is typically a distributed database with high throughput, lower latency, higher capacity, rich querying. Many of these properties are shared by blockchain. But, blockchain technology has its own advantages. Therefore, a fusion of both technologies called blockchain pipelining (Mcconaghy et al., 2016) can be adopted in the future.

Blockchain providing answers to big data limitations is the key focus in (Karafiloski & Mishev, 2017), where the concerns of big data technology viz., decentralization, trust, identity data ownership, data-driven decisions, etc., can be addressed using the power of blockchain through the seamless hybridization of these two technologies. This hybridization has potential applications in the decentralized protection of personal data, digital property, Internet of Things (IoT), healthcare, etc.

In (Kiyomoto & Hidano, 2017), the iKaaS platform is developed for the multi-cloud environment so as to share data and knowledge. Further, a revenue sharing process for the knowledge sharing, iKaaS, utilizes blockchain for registering and data processing.

The use of humungous number of online advertisements for manual exploration and analysis of human traffickers becomes unscalable. Therefore in (Portnoff, Huang, Doerfler, Afroz, & McCoy, 2017), a machine learning (ML) classifier is developed which employs stylometry for finding human traffickers in an automated and scalable method from multiple data sources. The advertisements data is scrapped from Backpage and blockchain transaction linking technique is employed to link specific purchases. The ML classifier yielded 90% True Positive Rate (TPR) and 1% False Positive Rate (FPR).

The increasing growth of data sources like the Internet led to the explosive growth of data for the evolution into big data age. Even though data growth is on the increase, the trust issue becomes the main concern of big data. Therefore, blockchain technology provides an answer to this problem by combining immutable and traceable features. In (Yue, Junqin, Shengzhi, & Ruijin, 2017), the authors provide a credible big data sharing model with decentralized and transparent blockchain technology.

Managing and tracking confidential user information shared by the providers is a difficult task in domains such as health care and land records maintenance. In (Banerjee & Joshi, 2017), to track this information, data privacy ontology with the capabilities of blockchain technology is developed for automated access-control and audit, that applies data privacy policies while sharing data to the third parties. The model can also be adopted by big data operators to automatically employ their privacy policies on data procedures and track the flow of the data.

d) Security

In one of the embodiments of any patent (US9672499B2, 2014), a data analytic system combined with the security features is developed for the hot wallet service refers to blockchain-based virtual currency wallet accessible via Internet. The data analysis system performs classification and/or credit scoring of the data from the profile database. This data analysis helps in detecting potential frauds in the hot wallet service.

In (Cai & Zhu, 2016), a study on the fraud detection in online businesses with a blockchain perspective is presented based on decision rules from the reputation system. Reputation systems are a kind of recommender systems (i.e. collaborative filtering systems) which are utilized in ratings of the online businesses. The robustness of blockchain-based reputation systems for rating fraud is presented in the last section of (Cai & Zhu, 2016).

Data and trust management remain the two main challenges of an intrusion detection system (IDS). Therefore in (Meng, Tischhauser, Wang, Wang, & Han, 2018), a review of the IDS (i.e. in particular anomaly-based techniques which use machine learning classifiers) employing blockchain technology is discussed to address these two challenges. Also, some of the challenges and limitations in the fusion of IDS with blockchain technology are also discussed in this review. In the above fusion, fraud data in the blockchain is the main source for the detection of frauds.

In (Juneja & Marefat, 2018), Stacked Denoising Autoencoder (SDA) is employed to detect anomalies of arrhythmia classification. This approach is combined with blockchain technology by giving access to the securely stored models and patient data on the chain. Increased accuracy results yielded by SDA are 99.15% and 98.55%.

In (Kapsammer et al., 2017), blockchain technology-based Volunteer Management System (VMS) is developed. The four layers of the system include trust layer, semantic layer, encouragement layer and market layer. The recommendation system is part of the encouragement layer which takes decisions based on rules-based profilers (which are developed based on web ontology language and description logic). The blockchain technology is employed to offer trust and confidence to the VMS.

According to (Pan & Yang, 2018), some of the cybersecurity challenges are data privacy, security, trust and trustworthiness. Thus, these challenges provide us to ponder on the technologies which can address them. they observed that artificial intelligence, machine learning and blockchain are some of the emerging technologies which offer us opportunities to address the challenges in the areas of cybersecurity.

Blockchain platform for clinical trial and precision medicine is developed in (Shae & Tsai, 2017), where analytics is employed (i.e. keyword-based analytics, semantic similarity model). The medical blockchain is developed to store the stroke clinical medical data from Chinese medical university hospital and the Taiwan health insurance database. For accessing the data securely blockchain technology is utilized.

Analytics for Blockchain

a) Security and Privacy

There are millions of nodes in public blockchain. Usually, some of the nodes attempt to cheat in the network for illegal interests and have an anomalous behaviour pattern. In order to study the behaviour pattern of the blockchain nodes and to increase the security, in (Huang et al., 2017) proposed a behaviour pattern-based clustering (BPC) algorithm to cluster 1321 nodes of stock trading blockchain. BPC is compared with classical clustering methods like density-based DBSCAN and Hierarchical Clustering (HIC). BPC applied a Dynamic Time Warping based distance measure for the clustering. BPC-based approach outperformed the classical methods of clustering.

Cyber-criminal entities may tend to utilize blockchain for their own benefits. To overcome these activities, in (Sun Yin & Vatrapu, 2017) developed a machine learning based cybercriminal entities estimation. Thirteen supervised machine learning classifiers were employed on the dataset of 854 samples of 12 classes out of which 5 classes are of cyber-criminal activity. Four classifiers based on bagging and gradient boosting prevailed over other classifiers with accuracies of 77.38%, 76.47%, 78.46% and 80.76%. After classification, these models were used to classify 100,000 uncategorized samples.

The key problem facing blockchain is the automatic clustering of the behaviour patterns of the blockchain nodes. In (Ermilov, Panov, & Yanovich, 2017), a greedy additive clustering technique is proposed for the automatic address clustering in the blockchain. This algorithm has an advantage of utilizing off-chain information from the Internet. This approach also avoids mistaken merging of the clusters and proves its superiority over the basic approach of clustering with blockchain data alone.

Reducing the high degree of anonymity in blockchain using supervised machine learning techniques is the primary goal of the proposed work in (Harlev, Sun Yin, Langenheldt, Mukkamala, & Vatrapu, 2018). Data consists of 434 entity samples of 10 different categories. Gradient boosting algorithm achieved an accuracy of 77% and an F1 score of 0.75. Though breaking the anonymity of the blockchain is an attack on privacy, but it has huge societal implications and trust enhancement of the technology.

In (Bogner, 2017), visualization and interactive querying are added to detect anomalies in blockchain technology. JSON-based search and analytics engine called Elasticsearch as well as visualization front-end called Kibana are utilized in developing the machine learning based anomaly detection system. This system, with the help of visualizations generates automated alerts and perform analysis.

Swan in (Swan, 2015, 2018), introduced the concept of deep learning chains. The first reason for the fusion of deep learning and blockchain technology is to make probabilistic deductions using computation graphs. deep learning algorithms can set fees and sense duplicitous actions in the blockchains and blockchains can be utilized for security, tracing, and compensation of deep nets.

b) Optimization

In (Hussein et al., 2018), a Genetic Algorithm (GA) -based approach to optimize the request selection is proposed. The GA assists in the process flow and assures controlling the requests. This ensures the overall stability of the system while reducing the latency of the requests as well. Also, a Discrete Wavelet Transform is employed to enhance the security of the system.

The transaction data in blockchain is formulated as a big data analytics problem. The irrelevant data and poorly optimized traceability are two prominent issues in the blockchain. Therefore in (Chen, 2018), a Takagi-Sugeno Fuzzy cognitive map together with artificial neural network is developed as a traceability chain algorithm. The goal is to reach a traceability decision rather than a consensus decision. Thus, this algorithm reduces the mining latency and arrives at optimized decisions.

c) Complexity

In (Göbel, Keeler, Krzesinski, & Taylor, 2016), the effects of propagation delays in the context of selfish mine strategy on the blockchain is studied. Markov chain is employed to detect the mining block-hiding behaviour by monitoring the rate of production of orphan blocks. By doing so the rate of production behaviour and hence the revenue generation of both honest miners and selfish miners is studied. In future, this study would result in the discovery of tomographic techniques for the network.

Because blockchain is a public ledger where transactions are registered it is an example of big data where straightforward visualization is not informative. Therefore in (Bistarelli & Santini, 2017), the authors employed from visual analytics techniques to analyse the transactions and filter out undesired data. The focus is to view transactions as grouped into disconnected 'islands' for more concentration on these groups.

Figure 3. Journal articles per year

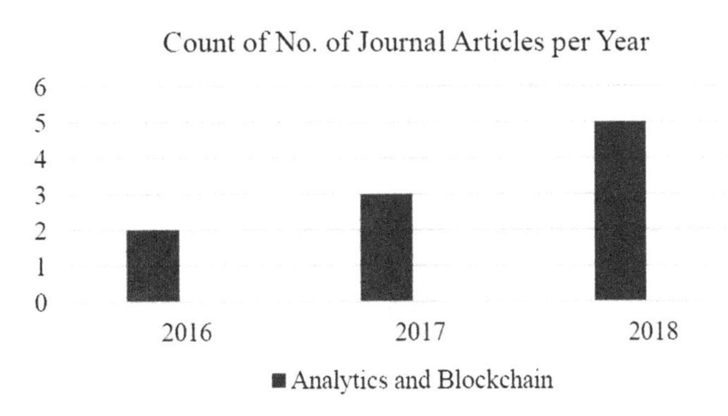

Blockchain has architectural requirements which create trade-offs between non-functional requirements like performance, cost and security. Therefore in (Yasaweerasinghelage, Staples, & Weber, 2017), architectural performance modelling and simulation tools are utilized to predict the latency of blockchain-based systems with analytical solvers or simulations (i.e. for example Palladio). In this work (Yasaweerasinghelage et al., 2017), issues of a number of confirmation blocks and inter-block times are addressed. Relative error of the median response time (viz. is mostly 10%) is computed, which in turn helps in the architectural decision-making.

Law can be provided as a service in blockchain by employing analytical solutions. In (Wasim, Ibrahim, Bouvry, & Limba, 2017), Probability Factor Model (PFM) is implemented over the blockchain to identify potential breaches in Service Level Agreements (SLA) and finding out possible recurrences of the same. Complex operations performed on the Redis, MongoDB, as well as message brokers, are likely to cause breaches. These potential breaches are identified by factor analysis and stochastic modelling of PFM.

The blockchain is a rich and increasing source of valued information which is difficult to interpret. Therefore in (Bartoletti, Lande, Pompianu, & Bracciali, 2017), authors proposed a Scala library-based views and analysis for internal data stored within blockchain as well as data stored on external data sources like SQL or NoSQL. In the future, information flow, forking and attacks can also be analysed.

Analysis of the Surveyed Articles

In this section, detailed descriptive analytics of the surveyed articles is performed. The articles on ***'Analytics with Blockchain'*** theme are more in number than ***'Analytics for Blockchain'*** theme. The reason could be that researchers have not yet fully realized the potential and power of analytical algorithms thereby not attempting to solve various operational issues on blockchain using analytics.

Figure 3 illustrates that journal articles are more in number in 2018 than in the previous years. Figure 4 illustrates that the conference articles are more in number in 2016. Also, the 'Analytics with Blockchain' theme has more articles in all the years compared to the 'Analytics for Blockchain' theme.

The articles per publisher and the trend of articles published per year are depicted in Figures 5 and 6. The list is topped by IEEE. IEEE has the greatest number of conference articles and Elsevier has the greatest number of journal articles. No journal articles were published in ACM and no conference articles were published in Springer.

Figure 4. Conference articles per year

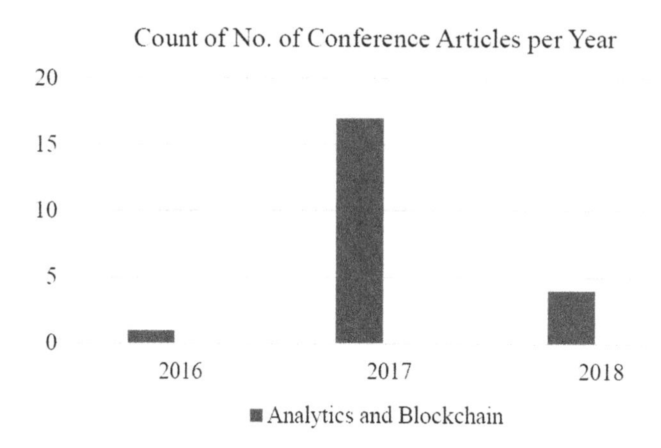

Figure 5. Articles per publisher

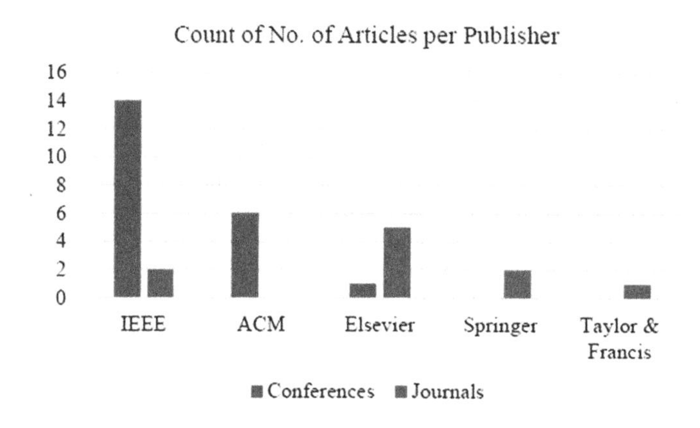

Figure 6. Trend of articles published per year

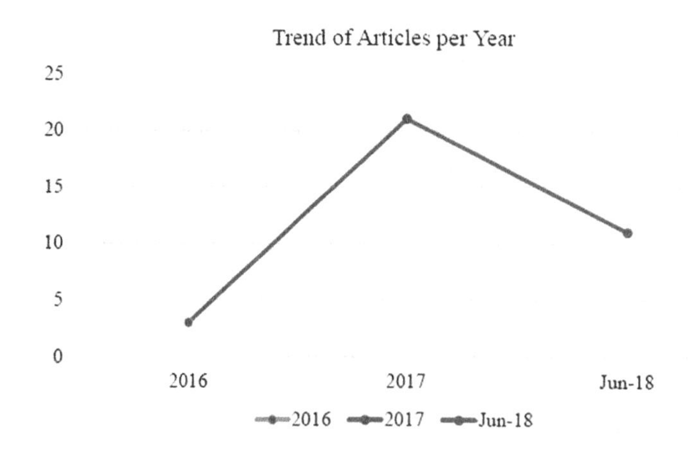

Analytical solutions employed for/with blockchain are summarised in Table 1 by providing the count and references of the analytical sub-categories.

Table 1. Count and references of articles in the survey by analytical subcategory

Analytics Sub-categories	Count	References
Machine Learning	12	(Bartoletti et al., 2017; Ermilov et al., 2017; Göbel et al., 2016; Harlev et al., 2018; Huang et al., 2017; Juneja & Marefat, 2018; Meng et al., 2018; Pokrovskaia, 2017; Portnoff et al., 2017; Sun Yin & Vatrapu, 2017; Xu et al., 2017; US9672499B2, 2014)
Deep Learning	2	(Swan, 2015, 2018)
Natural Language Processing	2	(Laskowski & Kim, 2016; C. Zhang et al., 2017)
Big Data Analytics	8	(Banerjee & Joshi, 2017; Bartoletti et al., 2017; Karafiloski & Mishev, 2017; Kiyomoto & Hidano, 2017; Roman-Belmonte et al., 2018; Shae & Tsai, 2017; Shih et al., 2018; Yue et al., 2017)
Fuzzy Sets	1	(Larios-Hernández, 2017)
Optimization	5	(Göbel et al., 2016; Hussein et al., 2018; Munsing et al., 2017; Yasaweerasinghelage et al., 2017; T. Zhang et al., 2018)
Visualization	2	(Bistarelli & Santini, 2017; Bogner, 2017)
Artificial Intelligence (Recommender systems)	3	(Cai & Zhu, 2016; Kapsammer et al., 2017; Pan & Yang, 2018)

Table 2. Count of articles for the theme 'blockchain with analytics'

Blockchain With Analytics	No. of Patents	No. of Journals	No. of Conferences	Total
Ease and Cost of Implementation-based	-	3	4	7
Crypto-Currency-based	-	-	1	1
Customer Data-based	-	1	5	6
Security-based	1	2	4	7
Total	***1***	***6***	***14***	***21***

Table 3. Count of articles for the theme 'analytics for blockchain'

Analytics For Blockchain	No. of Patents	No. of Journals	No. of Conferences	No. of Books/Chapters	Total
Security	-	1	4	-	5
Privacy	-	-	-	2	2
Optimization	-	2	-	-	2
Complexity	-	1	4	-	5
Total	**-**	***4***	***8***	***2***	***14***

Table 4. References of articles for the theme 'blockchain with analytics'

Analytics with Blockchain	Ref. of Patents	Ref. of Journals	Ref. of Conferences
Ease and Cost of Implementation-based	-	(Larios-Hernández, 2017; Xu et al., 2017; T. Zhang et al., 2018)	(Munsing et al., 2017; Pokrovskaia, 2017; Shih et al., 2018; C. Zhang et al., 2017)
Crypto-Currency-based	-	-	(Laskowski & Kim, 2016)
Customer Data-based	-	(Roman-Belmonte et al., 2018)	(Banerjee & Joshi, 2017; Karafiloski & Mishev, 2017; Kiyomoto & Hidano, 2017; Portnoff et al., 2017; Yue et al., 2017)
Security-based	(US9672499B2, 2014)	(Cai & Zhu, 2016; Meng et al., 2018)	(Juneja & Marefat, 2018; Kapsammer et al., 2017; Pan & Yang, 2018; Shae & Tsai, 2017)

Tables 2 and 3 present the count of the research articles in each of the theme. Tables 4 and 5 present the references of the research articles in each of the theme. The issues addressed in each of the themes of the survey are presented in Tables 2-5.

Discussion and Future Directions

As we can see in this survey, analytical techniques are making quite an impact for the blockchain technology as well together with it. Our survey presented in this chapter shows how analytical solutions play a vital role for/with blockchain technology, which in the past, already shown their importance in other technologies and application areas. In the context of this survey, we make some key discussion points of the open areas and future directions for further research:

- Less work is reported in the theme of ***'Analytics for Blockchain'.*** There are few reported journal articles in the above-mentioned theme that too in the areas of security, privacy and complexity. Therefore, these areas can also be fertile ground for journal articles. Secure multi party computation, privacy preserving predictive analytics can be explored further.

Table 5. References of articles for the theme 'analytics for blockchain'

Analytics For Blockchain	Ref. of Patents	Ref. of Journals	Ref. of Conferences	Ref. of Books/ Chapters
Security	-	(Huang et al., 2017)	(Bogner, 2017; Ermilov et al., 2017; Harlev et al., 2018; Sun Yin & Vatrapu, 2017)	-
Privacy	-	-	-	(Swan, 2015, 2018)
Optimization	-	(Chen, 2018; Hussein et al., 2018)	-	-
Complexity	-	(Göbel et al., 2016)	(Bartoletti et al., 2017; Bistarelli & Santini, 2017; Wasim et al., 2017; Yasaweerasinghelage et al., 2017)	-

- These issues considered in each of the themes are not exhaustive, as we are limited by the availability of research articles. There will also be other issues i.e. related to security or technology optimization and any other key focus areas in each of the themes which can be researched further.
- In future, issues and challenges in other technologies like big data, 5G wireless networks, IoT, edge analytics etc., can be efficiently solved using the fusion of analytical solutions with blockchain technology (N. Zhang et al., 2018).
- Security issues in blockchain such as 51% vulnerability, private key security, criminal activity, vulnerabilities in smart contract, etc., that are highlighted in (Li et al., 2017) apart from what is discussed in this survey can also be solved in the future by employing the analytical solutions for blockchain.
- Less work is reported in the application of optimization techniques for improving the blockchain implementation. Here both classical optimization (derivative-based) and Swarm or Evolutionary Computing (derivative-free) techniques can be applied for improving blockchain implementation, based on the objective function(s) and nature of the search space we are solving.
- Various issues related to latency, network, cost reliability, availability and privacy in blockchain play an important role in the operation of blockchain. These issues can be solved using prescriptive analytical solutions such as optimization methods, for improving the overall performance of the blockchain.
- Data arising from blockchain operation can be mined using data and text mining techniques for better solutions to the customer as well improving the performance of the blockchain. For example, data from the bitcoin trading can be fused with data mining techniques to predict the price volatility of the crypto-currencies. Here algorithmic trading can be explored and exploited with equal felicity as in the case of traditional currencies.
- Mining or solving a puzzle to add blocks is an important task in blockchain where huge computational resources as well time is spent. To overcome this issue various analytical solutions like deep learning, EC-based optimization, etc., can be employed to effectively and efficiently solve the mining task in blockchain.
- Ease, trust and cost of doing business are improved when analytical solutions are fused with blockchain. This, in turn, adds to the revenue and improves the Return-on-Investment (ROI) of the business.
- Data flow within blockchain can be regulated, which has implications for the areas of big data. Optimal resource allocation and management can also be accomplished in the blockchain using optimization techniques in the future.
- Analytical solutions have long been applied in/with cloud computing (Ravi, Khandelwal, Krishna, & Ravi, 2018), which if adopted in the blockchain technology will make the blockchain cloud ready thereby imparting all the benefits that cloud computing brings along.
- Maintaining land records registry, of a city for realty purposes in a fool-proof and tamper-proof manner requires blockchain so that illegal activities of selling the property to multiple parties can be caught with irrefutable evidence. Here again, analytics plays a spectacular role in identifying and mitigating the frauds (Cai & Zhu, 2016; Meng et al., 2018).
- Sentiment analysis can be performed on the reviews and feedback posted by the user community in the social media about the issues and utility of blockchain and/or crypto-currencies.

- One more area fertile to explore is the impact of GDPR (*Regulations Regulation (EU) 2016/679 Of The European Parliament And Of The Council,* 2016) on the use of blockchain technology for customer data using analytical techniques
- Hardware reliability optimization studies can be performed on the blockchain networks using evolutionary multi-objective/single objective optimization.
- Deep reinforcement learning enabled autonomous vehicles can play a stellar role in the analytics driven blockchain for secure maintenance of land records, issuing letter of credit in banks etc.

CONCLUSION

Of late, blockchain emerged as an exciting area with phenomenal applications in diverse fields. However, we discovered that all three flavours of analytics can be applied with great felicity in solving the limitations of blockchain, thereby making it a smart blockchain. We surveyed the articles published during 2016-2018 by keeping in focus the two proposed themes of the surveys i.e. ***'Analytics with Blockchain'*** and ***'Analytics for Blockchain'.*** This first-of-its-kind survey focuses on the four issues each, in real-world applications where blockchain is employed in conjunction with analytical solutions, and for blockchain where analytical solutions are applied. These analytical solutions for/with blockchain offer a benefit of revenue, security, ease of doing business, etc. This survey will be useful for practitioners, researchers and academicians in the blockchain domain. Discussion and future directions presented in this survey will also be helpful for budding researchers to choose their research and application area.

REFERENCES

Banerjee, A., & Joshi, K. P. (2017). Link before you share: Managing privacy policies through blockchain. In *2017 IEEE International Conference on Big Data (Big Data)* (pp. 4438–4447). Boston, MA: IEEE. 10.1109/BigData.2017.8258482

Bartoletti, M., Lande, S., Pompianu, L., & Bracciali, A. (2017). A general framework for blockchain analytics. In *Proceedings of the 1st Workshop on Scalable and Resilient Infrastructures for Distributed Ledgers - SERIAL '17* (pp. 1–6). New York, NY: ACM Press. 10.1145/3152824.3152831

Bistarelli, S., & Santini, F. (2017). Go with the -Bitcoin- Flow, with Visual Analytics. In *Proceedings of the 12th International Conference on Availability, Reliability and Security - ARES '17* (pp. 1–6). New York, NY: ACM Press. 10.1145/3098954.3098972

Bogner, A. (2017). Seeing is understanding – anomaly detection in blockchains with visualized features. In *UBICOMP / ISWC* (pp. 5–8). Maui, HI: ACM; doi:10.1145/3123024.3123157

Cai, Y., & Zhu, D. (2016). Fraud detections for online businesses: A perspective from blockchain technology. *Financial Innovation*, *2*(1), 20. doi:10.118640854-016-0039-4

Chen, R.-Y. (2018). A traceability chain algorithm for artificial neural networks using T–S fuzzy cognitive maps in blockchain. *Future Generation Computer Systems*, *80*, 198–210. doi:10.1016/j.future.2017.09.077

Di Pierro, M. (2017). What Is the Blockchain? *Computing in Science & Engineering*, *19*(5), 92–95. doi:10.1109/MCSE.2017.3421554

Ermilov, D., Panov, M., & Yanovich, Y. (2017). Automatic Bitcoin Address Clustering. In *2017 16th IEEE International Conference on Machine Learning and Applications (ICMLA)* (pp. 461–466). IEEE. 10.1109/ICMLA.2017.0-118

Fernandez-Carames, T. M., & Fraga-Lamas, P. (2018). A Review on the Use of Blockchain for the Internet of Things. *IEEE Access: Practical Innovations, Open Solutions*, *6*, 32979–33001. doi:10.1109/ACCESS.2018.2842685

Göbel, J., Keeler, H. P., Krzesinski, A. E., & Taylor, P. G. (2016). Bitcoin blockchain dynamics: The selfish-mine strategy in the presence of propagation delay. *Performance Evaluation*, *104*, 23–41. doi:10.1016/j.peva.2016.07.001

Harlev, M. A., Yin, S. H., Langenheldt, K. C., Mukkamala, R., & Vatrapu, R. (2018). Breaking Bad: De-Anonymising Entity Types on the Bitcoin Blockchain Using Supervised Machine Learning. In *Proceedings of the 51st Hawaii International Conference on System Sciences* (pp. 3497–3506).

Hilton Waikoloa Village, Hawaii: Association of Information Systems. Retrieved from https://scholarspace.manoa.hawaii.edu/handle/10125/50331

Hölbl, M., Kompara, M., Kamišalić, A., & Nemec Zlatolas, L. (2018). A Systematic Review of the Use of Blockchain in Healthcare. *Symmetry*, *10*(10), 470. doi:10.3390ym10100470

Huang, B., Liu, Z., Chen, J., Liu, A., Liu, Q., & He, Q. (2017). Behavior pattern clustering in blockchain networks. *Multimedia Tools and Applications*, *76*(19), 20099–20110. doi:10.100711042-017-4396-4

Hussein, A. F., ArunKumar, N., Ramirez-Gonzalez, G., Abdulhay, E., Tavares, J. M. R. S., & de Albuquerque, V. H. C. (2018). A medical records managing and securing blockchain based system supported by a Genetic Algorithm and Discrete Wavelet Transform. *Cognitive Systems Research*, *52*, 1–11. doi:10.1016/j.cogsys.2018.05.004

Juneja, A., & Marefat, M. (2018). Leveraging blockchain for retraining deep learning architecture in patient-specific arrhythmia classification. In *2018 IEEE EMBS International Conference on Biomedical & Health Informatics (BHI)* (pp. 393–397). IEEE. 10.1109/BHI.2018.8333451

Kapsammer, E., Kimmerstorfer, E., Pröll, B., Retschitzegger, W., Schwinger, W., & Schönböck, J. … Rets-Chitzegger, W. (2017). iVOLUNTEER-A Digital Ecosystem for Life-long Volunteering *. In iiWAS. Salzburg, Austria: ACM. doi:10.1145/3151759.3151801

Karafiloski, E., & Mishev, A. (2017). Blockchain solutions for big data challenges: A literature review. In *IEEE EUROCON 2017 -17th International Conference on Smart Technologies* (pp. 763–768). IEEE. 10.1109/EUROCON.2017.8011213

Kiyomoto, S., & Hidano, S. (2017). Knowledge sharing framework for iKaaS platform. In *2017 2nd International Conference on Knowledge Engineering and Applications (ICKEA)* (pp. 146–150). IEEE. 10.1109/ICKEA.2017.8169919

Larios-Hernández, G. J. (2017). Blockchain entrepreneurship opportunity in the practices of the unbanked. *Business Horizons*, *60*(6), 865–874. doi:10.1016/j.bushor.2017.07.012

Laskowski, M., & Kim, H. M. (2016). Rapid Prototyping of a Text Mining Application for Cryptocurrency Market Intelligence. In *2016 IEEE 17th International Conference on Information Reuse and Integration (IRI)* (pp. 448–453). IEEE. 10.1109/IRI.2016.66

Li, X., Jiang, P., Chen, T., Luo, X., & Wen, Q. (2017). A survey on the security of blockchain systems. *Future Generation Computer Systems*. doi:10.1016/j.future.2017.08.020

Mcconaghy, T., Marques, R., Müller, A., De Jonghe, D., Mcconaghy, T., & Mcmullen, G. … Granzotto, A. (2016). *BigchainDB: A Scalable Blockchain Database*. Berlin, Germany. Retrieved from http://www.noql.com

Meng, W., Tischhauser, E. W., Wang, Q., Wang, Y., & Han, J. (2018). When Intrusion Detection Meets Blockchain Technology: A Review. *IEEE Access: Practical Innovations, Open Solutions*, *6*, 10179–10188. doi:10.1109/ACCESS.2018.2799854

Merkle, R. C. (1979). *US4309569A*. USPTO. Retrieved from https://patents.google.com/patent/US4309569

Munsing, E., Mather, J., & Moura, S. (2017). Blockchains for decentralized optimization of energy resources in microgrid networks. In *2017 IEEE Conference on Control Technology and Applications (CCTA)* (pp. 2164–2171). Mauna Lani, HI: IEEE. 10.1109/CCTA.2017.8062773

Nakamoto, S. (2008). Re: Bitcoin P2P e-cash paper. Retrieved August 24, 2018, from https://satoshi.nakamotoinstitute.org/emails/cryptography/11/

Nakamoto, S. (2009). Bitcoin: A Peer-to-Peer Electronic Cash System. Retrieved August 23, 2018, from www.bitcoin.org

Pan, J., & Yang, Z. (2018). Cybersecurity Challenges and Opportunities in the New "Edge Computing + IoT" World. In *Proceedings of the 2018 ACM International Workshop on Security in Software Defined Networks & Network Function Virtualization - SDN-NFV Sec'18* (pp. 29–32). New York, NY: ACM Press. 10.1145/3180465.3180470

Pokrovskaia, N. N. (2017). Tax, financial and social regulatory mechanisms within the knowledge-driven economy. Blockchain algorithms and fog computing for the efficient regulation. In *2017 XX IEEE International Conference on Soft Computing and Measurements (SCM)* (pp. 709–712). IEEE. 10.1109/SCM.2017.7970698

Portnoff, R. S., Huang, D. Y., Doerfler, P., Afroz, S., & McCoy, D. (2017). Backpage and Bitcoin. In *Proceedings of the 23rd ACM SIGKDD International Conference on Knowledge Discovery and Data Mining - KDD '17* (pp. 1595–1604). New York, NY: ACM Press. 10.1145/3097983.3098082

Ravi, K., Khandelwal, Y., Krishna, B. S., & Ravi, V. (2018). Analytics in/for cloud-an interdependence: A review. *Journal of Network and Computer Applications*, *102*, 17–37. doi:10.1016/j.jnca.2017.11.006

Regulations Regulation (EU) 2016/679 Of The European Parliament And Of The Council of 27 April 2016 on the protection of natural persons with regard to the processing of personal data and on the free movement of such data, and repealing Directive 95/46/EC (General Data Protection Regulation). (2016). Retrieved from https://eur-lex.europa.eu/legal-content/EN/TXT/PDF/?uri=CELEX:32016R0679

Roman-Belmonte, J. M., De la Corte-Rodriguez, H., & Rodriguez-Merchan, E. C. (2018). How blockchain technology can change medicine. *Postgraduate Medicine*, *130*(4), 420–427. doi:10.1080/00325481.2018.1472996 PMID:29727247

Salah, K., Rehman, M. H. U., Nizamuddin, N., & Al-Fuqaha, A. (2019). Blockchain for AI: Review and Open Research Challenges. *IEEE Access: Practical Innovations, Open Solutions*, *7*, 10127–10149. doi:10.1109/ACCESS.2018.2890507

Shae, Z., & Tsai, J. J. P. (2017). On the Design of a Blockchain Platform for Clinical Trial and Precision Medicine. In *2017 IEEE 37th International Conference on Distributed Computing Systems (ICDCS)* (pp. 1972–1980). IEEE. 10.1109/ICDCS.2017.61

Shih, D.-H., Shih, P.-Y., & Wu, T.-W. (2018). An infrastructure of multi-pollutant air quality deterioration early warning system in spark platform. In *2018 IEEE 3rd International Conference on Cloud Computing and Big Data Analysis (ICCCBDA)* (pp. 648–652). IEEE. 10.1109/ICCCBDA.2018.8386595

Swan, M. (2015). *Blockchain : blueprint for a new economy*. Sebastopol, CA: O'Reilly; Retrieved from https://books.google.co.in/books?hl=en&lr=&id=RHJmBgAAQBAJ&oi=fnd&pg=PR3&ots=XQtFD1ZSe0&sig=Dkx_avuMXVeCtYv5LD-6N_TG7uQ&redir_esc=y#v=onepage&q&f=false

Swan, M. (2018). Blockchain for Business: Next-Generation Enterprise Artificial Intelligence Systems. *Advances in Computers*, *111*, 121–162. doi:10.1016/bs.adcom.2018.03.013

Wasim, M. U., Ibrahim, A. A. Z. A., Bouvry, P., & Limba, T. (2017). Law as a service (LaaS): Enabling legal protection over a blockchain network. In *2017 14th International Conference on Smart Cities: Improving Quality of Life Using ICT & IoT (HONET-ICT)* (pp. 110–114). IEEE. 10.1109/HONET.2017.8102214

Xu, C., Wang, K., & Guo, M. (2017). Intelligent Resource Management in Blockchain-Based Cloud Datacenters. *IEEE Cloud Computing*, *4*(6), 50–59. doi:10.1109/MCC.2018.1081060

Yang, D., Kou, L., & Liu, A. (2014). *US9672499B2*. USPTO. Retrieved from https://patents.google.com/patent/US9672499B2/en

Yasaweerasinghelage, R., Staples, M., & Weber, I. (2017). Predicting Latency of Blockchain-Based Systems Using Architectural Modelling and Simulation. In *2017 IEEE International Conference on Software Architecture (ICSA)* (pp. 253–256). Gothenburg, Sweden: IEEE. 10.1109/ICSA.2017.22

Yin, S. H., & Vatrapu, R. (2017). A first estimation of the proportion of cybercriminal entities in the bitcoin ecosystem using supervised machine learning. In *2017 IEEE International Conference on Big Data (Big Data)* (pp. 3690–3699). IEEE. 10.1109/BigData.2017.8258365

Yli-Huumo, J., Ko, D., Choi, S., Park, S., & Smolander, K. (2016). Where Is Current Research on Blockchain Technology?—A Systematic Review. *PLoS One*, *11*(10). doi:10.1371/journal.pone.0163477 PMID:27695049

Yue, L., Junqin, H., Shengzhi, Q., & Ruijin, W. (2017). Big Data Model of Security Sharing Based on Blockchain. In *2017 3rd International Conference on Big Data Computing and Communications (BIG-COM)* (pp. 117–121). IEEE. 10.1109/BIGCOM.2017.31

Zhang, C., Romagnoli, A., Zhou, L., & Kraft, M. (2017). From Numerical Model to Computational Intelligence: The Digital Transition of Urban Energy System. *Energy Procedia*, *143*, 884–890. doi:10.1016/j.egypro.2017.12.778

Zhang, N., Yang, P., Ren, J., Chen, D., Yu, L., & Shen, X. (2018). Synergy of Big Data and 5G Wireless Networks: Opportunities, Approaches, and Challenges. *IEEE Wireless Communications*, *25*(1), 12–18. doi:10.1109/MWC.2018.1700193

Zhang, T., Pota, H., Chu, C.-C., & Gadh, R. (2018). Real-time renewable energy incentive system for electric vehicles using prioritization and cryptocurrency. *Applied Energy*, *226*, 582–594. doi:10.1016/j.apenergy.2018.06.025

Chapter 4
The Significance of the Study of the Brain's Hippocampus for the Progress of Biologically-Inspired Computational Systems

Konstantinos Domdouzis
https://orcid.org/0000-0003-3679-3527
Sheffield Hallam University, UK

ABSTRACT

The field of Artificial Intelligence faces unprecedented progress. It is expected that the use of Artificial Intelligence to different sectors of science and economy will be increased. This is also shown by the fact that at the moment, Artificial Intelligence is characterised by popularity which is proven through the constant presentations on the news. This chapter shows how the study of the brain's hippocampus can further progress the field of Artificial Intelligence. The chapter presents indicative examples of the literature that show how the study of the hippocampus can lead to the development of specific applications. It also shows the impact to the development of biologically-inspired systems through the analysis of specific capabilities of the hippocampus. A number of conclusions are drawn in relation to the significance of the study of the brain's hippocampus for the development of new applications.

INTRODUCTION

The hippocampus is an important area of the brain that contributes significantly to memory and navigation. Alzheimer's disease and other forms of dementia can affect and even damage the specific area. The hippocampus is also closely related to epilepsy, schizophrenia and post-traumatic stress disorder (The University of Hong Kong, 2017). The complexity of hippocampus can be shown by the fact that hippocampal place neurons do not only represent current location but also, they fire in sequences and these sequences appear to simulate past and future spatial trajectories. The firing sequences match the structure of a complex maze (Jeffery & Casali, 2014). Furthermore, the role of hippocampus in com-

DOI: 10.4018/978-1-7998-2112-0.ch004

plex brain networks and especially how it integrates with other brain regions is not well understood by scientists (The University of Hong Kong, 2017).

An example of the complexity of the anatomy and of the operations of hippocampus is spatial navigation. Spatial navigation is based on a network of interconnected networks that include the hippocampus, the prefrontal cortex and the basal ganglia. The exploration of the flows of information between these structures will reveal the processes involved to learning complex navigational tasks (Chersi & Burgess, 2015). The hippocampus is long known to be associated with episodic memory (Scoville & Milner, 2000) and the discovery of place cells play a significant role in spatial memory (O'Keefe & Nadel, 1978).

During spatial navigation, cognitive maps are developed. The instantiation of the cognitive maps is done by place, grid, border and head direction cells that are located mainly in the hippocampus. The human hippocampus and the entorhinal cortex support map-like spatial codes while parahippocampal and retrosplenial cortices allow cognitive maps to be related to specific environmental landmarks. The hippocampal and entorhinal spatial codes are used in coordination with the frontal lobe in order to develop navigational routes (Epstein et al., 2017).

The analysis of the spatial navigation capabilities of the hippocampus shows its complexity and this can be the basis for the re-evaluation of current AI systems and the development of new systems based exactly on this complexity. In recent years there has been increment in the complexity and quality of Artificial Intelligence systems. Examples of the evolution of Artificial Intelligence are the Deep Learning Theory, the Deep Reinforcement Learning, the Generative Adversarial Networks, the Lean and Augmented Data Learning, Probabilistic Programming, the Hybrid Learning Models, and the Automated Machine Learning (Rao et al., 2018). Since 2006, deep learning has emerged as a new era of machine learning research ((Bengio, 2009), (Hinton et al., 2006)). Deep Learning is a set of computational methods that allow an algorithm to program itself through learning from a large set of examples that show the desired behaviour (Gulshan et al., 2016). The deep learning techniques impact key areas of machine learning and artificial intelligence.

This paper aims to present how the study of the functions of the brain's hippocampus can result to the extraction of conclusions that will help in the development of more advanced Artificial Intelligence systems. The paper presents a range of indicative applications from different fields that are based on the characteristics of hippocampus. It then presents arguments on how specific characteristics of the function of the hippocampus can impact the development of biologically-inspired systems. Finally, a range of conclusions are drawn that show how the study of hippocampus can be the basis for the improvement of current biologically-inspired systems.

HIPPOCAMPUS-BASED COMPUTATIONAL APPLICATIONS

Neuroscience and Artificial Intelligence are interconnected since neuroscience can be a source of inspiration for the development of new algorithms to be used in Artificial Intelligence and also, by validating existing Artificial Intelligence approaches. According to neuroscience, intelligent behaviour is based on multiple memory systems (Tulving, 1985). These systems include both reinforcement-based mechanisms and instance-based mechanisms (Gallistel & King, 2009). Instance-based mechanisms are related to episodic memory (Tulving, 2002) and it is often related to circuits in the medial temporal lobe including the hippocampus (Squire et al., 2004). Neuroscience was important in the emergence of the field of Reinforcement Learning. Adolescents show better reinforcement learning and an enhanced link between reinforcement learning and episodic memory. This can be shown through the use of a probabilistic reinforcement learning task combined with reinforcement learning and fMRI. Their findings reveal a significant role of hippocampus in reinforcement learning in adolescence and suggest that reward sensitivity in adolescence is related to the different adaptive ways of how adolescents learn from experience (Davidow et al., 2016). Specifically, the stratium and the hippocampus collaborate in order to support episodic encoding and reinforcement learning ((Adcock et al., 2006), (Bunzeck et al., 2010), (Wimmer & Shohamy, 2012)).

Reinforcement Learning is an Artificial Intelligence field with many applications ranging from finance to robotics. Reinforcement Learning can increase the adaptability of robotics systems and this capability is very important in the interaction with a complex and dynamic environment (Polydoros & Nalpantidis, 2017). There are a number of applications of Reinforcement Learning in Robotics, in Data Science, in Education and in Health. For example, in Robotics, reinforcement learning is used in process planning, demand forecasting, fleet logistics and production coordination. Reinforcement Learning is also used in autonomous vehicles, smart grid and machine tuning while it finds significant applications in predictive maintenance, inventory monitoring, and supply chain risk management (Lorica, 2017).

An interesting application of Reinforcement Learning in High Speed Rail Cognitive Radio is presented by Wu et al. (2016). The application of Cognitive Radio to high-speed rail can result to problems, such as frequent channel switch. The Cognitive Radio (CR) technology can be embedded into the base station which is called the Cognitive Base Station (CBS). The aim of the Cognitive Radio is to achieve the best available spectrum without interfering with the licensed user. The CBS proposed by Wu et al. (2016) has four spectrum management functions: spectrum sensing, spectrum mobility, spectrum decision and spectrum sharing ((Chkirbene & Hamdi, 2015), (Lee & Akyildiz, 2012). The CR selects the most available spectrum depending on the individual users' requirements. CBS implements reinforcement learning to adapt to the different environmental conditions. The use of Reinforcement Learning in the Cognitive Base Station can reduce frequency hopping and also, impact positively the performance of the CBS (Wu et al., 2016).

The Intelligent Soft Arm Control (ISAC) was developed by the Cognitive Robotics Lab of Vanderbilt University. This is a cognitive robotic system and it is used as a research platform for human-humanoid interaction for robotic cognitive systems. The lab has analysed the impact of the use of an episodic memory module in terms of the robot performance and computational resources. ISAC is very important in the generation of intelligent behaviours in humanoid robots. Computation on ISAC takes place within the framework of the Intelligent Machine Architecture (IMA), a true multi-agent system designed to promote code-reuse (Alford et al, 1999). ISAC's memory is divided into three classes: Short-Term Memory (STM), Long-Term Memory (LTM), and the Working Memory System (WMS). The STM holds information about the current environment while the LTM holds learned behaviours, semantic and perceptual knowledge and past experiences. The WMS holds task specific STM and LTM information and provides the information flow to the cognitive processes during the realisation of a task (Dodd & Gutierrez, 2005).

RatSLAM is a robotic navigation system, the operation of which is based on models of the hippocampus of rodents. RatSLAM has managed significant contributions in the field of Simultaneous Localization and Mapping (SLAM). These contributions showed a biologically-inspired mapping system can compete with a probabilistic robot mapping system (Milford et al., 2016). RatSLAM is useful in the development of maps of the real-world environment. The algorithm used by RatSLAM is based on the analysis of spatial encoding in rat brains in order to perform and recall of maps. The main elements

Figure 1. ISAC Cognitive Robot Architecture
Adapted from (Dodd & Gutierrez, 2005)

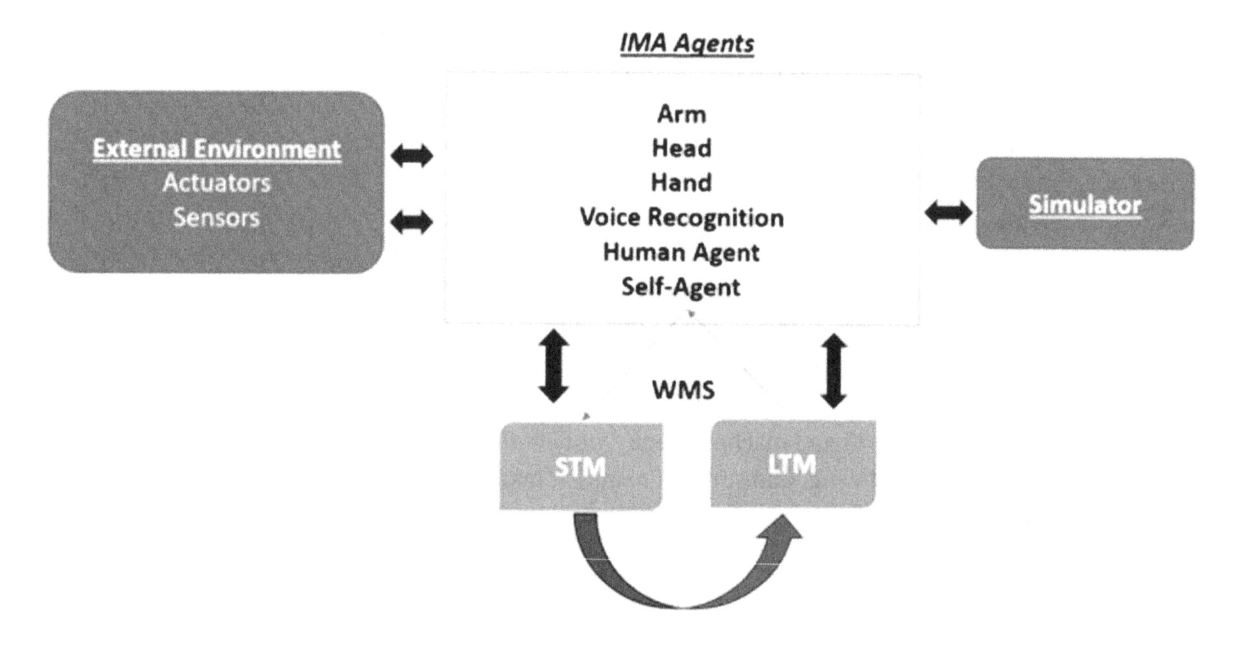

of the RatSLAM system are the pose cells, the local view cells and the experience map. The activity of the pose cells is updated by self-motion cues and it is calibrated by local view. The self-motion cues are responsible for path integration in the pose cells while the external cues trigger local view cells that are associated with pose cells through associative learning (Milford & Wyeth, 2008). The local view cells represent learned unique scenes of the environment. The experience map is a topological representation that encodes the pose cells and local view cells in nodes and links (Heath et al., 2013). Furthermore, it can integrate the data from the pose cells, the local view cells and the self-motion cues and develop a set of spatial experiences. These positions of these experiences and the transition from one experience to another can be integrated to a semimetric map (Milford & Wyeth, 2008).

Whisker-RatSLAM is a 6D extension of the RatSLAM and it focuses on object-recognition through the generation of point clouds of objects based on data received by an array of biomimetic whiskers. ViTa-SLAM harnesses both vision and tactile information for performing SLAM. It provides a mechanism for fusing non-unique tactile and unique visual data (Struckmeier et al., 2015).

Hirel at al. (2011) present a biologically-inspired model of hippocampal spatio-temporal learning and prefrontal cortex that can be used in robotic applications that have spatio-temporal constraints. Such applications are related to place-preference tasks and automated-security patrols.

Figure 2 shows the neural network for the associative learning. Sensory information is computed and a winner-take-all (WTA) competition is realised between neurons. Any change in the winning state triggers the spectral timing memory. When a new sensory event happens, the association of this new information with the spectral timing memory of the last event is learned by state prediction cells. In the preliminary phase, the robot explores the environment through the use of a camera which allows the autonomous learning and recognition of visual landmarks. New visual place cells which are represented as a constellation of visual landmarks are autonomously learned. The robot learns a topological repre-

Figure 2. Hippocampus' Associative Learning Architecture
Adapted from (Hirel et al., 2011)

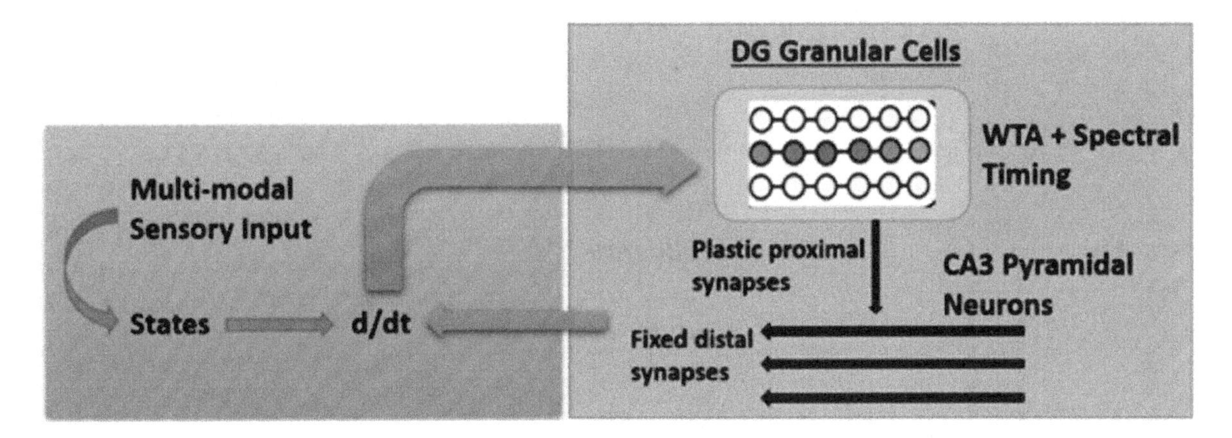

sentation of its environment. The next step is to teach the robot how to select specific places. This task is characterised by a number of behaviours that are controlled by sensory events. The training of the robot is achieved through the interaction with humans. A leash which is placed to an artificial neck is used to guide the robot. A camera is mounted on a pan-tilt system allows the recognition of visual landmarks. Visual place cells are represented as constellations of visual landmarks. During learning, the robot has no knowledge of its environment and it is guided by a human using the leash. The trajectory forms a loop that the robot must realise. When the loop is realised, the robot is back at the starting point. In this case, the leash is removed from the robot and the robot is free to reproduce the trajectory (Hirel et al., 2011).

Molecular communication enables the communication among nanomachines, such as biosensors and bio-actuators (Suzuki et al., 2014). A molecular communication system propagates information in terms of chemical signals from transmitting bio-nanomachines to receiving bio-nanomachines (Farhan & Mushfiq, 2015). A significant application of molecular communication is Body Area Nano Networks (BANNs) where nanoscale machines develop networks with each other in order to perform biomedical tasks, such as physiological sensing and neural signal transduction. A neuron-based BANN includes a number of nanomachines (e.g. sensors) and a network of neurons that form a particular topology. This topology allows nanomachines to communicate with each other. Two methods can be used for the specific topology. The first method is based on the development of neurons into 3D topology patterns and the second method is based on the use of Neuronal Time Division Multiple Access (TDMA) as a communication method among neurons. The assembly method uses silica beads as growth surface and bead-bead contacts as geometrical constraints on neuronal connectivity. A web lab experiment verified the assembly method with cells from the hippocampus (Suzuki et al., 2014).

Three types of BANNs are presented in Figure 3. These types include the neuro-sensory BANNs, the neuro-control BANNs, and the neuro-transduction BANNs. A neuro-sensory BANN includes sensor nanomachines that transmit sensory data to the sink nanomachine. This nanomachine serves as a sub-dermal or epidermal interface that connects a neuron-based BANN with external devices and it converts electrochemical signals to electromagnetic or electrostatic ones. Sensory data are then transmitted mobile devices such as a smartphone or a tablet. A neuro-control BANN includes a subdermal/epidermal interface node that receives signals from mobile devices and transmits these signals to actuator nanomachines.

Figure 3. Molecular communication using BANNs
Adapted from Suzuki et al. (2014)

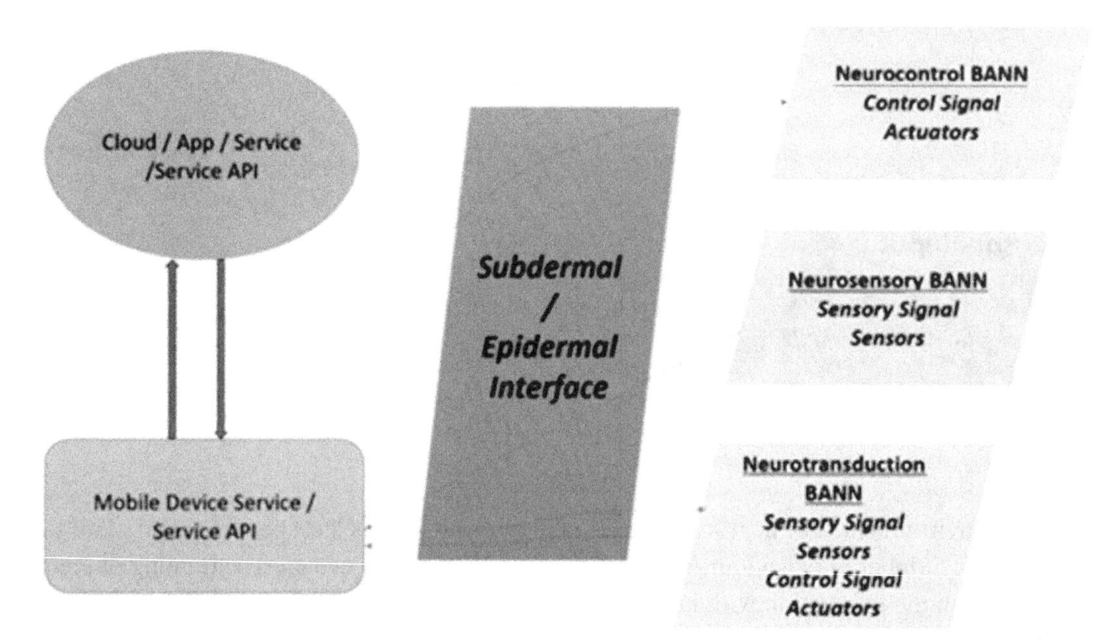

The actuator nanomachines are equipped with pumps that release drug molecules or neurostimulators to a nerve. A neuro-transduction BANN is a hybrid of a neuro-sensory and a neuro-control BANN. A neuro-transduction BANN has sensor nanomachines that transmit sensory signals to a subdermal/epidermal interface node that combines a sink nanomachine with a control signalling nanomachine. The interface node transduces an incoming sensory signal to a control signal and it transmits the control signal to actuator nanomachines [Suzuki et al. (2014)]. A BANN can be used to connect a nerve of the spinal cord to a prosthetic device (Chae et al., 2009).

The hippocampus is related to the formulation of neural mechanisms that are responsible for decision making. There is also potential involvement of the hippocampus in value-based decision making, especially during model-based reinforcement learning. This is very useful for Neuroeconomics which is field that seeks to explain human decision making in Economics. In order to examine how the process of value-related information is realised by the hippocampus, the task of dynamic foraging by the hippocampus was analysed using a reinforcement learning model. The study showed that the CA1 neurons of the hippocampus converged the signals that were necessary to evaluate the outcome of a specific event (Shanghai Neuroeconomics, 2017). Bhui (2018) shows the significance of the hippocampus and other medial temporal lobe structures in the realisation of case-based reasoning through the initiation of stimulus and reward associations. The comprehension of these operations of the hippocampus can reveal how case-based calculations can contribute to or interfere with model-free and model-based control (Bhui, 2018). Case-based reasoning is very important in risk and knowledge management (Dash Wu et al., 2014).

The emulation of biological synapses that realise memory and learning functions is an important step towards the realisation of bio-inspired systems. The development of artificial synaptic devices is based on the use of inorganic materials and semiconductor devices (Park & Lee, 2017). A number of neuromorphic computing projects have also been developed based on the use of analog and digital technologies. For example, DARPA has funded the SyNAPSE (Systems of Neuromorphic Adaptive Plastic Scalable Electronics) program which aims to develop an artificial brain that simulates the mammalian brain. IBM developed the TrueNorth chip which uses 5 billion transistors that make up 256 million synapses. A highly-parallel neuromorphic computer, the SpiNNaker (Spiking Neural Network Architecture), has been developed by the University of Manchester and it simulates the behaviour of 1 million neurons. The IBM BlueGene supercomputer is used to simulate sections of the brain of the rat (Upadhyay et al., 2016). Xu et al. (2016) show how artificial synapses in a two-terminal structure of substrate/buffer-capped conducting polymer electrode can emulate biological synapses. Brain plasticity is the ability-flexibility of the brain to change over time. Studies on the hippocampus of the rat show that the aging of the hippocampus results to a loss of the functional synapses in the area CA1 of the hippocampus, a decrement in the NMDA-receptor-mediated responses at performant path synapses, and a modification of Ca^{2+} regulation in area CA1. These changes may affect the synaptic plasticity (Rosenzweig & Barnes, 2003). The fabrication of artificial synapses can lead to better analysis of brain plasticity and the development of new methods for the therapy of diseases, such as dementia.

The above indicative examples from the literature show how the study of the brain's hippocampus can lead to the development of new applications in fields such as robotics. The hippocampus offers a range of capabilities and each of these capabilities can be the basis for the development of advanced computational and robotic systems. The fact that some of these applications currently exist prove the significance of the hippocampus and this is shown in the next section where the impact of the study of hippocampus in the development of biologically-inspired systems is shown. The complexity of the anatomy of the hippocampus and its operation means that more applications could be developed by analysing further this anatomy and associating it with existing computational systems.

IMPACT ON THE DEVELOPMENT OF BIOLOGICALLY-INSPIRED SYSTEMS

There are specific capabilities and functions of the hippocampus that can be useful for the further progress of current Artificial Intelligence systems. The hippocampus is capable of developing cognitive maps. The key features of processing in the hippocampus support a flexible Model-based Reinforcement Learning (MBRL) mechanism for spatial navigation. A computational MBRL framework includes characteristics that are related to the computational properties of the hippocampus, such as the hierarchical spatial representation and the remapping of place cells. These characteristics allow the easy adaptation to the dynamic conditions and the learning of multiple tasks (Barzov, 2017). The hippocampus includes place cells that are found in the CA1 and CA3 areas. Grid cells are found in the medial entorhinal cortex and in the pre- and para-subiculum while head-direction calls are found in the pre-subiculum and in the deep layers of the medial entorhinal cortex. Boundary cells that are found in the subiculum and in the entorhinal cortex typically fire at a specific distance from an environmental boundary. The parietal cortex includes a trajectory neuron. The hippocampus provides a cognitive map which includes information related to goal-directed decision making while the striatum learns associations between stimuli and responses. Sensory inputs are received by both the hippocampus and the striatum. The striatum only learns when a reward signal is given while the hippocampus has to receive additional information about the direction of the head. Each area produces as an output an estimated optimal action which is then compared and chosen by the prefrontal cortex (Chersi & Burgess, 2015).

The need for re-evaluating current biologically-inspired systems is imminent. A new type of intelligent computing methods have been developed in order to surpass the limitations posed by traditional Artificial Intelligence methods. The intelligent computing methods are called bioinspired intelligent algorithms (BIAs) and they are characterised by significant progress in the combining neuroscience-related and computing-related knowledge and implementing it to advanced fields, such as robotics (Ni et al., 2015). Based on the behavioural characteristics of organisms, a number of BIAs have been developed. Most of these algorithms are inspired from the foraging behaviours of organisms, such as the ant colony algorithm, the bee colony algorithm, and the fish swarm algorithm. There are also other algorithms such as the monkey climbing algorithm which simulates the climbing behaviour of monkeys and the bacterial chemotaxis algorithm which simulates the chemotaxis behaviour of bacteria (Ni et al., 2015). Putintsev et al. (2015) present the development of hybrots. The operation of hybrots is based on the interaction between a living neuron and a machine that acts as an artificial body. The use of such systems allows the control of the exchange of data between the brain and the body and the limitation of the complexity of this exchange. Hybrots are used in the study of Artificial Neural Networks and especially of their adaptation capabilities (Putintsev et al., 2015).

Song & Giszter (2011) describe a system which includes a live rat and a robot. They used rats based on specific criteria, such as nervous activity and movement. Nichrome tetrode wires were implanted into the motor cortex of the rats. There was reception of signals by the tetrodes in the frequency range from 300 Hz to 7.5 kHz and bursts of neuronal activity (spikes) were detected. The total number of spikes at a specific time interval (100ms) was used as the control signal for the robot.

The hippocampus contributes significantly in associative memory binding (Davachi, 2006; Eichenbaum & Cohen, 2001). This allows the long-term encoding of the co-occurrences of people, objects and locations with the spatial, temporal and interactions among them (Konkel & Cohen, 2009). It also allows the encoding of representations of relationships among events across time. The hippocampus is also characterised by representational flexibility which allows the reconstruction and recombination of information (Eichenbaum & Cohen, 2001). Hippocampal relational representations support flexible cognition. The role of the hippocampus in relational binding supports our ability to reconstruct richly detailed, multimodal memories of our remote past and our ability to imagine events of our distant futures (Hilliard et al., 2016). The hippocampus integrates and transforms memories that overlap into relational schemas (McKenzie et al., 2014). Recent research on how the hippocampus represents memories shows that there is competition between pattern completion of a new experience to a previously stored representation and pattern separation mechanism that separates overlapping memories in order for any interference between representations of memories to be reduced (Deng et al., 2013, Colgin et al., 2010). Little is known about the encoding and organization of memories in the hippocampus and how the mechanisms of pattern completion and separation operate in this context (McKenzie et al., 2014).

A network of strongly interconnected structures that include the hippocampus, the prefrontal cortex and the basal ganglia is used for the realisation of spatial navigation. The identification of how information flows between these structures could reveal the processes that characterise the navigational tasks in the brains of the mammals. There are strong excitatory connections between the hippocampus and the medial prefrontal cortex (mPFC). In the basal ganglia (BG), the striatum receives connections from both the hippocampus and the mPFC. Bilateral connections exist between the BG and mPFC (Groenewegen, Wright, & Uylings, 1997). An auto-associative network function of the CA3 region of the hippocampus supports mnemonic functions. Specifically, the CA3 region is responsible for the acquisition and encoding of spatial data of the short-term memory. The encoding of information can be used for the development of relational representations (Kesner, 2007).

The analysis of how the brain integrates different types of learning and how it coordinates these different types is important in comprehending the complex operations of human and animal brains. The hippocampus enhances the adaptation of the brain to different situations by making predictions of the future states. These predictions are based on the representation of successor states of a current state. In other words, the hippocampus creates a predictive map. This approach includes two different types of algorithms that are part of reinforcement learning and they are the 'model-based' and the 'model-free' algorithms. Model-based algorithms learn models of the environment that can be later simulated while model-free algorithms are based on the direct experience with the environment (Stachenfeld et al., 2017).

Pigeons are known for their ability to return to home territory even if they have been released miles away from it. This ability is based on the pigeon's vision and their hippocampus which is used to construct an environmental spatial map. This ability has not yet been achieved by modern-day UAVs. Modern-day UAVs rely on GPS for their navigation. Current GPS technologies are not so advanced in order to provide precise information and the reason is because GPS signals can be affected by surrounding structures. Furthermore, the hippocampi of different organisms adapt to different environmental conditions. For example, the rat's hippocampus is adapted to an environment which includes tunnels and corridors while the pigeon's environment is adapted to the flying environment. Based on these facts, the operation of a robot depends on the quantity of the spatial cues of the environment it operates in. The most important visual features can be used to detect very important locations in the environment (Oyekan, 2015).

Neuroscience can provide new analytic tools for the comprehension of computation in AI systems. The products of Artificial Intelligence are usually 'black boxes' as the nature of computations that occur to them are not well comprehended. The application of neuroscience tools to AI systems will provide insights to these systems. This is called 'virtual brain analytics' (Hassabis et al., 2017). The visualisation of brain states through dimensionality reduction is used in neuroscience and has been applied to neural networks (Zahavy et al., 2016).

The role of the hippocampus is defined by the relational organisation and flexibility of the cognitive maps. The hippocampus develops multiple navigational strategies as well as other spatial and non-spatial knowledge domains that focus on relational memory organisation. In general, the brain acts as a statistician that applies probabilistic learning in order to develop an hierarchical model of the world (Barzov, 2017). The auditory system supported by a network of auditory, cortical, hippocampal, and frontal sources scans the environment and it efficiently represents complex stimulus statistics and it responds to the emergence of regular patterns (Barascud et al., 2016).

Hippocampus supports rapid behavioural adjustment and experience replay which supports interleaved training of deep neural networks (Hassabie et al., 2017). Interleavers for DNNs use sparsity in order to lower memory and computational requirements (Dey et al., 2018). DNNs are useful for speech processing and autonomous vehicles. Modern DNNs are characterised by million of parameters (Khrizhevsky et al., 2012) which make the networks difficult to implement in hardware. A sparse network offered by interleavers can reduce the number of these parameters (Han et al., 2016). Rolls (2013) suggests that the hippocampal area CA3 acts as an auto-associative memory which is capable of pattern completion while the dentate gyrus is responsible for pattern separation. If there is no similarity among patterns, then a new input will be stored as a new memory (Kumaran et al., 2016). In this case, the retrieval of complete memories from partial cues can be achieved realising operations similar to database retrieval (Marblestone et al., 2016). The enhancement of database technologies can impact the field of robotics as databases that simulate more accurately the human perception can be embedded to robots. Also, the way information are inter-related to each other and the simulation of these inter-relationships in a computational manner can reveal some of the more advanced functionalities of human perception and the way human personalities can adapt to the dynamic changes of their environments. This is very important for the field of behavioural sciences.

Adaptability is a great characteristic of the brain's hippocampus. Modern Artificial Intelligence systems are characterised by adaptability to the dynamic conditions of their environment, however this adaptability constantly progresses. Further and more detailed examination of the characteristics and the functions of the hippocampus will lead to the extraction of useful conclusions on how the data flow in the brain is realised and how the connections among these data are achieved. There are serious reasons that explain the significance of the study of the brain's hippocampus and how it can contribute to the development of current Artificial Intelligence systems.

FUTURE RESEARCH DIRECTIONS

The exact steps of the ways the hippocampus generate adaptiveness and how this affects the brain's situational awareness need to be clarified. This will lead to the re-evaluation of the current properties and capabilities of biologically-inspired systems. One of the foci of Synthetic Biology is the development of artificial cells. This type of cells can contain the same biological material as biological cells. Artificial cells can be of macro-, micro-, nano- and molecular dimensions. There are also unlimited possibilities in variations in the membranes and contents of these cells (Chang, 2007). Artificial cells play an important role in cell and organ transplantation in regenerative medicine (Chang, 2005). Bionic organ science allows the reproduction of the organ architecture and function through a complex interplay of cellular and mechanical elements. Some bio-artificial organs may be used for the replacement of anatomical defects while others allow the compensation of failing organ functions (Famulari et al., 2003). The study of the hippocampus can provide useful conclusions on the situational awareness capabilities of the cells. Specifically, it would be interesting to check how artificial intelligence can be embedded into current organ transplantation technologies and improve the situational awareness of artificial cells. In this case, improvements can be made to the current bio-robotic technologies that could be used for organ transplantations. Other areas that can be benefited from the study of the hippocampus and how this study is related to the enhancement of current biologically-inspired systems is the prevention of cybercrime and the re-organization of supply chains in different fields of engineering such as aeronautics and construction management. The tracking capabilities of the hippocampus can be the basis for the development of more accurate supply chains which will correspond to the needs of their environments. Efficient planning characterises a successful supply chain. Specific characteristics of efficient planning are concurrency, forecasting and optimization and these are characteristics that can be further studied through the study of the hippocampus (Gaus et al., 2018). In relation to cyber security systems, situational awareness plays an important role in the development of advanced cyber-crime prevention systems. This will lead to more accurate prediction of extreme crimes which in turn leads to cyber resilience. Cyber resilience refers to the ability to prepare for, respond to and recover from cyber-attacks.

CONCLUSION

The capability of the hippocampus to adapt to constantly changing conditions is considered significant in the development of up-to-date Artificial Intelligence systems. There is an on-going attempt to further enhance current AI systems and the different capabilities of the hippocampus as shown in the previous sections of this paper can contribute towards this direction. The study of the hippocampus can help in the refinement and clarification of the type and amount of data that should be used by biologically-inspired systems and also, provide conclusions on the exact structural elements and their topologies that should be embedded to a biologically-inspired system. The hippocampus adjusts the adaptiveness of the brain and this ability needs to be further examined so that new ways of comprehending adaptiveness are generated. These new ways can in turn produce conclusions on how adaptiveness can be embedded to biologically-inspired systems. Managing adaptiveness can allow a biologically-inspired system to better understand its environment and interact with other systems that belong to this environment. The further comprehension of adaptiveness can enable the development of systems that are characterised by enhanced situational awareness and this could optimize for example different types of supply chains, the operation of bio-robotic devices, the way environmental pollution is perceived by a biologically-inspired system, the way an autonomous vehicle perceives it surroundings or even how a robotic teaching assistant interacts with its human students.

Furthermore, the detailed analysis of how data are represented and connected to hippocampus can lead to the better comprehension of diseases, such as Parkinson's and Alzheimer's or even schizophrenia. These neurodegenerative disorders are associated with genetics, age, geographic location and gender. The study of the hippocampus through the development of more advanced AI systems will result to the examination of other factors that are associated with these neurogenerative disorders. Furthermore, the way data are represented in the hippocampus and how they are related to each other can lead to better conclusions on how data could be represented and related in database technologies, especially those that are based on non-relational concepts. There are some challenges that also need to be considered. Especially for the field of Synthetic Biology, many areas of it are still unknown. Also, it is possible that many aspects of biological systems are not open to the integration with automated systems. This argument is further enhanced by the complexity of these systems and this complexity does not allow these systems to be more open to the integration with automated systems.

REFERENCES

Adcock, R. A., Thangavel, A., Whitfield-Gabrieli, S., Knutson, B., & Gabrieli, J. D. E. (2006). Reward-motivated learning: Mesolimbic activation precedes memory formation. *Neuron*, *50*(3), 507–517. doi:10.1016/j.neuron.2006.03.036 PMID:16675403

Alford, W. A., Rogers, T., Wilkes, D. M., & Kawamura, K. (1999). Multi-Agent System for a Human-Friendly Robot. In *Proceedings of the 1999 IEEE International Conference on Systems, Man, and Cybernetics (SMC '99)*, 1064-1069, Tokyo, Japan. 10.1109/ICSMC.1999.825410

Ball, D., Heath, S., Wiles, J., Wyeth, G., Corke, P., & Milford, M. (2013). OpenRatSLAM: An open source brain-based SLAM system. *Autonomous Robots*, *34*(3), 149–176. doi:10.100710514-012-9317-9

Barascud, N., Pearcec, M. T., Griffiths, T. D., Friston, K. J., & Chaita, M. (2016). Brain responses in humans reveal ideal observer-like sensitivity to complex acoustic patterns. *Proceedings of the National Academy of Sciences of the United States of America (PNAS)*, E616-E625 10.1073/pnas.1508523113

Barzov, Y. (2017). *Hacking Hippocampus: the Next Frontier for Machine Learning and beyond.* Retrieved September 14, 2019 from https://towardsdatascience.com/machine-learning-as-hacking-of-the-brain-6aab8c4a9e7d

Bengio, Y. (2009). Learning deep architectures for AI. *Foundations and Trends in Machine Learning*, *2*(1), 1–127. doi:10.1561/2200000006

Bhui, R. (2018). Case-Based Decision Neuroscience: Economic Judgment by Similarity. Retrieved July 20, 2019 from https://scholar.harvard.edu/files/rbhui/files/case-based_decision_neuroscience.pdf

Bunzeck, N., Dayan, P., Dolan, R. J., & Duzel, E. (2010). A common mechanism for adaptive scaling of reward and novelty. *Human Brain Mapping*, *31*(9), 1380–1394. doi:10.1002/hbm.20939 PMID:20091793

Chae, M. S., Yang, Z., Yuce, M. R., Hoang, L., & Liu, W. (2009). A 128-Channel 6 mW Wireless Neural Recording IC With Spike Feature Extraction and UWB Transmitter. *IEEE Transactions on Neural Systems and Rehabilitation Engineering*, *17*(4), 312–321. doi:10.1109/TNSRE.2009.2021607 PMID:19435684

Chang, T. M. (2005). The role of artificial cells in cell and organ transplantation in regenerative medicine. *Panminerva Medica*, *47*(1), 1–9. PMID:15985972

Chang, T. M. (2007). 50th anniversary of artificial cells: Their role in biotechnology, nanomedicine, regenerative medicine, blood substitutes, bioencapsulation, cell/stem cell therapy and nanorobotics. *Artificial Cells, Blood Substitutes, and Immobilization Biotechnology*, *35*(6), 545–554. doi:10.1080/10731190701730172 PMID:18097783

Chersi, F., & Burgess, N. (2015). The Cognitive Architecture of Spatial Navigation: Hippocampal and Striatal Contributions'. *Neuron*, *88*(1), 64–77. doi:10.1016/j.neuron.2015.09.021 PMID:26447573

Chkirbene, Z., & Hamdi, N. (2015). A survey on spectrum management in cognitive radio networks. *International Journal of Wireless and Mobile Computing*, *8*(2), 153–165. doi:10.1504/IJWMC.2015.068618

Colgin, L. L., Leutgeb, S., Jezek, K., Leutgeb, J. K., Moser, E. I., McNaughton, B. L., & Moser, M.-B. (2010). Attractor-map versus autoassociation basedattractor dynamics in the hippocampal network. *Journal of Neurophysiology*, *104*(1), 35–50. doi:10.1152/jn.00202.2010 PMID:20445029

Dash Wu, D., Chen, S.-H., & Olson, D. L. (2014). Business intelligence in risk management: Some recent progresses. *Information Sciences*, *256*, 1–7. doi:10.1016/j.ins.2013.10.008

Davachi, L. (2006, December). Item, context and relational episodic encoding in humans. *Current Opinion in Neurobiology*, *16*(6), 693–700. doi:10.1016/j.conb.2006.10.012 PMID:17097284

Davidow, J. Y., Foerde, K., Galvan, A., & Shohamy, D. (2016). An Upside to Reward Sensitivity: The Hippocampus Supports Enhanced Reinforcement Learning in Adolescence. *Neuron*, *92*(1), 93–99. doi:10.1016/j.neuron.2016.08.031 PMID:27710793

Deng, W., Mayford, M., & Gage, F. H. (2013). Selection of distinct populations of dentate granule cells in response to inputs as a mechanism for pattern separation in mice. *eLife*, *2*. doi:10.7554/eLife.00312 PMID:23538967

Dey, S., Beerel, P. A., & Chugg, K. M. (2018). Interleaver Design for Deep Neural Networks, *arXiv:1711.06935v2*.

Dodd, W., & Gutierrez, R. (2005). *The role of episodic memory and emotion in a cognitive robot.* In *Proceedings of the IEEE International Workshop on Robot and Human Interactive Communication (ROMAN)*. IEEE. Nashville, TN 10.1109/ROMAN.2005.1513860

Eichenbaum, H., & Cohen, N. J. (2001). *From Conditioning to Conscious Recollection: Memory Systems of the Brain*. Oxford, UK: Oxford University Press.

Epstein, R. A., Patai, E. Z., Julian, J. B., & Spiers, H. J. (2017). The cognitive map in humans: Spatial navigation and beyond. *Nature Neuroscience*, *20*(11), 1504–1513. doi:10.1038/nn.4656 PMID:29073650

Famulari, A., De Simone, P., Verzaro, R., Iaria, G., Polisetti, F., Rascente, M., & Aureli, A. (2003). Artificial Organs as a Bridge to Transplantation. *Artificial Cells, Blood Substitutes, and Biotechnology*, *31*(2), 163–168. doi:10.1081/BIO-120020174 PMID:12751836

Farhan, F., & Mushfiq, M. (2015). Biological Cell and Molecular Communication Technology: Overview and Challenges. *European Scientific Journal*, *3*, 56–60.

Gallistel, C., & King, A. P. (2009). *Memory and the Computational Brain: Why Cognitive Science will Transform Neuroscience*. Wiley-Blackwell. doi:10.1002/9781444310498

Gaus, T., Olsen, K., & Deloso, M. (2018). *Synchronizing the digital supply network Using artificial intelligence for supply chain planning*. Retrieved May 10, 2019, from https://www2.deloitte.com/us/en/insights/focus/industry-4-0/artificial-intelligence-supply-chain-planning.html

Groenewegen, H. J., Wright, C. I., & Uylings, H. B. (1997). The anatomical relationships of the prefrontal cortex with limbic structures and the basal ganglia. *Journal of Psychopharmacology (Oxford, England)*, *11*(2), 99–106. doi:10.1177/026988119701100202 PMID:9208373

Gulshan, V., Peng, L., Coram, M., Stumpe, M. C., Wu, D., Narayanaswamy, A., ... Webster, D. R. (2016). Development and Validation of a Deep Learning Algorithm for Detection of Diabetic Retinopathy in Retinal Fundus Photographs. *Journal of the American Medical Association*, *316*(22), 2402–2410. doi:10.1001/jama.2016.17216 PMID:27898976

Hassabis, D., Kumaran, D., Summerfield, C., & Botvinick, M. (2017). Neuroscience-Inspired Artificial Intelligence. *Neuron*, *95*(2), 245–258. doi:10.1016/j.neuron.2017.06.011 PMID:28728020

Hilliard, C., Wagner Cook, S., & Duff, M. C. (2016). Hippocampal declarative memory supports gesture production: Evidence from amnesia. *Cortex*, *85*, 25–36. doi:10.1016/j.cortex.2016.09.015 PMID:27810497

Hinton, G., Osindero, S., & The, Y. (2006). A fast learning algorithm for deep belief nets. *Neural Computation*, *18*(7), 1527–1554. doi:10.1162/neco.2006.18.7.1527 PMID:16764513

Hirel, J., Gaussier, P., & Quoy, M. (2011). Biologically inspired neural networks for spatio-temporal planning in robotic navigation tasks. In *Proceedings of the 2011 IEEE International Conference on Robotics and Biomimetics*, December 7-11, 2011. Phuket, Thailand IEEE. 10.1109/ROBIO.2011.6181522

Jeffery, K., & Casali, G. (2014). Hippocampal Neurons: Simulating the Spatial Structure of a Complex Maze. *Current Biology*, *24*(14), R643–R645. doi:10.1016/j.cub.2014.06.001 PMID:25050959

Kesner, R. P. (2007). Behavioral functions of the CA3 subregion of the hippocampus. *Learning & Memory (Cold Spring Harbor, N.Y.)*, *14*(11), 771–781. doi:10.1101/lm.688207 PMID:18007020

Konkel, A., & Cohen, N. J. (2009). Relational memory and the hippocampus: Representations and methods. *Frontiers in Neuroscience*, *3*(2), 166–174. doi:10.3389/neuro.01.023.2009 PMID:20011138

Kumaran, D., Hassabis, D., & McClelland, J. L. (2016). What learning systems do intelligent agents need? complementary learning systems theory updated. *Trends in Cognitive Sciences*, *20*(7), 512–534. doi:10.1016/j.tics.2016.05.004 PMID:27315762

Lee, W. Y., & Akyildiz, I. F. (2012). Spectrum-aware mobility management in cognitive radio cellular networks. *IEEE Transactions on Mobile Computing*, *11*(4), 529–542. doi:10.1109/TMC.2011.69

Lorica, B. (2017). Practical applications of reinforcement learning in industry. Retrieved May 25, 2019, from https://www.oreilly.com/ideas/practical-applications-of-reinforcement-learning-in-industry

Marblestone, A. H., Wayne, G., & Kording, K. P. (2016). Toward an Integration of Deep Learning and Neuroscience. *Frontiers in Computational Neuroscience*, 1–41. PMID:27683554

McKenzie, S., Frank, A. J., Kinsky, N. R., Porter, B., Rivière, P. D., & Eichenbaum, H. (2014). Hippocampal Representation of Relatedand Opposing Memories Develop within Distinct, Hierarchically Organized Neural Schemas. *Neuron*, *83*(1), 202–215. doi:10.1016/j.neuron.2014.05.019 PMID:24910078

Milford, M., Jacobson, A., Chen, Z., & Wyeth, G. (2016). RatSLAM: Using Models of Rodent Hippocampus for Robot Navigation and Beyond. In M. Inaba, & P. Corke (Eds.), *Robotics Research. Springer Tracts in Advanced Robotics* (Vol. 114). Springer International Publishing. doi:10.1007/978-3-319-28872-7_27

Milford, M. J., & Wyeth, F. G. F. (2008). Mapping a Suburb With a Single Camera Using a Biologically Inspired SLAM System. *IEEE Transactions on Robotics*, *24*(5), 1038–1053. doi:10.1109/TRO.2008.2004520

Neuroeconomics, S. (2017). Hippocampus and Decision Making. Retrieved June 17, 2019, from http://www.shanghai-neuroeconomics.org/2017-sh-colloq/2017/5/2/hippocampus-and-decision-making

Ni, J., Wu, L., Fan, X., & Yang, S. X. (2015). Bioinspired Intelligent Algorithm and Its Applications for Mobile Robot Control: A Survey. *Computational Intelligence and Neuroscience*, *2016*. PMID:26819582

O'Keefe, J., & Nadel, L. (1978). *The Hippocampus as a Cognitive Map*. Oxford University Press.

Oyekan, J. (2015). A vision-based terrain morphology estimation model inspired by the avian hippocampus. *Digital Communications and Networks*, *1*(2), 134–140. doi:10.1016/j.dcan.2015.04.002

Park, Y., & Lee, J.-S. (2017). Artificial Synapses with Short- and Long-Term Memory for Spiking Neural Networks Based on Renewable Materials. *ACS Nano*, *11*(9), 8962–8969. doi:10.1021/acsnano.7b03347 PMID:28837313

Polydoros, A. S., & Nalpantidis, L. (2017). Survey of Model-Based Reinforcement Learning: Applications on Robotics. *Journal of Intelligent & Robotic Systems*, *86*(2), 53–173. doi:10.100710846-017-0468-y

Putintsev, N. I., Vishnevskya, O. V., & Vityaev, E. E. (2015). Development of Artificial Cognitive Systems Based on Models of the Brain of Living Organisms. *Russian Journal of Genetics: Applied Research*, *5*(6), 589–600. doi:10.1134/S207905971506012X

Rao, A., & Voyles, J. R. (2017). Top 10 artificial intelligence (AI) technology trends for 2018. Retrieved June 2, 2019, from http://usblogs.pwc.com/emerging-technology/top-10-ai-tech-trends-for-2018/

Rolls, E. T. (2013). The mechanisms for pattern completion and pattern separation in the hippocampus. *Frontiers in Systems Neuroscience*, *7*, 74. doi:10.3389/fnsys.2013.00074 PMID:24198767

Rosenzweig, E. S., & Barnes, C. A. (2003). Impact of aging on hippocampal function: Plasticity, network dynamics, and cognition. *Progress in Neurobiology*, *69*(3), 143–179. doi:10.1016/S0301-0082(02)00126-0 PMID:12758108

Scoville, W. B., & Milner, B. (2000). Loss of recent memory after bilateral hippocampal lesions, 1957. *The Journal of Neuropsychiatry and Clinical Neurosciences*, *12*(1), 103–113. doi:10.1176/jnp.12.1.103-a PMID:10678523

Song, W., & Giszter, S. F. (2011). Adaptation to a cortex-controlled robot attached at the pelvis and engaged during locomotion in rats. *The Journal of Neuroscience*, *31*(8), 3110–3128. doi:10.1523/JNEUROSCI.2335-10.2011 PMID:21414932

Squire, L. R., Stark, C. E., & Clark, R. E. (2004). The medial temporal lobe. *Annual Review of Neuroscience*, *27*(1), 279–306. doi:10.1146/annurev.neuro.27.070203.144130 PMID:15217334

Stachenfeld, K. L., Botvinick, M. M., & Gershman, S. J. (2017). The hippocampus as a predictive map. *Nature Neuroscience*, *20*(11), 1643–1653. doi:10.1038/nn.4650 PMID:28967910

Struckmeier, O., Tiwari, K., Salman, M., Pearson, M. J., & Kyrki, V. (2019). ViTa-SLAM: A Bio-inspired Visuo-Tactile SLAM for Navigation while Interacting with Aliased Environments, arXiv:1906.06422v4. Retrieved August, 01, 2019, from https://arxiv.org/pdf/1906.06422.pdf

Suzuki, J., Balasubramaniam, S., Pautor, S., Meza, V. D. P., & Koucheryavy, Y. (2014). A Service-Oriented Architecture for Body Area NanoNetworks with Neuron-based Molecular Communication. *Mobile Networks and Applications*, *19*(6), 707–717. doi:10.100711036-014-0549-0

The University of Hong Kong. (2017). New functions of hippocampus unveiled: Scientists achieve major breakthrough in untangling mysteries of the brain. Retrieved July 12, 2019, from www.sciencedaily.com/releases/2017/09/170929093215.htm

Tulving, E. (1985). How many memory systems are there? *The American Psychologist*, *40*(4), 385–398. doi:10.1037/0003-066X.40.4.385

Tulving, E. (2002). Episodic memory: From mind to brain. *Annual Review of Psychology*, *53*(1), 1–25. doi:10.1146/annurev.psych.53.100901.135114 PMID:11752477

Upadhyay, N. K., Joshi, S., & Yan, J. J. (2016). Synaptic electronics and neuromorphic computing. *Science China. Information Sciences*, *59*(6). doi:10.100711432-016-5565-1

Wimmer, G. E., & Shohamy, D. (2012). Preference by association: How memory mechanisms in the hippocampus bias decisions. *Science*, *338*(6104), 270–273. doi:10.1126cience.1223252 PMID:23066083

Wu, Q.-T., Wu, C., & Wang, Y. M. (2016). *Application of Reinforcement Learning on High-Speed Rail Cognitive Radio*. Paper presented at the 2016 International Conference on Artificial Intelligence: Techniques and Applications (AITA 2016). Shanghai, China

Xu, W., Cho, H., Kim, Y. H., Kim, Y. T., Wolf, C., Park, C. G., & Lee, T. W. (2016). Organometal Halide Perovskite Artificial Synapses. *Advanced Materials*, *28*(28), 5916–5922. doi:10.1002/adma.201506363 PMID:27167384

Zahavy, T., Zrihem, N. B., & Mannor, S. (2016). Graying the black box: understanding DQNs, *arXiv:160202658*

ADDITIONAL READING

Andersen, P., Morris, R., Amaral, D., Bliss, T., & O'Keefe, J. (Eds.). (2006). *The Hippocampus Book*. New York, NY, US: Oxford University Press. doi:10.1093/acprof:oso/9780195100273.001.0001

Bera, R. K. (2018). Synthetic Biology, Artificial Intelligence, and Quantum Computing. *Genetic Engineering Technology and Synthetic Biology*. IntechOpen Retrieved October 10, 2019 from: https://www.intechopen.com/online-first/synthetic-biology-artificial-intelligence-and-quantum-computing

Fan, S. (2018). *DeepMind's New Research on Linking Memories, and How It Applies to AI*. Retrieved September 15, 2019 from: https://singularityhub.com/2018/09/26/deepminds-new-research-on-linking-memories-and-how-it-applies-to-ai/

Nilsson, N. J. (2009). *The Quest for Artificial Intelligence - A History of Ideas and Achievements*. New York, NY: Cambridge University Press. doi:10.1017/CBO9780511819346

Wibral, M., Lizier, J. T., & Priesemann, V. (2015). Bits from brains for biologically inspired computing. *Frontiers in Robotics and AI*, *2*, 1–25. doi:10.3389/frobt.2015.00005

Yamakawa, H. (2012). Hippocampal Formation Mechanism Will Inspire Frame Generation for Building an Artificial General Intelligence. In J. Bach, B. Goertzel, & M. Iklé (Eds.), Lecture Notes in Computer Science: Vol. 7716. *Artificial General Intelligence. AGI 2012*. Berlin: Springer. doi:10.1007/978-3-642-35506-6_37

KEY TERMS AND DEFINITIONS

Body Area Nano Networks (BANNs): A miniature network of sensors that operate on or around the human body.

CA (Cornu Ammonis or Ammon's Horn): This is Hippocampus Proper which is a major region of the Hippocampus. It includes four sub-regions, namely CA1, CA2, CA3 and CA4.

Cortex: The outer layer of an organ.

DARPA (Defence Advanced Research Projects Agency): This is an agency of the United States Department of Defence that is used for the development of advanced technologies used by the military.

Hippocampus: A major element of the mammalian brain used for spatial navigation and transformation of short-term memory to long-term memory.

ISAC (Intelligent Soft Arm Control): A cognitive robotic platform used for human-computer interaction.

RatSLAM (Rat - Simultaneous Localization and Mapping): A robotic navigation system used for the development of maps of the environment.

SpiNNaker (Spiking Neural Network Architecture): This is a supercomputer used at the University of Manchester in order to mimic how the human brain functions.

SyNAPSE (Systems of Neuromorphic Adaptive Plastic Scalable Electronics): A DARPA program that aims to develop a cognitive artificial brain based on the functions of the mammalian brain.

Chapter 5
Monitoring, Predicting, and Optimizing Energy Consumptions:
A Goal Toward Global Sustainability

Pedro J. S. Cardoso
https://orcid.org/0000-0003-4803-7964
Universidade do Algarve, Portugal & LARSyS, Institute for Systems and Robotics, Lisbon, Portugal

Jânio Monteiro
https://orcid.org/0000-0002-4203-1679
Universidade do Algarve, Portugal & INESC-ID, Lisbon, Portugal

Cristiano Cabrita
Universidade do Algarve, Portugal & Centre of Intelligent Systems, IDMEC, Portugal

Jorge Semião
https://orcid.org/0000-0002-7667-7910
Universidade do Algarve, Portugal & INESC-ID, Lisbon, Portugal

Dario Medina Cruz
https://orcid.org/0000-0001-9465-0845
Universidade do Algarve, Portugal

Nelson Pinto
https://orcid.org/0000-0002-8041-9199
Universidade do Algarve, Portugal

Célia M.Q. Ramos
https://orcid.org/0000-0002-3413-4897
Universidade do Algarve, Portugal & CIEO, Portugal

Luís M. R. Oliveira
Universidade do Algarve, Portugal & CISE, Portugal

João M. F. Rodrigues
https://orcid.org/0000-0002-3562-6025
Universidade do Algarve, Portugal & LARSyS, Institute for Systems and Robotics, Lisbon, Portugal

DOI: 10.4018/978-1-7998-2112-0.ch005

ABSTRACT

Energy consumption and, consequently, the associated costs (e.g., environmental and monetary) concern most individuals, companies, and institutions. Platforms for the monitoring, predicting, and optimizing energy consumption are an important asset that can contribute to the awareness about the ongoing usage levels, but also to an effective reduction of these levels. A solution is to leave the decisions to smart system, supported for instance in machine learning and optimization algorithms. This chapter involves those aspects and the related fields with emphasis in the prediction of energy consumption to optimize its usage policies.

INTRODUCTION

A recent study by the International Energy Agency (2019) shows the continuous increase of the global energy consumption which, in 2018, nearly doubled the average of the growth between 2010 and 2017. This growth was driven by a demanding global economy and population needs such as the higher consumptions in the heating and cooling of buildings which, for instance, in 2012 accounted for around 50% of the consumptions in the European Union (EU Publications, 2016). This led to an increasing demand of energy with significant growth in the consumption of renewable and non-renewable energy sources, with CO_2 emission raising almost 2% between 2017 and 2018, as a consequence. In 2018, electricity demand was accountable for more than 50% of the energy needs (International Energy Agency, 2019), contributing to two-thirds of emissions growth. Nevertheless, there are some good news as emissions declined for some of the most industrialized countries, such as Germany, Japan, Mexico, France, and the United Kingdom, and renewables incorporation increased by 4% in 2018, accounting for almost one-quarter of global's energy demand growth.

Globally, more than 40% of total energy consumption and 30% of total CO_2 emissions are associated to buildings. In the European Union, buildings also represent 40% of total energy consumption and approximately 36% of greenhouse gas emissions (Ahmad, Mourshed, Mundow, Sisinni, & Rezgui, 2016; Ahmad, Mourshed, & Rezgui, 2017), being urgent the contraction or, at least, a stabilization of those values. Some measures to reduce the energy needs must therefore be taken, supported on consuming politics, buildings' improvements, consumers' awareness, or intelligent buildings promotion. For instance, artificial intelligence techniques can be used to analyze the history of energy consumption of buildings and to predict its future consumption levels. For instance, the integration of predictive power consumption models is one essential component in energy control and operating strategies (Pinto *et al.*, 2019). This can contribute to the minimization of the energy costs of buildings, using Energy Management Systems (EMS) capable of scheduling the loads according with the generation levels or tariff rates. It can also be used in the development of decision support systems, that inform users about the best solutions to deal with a certain consumption pattern, and/or the implementation of automated systems for the detection of excessive consumption levels. All these solutions require the development and/or usage of prediction algorithms but, accurate forecasting of consumption is a challenge and a very complex task, not only because these are typically non-linear systems, but also correlate to seasonal variations and atmospheric conditions, while being dependent on users' habits.

In many countries, renewable energy production already represents an important percentage of the total energy that is generated in electrical grids. In order to reach higher levels of integration, demand side management measures are required. In fact, different from the traditional electrical grids, where at any given instant the production is adjusted to meet the demand, when using renewable energy sources, the demand must be adapted in accordance with the generation levels, since these cannot be controlled. EMS can have here a major role to monitor both the generation and consumption patterns, and to control electrical appliances, alleviating users from the burden of individual control of each appliance. Therefore, a platform for the monitoring, prediction, and optimization of consumptions in buildings is an asset that can contribute to the improvement of today's reality.

This chapter focuses aspects of a possible platform for the monitoring, prediction, and optimization of consumptions with special interest in the prediction of energy consumptions, in order to optimize its usage policies. The chapter will start by doing a thoughtful state-of-the art analysis on the main addressed fields, such as the global energy needs, detailing the European case. Then the intelligent energy management will be addressed namely, the energy management in general buildings, the implication of the introduction of electric vehicles in the available electrical networks, and the aspects of energetical sustainability in touristic buildings and activities. Later, a section will be devoted to the analysis of machine learning (ML) methods (e.g., for energy consumption prediction and anomaly detection), referring some state-of-the-art algorithms. Solutions, recommendations, and conclusions will end the chapter.

BACKGROUND

Global Trends on Energy Production and Consumption

Energy consumption has been steadily growing at least since 1990 (the exception was 2009), passing from around 8,761 Mtoe (million tons of oil equivalent) to 14,126 Mtoe, according with data published by Enerdata (2018) – see Figure 1. This growth corresponds to a global increase of more than 60%, with some regions more than doubling their consumptions, such as Middle East (253%), Asia (172.9%), or Africa (110.7%) – see Table 1. Although the Middle East and Africa represent a small portion of the global consumption, those values show a trend accompanied by the five major emerging national economies, usually designated as BRICS (Brazil, Russia, India, China, and South Africa), which had a growth of 127.6%. Other groups of countries instead showed a more bearable growth, see for example the G7 (8.8%) or North America (17.4%).

Breaking up consumptions by nations (see Table 2), considering the top 10 consumers in 2017, the major impact comes from China which consumes 22% of the World's energy, corresponding to a growth of 256.6% from 1990 to 2017. In 2017, the top 10 consumers total almost 63% of the World's consumption, showing some of them (e.g., China or India) a tendency to continue their steep increase as is observable in Figure 2. Nevertheless, in the top 10 of consumers, some countries diminished their energy consumption in the same period, as are the cases of Russia (-15.6%), Japan (-2.1%), and Germany (-11.4%).

The scenario is not much different when it comes to the specific case of electricity domestic consumption, as it can be seen in Figure 3. The values show that worldwide the domestic electricity consumption more than doubled from 1990 and 2017, with its major growth in the Asian continent, with the BRICS group of countries following a very similar pattern of increase.

In a more in-depth analysis (see Table 3), the top 10 domestic electricity consumer countries are again led in 2017 by China with 25.8% of the World's domestic electricity consumption, and almost 1000.0% growth since 1990, having actually exceeded that of the United States in 2010 (see Figure 4). Other countries in the top 10 have also more than doubled their domestic energy consumption in the

Figure 1. Energy consumption from 1990 to 2017 by groups of countries

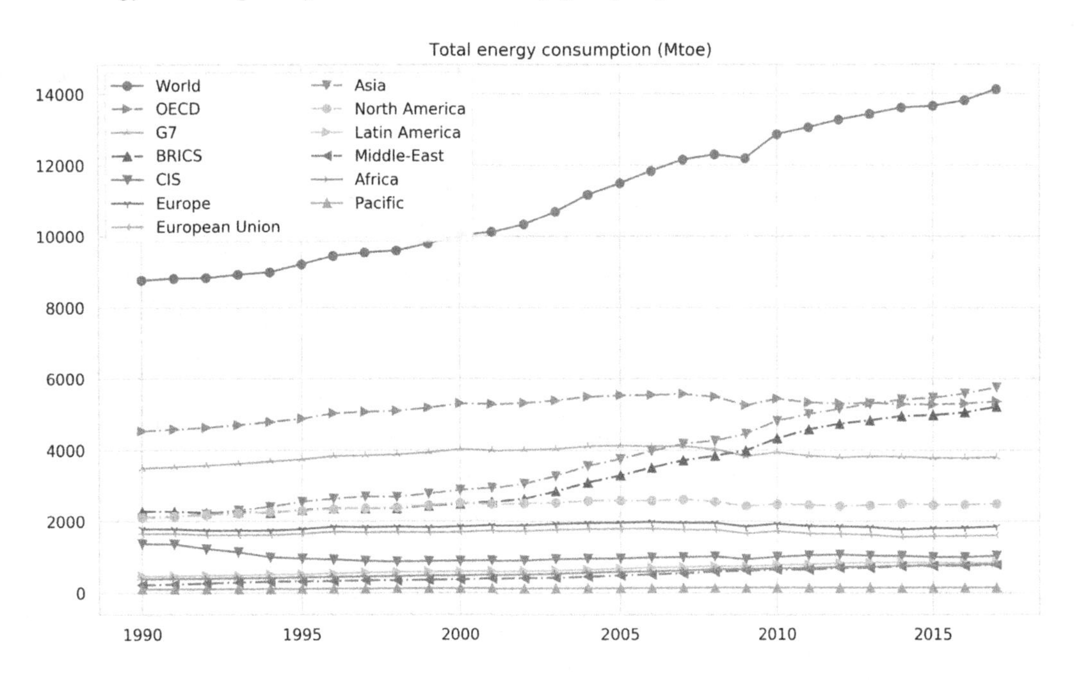

Table 1. Energy consumption (in Mtoe) from 1990 to 2017 by groups of countries

	Consumption in 1990	Consumption in 2017	Mean growth by year (1990 - 2017)	% of growth (1990 - 2017)	% of global consumption (in 2017)
World	8,760.9	14,125.6	198.7	61.2%	100.0%
Asia	2,108.5	5,754.9	135.1	172.9%	40.7%
OECD	4,531.2	5,363.2	30.8	18.4%	38.0%
BRICS	2,290.6	5,213.6	108.3	127.6%	36.9%
G7	3,496.2	3,804.1	11.4	8.8%	26.9%
North America	2,120.8	2,488.8	13.6	17.4%	17.6%
Europe	1,786.1	1,856.6	2.6	3.9%	13.1%
European Union	1,655.0	1,610.8	-1.6	-2.7%	11.4%
CIS	1,372.8	1,037.1	-12.4	-24.5%	7.3%
Latin America	462.8	846.7	14.2	83.0%	6.0%
Africa	382.2	805.4	15.7	110.7%	5.7%
Middle-East	222.5	785.6	20.9	253.0%	5.6%
Pacific	103.7	154.1	1.9	48.6%	1,1%

Table 2. Energy consumption (in Mtoe) for the top 10 consumers in 2017

	Consumptions in 1990	Consumption in 2017	Mean growth by year (1990 - 2017)	% of growth (1990 - 2017)	% of global consumption (in 2017)
China	870.7	3,104.9	82.7	256.6%	22.0%
United States	1,909.9	2,201.4	10.8	15.3%	15.6%
India	305.7	933.9	23.3	205.5%	6.6%
Russia	881.9	744.1	-5.1	-15.6%	5.3%
Japan	438.4	429.0	-0.3	-2.1%	3.0%
Germany	354.9	314.3	-1.5	-11.4%	2.2%
South Korea	94.0	295.8	7.5	214.8%	2.1%
Brazil	140.6	290.7	5.6	106.8%	2.1%
Canada	210.9	287.4	2.8	36.2%	2.0%
Iran	69.3	252.7	6.8	264.4%	1.8%

Figure 2. Energy consumption from 1990 to 2017 by the World's top 10 consumers in 2017

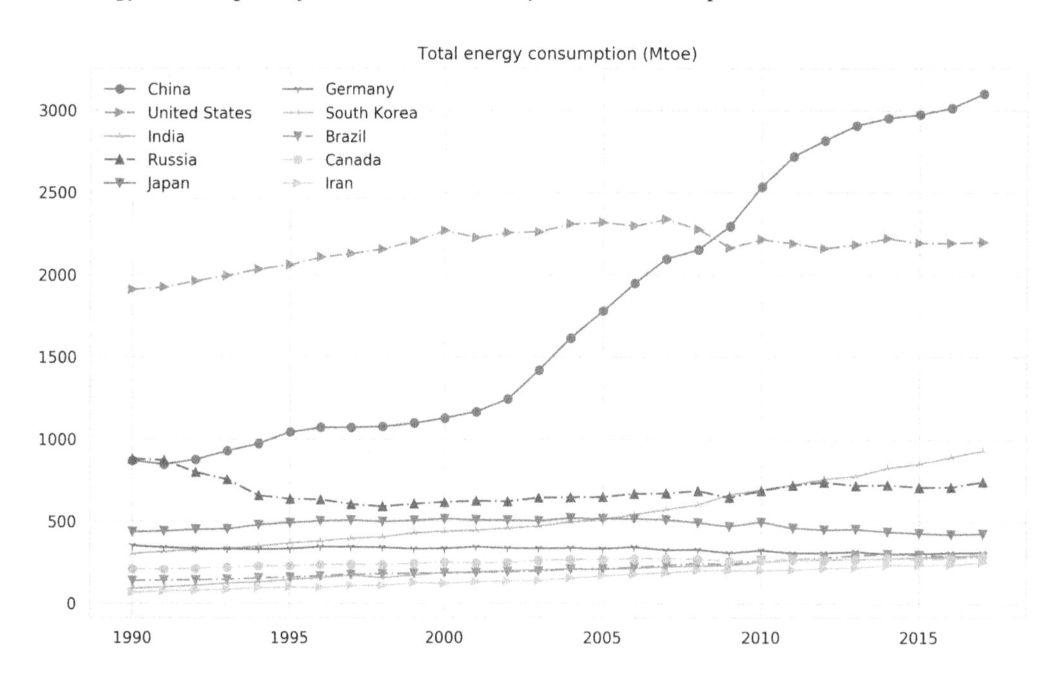

same period such as South Korea (466.0%), India (437.5%), or Brazil (142.0%). In the top 10, the only country which, according to Enerdata (2018), did not increase its electrical domestic consumption since 1990 was Russia with a decrease of 3.1% during those years.

However, there are some good news. The use of renewable sources of electricity is creating its path as it is shown in Table 4. Despite the outstanding growth in electricity consumption, the percentage of renewable production has steadily increased in most regions between 1990 and 2017, being the exceptions Latin America (which passed from 66.1% to 56.0%, a -10.1% variation), Africa (-0.3% variation)

Figure 3. Electricity domestic consumption (in TWh) from 1990 to 2017 by groups of countries

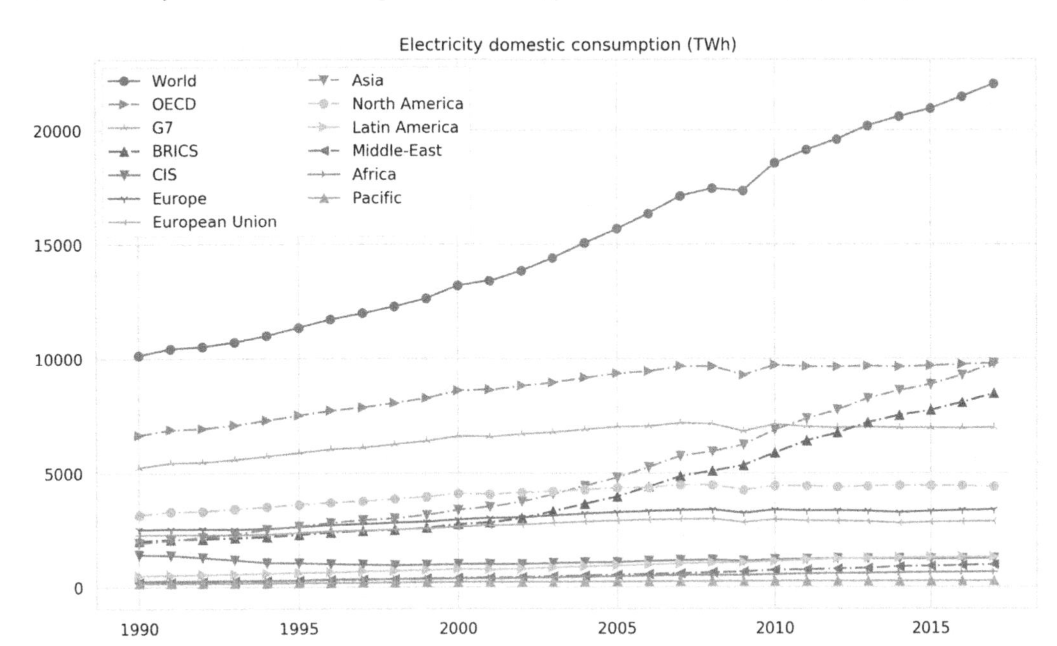

Table 3. Domestic Electricity consumption for the top 10 consumers in 2017

Country	*Consumption in 1990 (TWh)*	*Consumption in 2017 (TWh)*	*Mean growth by year from 1990 to 2017*	*% of growth from 1990 to 2017*	*% of global consumption in 2017*
China	534.3	5,683.4	190.7	963.8%	25.8%
United States	2,712.6	3,807.7	40.6	40.4%	17.3%
India	215.0	1,155.8	34.8	437.5%	5.2%
Japan	780.6	1,018.7	8.8	30.5%	4.6%
Russia	917.4	888.8	-1.1	-3.1%	4.0%
Canada	433.0	571.6	5.1	32.0%	2.6%
South Korea	94.4	534.2	16.3	466.0%	2.4%
Germany	481.0	531.0	1.9	10.4%	2.4%
Brazil	215.8	522.1	11.3	142.0%	2.4%
France	323.3	445.2	4.5	37.7%	2.0%

and the Middle-East (-2.7% variation). Nevertheless, Latin America continues to be the region with the highest percentage of electricity coming from renewable sources. Other regions stand out for their growth in the percentage of renewable electricity such as Europe (from 18.1% in 1990 to 33.6% in 2017) and, in particular, the European Union (from 12.6% in 1990 to 30.2% in 2017). The G7 (Canada, France, Germany, Italy, Japan, United Kingdom, and United States) had a variation of almost 10.0%, having 24.5% of its electrical energy produced over renewable sources. This evolution is many times led

Figure 4. Domestic electricity consumption (in TWh) since 1990 by the top 10 consumer countries in 2017

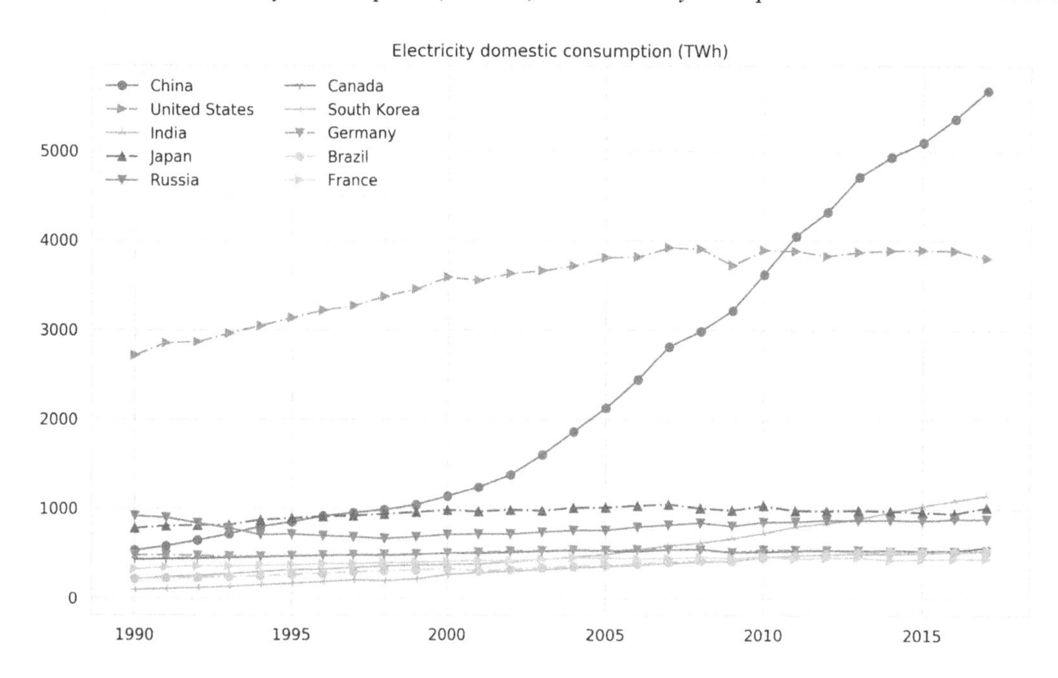

by countries which did not appear in the previous analysis, such as Norway, Colombia, New Zealand, Venezuela, Sweden, Chile, Portugal, and Romania – see the top 10 countries in terms of percentage of electricity produced in 2017 in Figure 5. Fundamental to the path toward global sustainability, several countries follow this route as stated by many authors (Bhattacharya, Paramati, Ozturk, & Bhattacharya, 2016; Gullberg, Ohlhorst, & Schreurs, 2014; Kitzing, Mitchell, & Morthorst, 2012; Krajačić, Duić, & Carvalho, 2011; Mustapa, Peng, & Hashim, 2010; Pinson, Mitridati, Ordoudis, & Ostergaard, 2017; Schiermeier, 2013).

Europe's Energy Consumption Goals: An Example

In 2016, the European Commission presented several measures aimed at providing a stable legislative framework needed to facilitate the energy transition. In this context, the EU Regulation 2018/1999 established that all member states should prepare and submit to the European Commission a national energy and climate plan (NECP) with a medium-term perspective (2021-2030) but aiming to achieve the goals set for 2050.

In the context of energy, looking into a 2050 horizon, the carbon neutrality perspective allows us to anticipate the following changes, as mentioned in (Direção-Geral de Energia e Geologia, 2018): (i) an energy transition based on the complete decarbonization of the energy sector, with the totality of the electric energy being generated using renewable energy sources, which in turn will require a significant rethinking of the transmission and distribution grids, including storage capacities, decentralized production, digitization and interconnections; (ii) a strong focus on energy efficiency, a cross-cutting theme for all sectors with a strong emphasis on industry, residential, services, and mobility; (iii) a complete decarbonization of the road and rail transport sectors, including the implications in terms of technologies and its substitution (with a clear focus on electric mobility, smooth mobility, and shared mobility),

Table 4. Percentage of renewable electricity production in 1990 and 2017 by groups of countries

	% of renewables in 1990	% of renewables in 2017	Variation from 1990 to 2017
Latin America	66.1%	56.0%	-10.0%
Europe	18.1%	33.6%	15.5%
America	25.2%	32.1%	6.9%
European Union	12.6%	30.2%	17.6%
BRICS	24.2%	26.1%	1.9%
OECD	17.7%	25.6%	7.9%
Pacific	22.6%	24.9%	2.4%
World	19.7%	24.8%	5.1%
G7	15.1%	24.5%	9.4%
North America	18.5%	24.4%	5.9%
Asia	17.5%	21.4%	3.9%
Africa	18.4%	18.1%	-0.3%
CIS	13.7%	17.4%	3.7%
Middle-East	4.9%	2.1%	-2.7%

Figure 5. Percentage of electricity produced from renewable sources between 1990 and 2017, top 10 countries

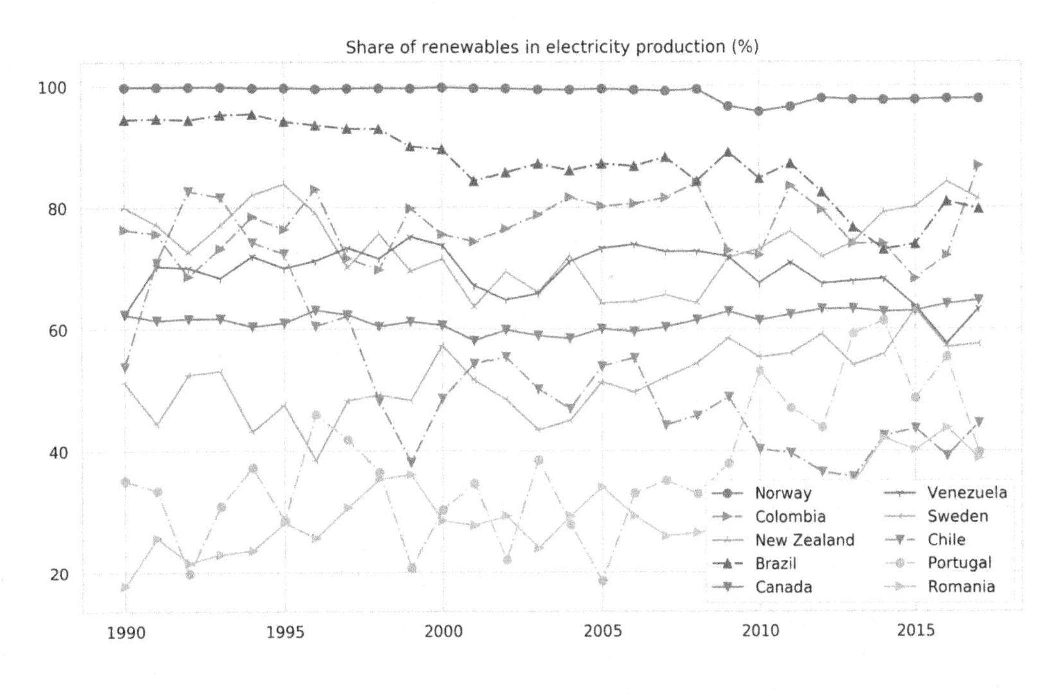

but also in what relates to the models of territorial organization of cities, their implications in terms of mobility needs, and in terms of collective mobility versus individual mobility; and (iv) a strong focus in innovation and in the creation of new business models in the industrial sectors, including agro-food, reinforcing the perspectives of the circular economy and of the "industry 4.0", supporting innovative, efficient, greener, and zero emission solutions in the next 30 years.

As above, a carbon-neutral society based on a circular economy, which conserves resources at its highest economic value, is also creating employment (but more qualified), wealth (but more sustained), and well-being (but more shared).

In the case of Portugal, for instance, the objective of the NECP is to be neutral in emissions of greenhouse gases (GHG) by 2050. In order to achieve the objective of carbon neutrality in 2050, a GHG emission reduction between 85.0-90.0% will be required. However, the largest reduction of emissions will have to be achieved in the 2020-2030 decade, with decreases ranging from 45.0% to 55.0%. As defined in (Direção-Geral de Energia e Geologia, 2018), the Portuguese goals for the 2030 horizon are:

1. Achieve a reduction of CO_2 emissions, compared to 2005, by 17.0%;
2. In 2030, 47.0% of energy consumption must result from renewable energy sources (percentage of the gross final energy consumption) distributed by: (i) a weight of 80.0% of renewable energy in the consumption of electricity; (ii) a weight of 20.0% of renewable energy in transportations; (iii) a weight of 38.0% of renewable energy in cooling and heating systems; (iv) increase energy efficiency by 35.0% (percentage of reduction of gross primary energy); and (v) increase the electrical interconnections with Europe in 15.0%.

As described in (Bernardo, 2019; Gonçalves, 2019) some of the more specific goals of NECP include: (i) to generate practically the same amount of solar energy as wind energy by 2030; (ii) introduce heat pumps in 15.0% of homes; (iii) achieve a reduction of 82.0% in the urban solid waste that is deposited in landfills; (iv) reduce between 4.0% and 9.0% of the urban solid waste that is produced per capita; (v) assure almost all the electricity needed by light goods vehicles; (vi) achieve a ratio of 33.0% in electric mobility; and (vii) reconfigure the energy market.

Being somehow global, these objectives are at the same time promising and challenging. Among them, the objective that 80.0% of consumption should result from renewable energy sources cannot be met by simply generating more energy. In fact, as we cannot control generation from sources like wind and solar, storage and demand side management measures are required to assure the real time equilibrium between generation and consumption. That equilibrium is also very challenging because these renewable energy sources are not only known as intermittent, but also because they are being placed, in a distributed manner, at the periphery of the electrical grid.

So, instead of adjusting the generation according with the consumption, in demand side management, the consumption is adjusted according with the generation. However, demand side management, requires communication systems, to actively inform the loads when they should work. They also require optimization algorithms and machine learning algorithms. Optimization algorithms are used to schedule the working periods of loads according with the tariff rates, and machine learning algorithms are used to predict future consumption and generation levels. These are the stepping-stones of what is called smart grids (Entidade Reguladora dos Serviços Energéticos, 2019; Farhangi, 2010; Tuballa & Abundo, 2016).

The next section will be devoted to intelligent energy management, with emphasis in (general) buildings, touristic buildings, and electric vehicles, with the objective of attaining sustainable energy consumptions activities.

INTELLIGENT ENERGY MANAGEMENT

Intelligence at the Disposal of Managing Energy Consumption in Buildings

Seen as an electricity network based on digital technology, a smart grid is primarily used to supply electricity to consumers via two-way digital communication. The smart grid comprises a system which allows to monitor, analyze, control, and communicate within the energy supply chain (Ghasemi, Shayeghi, Moradzadeh, & Nooshyar, 2016). In its core, the grid consists of controls, sensors, computers, automation, smart buildings, and equipment which will employ technologies in order to work with the electrical network, responding digitally to the quickly changing energy demands of the users.

In the literature, a wide variety of optimization techniques were already developed to solve the different challenges placed by smart grids. Momoh (2012) gives a classification of the optimization solutions in five groups, namely: (i) decision systems, (ii) static methods, (iii) adaptive dynamic programming, (iv) evolutionary programming, and (v) intelligent systems. In some cases, these techniques can also be combined to form hybrid solutions. Among these solutions, the evolutionary computation family of methods is currently widely used to solve a large range of problems (J. C. Bansal, Singh, & Pal, 2019). In particular, the genetic algorithms (GA) are highly relevant for its industrial applications, because they are capable of handling problems with non-linear constraints, multiple objectives, and dynamic components properties that frequently appear in real-world problems (Roeva, Fidanova, & Paprzycki, 2013). In the microgrid's scope they have been applied on profit-based optimal generation scheduling (Muhamad Razali & Hashim, 2010), on operation planning of multi smart microgrids (Nagasaka *et al.*, 2012), on multi-objective environmental/economic power dispatch (Mohamed & Koivo, 2008), on partitioning a distribution network with the aim to minimize the energy exchange among the microgrids (Korjani, Facchini, Mureddu, & Damiano, 2017), for high power loss sensitivity identification on capacitor placement and performance improvement in electrical power distribution systems (Mahdavian, Kafi, Movahedi, & Janghorbani, 2017), and even on the optimization of control parameters of hierarchical control to improve the power quality of the microgrid (Wang *et al.*, 2018).

In this context, smart grids create an exceptional opportunity for the support of the smart net zero energy buildings development (Irfan, Abas, & Saleem, 2019) and offer the step towards the internet of things (IoT) for the energy and building industry (Irfan *et al.*, 2019). As a result of combining novel concepts, models and ancillary services from generation, transmission and distribution right into customer devices they can provide energy more securely, economically and efficiently and can also enable participants of the electricity market to adjust their bidding strategies based on demand-side management (DSM) models, in order to maximize profits given an exact and strong estimate of future electricity price estimate. Any program that provides an economic model for the customers, with the aim of regulating the quantity and time of energy consumption can be addressed as DSM. However, in order to use the energy that is produced at a certain time, users need to allow the adjustment of the working periods of their electrical appliances. This can be done through a manual control of equipment, machines or lighting systems, or preferably through automated mechanisms, using energy management systems. EMS perform the control of electrical devices, reducing user intervention by running an optimization algorithm that decides when electrical devices should work (Keles *et al.*, 2015). To accomplish this, EMS take into account several factors, including: (i) user restrictions, (ii) generation levels, (iii) tariff rates, (iv) contracted power level, and (v) electrical circuit limitations (Eggink & Reinders, 2017). Based on these parameters and restrictions, EMS then run optimization algorithms that search for a solution that minimizes costs.

When the EMS software is equipped with scheduling optimization algorithms some researchers refer to them as intelligent EMS (Yu, Chen, & Tong, 2016). The underlying scheduling problem is then closely related to the classical deadline scheduling problem first considered by Liu and Layland (1973) where, generally, jobs arrive with different sizes and deadlines, are processed by several identical processors, each one capable of processing one job at a time, possibly preempted without cost.

With the purpose of reducing greenhouse gases, national governments are promoting the replacement of traditional fuels with renewable power sources. Due to its capability of being easily installed on roofs, facades or courtyards, renewable power sources based on photovoltaic (PV) technology have quickly expanded in both rural and urban areas, with its capacity rapidly increasing. Furthermore, most PV installations are grid-connected and located close to the final users. By limiting the amount of energy based on non-renewable sources drawn from the distribution grid one will end up by maximizing self-consumption. In this context, building energy management systems (BEMS) have the key role to properly optimize generation and consumption. BEMS possess the capability to exchange information in real-time with the building's local management systems in order to coordinate multiple actions, while considering the status of the entire building. Traditionally, to implement the control strategy, monitored data are collected periodically and stored at a server for further offline analysis (Barchi, Miori, Moser, & Papantoniou, 2018). Other solutions exist, such as the control solution presented by Barchi *et al.* (2018) which uses a battery energy storage system (BESS) that operates in three distinct modes, allowing multiples combinations of power flow between the local systems. BESS maximizes both the exploitation and self-consumption of power generated by PV technology.

Further, energy management in buildings is closely related to the energy required to maintain a desired minimum comfort level for the occupants. The perception of comfort is related to several environmental factors such as lighting, temperature, and air quality, but most of the research is focused on indoor thermal comfort maximization. In this context, Ferreira, Ruano, Silva, & Conceição (2012), Ma *et al.* (2012) and Frătean & Dobra (2018) have demonstrated that energy prediction and control strategies based on model-based predictive control can improve building energy efficiency and reduce building energy costs. Specifically, Ferreira *et al.* (2012) employed a discrete model-based predictive control methodology, consisting of three major components: the predictive models, implemented by radial basis function neural networks; the cost function devised to minimize energy consumption and maintain thermal comfort (using the predicted mean vote index); and the optimization method, a discrete branch and bound approach. They advocate that the good estimation accuracy over wide ranges of the input variables was determined by the accurate planning and estimation of the on-line control models, which resulted in energy savings of around 50.0%. In similar fashion, the work by Papas, Estibals, Ecrepont, & Alonso (2018) gives an overview on the application of efficient energy networks, which build up on the comparison of three different modeling techniques to optimize both thermal and electrical consumption of a smart building in France. This work explored the dynamic thermal simulation, a "black box" model using artificial neural networks (ANN) for the simulation of energy system parameters, and a global model developed in Matlab Simulink incorporating all systems of consumption in interaction with the building.

Given the previous approaches, accurate and effective forecast of tariff rates and load demand in smart grids is then an important challenge as they play a decisive part in the predictive based scheduling approaches. A broad range of research papers are dedicated to compare the performance of forecasting methods like ANN and random forests (RF) (Ahmad *et al.*, 2017), autoregressive integrated moving average model (ARIMA) (Box, Jenkins, Reinsel, & Ljung, 2016; Jeihoonian, Ghaderi, & Piltan, 2010), particle swarm optimization (PSO) and multi-objective genetic algorithm (MOGA) (Litchy & Nehrir,

2014), recurrent neural networks (Leonori, Rizzi, Paschero, & Mascioli, 2018), or hybrid evolutionary fuzzy neural network (Minhas, Khalid, & Frey, 2017). In this context, Ghasemi *et al.* (2016) proposed a hybrid algorithm for simultaneous forecast of price and demand that uses a set of effective tools in the preprocessing part, a forecast engine, and a tuned algorithm. On the other hand, Ahmad *et al.* (2017) compared the performance of the widely-used feed-forward back-propagation ANN with RF for predicting the hourly heating, ventilation, and air conditioning (HVAC) energy consumption of a hotel. Incorporating social parameters, such as the numbers of guests, marginally increased prediction accuracy in both cases. Despite performing marginally better than RF, the latter features ease of tuning, which offers ensemble-based algorithms an advantage for dealing with multi-dimensional complex data, typical in buildings. Recently, Cabrita, Monteiro, & Cardoso (2019) proposed a GA based on a predictive smart energy management scheduling approach for household appliances that achieves minimum consumption costs and reduction in peak load at real-time. A scenario of self-consumption is assumed where the surplus from local power generation can be sold to the grid, with the presence of appliances that can be shiftable from peak hours to off-peak hours.

Integration of Electric Vehicles in the Energetic Management

The large-scale adoption of electric vehicles (EVs) present an opportunity to provide electric energy storage (EES)-based ancillary services. However, they can increase load side uncertainties while requiring significant amounts of energy and charging power, which can impact the distribution grid. For instance, Di Paolo (2018) and Veldman & Verzijlbergh (2015) studied the potential impact of integrating EV charging on the distribution grid. Di Paolo (2018) results showed that in terms of EV plug-in time, 80% of the time the EV was connected for less than 2 hours, while less than 1% of the time EV users plugged in their EVs for more than 5 hours. Further, energy peak brought by EV charging from the grid was between 3 and 5 pm being most of the energy drawn to EVs during daytime. On the other hand, Veldman & Verzijlbergh (2015) assessed the effect of wind energy on electricity prices in a scenario with high wind production penetration. They concluded that when EVs base their charging schedules on electricity price, which is for a significant part determined by the instantaneous production of wind energy, this leads to high peaks in network load and, consequently, high network investment costs; when EV charging is controlled with the objective of minimizing peaks in network load, far less reinforcements are needed.

Alternatively, energy stored in EVs can be used as an auxiliary source of electrical energy storage (Kang, Duncan, & Mavris, 2013), which in smart grids can be used to regulate voltage profiles and power of the network, and compensate for fluctuations in renewable energy generation (Dallinger, 2013). This fact is even more decisive with the high ranges and resulting enhanced battery capacities of EVs. Thus, even without considering reverse charging, EVs can be seen as a challenge and, at the same time, as a storage solution.

Sustainable charging of EV vehicles is only practical if the electricity used to charge them comes from sustainable sources. The work from Van Der Meer, Mouli, Mouli, Elizondo, & Bauer (2018) proposed an EMS that incorporated PV power influx forecasting while being capable of optimally planning and allocating power in a parking place. The developed EMS consists of two components: an ARIMA to predict PV power production and a mixed-integer linear programming (MILP) framework that optimally allocates power to minimize charging cost. The optimal way to charge EVs is to schedule the charging by taking into consideration the EV user preferences, local renewable generation, distribution network, and energy prices from the market. There are innumerous examples of algorithms that exploit the elasticity of electric vehicle loads to fill the valleys in electric load profiles such as the works of Gan, Topcu, & Low (2013), Grammatico (2016), or Sortomme, Hindi, MacPherson, & Venkata (2011), just to name a few. Sortomme *et al.* (2011) used a centralized control strategy to optimize various objectives (non-simultaneously) like minimizing distribution network power loss, minimizing load variance, maximizing load factor, and maximizing supportable EV penetration level. Their results showed that using load factor or load variance as the objective function brings about additional benefits of reduced computation time and problem convexity when compared to optimizing system losses. Chen, Ji, & Tong (2012) proposed an adaptation of the earliest departure first (EDF) algorithm to charging EV vehicles. Although the algorithm already considered local production from renewable sources, it treated the variability of renewable sources considering the option of having non-conclusion penalties when the reserve is not assured. Later, Monteiro & Nunes (2015) modified the EDF algorithm to consider two energy components, one mandatory and one optional which varies according to the variable nature of renewable sources. The optimization methodology employed linear programming (LP) and attained a considerable cost reduction, low computational complexity, and processing times, while achieving a high ratio of renewable energy usage. García-Álvarez, González, & Vela (2015) have addressed the problem of scheduling the charging of electric vehicles through the use of a GA for a real environment in which a number of vehicles may require charge from an electric system installed in a garage/parking place. In the considered scenario, there are several constraints, such as load factor and imbalance constraints among the three phase power lines of the electric feeder. The GA operators were specifically designed to efficiently deal with the scheduling problem. Instead, Yu *et al.* (2016), used a strategy based on threshold admission and greedy scheduling (TAGS) policies, online algorithms aimed at maximizing the overall operation profit in a scenario where the power dispatch module draws power from a mix of storage, local renewable EVs, and purchased power from the grid.

When fully centralized control is not possible, one can iteratively adjust the price of charging and the charging profile of the consumers in the quest for solutions that minimize electricity generation costs by scheduling EV demand to fill the overnight non-EV demand "valley" (Grammatico, 2016; Z. Ma, S. Callaway, & Hiskens, 2013). The former favors optimization of charging profiles through a day-ahead negotiation process between the utility and EV based on the prediction of load profiles. Likewise, Grammatico (2016) introduced a unifying framework employing a dynamic control law which promoted faster charging/discharging schedules towards a convenient equilibrium. Their strategy assumed that the plugin electric vehicles (PEV) agents determine their optimal charging strategy with respect to an incentive signal (e.g., the electricity price forecast), and the average among these resulting strategies is used to estimate the total demand over the charging horizon. In turn, an updated incentive signal is computed based on such average PEV demand, broadcast to the whole PEV population, and then the process is repeated. Research conducted by Liu, Phanivong, Shi, & Callaway (2019) and Ma, Zou, Ran, Shi, & Hiskens (2016) focused on filling the overnight load "valley" where the non-EV electricity demand is at its lowest while satisfying both local charging needs like battery degradation and distribution network constraints (e.g., overloading). More recently, a MILP which also regarded battery degradation minimization for charging a fleet of EVs from PV based on a receding-horizon model predictive EMS system was introduced by Mouli, Kefayati, Baldick, & Bauer (2019). The system included: (i) the charging of EV from PV, using time of use tariffs to sell PV power and charge EV from the grid, (ii) implementation of vehicle-to-grid (V2G) for grid supporting, (iii) using EV to offer ancillary services in the form of reserves, (iv) considering distribution network capacity constraints, and (iv) scheduling of the connection of a single electric vehicle supply equipment (EVSE) to several EV. The authors concluded that a large portion of the V2G revenues observed derived from increased regulation services offered rather than from V2G energy sales.

In summary, studies on EV charging control roughly fall into three categories (Gan *et al.*, 2013): (i) time-of-use price, (ii) centralized control, and (iii) decentralized control. Category (i) algorithms tend to explore the effect of higher electricity price during peak hours on shifting EV load. Category (ii) require a centralized infrastructure to collect information from all the EVs and centrally optimize over their charging profiles. The complexity of this centralized optimization increases with the number of EVs. Lastly, and potentially enabled by technologies such as EMS, home area networks, and advanced EV chargers, decentralized control, (iii), introduces load-side participation into the power market and has been attracting more attention recently.

Energy Consumption in Touristic Buildings and Sustainable Tourism

Energy is one of the most used resources in the tourism activity (Pan *et al.*, 2018), having a great impact on environment due to the energy required in the building's construction, to the high energetical costs associated to the transports of tourists or, in general, to the necessary power to run the full touristic operation (e.g., heating or cooling environments), to name just a few examples. In addition to the required high energy levels, touristic regions are in general at great stress due other environmental worries such as the waste production or the water usage, which led to the materialization of environmental concerns and the emergence of sustainable development goals (World Tourism Organization, 2019a, 2019b).

Associated to the mentioned concerns, the concept of "sustainable tourism" emerged as a new paradigm, being characterized by a development without increasing the production of materials or the energy consumption and the capacity to somehow generate new energy (Hall, 2010). The relation between sustainable tourism and renewable energy, also called green energy, is very narrow and delicate, causing deep changes in the consumption pattern of the tourism sector. Furthermore, the relevancy of the "green" politics leveraged the tourism industry to lead a number of innovative sustainable energy initiatives, highlighting the use of energy efficient transports, the use of renewable energy in detriment of others, and the definition of indicators and actions that contribute to the energy efficiency of hotels (Pan *et al.*, 2018).

In this sustainable tourism context, hotels can be transformed in green buildings, where energy, water and waste are the most relevant criteria to get a green hotel certification (Teng, Horng, Hu, Chien, & Shen, 2012). In the energy case there are a set of green practices that are consistent with energy saving and use renewable energy, such as (Pan *et al.*, 2018; Teng *et al.*, 2012): reuse of towels by the guests, installing energy-efficient lights to reduce energy consumption without affecting hotel functionality, usage of energy certified products, optimally adjusting air-conditioning systems, installing auto-sensing systems for energy supply, usage of energy-efficient refrigeration facilities and measures, development of performance indicators and optimize measures to reduce energy consumption in a response to the weather changes and usage patterns, inclusion of wireless sensors and electronic devices that permit the internet of things, acquire a real-time energy consumption monitoring system, and develop a dashboard energy consumption with performance indicators to control and manage the hotel consumption. In general, a technological environment associated with the Internet of Everything.

The success of the hotel industry depends on the customer's satisfaction, such that green practices applied to the industry has contributed to increase the clients' satisfaction (Moise, Gil-Saura, & Ruiz-

Molina, 2018), which provided revenues and profits (Yadav, Kumar Dokania, & Swaroop Pathak, 2016), and clients' loyalty (Dominici & Guzzo, 2010; Han & Kim, 2010). The customers satisfaction also contributes to increase the hotel's brand and the "green hotel" label has become a marketing strategy, attracting customers (Pizam, 2009), since the image of a green hotel, concerned with the environment and sustainability, contributes to the decision-making process of the tourist as well as conducts the behavioral intentions at the moment of choice (Chan, 2013a). In this sense, customer's interest in green hotels created opportunities for hotels to promote their green products and services as it enhances customer satisfaction and hoteliers began to include in their management processes the strategies associated with green marketing, which can be defined as the management process responsible for identifying, anticipating, and satisfying the requirements of customers and society in a profitable and sustainable way.

As pointed out by Chan (2013a, 2013b), in addition to customer perspectives, green marketing strategies applied to hotels also have relevance in the perspective of managers, creating a green image, which can increase the hotel's competitiveness and help define its position in a different way, taking in consideration the achievement of new segments that they want to captivate.

For other authors, such as Mejia (2019) and Punitha & Mohd Rasdi (2013), the application of numerous green practices can place tourism at the forefront of such sustainable transformation, focusing on hospitality; the combination of technology, sensors, IoT, and bigdata; interests among consumers; increased satisfaction and loyalty; hotel managers increase profits and improve the image; contribution to deepening the concepts associated with green marketing since it allows the monitoring and control of spending, mainly through the definition of associated indicators, for performance analysis and certification; thus creating hotels with social responsibility, and concerned with sustainability objectives.

PREDICTING ELECTRICAL CONSUMPTIONS IN BUILDINGS

This section describes some prediction methods which are applicable to predict energy consumption. As any ML book will tell you, in its fundamental form, ML is backed in mathematical formulations and computer science techniques to build models from given data, which are then applied to new and unseen data, in order to predict new outcomes (Alpaydin, 2016; Domingos, 2015; Witten, Frank, Hall, & Pal, 2016). Although many other approaches could be explored, this section will briefly explore autoregressive (autoregressive with Exogenous input, autoregressive moving average with exogenous input, autoregressive integrated moving average with exogenous input, and seasonal autoregressive integrated moving average with exogenous input), Bayesian linear regression, decision forest regression, and boosted decision tree regression. The reader should notice many other methods exist which have also proven their worth in many regressions problems, such as, ordinary least squares, Ridge regression, Lasso regression, support vector regressor, or *k*-nearest neighbors (Bonaccorso, 2017; Dangeti, 2017; Kim, Koh, Lustig, Boyd, & Gorinevsky, 2007).

Autoregressive Methods

Autoregressive methods are widely used to deal with time series (Box *et al.*, 2016). In their essence, to predict the value in the next time step(s), they use observations from previous time steps as input to build a regression equation. The choice of an autoregressive model depends on the compromise between the simplicity of the model and the properties of a time series. If a time series is stationary (Box *et al.*,

2016) it is very common to choose the autoregressive with exogenous input (ARX) model and/or the autoregressive moving average with exogenous input (ARMAX) model. ARMAX is more complex than ARX being able to deal with stationary time series. If a time series is non-stationary, the autoregressive integrated moving average with exogenous input (ARIMAX) and/or the seasonal autoregressive integrated moving average with exogenous input (SARIMAX) can be used. Both models, ARIMAX and SARIMAX, can deal with stationary and non-stationary series. However, if seasonal series have seasonal elements the best option may be SARIMAX.

The autoregressive methods can be used, for instance, to predict energy consumption in buildings such as in (Bourdeau, Zhai, Nefzaoui, Guo, & Chatellier, 2019). Campos & da Silva (2016) conducted an analysis to predict the load demand in the context of smart grids, using ARX, ANN, and ANN combined with a GA (ANN-GA). In this analysis, the ARX presented lower execution time and higher mean absolute percentage error value when compared to ANN and ANN-GA. In the same year, Amara, Agbossou, Dube, Kelouwani, & Cardenas (2016) developed a hybrid model to predict the electricity demand according to outside temperature and compared it with the ARMAX model. ARMAX presented higher prediction errors than the hybrid model. Goswami, Ganguly, & Kumar Sil (2018) used ARIMA and ARIMAX to predict load demand in West Bengal. Climatic variations were considered as exogenous variables for the ARIMAX model. In the same work, ARMAX presented lower mean absolute error values in all scenarios when compared to ARIMA. Cai, Pipattanasomporn, & Rahman (2019) made a comparison between the SARIMAX, the gated recurrent neural network (GRNN), and the gated convolutional neural network (GCNN) regarding performance in predicting the load level in buildings. In this same work, the SARIMAX presented less execution time, but with less accuracy when compared to the other methods.

Bayesian Linear Regression

As the name suggests, Bayesian linear regression (BLR) (Bishop & Tipping, 2003) is a method associated with linear regression in which statistical analysis is performed within the context of Bayesian inference. Bayesian inference is a statistical method in which Bayes' theorem is used to update the probability of a hypothesis as more evidence or information becomes available (Islikaye & Cetin, 2018). The Bayes' Theorem is defined, in a simplified way, by the expression

$$P(y \mid \boldsymbol{x}) = \frac{P(y).P(\boldsymbol{x} \mid y)}{P(\boldsymbol{x})},$$

where $x=\{x_1,\ldots,x_n\}$ are the input variables, y is the output variable, $P(y|x)$ is the posterior probability of y given the evidence x, $P(y)$ is the prior probability of y, $P(x|y)$ is called the likelihood of evidence x if hypothesis y is true, and $P(x)$ is the prior probability of x in which the evidence itself is true.

For example, in (Tehrani, Khan, & Crawford, 2016), a BLR model, formulated in a recursive structure, was proposed to predict the base level of energy consumption demand in a smart grid, where real-time consumption forecasting is important for energy distribution network operators and market participants. In this same work, the results of the proposed BLR model showed a very good hourly load forecast. Later, Yuan *et al.* (2017) used a Bayesian approach to predict the energy demand in some regions of China. The results of the Bayesian approach showed to be satisfactory for an energy forecast, considering the model uncertainty, regional heterogeneity, and cross-sectional dependence. Also, in the same work, the model served as an incentive to develop some energy policies in the same regions of China. Islikaye & Cetin (2018) used a BLR method to predict the total load electrical power output of a base of operations in a power plant, where it were considered as input variables: ambient temperature, atmospheric pressure, relative humidity, and exhaust vacuum. The proposed BLR model presented satisfactory results, but when compared to other methods, in this same work, it was not the one with better performance.

Decision Forest Regression

The decision forest (DF) is an associative method based on *T* decision trees (Hartshorn, 2016). Trees have a structure constituted by nodes and branches, organized hierarchically. The nodes are classified as internal nodes (split node) and terminal nodes (leaves) (Ahmad *et al.*, 2017), as depicted in Figure 6. Depending on how trees are built, there are three major classification for DF: bagging, random decision forests, and boosting ensemble.

The bagging case uses a different random subset of the original data set for each tree in the ensemble. By default, a 100% sampling rate with substitution for each tree, which means that some of the original instances will be repeated and others left out. In a random DF (RDF) the bagging is extend by only considering a random subset of the input fields (features) at each split. For instance, in (Criminisi, Shotton, & Konukoglu, 2012) DF are applied to a regression problem and each bootstrap sample generates a tree in a greedy manner until it reaches a stopping criterion, such as the depth or maximum level of decision trees or minimum number of samples per leaf. For such, the samples S_j arriving at each node are divided into two subsets, S_j^L and S_j^R, taking into account a randomly selected subset x_j of the set of independent variables *x* (features), whose size is fixed in all the nodes. In binary trees the following properties must be met (Criminisi *et al.*, 2012): $S_j = S_j^L \cup S_j^R$, $S_j^L \cap S_j^R = \varnothing$, $S_j^L = S_{2j+1}$ and $S_j^R = S_{2j+2}$. In the regression case, the splitting of the samples arriving at each node *j* is optimized by selecting the value of the independent variable x_j where the division of the samples minimizes, for instance, the quadratic error of each node.

After properly training the model (including data cleaning, selection of the appropriate set of features, model's parameter optimization, cross-validation, etc.), given a new observation for which it is required a prediction, the task of each node of each DF regression tree is to decide to which branch the observation should be directed, until its sample reaches a leaf. Then, the prediction made for the same sample is given by the calculation of the average of the training samples that are in the said leaf. In both bagged and RDF trees are independent and can be trained in parallel.

Boosted decision tree regression (BDTR) technique is somehow different from the previous ones (Hartshorn, 2017). Individual trees in a boosted tree differ from trees in bagged or random forest ensembles since they do not try to predict the objective field directly. Instead, learners are "taught" sequentially, which implies that trees cannot be trained in parallel. Earlier learners fit simple models to the data and latter learners try to correct mistakes made in previous iterations, as the construction of the trees minimizes a suitable predefined loss function at each step (Goon, 2018; Gupta, Shrivastava, Khosravi, & Panigrahi, 2016).

Gerossier, Girard, & Kariniotakis (2019) used a DF to predict the consumption scenarios for the following day of a fleet of EVs. Their work used real consumption information of the EV measured in a loading station. In (Johannesen, Kolhe, & Goodwin, 2019) several regression methods were analyzed, namely a random forest regressor, a *k*-nearest neighbor regressor, and a linear regressor with the objective

Figure 6. Outline of a decision forest

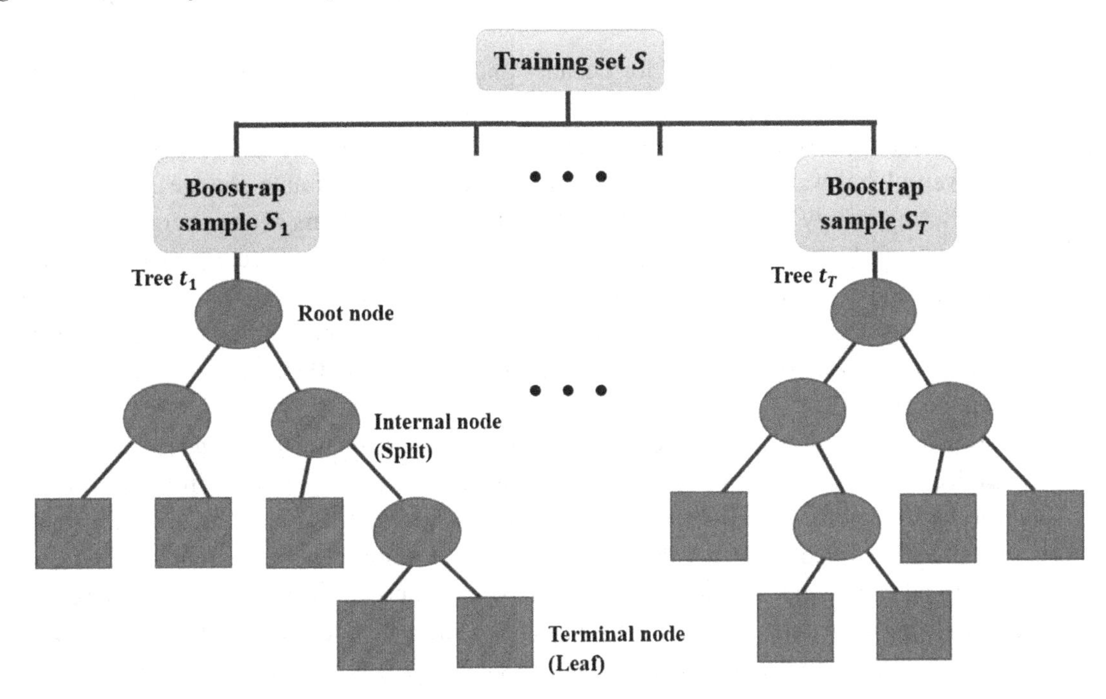

of predicting electric loads in urban regions, considering the season, day of the week, and the weather. The random forest regressor presented better short-term (30 min) electric load prediction results and the *k*-nearest neighbor regressor presented better results for the long-term load prediction (24 h). Touzani, Granderson, & Fernandes (2018) proposed a model based on gradient boosted machine to predict the energy consumption of commercial buildings. To obtain the performance of the model, a dataset of 410 commercial buildings was used and the results of the metrics presented optimal results, when compared with a random forest algorithm and a linear regression model (designated time of week and temperature – TOWT). The BDTR in (A. Bansal, Rompikuntla, Gopinadhan, Kaur, & Kazi, 2015; Burges, 2010) uses an efficient implementation called multiple addictive regression tree (MART). MART is a method based on boosting where the result is a linear combination of the outputs of a set of regression trees. Bansal *et al.* (2015) used MART to predict power consumption for smart meters in order to create custom electricity plans for home users based on usage history. A model called the instance-based transfer learning embedded gradient boosting decision trees (IBT-GBDT) was proposed in (L. Cai, Gu, Ma, & Jin, 2019) to predict generations of wind energy. The IBT-GBDT model presented better prediction accuracy results when compared to other models based on gradient boosting decision trees. In (Lingamallu & Garg, 2019), the gradient boosting regression trees were used to predict the energy consumption used in the cooling of a building, allowing a better analysis and control of energy demand in real time. In this work, in each iteration the prediction consisted of a single step forward and the mean absolute percentage error was as low as 14.7%.

SOLUTIONS, RECOMMENDATIONS and FUTURE RESEARCH DIRECTIONS

World's energy needs are steeply growing for decades and will continue for many more. This growth implies large ecological costs as many of the energy is the product of the fossil fuels burning, resulting in air and water pollution. Responsible for large amounts of CO_2 emissions, this energy policies/needs lead us to a proved global warming, having as consequence serious climate changes. Therefore, the usage of non-polluting energy sources is a major goal since the reductions of the energy requirements seems to be out of control. The present major sources of non-polluting energy (e.g., solar and wind) are intermittent sources and, to avoid large arrays of batteries or other energy storage facilities, require the usage of the energy as it is produced. Furthermore, the decentralized production of energy, resulting in the so called microgrids, also pose other optimization problems as is the need to trade the produced energy between the grids.

A contribution to the solution of the aforementioned problems is to characterize and predict consumptions and optimize movable charges location in time. This prediction can be done from small building (micro-sites) to large regions (macro-sites), from small to large electrical charges. Industry, tourism, transports, etc. all have a role in this environment. In short, the development of smart automated systems supported in artificial intelligence, and in machine learning techniques in particular, will play a major role. Their capacity to learn from data, to infer new solutions, to simulate occurrences, to detect anomalies, to optimize resources, etc. will replace human decision as it is doing, for instance, in automated driving.

Therefore, there are some typical challenges in the machine learning field of research. A first one is in the understanding of which processes need automation, which in an initial phase should be manual ordinary activities. A second issue relates to the data quality and quantity, being intrinsically related to the acquisition and transmission of it and consequently with fields such as the IoT/IoE (internet of everything). The development and implementation of known, similar or new methods, possibly to solve problems yet to be posed, will be a challenge. Finally, the lack of skilled resources might impose a serious drawback. These are some fields which need to be tackled as soon, and deeply, as possible, to avoid future bottlenecks in this emerging field.

CONCLUSION

Smart systems are here to stay in the energy management. The intelligent usage of energy can be addressed at several levels in order to achieve a more sustainable planet. In the case of electrical energy, several challenges are being posed, making this a very interesting research environment. Detecting electrical anomalies, predicting consumptions, optimizing charge disposable are just a few assets that can be addressed by information systems. This will produce environments where users will be alleviated from the burden of individually controlling each of their appliances, not only contributing to their wellbeing, but also to a better usage of renewable energy sources.

ACKNOWLEDGMENT

This work was supported by the Portuguese Foundation for Science and Technology (FCT), projects LARSyS (UID/EEA/50009/2019), CIAC (UID/Multi/04019/2019), CinTurs (UID/SOC/04020/2019),

CEFAGE (UID/ECO/04007/2019), CIEO, ACCES4ALL (SAICT-POL/23700/2016), IDMEC, under LAETA (UID/EMS/50022/2019); Portugal2020, CRESC2020, PO Norte 2020, FEDER; and by the European Union, under the FEDER and INTERREG programs, in the scope of the AGERAR (0076_ AGERAR_6_E) project project.

REFERENCES

Ahmad, M. W., Mourshed, M., Mundow, D., Sisinni, M., & Rezgui, Y. (2016). Building energy metering and environmental monitoring – A state-of-the-art review and directions for future research. *Energy and Building*, *120*, 85–102. doi:10.1016/j.enbuild.2016.03.059

Ahmad, M. W., Mourshed, M., & Rezgui, Y. (2017). Trees vs Neurons: Comparison between random forest and ANN for high-resolution prediction of building energy consumption. *Energy and Building*, *147*, 77–89. doi:10.1016/j.enbuild.2017.04.038

Alpaydin, E. (2016). *Machine Learning: the New AI*. The MIT Press.

Amara, F., Agbossou, K., Dube, Y., Kelouwani, S., & Cardenas, A. (2016). Estimation of temperature correlation with household electricity demand for forecasting application. In *Proceedings of IECON 2016 - 42nd Annual Conference of the IEEE Industrial Electronics Society*, 3960–3965. 10.1109/IECON.2016.7793935

Bansal, A., Rompikuntla, S. K., Gopinadhan, J., Kaur, A., & Kazi, Z. A. (2015). *Energy Consumption Forecasting for Smart Meters*. Retrieved from http://arxiv.org/abs/1512.05979

Bansal, J. C., Singh, P. K., & Pal, N. R. (Eds.). (2019). *Evolutionary and Swarm Intelligence Algorithms*. doi:10.1007/978-3-319-91341-4

Barchi, G., Miori, G., Moser, D., & Papantoniou, S. (2018). A Small-Scale Prototype for the Optimization of PV Generation and Battery Storage through the Use of a Building Energy Management System. In *Proceedings - 2018 IEEE International Conference on Environment and Electrical Engineering and 2018 IEEE Industrial and Commercial Power Systems Europe, EEEIC/I and CPS Europe 2018*, 1–5. 10.1109/EEEIC.2018.8494012

Bernardo, J. (2019). Plano Nacional Integrado Energia-Clima: Linhas de Atuação para o Horizonte 2021-2030.

Bhattacharya, M., Paramati, S. R., Ozturk, I., & Bhattacharya, S. (2016). The effect of renewable energy consumption on economic growth: Evidence from top 38 countries. *Applied Energy*, *162*, 733–741. doi:10.1016/j.apenergy.2015.10.104

Bishop, C., & Tipping, M. E. (2003). *Bayesian Regression and Classification* (Advances i, Vol. 190, pp. 267–285). Advances i, Vol. 190, pp. 267–285. Retrieved from https://www.microsoft.com/en-us/research/publication/bayesian-regression-and-classification/

Bonaccorso, G. (2017). *Machine Learning Algorithms*. Packt Publishing.

Bourdeau, M., Zhai, X. Q., Nefzaoui, E., Guo, X., & Chatellier, P. (2019). Modeling and forecasting building energy consumption: A review of data-driven techniques. *Sustainable Cities and Society*, *48*, 101533. doi:10.1016/j.scs.2019.101533

Box, G., Jenkins, G., Reinsel, G., & Ljung, G. (2016). Time series analysis: forecasting and control (5th Ed.) (D. Balding, N. A. C. Cressie, G. M. Fitzmaurice, G. H. Givens, H. Goldstein, G. Molenberghs, … S. Weisberg, Eds.). Hoboken, NJ: John Wiley & Sons.

Burges, C. J. C. (2010). *From RankNet to LambdaRank to LambdaMART: An Overview*. Retrieved from https://www.microsoft.com/en-us/research/publication/from-ranknet-to-lambdarank-to-lambdamart-an-overview/

Cabrita, C. L., Monteiro, J. M., & Cardoso, P. J. S. (2019). *Improving Energy Efficiency in Smart-houses by Optimizing Electrical Loads Management*. 1–6.

Cai, L., Gu, J., Ma, J., & Jin, Z. (2019). Probabilistic Wind Power Forecasting Approach via Instance-Based Transfer Learning Embedded Gradient Boosting Decision Trees. *Energies*, *12*(1), 159. doi:10.3390/en12010159

Cai, M., Pipattanasomporn, M., & Rahman, S. (2019). Day-ahead building-level load forecasts using deep learning vs. traditional time-series techniques. *Applied Energy*, *236*, 1078–1088. doi:10.1016/j.apenergy.2018.12.042

Campos, B. P., & da Silva, M. R. (2016). Demand forecasting in residential distribution feeders in the context of smart grids. In *Proceedings 2016 12th IEEE International Conference on Industry Applications (INDUSCON)*, 1–6. IEEE. 10.1109/INDUSCON.2016.7874464

Chan, E. S. W. (2013a). Gap analysis of green hotel marketing. *International Journal of Contemporary Hospitality Management*, *25*(7), 1017–1048. doi:10.1108/IJCHM-09-2012-0156

Chan, E. S. W. (2013b). Managing green marketing: Hong Kong hotel managers' perspective. *International Journal of Hospitality Management*, *34*, 442–461. doi:10.1016/j.ijhm.2012.12.007

Chen, S., Ji, Y., & Tong, L. (2012). Deadline scheduling for large scale charging of electric vehicles with renewable energy. In *Proceedings of the IEEE Sensor Array and Multichannel Signal Processing Workshop*, 13–16. 10.1109/SAM.2012.6250449

Criminisi, A., Shotton, J., & Konukoglu, E. (2012). Decision Forests: A Unified Framework for Classification, Regression, Density Estimation, Manifold Learning and Semi-Supervised Learning. In *Foundations and Trends® in Computer Graphics and Vision* (Foundation, Vol. 7, pp. 81–227). Retrieved from https://www.microsoft.com/en-us/research/publication/decision-forests-a-unified-framework-for-classification-regression-density-estimation-manifold-learning-and-semi-supervised-learning/

Dallinger, D. (2013). *Plug-in Electric Vehicles Integrating Fluctuating Renewable Electricity*. Kassel University Press.

Dangeti, P. (2017). *Statistics for Machine Learning*. Packt Publishing Ltd.

Di Paolo, M. (2018). Analysis of harmonic impact of electric vehicle charging on the electric power grid, based on smart grid regional demonstration project - Los Angeles. In *Proceedings 2017 IEEE Green Energy and Smart Systems Conference, IGESSC 2017. November 1–5.* doi:10.1109/IGESC.2017.8283460

Direção-Geral de Energia e Geologia. (2018). *Plano Nacional Integrado Energia e Clima 2021-2030.*

Domingos, P. (2015). *The master algorithm: How the quest for the ultimate learning machine will remake our world.* Basic Books.

Dominici, G., & Guzzo, R. (2010). Customer Satisfaction in the Hotel Industry: A Case Study from Sicily. *International Journal of Marketing Studies*, *2*(2). doi:10.5539/ijms.v2n2p3

Eggink, W., & Reinders, A. (2017). Design it with LSCs; an exploration of applications for Luminescent Solar Concentrator PV technologies. In *Proceedings 2017 IEEE 44th Photovoltaic Specialist Conference (PVSC)*, 2109–2113. 10.1109/PVSC.2017.8366790

Enerdata. (2018). *Global Energy Statistical Yearbook.*

Farhangi, H. (2010). The path of the smart grid. *IEEE Power & Energy Magazine*, *8*(1), 18–28. doi:10.1109/MPE.2009.934876

Ferreira, P. M., Ruano, A. E., Silva, S., & Conceição, E. Z. E. (2012). Neural networks based predictive control for thermal comfort and energy savings in public buildings. *Energy and Building*, *55*, 238–251. doi:10.1016/j.enbuild.2012.08.002

Frătean, A., & Dobra, P. (2018). The impact of control strategies upon the energy flexibility of nearly zero-energy buildings: Energy consumption minimization versus indoor thermal comfort maximization. *2018 IEEE International Conference on Automation, Quality and Testing, Robotics, AQTR 2018 - THETA 21st Edition, Proceedings*, 1–6. 10.1109/AQTR.2018.8402759

Gan, L., Topcu, U., & Low, S. H. (2013). Optimal decentralized protocol for electric vehicle charging. *IEEE Transactions on Power Systems*, *28*(2), 940–951. doi:10.1109/TPWRS.2012.2210288

García-Álvarez, J., González, M. A., & Vela, C. R. (2015, July). A genetic algorithm for scheduling electric vehicle charging. In *Proceedings of the 2015 Annual Conference on Genetic and Evolutionary Computation* (pp. 393-400). ACM. doi:10.1145/2739480.2754695

Gerossier, A., Girard, R., & Kariniotakis, G. (2019). Modeling and Forecasting Electric Vehicle Consumption Profiles. *Energies*, *12*(7), 1341. doi:10.3390/en12071341

Ghasemi, A., Shayeghi, H., Moradzadeh, M., & Nooshyar, M. (2016). A novel hybrid algorithm for electricity price and load forecasting in smart grids with demand-side management. *Applied Energy*, *177*, 40–59. doi:10.1016/j.apenergy.2016.05.083

Gonçalves, H. (2019). PNEC 2030 um "Admirável Mundo Novo" o que fazer?".

Goon, T. (2018). Difference between random forest and Gradient boosting Algo. Retrieved May 2, 2019, from https://www.linkedin.com/pulse/difference-between-random-forest-gradient-boosting-algo-goon

Goswami, K., Ganguly, A., & Kumar Sil, D. A. (2018). Comparing Univariate and Multivariate Methods for Short Term Load Forecasting. In *Proceedings 2018 International Conference on Computing, Power, and Communication Technologies (GUCON)*, 972–976. 10.1109/GUCON.2018.8675059

Grammatico, S. (2016). Exponentially convergent decentralized charging control for large populations of plug-in electric vehicles. In *Proceedings 2016 IEEE 55th Conference on Decision and Control, CDC 2016*, 5775–5780. 10.1109/CDC.2016.7799157

Gullberg, A. T., Ohlhorst, D., & Schreurs, M. (2014). Towards a low carbon energy future – Renewable energy cooperation between Germany and Norway. *Renewable Energy*, *68*, 216–222. doi:10.1016/j.renene.2014.02.001

Gupta, S., Shrivastava, N. A., Khosravi, A., & Panigrahi, B. K. (2016). Wind ramp event prediction with parallelized gradient boosted regression trees. In *Proceedings 2016 International Joint Conference on Neural Networks (IJCNN)*, 5296–5301. 10.1109/IJCNN.2016.7727900

Hall, C. M. (2010). Changing Paradigms and Global Change: From Sustainable to Steady-state Tourism. *Tourism Recreation Research*, *35*(2), 131–143. doi:10.1080/02508281.2010.11081629

Han, H., & Kim, Y. (2010). An investigation of green hotel customers' decision formation: Developing an extended model of the theory of planned behavior. *International Journal of Hospitality Management*, *29*(4), 659–668. doi:10.1016/j.ijhm.2010.01.001

Hartshorn, S. (2016). *Machine Learning With Random Forests And Decision Trees: A Visual Guide For Beginners*. Kindle Edition.

Hartshorn, S. (2017). *Machine Learning With Boosting: A Beginner's Guide*. Kindle Edition.

International Energy Agency. (2019). *Global Energy & CO2 Status Report 2018*.

Irfan, M., Abas, N., & Saleem, M. S. (2019). Net Zero Energy Buildings (NZEB): A Case Study of Net Zero Energy Home in Pakistan. In *Proceedings 4th International Conference on Power Generation Systems and Renewable Energy Technologies, PGSRET 2018*, (September), 1–6. 10.1109/PGSRET.2018.8685970

Islikaye, A. A., & Cetin, A. (2018). Performance of ML methods in estimating net energy produced in a combined cycle power plant. In *Proceedings 2018 6th International Istanbul Smart Grids and Cities Congress and Fair (ICSG)*, 217–220. 10.1109/SGCF.2018.8408976

Jeihoonian, M., Ghaderi, S. F., & Piltan, M. (2010). Modeling and comparing energy consumption in basic metal industries by neural networks and ARIMA. In *Proceedings 2010 International Conference on Computer Information Systems and Industrial Management Applications, CISIM 2010*, 171–175. 10.1109/CISIM.2010.5643670

Johannesen, N. J., Kolhe, M., & Goodwin, M. (2019). Relative evaluation of regression tools for urban area electrical energy demand forecasting. *Journal of Cleaner Production*, *218*, 555–564. doi:10.1016/j.jclepro.2019.01.108

Kang, J., Duncan, S. J., & Mavris, D. N. (2013). Real-time scheduling techniques for electric vehicle charging in support of frequency regulation. *Procedia Computer Science*, *16*, 767–775. doi:10.1016/j.procs.2013.01.080

Keles, C., Karabiber, A., Akcin, M., Kaygusuz, A., Alagoz, B. B., & Gul, O. (2015). A smart building power management concept: Smart socket applications with DC distribution. *International Journal of Electrical Power & Energy Systems*, *64*, 679–688. doi:10.1016/j.ijepes.2014.07.075

Kim, S.-J., Koh, K., Lustig, M., Boyd, S., & Gorinevsky, D. (2007). An Interior-Point Method for Large-Scale -Regularized Least Squares. *IEEE Journal of Selected Topics in Signal Processing*, *1*(4), 606–617. doi:10.1109/JSTSP.2007.910971

Kitzing, L., Mitchell, C., & Morthorst, P. E. (2012). Renewable energy policies in Europe: Converging or diverging? *Energy Policy*, *51*, 192–201. doi:10.1016/j.enpol.2012.08.064

Korjani, S., Facchini, A., Mureddu, M., & Damiano, A. (2017). A genetic algorithm approach for the identification of microgrids partitioning into distribution networks. In *Proceedings IECON 2017-43rd Annual Conference of the IEEE Industrial Electronics Society* (pp. 21-25). IEEE. doi:10.1109/IECON.2017.8216008

Krajačić, G., Duić, N., & Carvalho, M. da G. (2011). How to achieve a 100% RES electricity supply for Portugal? *Applied Energy*, *88*(2), 508–517. doi:10.1016/j.apenergy.2010.09.006

Leonori, S., Rizzi, A., Paschero, M., & Mascioli, F. M. F. (2018). Microgrid Energy Management by ANFIS Supported by an ESN Based Prediction Algorithm. *Proceedings of the International Joint Conference on Neural Networks, 2018*-July 1–8. 10.1109/IJCNN.2018.8489018

Lingamallu, M., & Garg, V. (2019). Very Short-Term HVAC Cooling Energy Forecasting for an Educational Building in Real-Time. *IOP Conference Series: Earth and Environmental Science*, *238*. 10.1088/1755-1315/238/1/012069

Litchy, A. J., & Nehrir, M. H. (2014). Real-time energy management of an islanded microgrid using multi-objective Particle Swarm Optimization. In *Proceedings 2014 IEEE PES General Meeting | Conference & Exposition*, 1–5. 10.1109/PESGM.2014.6938997

Liu, C. L., & Layland, W. (1973). Scheduling Algorithms for Multiprogramming in Hard-Real-Time Environment. *Journal of the Association for Computing Machinery*, *20*(1), 46–61. doi:10.1145/321738.321743

Liu, M., Phanivong, P. K., Shi, Y., & Callaway, D. S. (2019). Decentralized charging control of electric vehicles in residential distribution networks. *IEEE Transactions on Control Systems Technology*, *27*(1), 266–281. doi:10.1109/TCST.2017.2771307

Ma, Y., Borrelli, F., Hencey, B., Coffey, B., Bengea, S., & Haves, P. (2012). Model predictive control for the operation of building cooling systems. *IEEE Transactions on Control Systems Technology*, *20*(3), 796–803. doi:10.1109/TCST.2011.2124461

Ma, Z., Zou, S., Ran, L., Shi, X., & Hiskens, I. A. (2016). Efficient decentralized coordination of large-scale plug-in electric vehicle charging. *Automatica*, *69*, 35–47. doi:10.1016/j.automatica.2016.01.035

Ma, Z. S., Callaway, D., & Hiskens, I. (2013). Decentralized Charging Control of Large Populations of Plug-in Electric Vehicles. *IEEE Transactions on Control Systems Technology*, *21*(1), 67–78. doi:10.1109/TCST.2011.2174059

Mahdavian, M., Kafi, M. H., Movahedi, A., & Janghorbani, M. (2017). Improve performance in electrical power distribution system by optimal capacitor placement using genetic algorithm. In *Proceedings ECTI-CON 2017 - 2017 14th International Conference on Electrical Engineering/Electronics, Computer, Telecommunications, and Information Technology*, 749–752. 10.1109/ECTICon.2017.8096347

Mejia, C. (2019). Influencing green technology use behavior in the hospitality industry and the role of the "green champion.". *Journal of Hospitality Marketing & Management*, *28*(5), 538–557. doi:10.1080/19368623.2019.1539935

Minhas, D. M., Khalid, R. R., & Frey, G. (2017). Short term load forecasting using hybrid adaptive fuzzy neural system: The performance evaluation. In *Proceedings 2017 IEEE PES-IAS PowerAfrica Conference: Harnessing Energy, Information and Communications Technology (ICT) for Affordable Electrification of Africa, PowerAfrica 2017*, 468–473. 10.1109/PowerAfrica.2017.7991270

Mohamed, F., & Koivo, H. (2008). *Multiobjective genetic algorithms for online management problem of microgrid. 3*, 46–54.

Moise, M. S., Gil-Saura, I., & Ruiz-Molina, M. E. (2018). Effects of green practices on guest satisfaction and loyalty. *European Journal of Tourism Research*, *20*(20), 92–104.

Momoh, J. (2012). *Smart Grid: Fundamentals of Design and Analysis (Band 63 von IEEE Press Series on Power Engineering)*.

Monteiro, J., & Nunes, M. S. (2015). *A Renewable Source Aware Model for the Charging of Plug-in Electrical Vehicles*. (May), 51–58. doi:10.5220/0005459000510058

Mouli, G. R. C., Kefayati, M., Baldick, R., & Bauer, P. (2019). Integrated PV charging of EV fleet based on energy prices, V2G, and offer of reserves. *IEEE Transactions on Smart Grid*, *10*(2), 1313–1325. doi:10.1109/TSG.2017.2763683

Muhamad Razali, N. M., & Hashim, A. H. (2010). Profit-based optimal generation scheduling of a microgrid. *PEOCO 2010 - 4th International Power Engineering and Optimization Conference, Program and Abstracts*, 232–237. 10.1109/PEOCO.2010.5559244

Mustapa, S. I., Peng, L. Y., & Hashim, A. H. (2010). Issues and challenges of renewable energy development: A Malaysian experience. In *Proceedings International Conference on Energy and Sustainable Development: Issues and Strategies (ESD 2010)*, 1–6. 10.1109/ESD.2010.5598779

Nagasaka, K., Ando, K., Xu, Y. B., Takamori, H., Wang, J., Mitsuta, A., … Go, E. (2012). A research on operation planning of Multi Smart Micro grid. *International Journal of Advanced Mechatronic Systems*, 351–356.

Pan, S.-Y., Gao, M., Kim, H., Shah, K. J., Pei, S.-L., & Chiang, P.-C. (2018). Advances and challenges in sustainable tourism toward a green economy. *The Science of the Total Environment*, *635*, 452–469. doi:10.1016/j.scitotenv.2018.04.134 PMID:29677671

Papas, I., Estibals, B., Ecrepont, C., & Alonso, C. (2018). Energy Consumption Optimization through Dynamic Simulations for an Intelligent Energy Management of a BIPV Building. In *Proceedings 7th International IEEE Conference on Renewable Energy Research and Applications, ICRERA 2018, 5*, 853–857. 10.1109/ICRERA.2018.8566915

Pinson, P., Mitridati, L., Ordoudis, C., & Ostergaard, J. (2017). Towards fully renewable energy systems: Experience and trends in Denmark. *CSEE Journal of Power and Energy Systems, 3*(1), 26–35. doi:10.17775/CSEEJPES.2017.0005

Pinto, N., Cruz, D., Monteiro, J., Cabrita, C., Semião, J., & Cardoso, P. J. S. ... Rodrigues, J. M. F. (2019). IoE-Based Control and Monitoring of Electrical Grids. In Harnessing the Internet of Everything (IoE) for Accelerated Innovation Opportunities (pp. 57–82). doi:10.4018/978-1-5225-7332-6.ch003

Pizam, A. (2009). Green hotels: A fad, ploy or fact of life? *International Journal of Hospitality Management, 28*(1), 1. doi:10.1016/j.ijhm.2008.09.001

EU Publications. (2016). *Overview of support activities and projects of the European Commission on energy efficiency and renewable energy in the heating and cooling sector.* doi:10.2826/607102

Punitha, S., & Mohd Rasdi, R. (2013). Corporate Social Responsibility: Adoption of Green Marketing by Hotel Industry. *Asian Social Science, 9*(17). doi:10.5539/ass.v9n17p79

Reguladora dos Serviços Energéticos, E. (2019). *Proposta de Regulamentação das Redes Inteligentes de eletricidade.* Retrieved from http://www.erse.pt/pt/consultaspublicas/consultas/Paginas/70_1.aspx

Roeva, O., Fidanova, S., & Paprzycki, M. (2013). Influence of the population size on the genetic algorithm performance in case of cultivation process modelling. In *Proceedings 2013 Federated Conference on Computer Science and Information Systems,* (pp. 371–376). IEEE.

Schiermeier, Q. (2013). Renewable power: Germany's energy gamble. *Nature, 496*(7444), 156–158. doi:10.1038/496156a PMID:23579661

Sortomme, E., Hindi, M. M., MacPherson, S. D. J., & Venkata, S. S. (2011). Coordinated charging of plug-in hybrid electric vehicles to minimize distribution system losses. *IEEE Transactions on Smart Grid, 2*(1), 186–193. doi:10.1109/TSG.2010.2090913

Tehrani, N. H., Khan, U. T., & Crawford, C. (2016). Baseline load forecasting using a Bayesian approach. In *Proceedings 2016 IEEE Canadian Conference on Electrical and Computer Engineering (CCECE),* 1–4. 10.1109/CCECE.2016.7726749

Teng, C.-C., Horng, J.-S., Hu, M.-L., Chien, L.-H., & Shen, Y.-C. (2012). Developing energy conservation and carbon reduction indicators for the hotel industry in Taiwan. *International Journal of Hospitality Management, 31*(1), 199–208. doi:10.1016/j.ijhm.2011.06.006

Touzani, S., Granderson, J., & Fernandes, S. (2018). Gradient boosting machine for modeling the energy consumption of commercial buildings. *Energy and Building, 158*, 1533–1543. doi:10.1016/j.enbuild.2017.11.039

Tuballa, M. L., & Abundo, M. L. (2016). A review of the development of Smart Grid technologies. *Renewable & Sustainable Energy Reviews, 59*, 710–725. doi:10.1016/j.rser.2016.01.011

Van Der Meer, D., Mouli, G. R. C., Mouli, G. M. E., Elizondo, L. R., & Bauer, P. (2018). Energy Management System with PV Power Forecast to Optimally Charge EVs at the Workplace. *IEEE Transactions on Industrial Informatics*, *14*(1), 311–320. doi:10.1109/TII.2016.2634624

Veldman, E., & Verzijlbergh, R. A. (2015). Distribution grid impacts of smart electric vehicle charging from different perspectives. *IEEE Transactions on Smart Grid*, *6*(1), 333–342. doi:10.1109/TSG.2014.2355494

Wang, R., Wu, S., Wang, C., An, S., Sun, Z., Li, W., … Fu, M. (2018). *Optimized Operation and Control of Microgrid based on Multi-objective Genetic Algorithm*. 1539–1544. doi:10.1109/POWERCON.2018.8601845

Witten, I. H., Frank, E., Hall, M. A., & Pal, C. J. (2016). *Data Mining: Practical Machine Learning Tools and Techniques*. Morgan Kaufmann.

World Tourism Organization. (2019a). Tourism and the SDGs. Retrieved May 2, 2019, from https://icr.unwto.org/content/tourism-and-sdgs

World Tourism Organization. (2019b). Tourism for SDGs – Welcome To The Tourism For SDGs Platform! Retrieved September 20, 2005, from http://tourism4sdgs.org/

Yadav, R., Kumar Dokania, A., & Swaroop Pathak, G. (2016). The influence of green marketing functions in building corporate image. *International Journal of Contemporary Hospitality Management*, *28*(10), 2178–2196. doi:10.1108/IJCHM-05-2015-0233

Yu, Z., Chen, S., & Tong, L. (2016). An intelligent energy management system for large-scale charging of electric vehicles. *CSEE Journal of Power and Energy Systems*, *2*(1), 47–53. doi:10.17775/CSEEJPES.2016.00008

Yuan, X.-C., Sun, X., Zhao, W., Mi, Z., Wang, B., & Wei, Y.-M. (2017). Forecasting China's regional energy demand by 2030: A Bayesian approach. *Resources, Conservation, and Recycling*, *127*, 85–95. doi:10.1016/j.resconrec.2017.08.016

ADDITIONAL READING

Abdelaziz Mohamed, M., & Eltamaly, A. M. (2018). *Modeling and Simulation of Smart Grid Integrated with Hybrid Renewable Energy Systems. Studies in Systems Decision and Control.* Cham: Springer; doi:10.1007/978-3-319-64795-1

Beaulieu, A., de Wilde, J., & Scherpen, J. M. A. (Eds.). (2016). *Smart Grids from a Global Perspective: Bridging Old and New Energy Systems. Power Systems series.* Springer International Publishing; doi:10.1007/978-3-319-28077-6

Hatti, M. (Ed.). (2019). *Renewable Energy for Smart and Sustainable Cities: Artificial Intelligence in Renewable Energetic Systems. Lecture Notes in Networks and Systems* (Vol. 62). Springer International Publishing; doi:10.1007/978-3-030-04789-4

Kabalci, E., & Kabalci, Y. (Eds.). (2019). *Smart Grids and Their Communication Systems. Energy Systems in Electrical Engineering book series (ESIEE)*. Singapore: Springer; doi:10.1007/978-981-13-1768-2

Keyhani, A. (2016). *Design of Smart Power Grid Renewable Energy System*. Wiley-IEEE Press.

Komarnicki, P., Lombardi, P., & Styczynski, Z. (2017). *Electric Energy Storage Systems: Flexibility Options for Smart Grids*. Springer-Verlag Berlin Heidelberg; doi:10.1007/978-3-662-53275-1

Lund, H. (2014). *Renewable Energy Systems*. Elsevier.

Palensky, P., & Dietrich, D. (2011). Demand side management: Demand response, intelligent energy systems, and smart loads. *IEEE Transactions on Industrial Informatics*, *7*(3), 381–388. doi:10.1109/TII.2011.2158841

KEY TERMS AND DEFINITIONS

Energy Storage System: Essential part of a renewable power generation system, aims at suppling a smooth output power to the power grid by storing energy which is feed as needed.

Internet of Everything (IoE): Networked connection of people, data, process, and things. In other words, IoE extends IoT by including intelligent and robust communication between machines-to-people, machine-to-machine, people-to-machines and people-to-people.

Internet of Things (IoT): Dynamic global network infrastructure, with self-configuring capabilities based on standard and interoperable communication protocols, where a massive number of physical and virtual things have identities, physical attributes, and virtual personalities.

Machine Learning: Mechanisms that use datasets to find patterns and correlations in order to build models which will be applied to new data in order to predict its outcomes.

Microgrid: Localized group of electricity sources and loads that normally operates connected to and synchronous with the traditional wide area synchronous grid but can also disconnect to island mode.

Smart System: System incorporating sensing actuation and control in order to make decisions based on the available data.

Chapter 6
The Exploration of Autonomous Vehicles

Anthony J. Gephardt
Waynesburg University, USA

Elizabeth Baoying Wang
Waynesburg University, USA

ABSTRACT

This chapter explores the world of autonomous vehicles. Starting from the beginning, it covers the history of the automobile dating back to 1769. It explains how the first production automobile came about in 1885. The chapter dives into the history of auto safety, ranging from seatbelts to full-on autonomous features. One of the main focuses is the creation and implementation of artificial intelligent (AI), neural networks, intelligent agents, and deep Learning Processes. Combining the hardware on the vehicle with the intelligence of AI creates what we know as autonomous vehicles today.

INTRODUCTION

As technology grows exponentially greater every day, it starts to seep more and more into our everyday lives. First, it started with computers, smartphones, and even home products like our fridge and coffee maker. Now, it is starting to become a staple in the automotive industry. It started out simple with the steam engine being the turn of the century technology. Soon technology became even more advanced with gasoline engines, electric engines, stability control, power steering, and even electric windows. In 2018, technology has advanced so much that our automobiles are assisting our driving and even driving for us. Self-driving vehicles have officially arrived and are available to the public.

The rest of the chapter is organized as follows. An overview of the background and previous history leading to technology and hardware in section 2. Section 3 discusses the software and AI used in autonomous vehicles. Section 4 discusses deep learning and how the AI is taught. Finally, we discuss the benefits and future endeavors of autonomous vehicles in section 5.

DOI: 10.4018/978-1-7998-2112-0.ch006

BACKGROUND

In 1769, Nicolas-Joseph Cugnot built the first "automobile". Cugnot was a military engineer who had experimented with many different steam engines. The French army wanted a faster and more efficient way to transport its cannons, so Cugnot built the "fardier à vapeu". The vehicle had three wheels, weighed over 2.5 tons, and was powered by a boiler mounted above the front wheel. The weight distribution was so off, if there was not enough weight on the back it would tip forward. Add that the fire for the boiler had to be lit every 15 minutes, with the weight and slowness, the project was scrapped (The Library of Congress, n.d.).

Even though the first true automobile is up for debate, generally, Karl Friedrich Benz is credited for creating the first true automobile. In 1885 Benz built the "Benz Patent-Motorwagen" which was the first production automobile to be powered by a gasoline engine (Daimler, n.d.). The automobile was revolutionary for its time with a set of innovations never seen before. The two-stroke engine produced 2/3 horsepower which drove three wheels crafted out of steel and wood. No one had manufactured steel to be used for wheels before. The wheels even had solid rubber surrounding them, which was the turn of the century innovation that inspired tires today.

The car that truly revolutionized the industry, utilized technology, and became the best-selling car in American history: The Ford Model T. The Model T was produced from 1908 to 1927 by the Ford Motor Company (History.com Editors, 2010). It is known for its practicality and reliability. It was the car for the "common" people of America. It was able to be produced so widely due to Henry's revolutionary idea for an assembly line (PBS, n.d.). Ford found out that to produce a low-cost car, there were 4 principals that needed to be followed: interchangeable parts, continuous flow, division of labor, and reduce wasted effort. Following these, Ford was able to reduce the cost of his model T down to less than $300 from the original $850. The simple motorized belt that moved parts along for people to assemble truly started the automotive industry.

Ever since the invention of the car and the production line, technology has not ceased its endeavor to grow in the automotive industry. In fact, automakers are creating new and exciting technology to make driving easier and more importantly safer. Besides the assembly line, one of the best "primitive" technologies integrated into vehicles was the electric ignition in 1911. Cranking the car to start posed somewhat of a safety hazard. Once the engine would start, sometimes the car would jump forward, injuring whoever cranked the car (Jardine Motors Group, n.d.).

It wasn't until the 1950s that other safety-oriented technology arose. It started with power steering in 1951 (Riley, 2018). This technology used hydraulic power to increase the pressure on the wheels, making it a lot easier to turn a car. It is still used to this day with electronic integrated steering. In 1959, Nils Bohlin, a Volvo safety engineer invented the 3-point seatbelt we see today (Jardine Motors Group, n.d.). Anti-Lock Braking (ABS) was introduced in 1971 which prevents the wheels from locking up under harsh braking. That is why your brakes pulse when you "slam" on the brakes. Airbags made their debut to the public in 1988. They were very shoddy at first, but with technology have improved greatly. In 2018 we have airbags virtually all around the inside of the car for maximum safety.

TECHNOLOGIES IN AUTONOMOUS VEHICLES

Gateway Technologies

One of the major gateway technologies that influenced autonomous driving were parking sensors. There are two types of parking sensors: Ultrasonic and Electromagnetic parking sensors. Ultrasonic sensors send out waves that bounce back which sends information through an onboard computer to alert the driver how far away an object is from the car. Usually, these sensors have a maximum range of up to 20 meters (Senix, n.d.). Sadly, these sensors are not the best to use. The waves emitted cannot bounce off fabrics or plastic very well causing the sensor to become unreliable. These sensors can also be covered up by dirt or dust, also making them unreliable. Electromagnetic sensors use a transceiver strip that generates an electromagnetic field. The strip detects disruptions in the electromagnetic field, sends the information through an onboard computer, and alerts the driver. These sensors have a better advantage over Ultrasonic due to being able to detect moving objects. Autonomous vehicles rely heavily on these sensors to know what is in the vicinity of the vehicle. This helps the vehicle steer, stop, start, etc.

Another gateway technology was the release of the Global Positioning System (GPS) navigation to the public. Originally GPS was exclusive to the United States military and air force. It was launched in 1973 after many predecessors. It was released to the public in the early 1980s but was not put into automotive technology until 2000. This was due to the military scrambling the signals given off by the satellites, virtually making it unusable for the public even though it was released (Jardine Motors Group, n.d.). It wasn't until Bill Clinton told the military to stop that the public could use them. Autonomous cars usually rely on their GPS system along with sensors to control where it is going, how fast it is going, etc.

The final core hardware piece of an autonomous car (besides the internal computers) is the reverse camera which was released in 2002, outside of the U.S (Jardine Motors Group, n.d.). This is a camera located on the back of the vehicle. When the vehicle is put into reverse, the camera will turn on and appear on a screen somewhere in the vehicle. This allows the driver to see what is behind the car, which greatly increases safety. In 2018, backup cameras also have a trajectory, so the driver can see how far to turn the steering wheel. In 2018, this technology was utilized to put cameras all around the car, giving the driver a 360-degree view (Riley, 2018).

Levels of Automation

Utilizing computers, sensors, GPS, and cameras driver assist features and autonomy came to the automotive world. Driver assists are different than autonomy. Driver assists are assists, so this includes lane departure warnings, blind spot monitoring, cross traffic alerts, and even high beam control. All of these need the driver to work, autonomy does not. Autonomy is when the car can take over for the driver. There are five levels of autonomy according to the Society of Automotive Engineers (SAE)(Figure 1) (National Highway Traffic Administration, n.d.)

Level 0 – No Automation: The driver must perform all driving tasks, no automation whatsoever.
Level 1 – Driver Assistance: The driver is controlling the vehicle, but some features may help the driver (Ex: Blind Spot Monitoring).
Level 2 – Partial Automation: The driver is controlling the vehicle, but the vehicle has the ability to accelerate, stopping, and steering. The driver <u>must</u> be monitoring the environment.

Figure 1. Automation levels

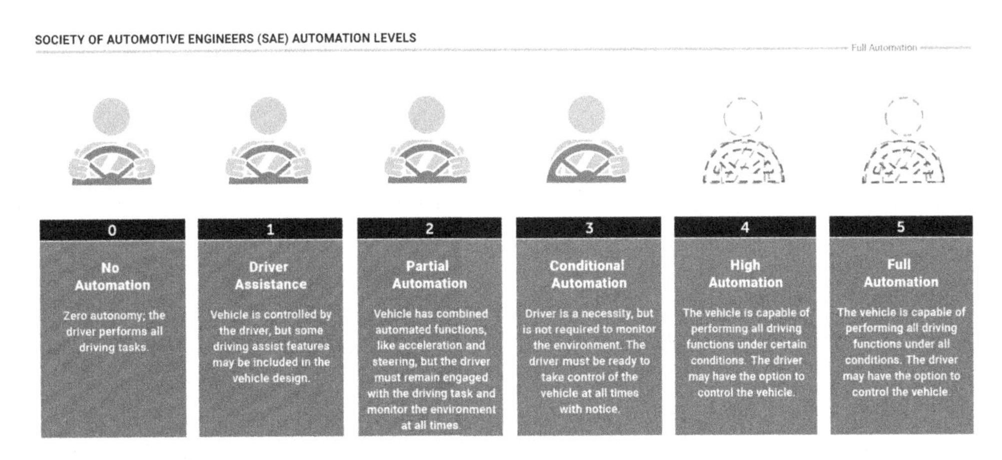

Level 3 – Conditional Automation: The driver is a necessity but <u>does not</u> have to be monitoring the environment. The driver must always be ready to take control of the vehicle.

Level 4 – High Automation: The vehicle is able to perform all driving tasks under <u>certain</u> conditions, but the driver can take over if desired.

Level 5 – Full Automation: The vehicle is able to perform all driving tasks under <u>all</u> conditions, but the driver can take over if desired.

Most vehicles mass produced by everyday manufactures like Ford or Audi are level 0 – 3 on the automation scale. These vehicles come with Blind Spot Monitoring, Active Lane assists, and even Adaptive Cruise Control. Toyota, Volvo, Google, Apple, Mercedes, Tesla, and Uber were the first companies to test and accept full automation. Google, Apple, and Uber use a very different automation system versus the normal automotive companies. When people think of autonomous vehicles, they think of the Volvo, Ford, or Toyota with a huge mount on top of the roof with a huge spinning sensor. These are essentially the same cameras and sensors that are built into Teslas or Mercedes but mounted on the roof. But, instead of using all cameras in tandem to create a 360-degree image, they use a Light Detection and Ranging (LIDAR) unit to provide a 360-degree image of the surrounding environment. This is a set of lasers that spin fast and measure the reflected light with a sensor. Using the information from the cameras and the LIDAR unit, the software inside the on-board computer can guide the car where it needs to go.

Tesla was one of the first companies in the world to mass-produce a fully autonomous vehicle, a level 5 on the automation scale (Tesla, n.d.). This was done by their best-selling vehicle since the start of Tesla: Model S. Model S uses 8 different cameras found all around the car, to give the computers a 360-degree view of what is around the car:

1. **Narrow Forward Camera** (250 meters of viewing distance) – This camera lets the car see far ahead on highways to adjust cruise control and speed.
2. **Main Forward Camera** (150 meters of viewing distance) – This is the main camera the car uses to sense surroundings, read signs, and detect pedestrians/traffic.
3. **Forward Side Cameras** (80 meters of viewing distance) – Cameras used for blind spot monitoring, reading signs, and detecting traffic/pedestrians.

Figure 2. Tesla Model S camera locations

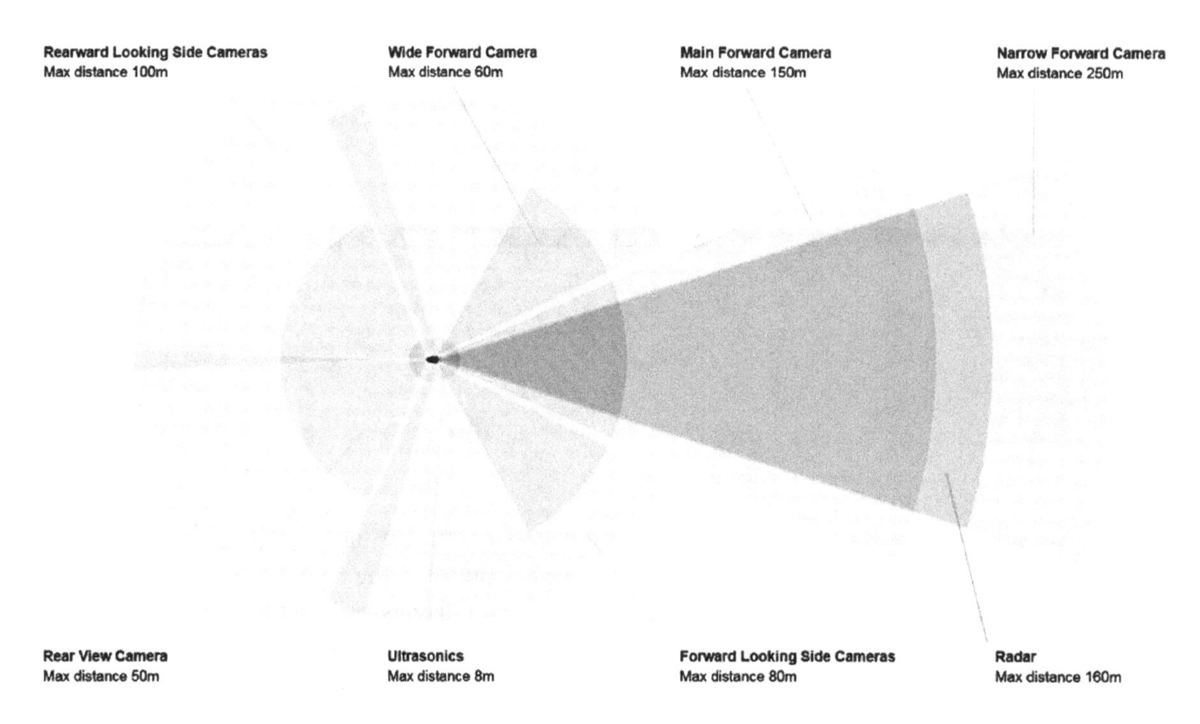

4. **Wide Forward Camera** (60 meters of viewing distance) - Camera used for blind spot monitoring, reading signs, and detecting traffic/pedestrians.
5. **Rearward Side Cameras** (100 meters of viewing distance) – Camera used for blind spot monitoring and backup information.
6. **Rear View Camera** (50 meters of viewing distance) – This is the main rear camera that lets the driver see when backing up, or when the car itself backs up.

Along with these 8 cameras, there are 12 ultrasonic sensors that pair with the cameras to complete "Tesla Vision". (Figure 2) (Castelluccio, 2018)

All this information taken in by the cameras is being processed by an onboard computer using Tesla's own deep neural network software. Tesla's software uses a Linux based kernel and is built with C++ and Python. With the Artificial Intelligent (AI) working with the sensors and cameras, the car is able to drive itself in what Tesla calls "Auto Pilot." Auto Pilot allows Teslas to: "match speed to traffic conditions, keep within a lane, automatically change lanes without requiring driver input, the transition from one freeway to another, exit the freeway when your destination is near, self-park when near a parking spot and be summoned to and from your garage." (Tesla, n.d.). Tesla's marketing encourages consumers to assume it is a level autonomy vehicle, it is not, it is at maximum potential an automation level 3. However, Elon Musk believes that by the middle of 2020, Teslas will be autonomy level 5 (Hawkins, 2019).

Even though Tesla is a very special case of autonomous cars, all full or semi-autonomous vehicles share the same core parts to make it actually autonomous. The vehicle needs sensors of some sort, whether that's ultrasonic or electromagnetic. Most autonomous vehicles use ultrasonic sensors because they are cheaper. The vehicle also needs radar or LIDAR. The vehicles need cameras to see where it is going, basically the "eyes" of the vehicle. Finally, the vehicle needs an onboard computer that uses AI, machine learning, and cloud management software.

SOFTWARE USED IN AUTONOMOUS VEHICLES

The software being run on the onboard computer is truly the "brains" of autonomous vehicles. Most of the time, this software is an advanced AI program that uses neural networks and deep learning to process the information given to it by the sensors and cameras. These are blocks upon blocks of code that all mesh together to tell the car what to do. These programs can be so complex and exclusive, auto manufacturers do not allow the public to see it. The program being run is what makes the vehicle autonomous, without it, sensors and cameras mean nothing.

Artificial Intelligence

Artificial intelligence is a relatively new technology that has just boomed in the past couple of years. It seems AI has been injected into everything we touch, from microwaves and thermostats to cameras and cars. This all came from two men who are regarded as the fathers of artificial intelligence: Marvin Minsky and John McCarthy. Both cognitive scientists started the development of the AI we know today.

Minsky mostly contributed to using neural networks with programming to create a machine that has "common sense". This was proven by Minsky creating SNARC (Stochastic Neural Analog Reinforcement Calculator). This machine is a neural net machine that has around 40 Hebb synapses connected to each other. There is a dial from 0-1, which is the probability signal. The synapses have a memory that detects if the signal comes in and if the signal goes out. Next to the synapse, a capacitor remembers if it received a signal or not and stores that data. A clutch is engaged, which the operator will "reward" the machine for engaging it. This machine is regarded as one of the first pioneering attempts in artificial intelligence. (Russel & Norvig, 2003)

Alongside Minsky, McCarthy was creating what we know as artificial intelligence, he even coined the term "Artificial Intelligence in 1955." In the early to late 1950s, McCarthy was creating a new type of programming language called Lisp. Lisp was primarily based on the Lambda calculus. "Lambda calculus can be called the smallest universal programming language of the world. Lambda calculus consists of a single transformation rule (variable substitution) and a single function definition scheme." (Rojas, 1998). It is the model that can be used to simulate a Turing machine. In its simplest form, lambda terms are built using three rules:

1. **x** – (A Variable) A character or string representing a parameter or mathematical/logical value
2. **(λx.M)** – (An Abstraction) Function definition, M representing the lambda term. X becomes bound in the expression.
3. **(M N)** – (An Application) Applying a function to an argument, M and N represent lambda terms.

Since Lisp was based off Lambda Calculus, it became the language of choice to program AI and logic programs in the 1960s. For a program published in the 1960s, it is still regarded as one of the most advanced languages for its time. It had easier program techniques like a "loop" macro and "reverse" function (see Figure 2) (Seibel, 2005). McCarthy had so much influence and early knowledge of the artificial intelligence world he founded the AI department of both MIT and Stanford University.

Minsky and McCarthy both "described artificial intelligence as any task performed by a program or a machine that, if a human carried out the same activity, we would say the human had to apply intelligence to accomplish the task." (Heath, 2018). Some examples of AI systems mimicking human tasks are problem-solving, planning, reasoning, learning, motion, and manipulation. With high-level AI, there are two categories of artificial intelligence: Narrow AI (Weak AI) and General AI. Narrow AI is the norm for today, what is going in our computers, phones, and appliances. It is a system that has been taught or learns how to carry out specific tasks without the need for specific programming of that task.

A great example is Siri, Google Assistant, or Cortana. All three of these assistants are programmed to do something a human can do, and learn how to improve on itself with speech, face, and screen recognition without having to be programmed specifically to do so. General AI is overall stronger, being able to think exactly like a human, think on its own, and have cognitive abilities. Google's Director of Engineering predicts that by 2029, an AI will be able to pass a Turing Test with flying colors and by 2045 humans will be able to connect their neocortex to a storage system with AI amplification.

Narrow AI is considered weak compared to general AI. However, it is the AI of choice for autonomous vehicles. Since narrow AI does one or two tasks, it makes sense for autonomous vehicles to use this type of AI. A car should not be making decisions as a human would, a car shouldn't be deciding where to eat for dinner. Companies want the AI to detect surroundings, cars, figures, etc. this allows the car to… be a car. Companies want the AI to behave in the sense of human seeing, hearing, and making decisions on where to turn, accelerate, or brake.

The AI learns many times throughout the day, every time you drive, every time OTHER people drive. It is a continuous cycle of new knowledge, which is called the Perception Action Cycle (Figure 3) (Gadam, 2018). It happens with the vehicle gathering data from its surroundings and feeding it to an

Figure 3. Action perception cycle

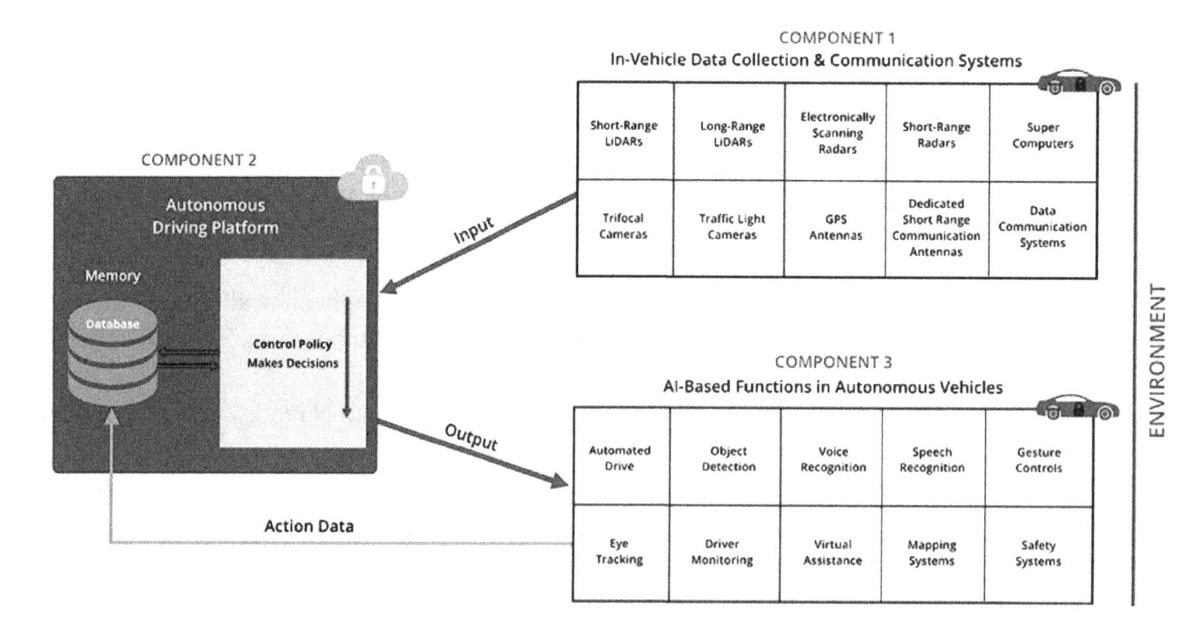

intelligent agent. This agent then makes decisions based on the information it was given and makes the vehicle perform specific actions in that environment. The more this cycle takes place, the more knowledge is gained by the agent, which improves accuracy in decision making.

There are three major parts to the Perception Action Cycle. The first part is data collection from sensors and communication systems. Autonomous vehicles use an abundance of sensors and cameras, as we said before. All this data is then processed and sent to the Autonomous Driving Platform, or better known as "The Cloud." The cloud is the second part of the cycle. It houses the intelligent agent the autonomous car is using to make decisions. It uses AI algorithms to make the decisions the car will act upon. The intelligent agent is also connected to a huge database that stores previous decisions, data from other cars, decisions form those cars and many other things. Combine this database with the real-time input coming from the vehicle, it aids the agent in making accurate decisions. This leads into the third part, which is the vehicle acting upon the decisions made by the intelligent agent. These decisions are what help the vehicle navigate safely through obstacles and traffic. These AI based functions also include eye-tracking and gesture controls.

Neural Networks

The Intelligent Agent cannot act on its own, it needs neural networks and deep learning processes to become "smarter" in its decision making. Neural networks date back to 1944 when Warren McCullough and Walter Pitts proposed the idea of an artificial type human brain. It was rejected surprisingly by the same man that created the first neural net machine, Marvin Minsky. It wasn't revived until the early 2000s when processors and graphics cards became much more powerful.

"A neural network is a type of machine learning which models itself after the human brain." (Hardesty, 2017). The net is made up of hundreds of thousands of densely interconnected nodes, the somewhat resemble a human brain. All of these nodes consist of layers, from bottom to top (Figure 4) (Ahire,

Figure 4. Diagram of a neural network

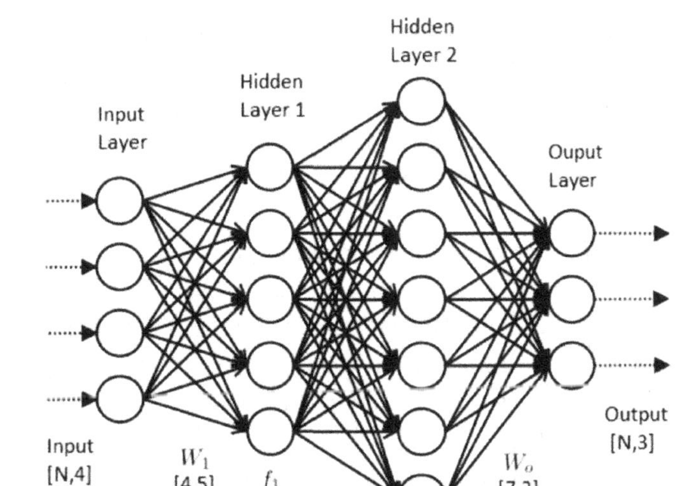

2018). Usually, the bottom layer is the input layer, where it takes the predefined or undefined information. It then goes through many "hidden" layers before it reaches the top layer, which is the output. "An individual node might be connected to several nodes in the layer beneath it, from which it receives data, and several nodes in the layer above it, to which it sends data." (Hardesty, 2017). As the information is being fed into the net, nodes will assign that information with a number, also known as a weight. The information then goes from layer to layer, with numbers being added to it. If the number is below an assigned threshold, it does not go to the next layer. If the number is above an assigned threshold, the node fires and keeps the data moving.

All of this means nothing unless the network is trained to do something. Whether that be to detect handwritten numbers or to detect pedestrians and cars in the street. The network is like the human brain, when you are a child, you must learn things like objects and behaviors. The same goes for the network when created it is an infant that must learn objects and behaviors. Teaching the network is called deep learning.

Deep Learning

Deep learning comes in three different methods: Supervised learning, Unsupervised learning, and Reinforced learning.

Supervised learning algorithm is the most commonly used when it comes to training a network. Supervised learning happens when there is a set of predefined images or numbers that have labels on them (also called ground truth labeling). The person teaching the network then knows how it should come out on the other side, but the machine does not. As the machine outputs correct or incorrect data, the weights are calibrated. The weights need to be calibrated until the network can output the desired result on a consistent basis. Relating to autonomous cars, the intelligent agent initially was fed hundreds of thousands, maybe millions of pictures of the surroundings of a car. The teacher knew what picture was of a person or a car or a tree and calibrated the weights until the agent was able to reliably detect and decipher what was in its surroundings.

Unsupervised learning is the opposite of supervised learning. There is no data set that is labeled going into the network. The data is fed into the network, blindly. Instead of a teacher adjusting the weights, the nodes do it themselves through a cost-benefit like program. This program tells the network how far off its answer was, and recalibrates the weights as needed. Another method is called cluster analysis. It "is used for exploratory data analysis to find hidden patterns or grouping in data. The clusters are modeled using a measure of similarity which is defined upon metrics such as Euclidean or probabilistic distance." (MathWorks, n.d.). There are five common clustering algorithms:

- Hierarchical Clustering: This is when the algorithm builds a multilevel hierarchy of clusters by creating a tree.
- K-Means clustering: This is when the algorithm partitions data into k distinct clusters depending on the distance from the center of the cluster.
- Gaussian Mixture Models: This is when the algorithm "models clusters as a mixture of multivariate normal density components."
- Self-organizing Maps: This combines with the neural network to learn topology and distribution of the data given.
- Hidden Markov Models: It uses previously observed data to recover a sequence of states.

Unsupervised learning would be used for teaching the agent "vision". It would see objects around it and learn what to avoid.

Finally, Reinforced learning is very different from the other two methods. Reinforced learning gives the network a data set, without labels, but instead of having a program calibrate weights when something is output wrong, it is reinforced with positive and negative feedback. Basically, the network is being rewarded for doing good and punished for being bad, as if it was a dog. We saw this happen with Minsky's neural net machine. There was a button the user would press to reward the machine for doing good, and one for it doing badly.

Nvidia is one of the pioneer companies that use their Graphics Processing Unit (GPU) power to create deep learning processes, neural networks, intelligent agents, and the AI we know today. Nvidia is also one of the leaders in using MATLAB and other programs to create a data set, feed it to a neural network, and translate it into autonomous vehicle actions. Nvidia posted an article on July 2017 explaining how the company programs three layers of deep learning in MATLAB.

Layer 1: This is the layer that information is fed into. After a data set is labeled, the network can use this information to learn. Figure 5 shows a picture already labeled with a truck, SUV, and car (Nehemiah & Jayaraman, 2017). This is just one of the millions of pictures the network would learn from.

Layer 2: This layer is called the middle layer or "Hidden Layer." "Middle layers are the core building blocks of the network, with repeated sets of convolutional, ReLU, and pooling layers." (Nehemiah & Jayaraman, 2017). Convolution pertains to the Convolutional Neural Network (CNN), which is a class of Artificial Neural Network (ANN). It uses convolutional layers to filter inputs for useful information. "The process involves combining input data (feature map) with a convolution kernel (filter) to form a transformed feature map." (Nehemiah & Jayaraman, 2017). ReLU pertains to the rectifier. "This is an activation function defined as the positive part of its argument: $f(x)=x^{+}=\max(0,x)$ where x is the input to the neuron." (Nehemiah & Jayaraman, 2017). "Pooling is a procedure that takes input over a certain area and reduces that to a single value (subsampling)."

Layer 3: This is the output layer, where all other layers connect. For these layers to work, the object detection has to be trained to identify things like pedestrians, trees, and roads. Once the Network has been taught and implemented into the Intelligent Agent, the agent can communicate with all vehicles connected to it. It will gather millions of pictures, continuously learn from them, and make better decisions for the vehicle to make. That is the foundation and brain which makes an autonomous vehicle autonomous.

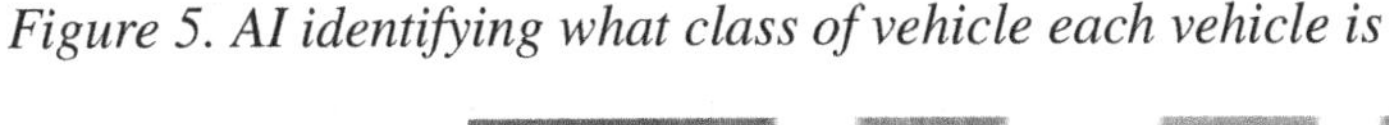

Figure 5. AI identifying what class of vehicle each vehicle is

Benefits and Challenges

After all, why would we want autonomous vehicles? What are the benefits of having a car that can drive itself? What are the challenges and problems?

One major benefit, and the main reason autonomous cars are becoming so popular: the roads will be safer. Every year, around 30,00 people die from traffic-related deaths in the U.S. alone. This could range from a fender bender on the highway, to flipping over in the winter. A car driving itself may be more "aware" of its surroundings than a human being. The car would not have blind spots, it would have cameras and sensors in those areas already. The computer could react faster than a human could if the car was about to collide and needed to apply brakes. According to Cadie Thompson of Business Insider: "In fact, if about 90% of cars on American roads were autonomous, the number of accidents would fall from 6 million a year to 1.3 million. Deaths would fall from 33,000 to 11,300, according to a study by the Eno Centre for Transportation." (Thompson, 2016).

Another major benefit would be reduced CO_2 emissions. Global warming and climate change are an extremely hot topic right now. At the current rate, the earth is going to heat above two degrees higher than it should be, which would cause mass destruction. With all these greenhouse gases in the air like methane, combined with CO_2, it is destroying the earth slowly. If most cars on the highways and in cities were autonomous, stop and go waves could be eliminated, promoting a constant, steady flow of traffic. This cuts down on congestion, which cuts down on CO_2 emissions. According to ASCE, "Americans currently spend more than 6.9 billion hours a year sitting in traffic." (Thompson, 2016). Not only would this cut down on CO_2, but save gas, which in turns cuts down on the need for fossil fuels. So far it is a win-win situation! Or is it?

With every pro, comes a con, and there a few challenges when it comes to autonomous vehicles. One of the major problems that come with autonomous vehicles is security. An autonomous vehicle is like another computer, smartphone, or tablet a consumer can own. It has a computer inside, an operating system of some sort, and is a sweet looking snack for a hacker. Hackers are one of the main hindrances for autonomous vehicles. There was a report done in 2015 about a hometown hacker, who hacked one of the first road going autonomous vehicles, while on the news. He then controlled it with a remote control, as if it was an R/C car. Currently, a few companies are on the upswing of cybersecurity like Tesla who have secured their network safely… for the time being.

Another problem is that autonomous vehicles are programmed to drive like humans. As mentioned earlier, the AI is trained as if it were a human. It is fed millions of images, taught right from wrong, and then sent out into the world. Professor Shrivastava from Arizona State University uses one of the most devastating cases to the autonomous car industry: the fatal accident of Uber in Arizona. In March 2018, a self-driving Uber car was driving down Mill Ave. when it struck a woman and killed her. He explains how the system was working like a human – it was too dark for the camera to see, so it didn't see an obstacle and kept going. If it was acting like a safety system, it would have detected an underlying obstacle and proceed accordingly. Uber is currently banned from testing in Arizona, causing their test drivers to be laid off.

FUTURE RESEARCH DIRECTIONS

Autonomous vehicles are emerging technologies. With advances in communication, controls and embedded hardware and software systems, Autonomous vehicles will be made more efficient and safer. Inter-vehicle communication will also be able to help with minimizing fuel used finding parking spaces. As more and more autonomous vehicles are used, more data will be gathered about driving patterns to make autonomous vehicles more intelligent and reliable.

CONCLUSION

Technology is growing exponentially every day. It enhances our lives tremendously from computers and smartphones to self-driving vehicles. This new breed of vehicles is becoming more popular every day. It started with simple safety features like back up cameras and parking sensors. It then expanded into driver assist features like lane departure or automatic braking. The automotive industry has now taken it a step further, combining safety features with AI, Intelligent Agents, Neural Networks, and Deep learning to create the autonomous vehicles we know today. Granted, there have been many hiccups along the way, but the benefits outweigh the cons by an extreme amount, pushing the automotive industry to prove autonomy at every corner. Who knows, maybe autonomous cars will be the norm someday and he driver's license may become obsolete.

REFERENCES

Administration, N. H. (n.d.). *Automated Vehicles for Safety*. Retrieved from National Highway Traffic Safety Administration: https://www.nhtsa.gov/technology-innovation/automated-vehicles-safety

Ahire, J. B. (2018, August 24). *The Artificial Neural Networks handbook: Part 1*. Retrieved from Medium: https://medium.com/coinmonks/the-artificial-neural-networks-handbook-part-1-f9ceb0e376b4

Castelluccio, M. (2018, October 3). *TESLA'S AUTOPILOT SEEN THROUGH THE WINDSHIELD*. Retrieved from SF Magazine: https://sfmagazine.com/technotes/october-2018-teslas-autopilot-seen-through-the-windshield/

Congress, T. L. (n.d.). *Who Invented the Automobile*. Retrieved from Everyday Mysteries: https://www.loc.gov/rr/scitech/mysteries/auto.html

Diamler. (n.d.). *Company History*. Retrieved from Diamler: https://www.daimler.com/company/tradition/company-history/1885-1886.html

Editors, H. (2010, April 26). *Model T*. Retrieved from HISTORY: https://www.history.com/topics/inventions/model-t

Gadam, S. (2018, April 19). *Artificial Intelligence and Autonomous Vehicles*. Retrieved from Medium: https://medium.com/datadriveninvestor/artificial-intelligence-and-autonomous-vehicles-ae877feb6cd2

Group, J. M. (n.d.). *The History of Tar Technology*. Retrieved from Jardine Motors Group: https://news.jardinemotors.co.uk/lifestyle/the-history-of-car-technology

Hardesty, L. (2017, April 14). *Explained: Neural networks*. Retrieved from MIT News: http://news.mit.edu/2017/explained-neural-networks-deep-learning-0414

Hawkins, A. J. (2019, April 22). *Here are Elon Musk's wildest predictions about Tesla's self-driving cars*. Retrieved from The Verge: https://www.theverge.com/2019/4/22/18510828/tesla-elon-musk-autonomy-day-investor-comments-self-driving-cars-predictions

Heath, N. (2018, February 18). *What is AI? Everything you need to know about Artificial Intelligence*. Retrieved from ZDNet: https://www.zdnet.com/article/what-is-ai-everything-you-need-to-know-about-artificial-intelligence/

Mathworks. (n.d.). *Unsupervised Learning*. Retrieved from Mathworks: https://www.mathworks.com/discovery/unsupervised-learning.html

Nehemiah, A., & Jayaraman, A. (2017, July 20). *Deep Learning for Automated Driving with MATLAB*. Retrieved from Nvidia Developer: https://devblogs.nvidia.com/deep-learning-automated-driving-matlab/

PBS. (n.d.). *Ford Installs First Moving Assembly Line 1913*. Retrieved from PBS: http://www.pbs.org/wgbh/aso/databank/entries/dt13as.html

Riley, C. (2018, October 5). *The Top 15 Automotive Innovations of All Time*. Retrieved from AutoWise: https://autowise.com/the-top-15-car-innovations-of-all-time/

Rojas, R. (1998). *A Tutorial Introduction to the Lambda Calculus*. Berlin, Germany: Freie Universität Berlin.

Russel, S., & Norvig, P. (2003). *Artificial Intelligence: A Modern Approach*. London, UK: Pearson Education.

Seibel, P. (2005). *Practical Common Lisp*. New York: Apress. doi:10.1007/978-1-4302-0017-8

Senix. (n.d.). *Ultrasonic Sensors FAQs*. Retrieved from Senix: https://www.senix.com/ultrasonic-sensor-faqs/

Tesla. (2018). *Autopilot*. Retrieved from Tesla: https://www.tesla.com/autopilot

Thompson, C. (2016, June 10). *The 3 biggest ways self-driving cars will improve our lives*. Retrieved from Business Insider: https://www.businessinsider.com/advantages-of-driverless-cars-2016-6

ADDITIONAL READING

Bourzac. (2016, March 28). *Bringing Big Neural Networks to Self-Driving Cars, Smartphones, and Drones*. Retrieved from IEEE Spectrum: https://spectrum.ieee.org/computing/embedded-systems/bringing-big-neural-networks-to-selfdriving-cars-smartphones-and-drones

Dettmers, T. (2015, November 3). *Deep Learning in a Nutshell: Core Concepts*. Retrieved from Nvidia Developer: https://devblogs.nvidia.com/deep-learning-nutshell-core-concepts/#pooling

Developer, N. (n.d.). *Convolutional Neural Network (CNN)*. Retrieved from Nvidia Developer: https://developer.nvidia.com/discover/convolutional-neural-network

Field, K. (2018, June 11). *Tesla Director Of AI Discusses Programming A Neural Net For Autopilot (Video)*. Retrieved from Clean Technica: https://cleantechnica.com/2018/06/11/tesla-director-of-ai-discusses-programming-a-neural-net-for-autopilot-video/

Insights, M. T. (2018, March 12). *What's Driving Autonomous Vehicles*. Retrieved from MIT Technology Review: https://www.technologyreview.com/s/609674/whats-driving-autonomous-vehicles/

MathWorks. (n.d.). *fasterRCNNObjectDetector*. Retrieved from MathWorks: https://www.mathworks.com/help/vision/ref/fasterrcnnobjectdetector.html;jsessionid=0dffcbceeef2b0710c74352a6d06

Renolds, C. W. (2006). *Steering Behaviors For Autonomous Characters*. Foster City: Sony Computer Entertainment America.

Seo, J. D. (2018, March 20). *Implementing Neural Network used for Self Driving Cars from NVIDIA with Interactive code [Manual Back Prop TF]*. Retrieved from Medium: https://towardsdatascience.com/implementing-neural-network-used-for-self-driving-cars-from-nvidia-with-interactive-code-manual-aa6780bc70f4

Silver, D. (2016, October 13). *C++ vs. Python for Automotive Software*. Retrieved from Medium: https://medium.com/self-driving-cars/c-vs-python-for-automotive-software-40211536a4ad

KEY TERMS AND DEFINITIONS

Artificial Intelligent (AI): A branch of computer science dealing with tasks that normally require human intelligence, such as visual perception, speech recognition, decision-making, and translation between languages.

Autonomous Vehicles: Is a vehicle that is capable of sensing its surroundings and moves with little or no driver assist.

Global Positioning System (GPS): A system of satellites, computers, and receivers that is able to determine the latitude and longitude of a receiver on Earth by calculating the time difference for signals from different satellites to reach the receiver.

Lisp: A computer programming language developed in late 1950s by John McCarthy at the Massachusetts Institute of Technology (MIT). It is based on a distinctive, fully parenthesized prefix notation.

Neural Network: Is inspired by the biological neural networks that constitute animal brains. It consists of a series of algorithms that discover underlying patterns in a dataset through a process that mimics the way the human brain operates.

Perception Action Cycle: Is an intuitive explanation of the technical setting of reinforcement learning. It is the circular flow of information that takes place between a moving object and its environment in form of a sensory-guided sequence of behavior.

Reinforced Learning: Is about taking suitable action to maximize reward in a particular situation. Reinforcement learning differs from the supervised learning in a way that in supervised learning there is training data whereas in reinforcement learning, there is no training dataset. Without training dataset, reinforcement learning has to learn from its experience.

Supervised Learning: Happens when a set of predefined images or numbers have labels on them. It maps an input to an output based on example input-output pairs.

Unsupervised Learning: Is the training of an artificial intelligence (AI) algorithm using a dataset that is neither classified nor labeled and allowing the algorithm to process the dataset and find patterns without guidance.

Chapter 7
Average Speed of Public Transport Vehicles Based on Smartcard Data

Vera Costa
https://orcid.org/0000-0003-0801-2113
INESC TEC, Faculdade de Engenharia, Universidade do Porto, Portugal

José Luís Borges
INESC TEC, Faculdade de Engenharia, Universidade do Porto, Portugal

Teresa Galvão Dias
https://orcid.org/0000-0001-6209-3626
INESC TEC, Faculdade de Engenharia, Universidade do Porto, Portugal

ABSTRACT

In public transport, traveler dissatisfaction is widespread, due to long waits and travel time, or the low frequency of the service provided. Public transport providers are increasingly concerned about improving the service provided. To improve public transport, detailed knowledge of the network and its weaknesses is necessary. An easy and cheap way to achieve this information is to extract knowledge from the data daily collected in a public transport network. Thus, this chapter focuses on data analysis resulting from the smartcard-based ticketing system. The main objective is to detect patterns of average speed for all days of the week and times of the day, along with pairs of consecutive stops. To perform the analyses, the average speed was deduced from ticketing data, and clustering methods were applied. The results show that it is possible to find segments with similar patterns and identify days and times with similar patterns.

DOI: 10.4018/978-1-7998-2112-0.ch007

INTRODUCTION

The development of cities led to an urban traffic increase. In turn, public transport providers and the government are encouraging the population to use public transport. This usage allows to save fuel, to reduce congestion, and it offers a safe, affordable, and convenient way to travel. However, improvements in public transport systems are a constant need. The introduction of Information and Communication Technologies with, for example, the implementation of smart cards and the vehicles' geolocation, and more recently, the use of smartphones enables the study and implementation of several service improvements through data analyses and optimization.

These technologies, as for example, Automatic Data Collection Systems (ADCS), created an opportunity to generate, at low marginal cost, large quantities of precise and disaggregate passenger trip data (Wilson, 2013), such as boarding and alighting times and locations, journey distances and collected fares. This data allows researchers to observe when and where the transactions took place and extract the footprints of many thousands of individuals (Gan et al., 2019). The exploration of massive real-time data collected from the ADCS provides an efficient way to disclose hidden mobility patterns and spatio-temporal regularities in urban mobility patterns (Pelletier, Trépanier, & Morency, 2011). In particular, information about public transport passenger behavior, such as travel purpose or activity (Gu & Mark, 2014), could be extracted. The availability of descriptive data about service usage enables Urban Public Transport (UPT) providers to optimize the transport service and manage their resources more efficiently (Giannopoulos, 2004). However, for this information to be useful to both public transport providers and the population, detailed and correct data exploration should be made.

Although many studies have been carried out to assess UPT service, most of them are not based on the data that is currently being collected. Instead, they are collected through counts (for example) and with resources to people who perform this collection. Thus, this made the data needed to be expensive and slow to collect. Consequently, the difficulties in obtaining data caused data to be updated only sporadically. The exploration of the data nowadays collected could give much more knowledge than the one that has been extracted. Thus, it is essential to define the type of information and knowledge that can be drawn from the data collected by different UPT providers and by using different technologies.

Having that in mind and considering the difficulties and limitations to have access to all data types, this study focuses on UPT travel data collected by smart cards in order to extract useful information. In particular, the analyses are performed to extract knowledge regarding the average speed of buses summarizing them into average speed vectors for clustering analysis.

In this paper, it is intended to answer the following questions:

- Is it possible to infer bus speed based on travel ticketing information?
- Are there bus speed patterns throughout the days of the week or time of the day? How can they be grouped?
- How can the extracted knowledge from these analyzes help UPT providers in decision making?

The chapter is structured as follows: the second section presents the background. The third section presents the main focus of the chapter, where 1) the material and methods used are described, 2) the results and discussion of the most significant results are presented, and 3) the issues, controversies, and problems are identified. In the fourth section, the solutions and recommendations are discussed. Future research directions are given in the fifth section, and, finally, the conclusion of the chapter is presented in the sixth section.

Background

Due to the need to participate in increasingly varied activities motivated by physiological, psychological, and economic necessity, mobility demand from people living in urban and metropolitan areas is continuously growing. This large number of activities implicates that people take complicated trips involving the use of various transport means or even several vehicles.

Transport systems provide access, mobility, and other benefits, and, at the same time, they put pressure on the human and natural environment. In this sense, the policy goals of several countries are targeted to make progress towards more sustainable transport systems and mobility patterns, and at the same time increasing the economic prosperity and quality of life. Additionally, a significant strand in policies intends to achieve higher usage of public transport and to influence modal shift. For this purpose, they expect to make available "good" public transports. However, "good" public transport has many attributes, including financial sustainability and an efficient provision of services with quality. To ensure the continuous improvement of delivered public transport services, the detailed knowledge of the UPT system is essential for focusing public transport agencies on their strategic goals.

The public transport performance evaluation can reflect various perspectives. There are many commonly-used public transport indicators, such as load factor and cost-per-vehicle-kilometer, measure operating efficiency. Other indicators indicate the user experience, such as rider comfort, travel speed, reliability, affordability, integration, and satisfaction. User-oriented indicators are essential for developing public transport systems that respond to user demands and, so, can attract more travelers.

More than 400 performance indicators are used in the transport industry today (Parks et al., 2010). Some examples of different indicators used in urban public transport are, among others, the regularity (Fu & Yang, 2007), the level of service (Fancello, Carta, & Fadda, 2014; Orth, Carrasco, Schwertner, & Weidmann, 2014; Polus & Shefer, 1984), the availability and reliability (Elms, 1998; Jasti & Ram, 2016; Turnquist & Blume, 1980).

A significant development in the last decade was the introduction of smartcards to supplement or even replace the magnetic stripe cards, which have become the industry standard over the previous years. More recently, in several cities, UPT has adopted the use of smartphones for ticket payment, journey planning, and information (P. M. Costa et al., 2016). These systems automatically and continuously collect the records of travelers' use of public transport because records of the fare payment can be regarded as travel records (Kusakabe & Asakura, 2014).

The use of smartphones as a payment system is new, and there are still a few works related to the data they collect. On the contrary, the use of smartcards has been of great interest to UPT providers and researchers, and there are a large number of works based on the use of this type of card.

With smart card systems, transport service providers are able to have access to larger volumes of personal travel data, to link those data to the individual card and/or traveler, to have access to continuous trip data covering long periods of time, and to know who their most frequent customers are (Bagchi & White, 2005). Thus, smart cards provide a means of significantly improving the quality of data in these respects, as well as obtaining more accurate estimates of the aggregate totals. Transport service providers can "reconstruct" the trips that people make over the day or longer, and examine travel behaviors that have been difficult to determine because of the deficiencies of existing data sources (Bagchi & White,

2005). Several studies have been developed to analyze smartcard data for various purposes, such as origin-destination matrices estimation (Munizaga & Palma, 2012; Nunes, Dias, & Falcão E Cunha, 2016), travel behavior analysis (Ali, Kim, & Lee, 2016; V. Costa, Fontes, Borges, & Dias, 2019; V. Costa, Fontes, Costa, & Dias, 2015b, 2015a; El Mahrsi, Come, Oukhellou, & Verleysen, 2017; Ferreira, Costa, Dias, & Falcão E Cunha, 2017; Jánošíkova, Slavík, & Koháni, 2014; Ma, Wu, Wang, Chen, & Liu, 2013; Weng, Liu, Song, Yao, & Zhang, 2018), and performance evaluation of public transport (Karim & Fouad, 2018; Kusakabe, Iryo, & Asakura, 2010; Zhou, Murphy, & Long, 2014). Performance evaluation of public transport is essential to understand the effectiveness of the system and the plans for its improvement. Different researchers have given a various number of measures to evaluate the performance of UPT. For example, Putra (2013), in his study, considers safety, accessibility, affordable tariff, capacity, regularity, swift and fast, on time, integration, efficiency, easiness, orderly, security, cozy, and low pollution. Other authors (Swathipriya, 2019) consider availability, convenience, comfort, mobility (travel speed), safety, affordability, ITS facility.

Focusing on mobility, in particular, the travel speed, most of the analyses in the literature focus on determining travel time between specific points, and overlook the speed of a vehicle on a particular type of link (Guessous, Aron, Bhouri, & Cohen, 2014), making relevant the development of a method for studying speed at different points in reference to the type of link and traffic conditions. Oskarbski et al. (2015) develop a method for determining the speed of buses in urban networks concerning traffic conditions using data from traffic control systems. In particular, these systems contain the location of vehicles in the road network. Vehicles send their position every 20 seconds on sections between stops, and every 10 seconds within a stop. Their results were promising and supported the development of a simplified speed model for all of the city's main roads with a 2x2 cross-section (dual carriageway with two lanes in each direction), reflecting the actual volume-capacity ratio of Gdynia (Poland). Authors consider two types of speed: the scheduled speed (relation between travel distance and travel time including all stops on the route) and technical speed (depends solely on the influence of local traffic conditions and vehicle characteristics).

In another study, Salonen et al. (2013) analyzed speed differences between transport systems. They consider that understanding different travel times for different modes of transport may be an effective way of the environmental and social sustainability of transport and land-use arrangements. In their literature review, they identified three travel time calculation models for both these travel modes: travel time by private car, travel time by public transport, and a door-to-door approach. The analysis performed in their work concludes that using conceptually corresponding models for car and UPT travel time calculations is the key to achieving a reliable analysis of modal accessibility disparity. A door-to-door approach in travel time calculations also makes the results truly comparable in absolute terms.

In cities without dedicated public transport lanes, buses must share road sections with other vehicles. Therefore, they experience the same delays caused by traffic lights, traffic control at junctions, or traffic incidents (Vasantha Kumar & Vanajakshi, 2013). Kumar et al. (2013) analyzed bus speed in India using GPS, considering two approaches: one based on the ratio of the section travel times of public transport to other vehicles, and other based on the quantifiable relationship between the public transport and other vehicles section travel times.

Wardman (2004) presents the valuations of public transport time relative to car travel time and on the valuations of the walking time, waiting time, and service headway associated with public transport use. According to the author, some studies indicate variations in travel time, depending on the choice of transport and the purpose of a trip. A key element to be determined is the variation in travel time component referred to as in-vehicle time, because regardless of the selected means of transport, the walk, and wait time must also be considered.

Although the analyzed studies focused on understanding the performance of public transportation with different types of data, studies relating the average speed of bus vehicles and smartcard are not found.

As mentioned, speed is an essential parameter in a transport network. It can be used to evaluate the quality of travel in a specific area. Thus, it is essential for understanding the speed of vehicles and vehicle streams for advanced traffic management systems. Additionally, the average speed allows estimating the travel time of UPT vehicles, which is used to define their frequency in a particular route. The increasing use of road networks implies noticeable deviations in public transport schedules, mainly in dense urban road networks.

MAIN FOCUS OF THE CHAPTER

Material and Methodology

To extract knowledge regarding the average speeds of buses summarizing them into average speed vectors for clustering analysis, in this chapter, three main steps were considered: (i) firstly, travel data was collected from smartcards (first subsection); (ii) secondly, the data was processed to deduce average speed (second subsection); and (iii) last, cluster analysis was performed in vectors of average speed of buses (third subsection). Such methodology was applied to a European medium size Metropolitan Area. Figure 1 presents an overview of the methodology followed. This methodology starts by collecting ticket validation data from a UPT network. Based on the information present in the collected data, the travel time between stops is estimated. The travel times need to be validated since data errors may occur. Thus, it is essential to perform data processing to avoid failures in the analysis conclusions.

Figure 1. Methodology

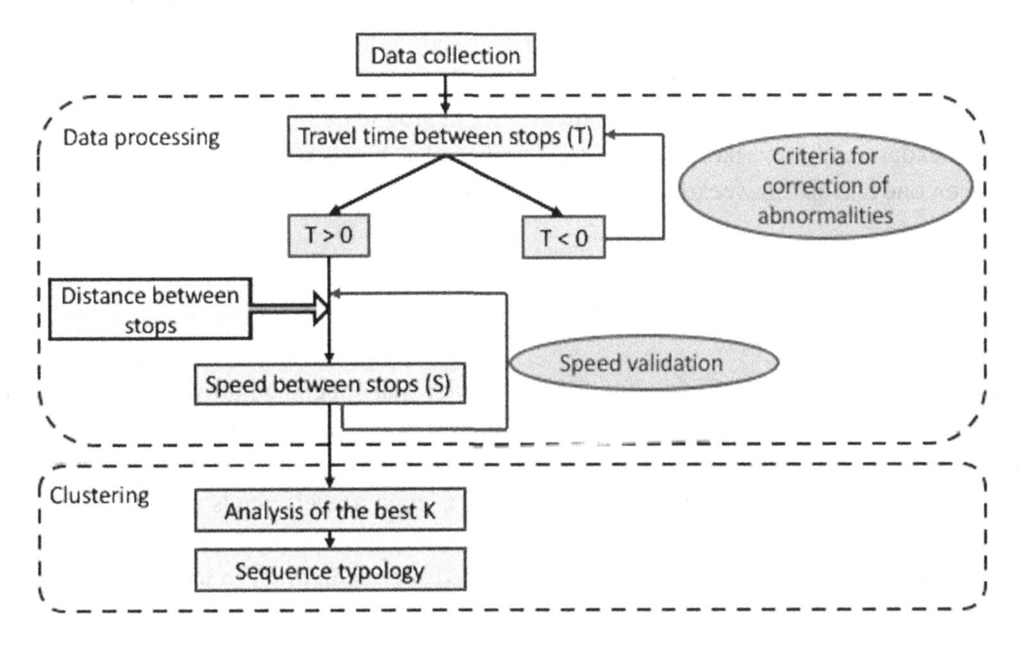

Based on the distance between stops and the deduced travel times between stops, it is possible to calculate the speed between those stops. These values are stored in vectors and after they will be used in clusters analysis. A detailed description of the main steps is described below.

Data Collection

The dataset used in this chapter is a sample of UPT´s travel data from the Metropolitan Area of Porto. The Smart Card Automated Fare Collection system of Porto, called the Andante system, was implemented in the metropolitan region in 2002 and emerged intending to promote the implementation of intermodality in the public transports of the city.

Metropolitan Area of Porto network covers an area of 1,575 km^2 and serves 1.75 million of inhabitants. It is composed of 126 bus lines (urban and regional), six metro lines, one cable line, three tram lines, and three train lines. This system is operated by 11 transport providers, from which Metro do Porto (metro system) and STCP (bus system) are the largest.

Porto network is based on an open and intermodal zonal system. The ticketing system uses a rechargeable intermodal smart card called Andante. There are two types of Andante transport tickets: Signature titles and Occasional titles. Signature titles have different groups of users where the charge depends, besides the journey length, also on the traveler age or economic conditions. While signatures cards can only be used by the cardholder, occasional titles can be used by different travelers (it has no personal information). Both cards are valid for a set of adjacent areas previously chosen by the passenger. Signature titles are valid for the charged month while occasional titles are valid within the limit ring acquired during a particular period, currently 1 hour for the minimum 2-zone ticket and longer as the number of valid zones increases. Thus, one journey may have one or more stages (validations), depending on the journey's period and the number of zones included in that journey.

For each traveler (i.e., for each Andante smart card), the information related to the boarding time (first boarding on the route), the line (or lines for each journey), and the stop (or stops for each journey) are available.

To analyze the average speed of the buses along their route based on smartcards data, this study uses data from validations of one Porto bus line and one direction during a month (January 2013), which corresponds to 99,073 validations (0.8% of the total number of validations).

The chosen line consists of 39 stops, starts the service at around 06:00 and ends at around 22:00, both during the weekdays and at weekends. Also, the number of vehicles per hour that carry out this route varies between one and six on weekdays, and between one and three on weekends.

Data Processing

Travel Time Between Stops (T)

Public transport is bound to predefined routes and schedules that depend on the time of day and the day of the week and are subject to frequent changes. Typical shortcomings in public transport travel time calculations are simplifying assumptions related to travel speeds along the route (Lei & Church, 2010). Due to the deficiency of detailed schedule information, average travel speeds are usually assigned to the entire route, ignoring differences between different parts of the route (Liu & Zhu, 2004). However, travel time between stops could help to adjust the provided service in order to improve it. To know the travel time between each pair of consecutive stops, several data sources may provide this information, either directly or indirectly, through GPS location data of bus vehicles, or by deduction from other data sources, respectively.

The recent development of data formats provided by Automated Vehicle Location (AVL), for example, opens up new opportunities, and electronic journey-planning services based on such data are now also provided for public transport users. Although the potential of these data formats and web-based services from the research point of view was anticipated over a decade ago (Martin et al., 2002), their use for research purposes has only recently begun (e.g. Eluru et al., 2012; Jäppinen et al., 2013; Lei and Church, 2010). Moreover, even though big public transport providers have quickly joined these data collection systems, small ones still do not have access to them. Thus, assuming that AVL systems are not accessible to all, this work focuses on the analysis of average speeds based on the ticketing data. Therefore, considering the data collected when validating tickets, it is possible to infer the timestamp from validations at two consecutive stops (x and $x+1$). Based on these timestamps, the travel time (T) between two stops (x and $x+1$) could be calculated as

$$T = t_{x+1}^{f} - t_{x}^{l}$$

where t_{x+1}^{f} is the timestamp of the first validation at stop $x+1$, and t_{x}^{l} is the timestamp of the last validation at stop x. Similarly, the time that a vehicle spends at each stop (dwell time (DT)) could be deduced. Thus,

$$DT = t_{x}^{l} - t_{x}^{f}$$

where t_{x}^{l} is the timestamp of the last validation at stop x, and t_{x}^{f} is the timestamp of the first validation at the same stop.

Consequently, the travel time of each trip (composed by N stops) could be written as

$$Travel\ time\ of\ each\ trip = \sum_{i=x}^{N-1} T_i + \sum_{i=x}^{N} DT_i$$

where T_i is the travel time between two consecutive stops and DT_i is the dwell time in each stop.

The deduction of travel times between pairs of bus stops should be positive. Otherwise, some criteria for the correction of abnormalities need to be used.

Speed Between Stops (S)

The speed (S) between two points is given by the division between the distance (D) traveled and the time (T) spent in the route, so that:

$$S = \frac{D}{T}$$

To calculate speed, it is necessary to know the distance traveled between each pair of stops. In this work, to obtain these distances, the Distance Matrix API from Google Maps Platform was used.

Distance Matrix API ("Google Maps Platform," 2019) provides the travel distance and the time for a matrix of origins and destinations. The API returns information based on the suggested route between two points, as calculated by the Google Maps API. It consists of a set of rows containing duration and/or distance values for each pair of points. For the calculation of distances, Distance Matrix API allows specifying the transportation mode to use, such as driving (by default), walking, bicycling, and transit. The definition of the mode is important due to the existence of exclusive bus lanes. Thus, the mode "transit" should be selected.

After the calculation of the speed of all pairs of stops, some data abnormalities should be verified, such as the maximum speed and the existence of NA (stops without ticket validation do not allow to infer the travel time between them, and the previous and next stops).

The speed data set between consecutive stops of a UPT line must then be stored in a vector. The set of vectors will be used in clustering analysis to find similarities along with the stops of each line or along the day.

Clustering

Analysis of the Best K

To perform clustering analysis, the number of clusters to be considered should be defined, since most algorithms need this value as an input. However, the correct choice of K is often ambiguous, with different interpretations. In this sense, the decision of the K value is frequently decided by the user.

Several methods can be used to determine what is the optimal number of clusters. Unfortunately, the answer is no definitive, because it depends on the method used for measuring similarities and the parameters used for partitioning. Some methods are, among others, the elbow method, the silhouette method, the gap statistic, and the clustering dendrogram (Kassambara, 2018).

The Elbow method focuses on the percentage of variance explained based on the number of clusters. The idea of this method is that one should choose several clusters, and adding another cluster does not give much better modeling. The percentage of explained variance by the clusters is plotted. The first clusters will add valuable information, but at some point, the marginal gain will drop dramatically and gives an angle in the graph (Purnima & Arvind, 2014), i.e., an elbow. Thus, the suggested K is that point.

The Silhouette method focuses on the quality of a clustering. The concept of silhouette width involves the difference between the within-cluster tightness and separation from the rest. Specifically, the silhouette width $s(i)$ for entity i is defined as (Rousseeuw, 1987):

$$s(i) = \frac{b(i) - a(i)}{\max\left(a(i), b(i)\right)}$$

where $a(i)$ is the average distance between i and all other entities of the cluster to which i belongs and $b(i)$ is the minimum of the average distances between i and all the entities in each other cluster.

Clustering can be characterized by the average silhouette width of individual entities. The largest average silhouette width, over different K, indicates the best number of clusters (Kodinariya & Makwana, 2013).

The gap statistic (Tibshirani, Walther, & Hastie, 2001) focuses on the comparison of the total within intra-cluster variation using different values of K. The K value that maximizes the gap statistic (i.e., that yields the most significant gap statistic) will be the optimal number of clusters.

Hierarchical clustering is usually used to understand the structure and relationships in the data better and based on them to decide the number of clusters that seems to be appropriate. A rule to decide the number of clusters is to look for the groups with the longest "branches." The shorter they are, the more similar they are.

The four described methods were used in the present study to define the best K. The more methods suggest the same number of clusters, the stronger the decision of K will be.

Sequence Typology

Sequences typology was carried out to highlight differences between days of the week, times of the day, and groups of stops, and thus to study the variability of the average speed according to sequences.

Euclidean distance between all sequences (or vectors) of average speeds was applied. The Euclidean distance between two points X and Y, with N dimensions, is calculated as

$$d(X,Y)=\sqrt{\sum_{i=1}^{N}(Y_i-X_i)^2}$$

The values of the Euclidean distance between all vectors of average speed were stored in a distance matrix, also called thc proximity matrix.

After that, a linkage criterion needs to be selected to determines the distance between sets of observations as a function of the pairwise distances between observations. Several methods are available such as single-linkage (or minimum), complete-linkage (or maximum), average-linkage (unweighted or weighted), centroid-linkage, and Ward's criterion. The result of the clustering can be visualized as a dendrogram, which shows the sequence of cluster fusion and the distance at which each fusion took place.

To visualize the behavior of sequences and corresponding clusters, heat maps in R was performed. Heat maps allow us to visualize clusters of samples and features simultaneously. They start with hierarchical clustering for the rows and the columns of the matrix. The columns or rows of the matrix are re-ordered according to the hierarchical clustering result, putting similar observations close to each other. The blocks of 'high' and 'low' values are adjacent to the data matrix. After, a color scheme is applied for the visualization, and the data matrix is displayed. Visualizing the data matrix in this way can help to find the variables that appear to be representative for each sample cluster.

RESULTS AND DISCUSSION

The results of the application of the proposed methodology are presented in four parts: the timestamps and speed validation rules are defined, the average speeds of buses are grouped by hour, then sequences of average speed between stops is studied in order to perform clusters, and finally, intragroup variability is studied within each group at a disaggregated level.

Figure 2. Average speed of Wednesdays of January 2013 between 9:00:00 and 10:00:00

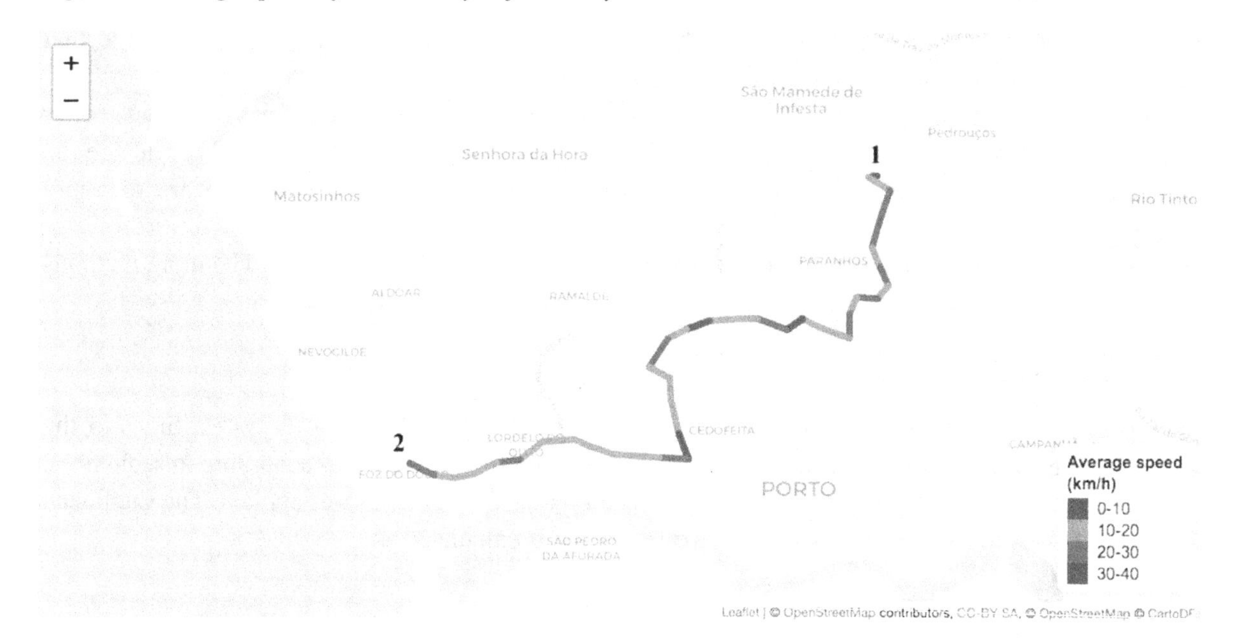

Timestamps and Speed Validation Rules

The inference of travel time between consecutive pairs of stops showed some anomalies since cases of negative times were obtained. This can occur because i) the bus vehicles have two ticket validation machines, each one with different times, or ii) the communication of the information with the server is slow and not sequential. In order to address these errors and to proceed with the analysis, one criterion was defined: it was considered the first validation of each stop and the last validation of the previous stop that had less time than the first validation of the next stop. This criterion allows us to remove cases of negative times between stops, and there is less risk of data loss.

After the inference of travel times, and the collection of distances, the calculation of speed is performed. Some incongruences are found, such as high-speed values that are impossible to achieve by bus and in the urban area. Thus, two validation rules were defined:

1) the assumed speed is at most 80 km/h. After the deduction of speed between consecutive pairs of stops, it was found that there were high speed-values, probably due to some errors in travel times. However, the maximum speed that the buses in the transport network that served as a case study in this chapter reach are 80 km/h, it was considered that no speed could exceed this value;
2) the missing average speed in a segment is equal to the maximum verified in the corresponding day and segment. After grouping the average speed by hours, the are some segments (pairs of consecutive stops) with NA values. This occurs because at least one stop of the segment does not have ticketing validations in the corresponding period. Thus, if the vehicle does not stop to allow boarding in one stop, it was assumed that it is possible to reach the maximum speed attained in that day and in the same segment.

Figure 3. Average speed of Fridays of January 2013 between 13:00:00 and 14:00:00

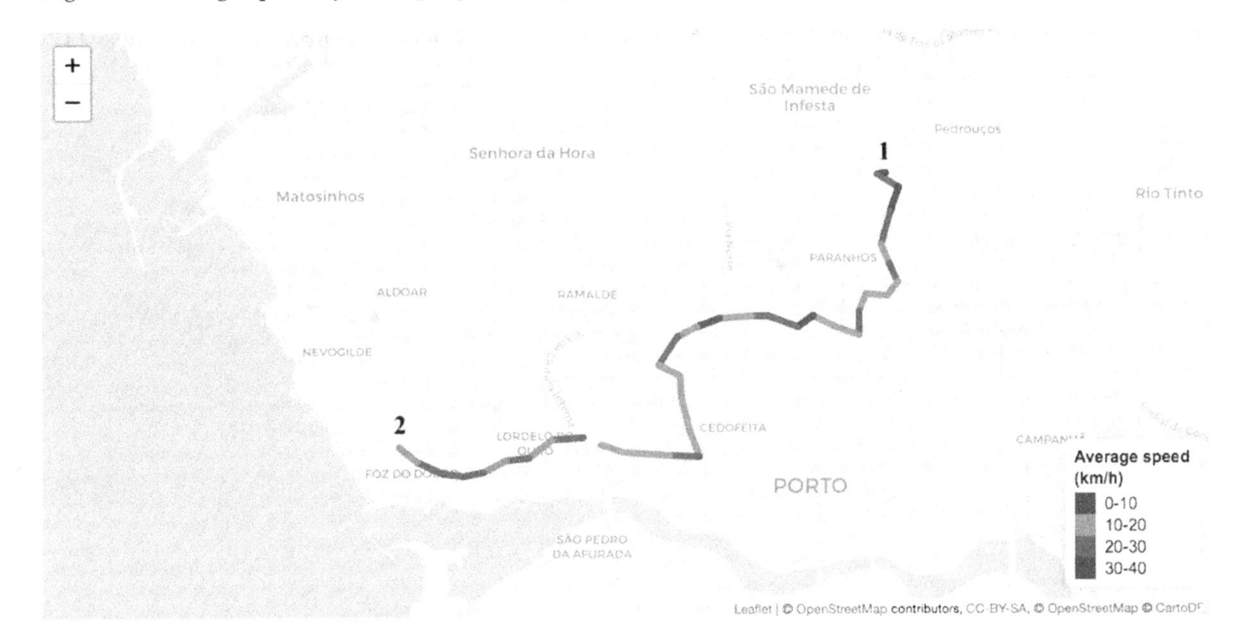

Average Speed

Depending on the day of the week and the time of the day, a different number of vehicles depart to perform the route selected in this study. Each vehicle has an average speed between two stops that may depend, among other factors, on the traffic in that route segment. The average speed is grouped by hours (rounded minutes and seconds to the hour). Thus, 8:00:00 includes all vehicles with departing from 7:30:00 to 8:29:59, and consequently, the average speed in 8:00:00 is the average speed of all vehicles grouped.

Figures above show two examples of the average speed along the route and direction in the study: Figure 2 is relative to all Wednesdays of the month that starts between 9:00:00 and 10:00:00, and Figure 3 relative to all Fridays of the month that starts between 13:00:00 and 14:00:00. Travels start in point 1 and finish in point 2.

The average speed at the beginning of this bus line mostly ranges between 0 and 10 km/h, which can be explained by the existence of traffic lights in the first segment. The same happens in other segments since this bus line circulates in a central area of the city where there are several traffic lights and a large number of vehicles that often cause congestion.

However, in some times of the day, vehicles can reach an average speed near to 40 km/h as it is possible to verify on Fridays between 13:00:00 and 14:00:00. This is the lunch period, and despite there is some traffic due to many people going to lunch at home or the restaurant, the early morning period is generally worse (as verified on Wednesdays between 9:00:00 and 10:00:00). On Wednesdays between 9:00:00 and 10:00:00, several segments have average speed lower than Fridays between 13:00:00 and 14:00:00. Nevertheless, some segments seem to be calm in terms of traffic, and consequently, the average speed is high in both periods.

Each day of the week and time of the day that has at least one vehicle of a bus departing for the studied line originates one vector with as many instances as the number of stops of the line. Thus, 112 vectors, each one with 35 instances, are generated.

Figure 4. Elbow method

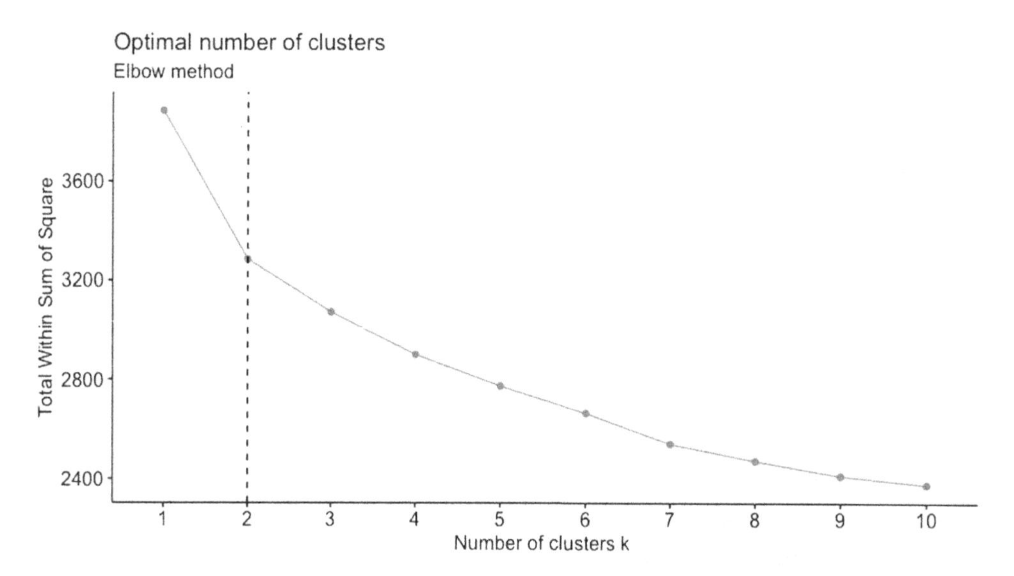

Figure 5. Silhouette method

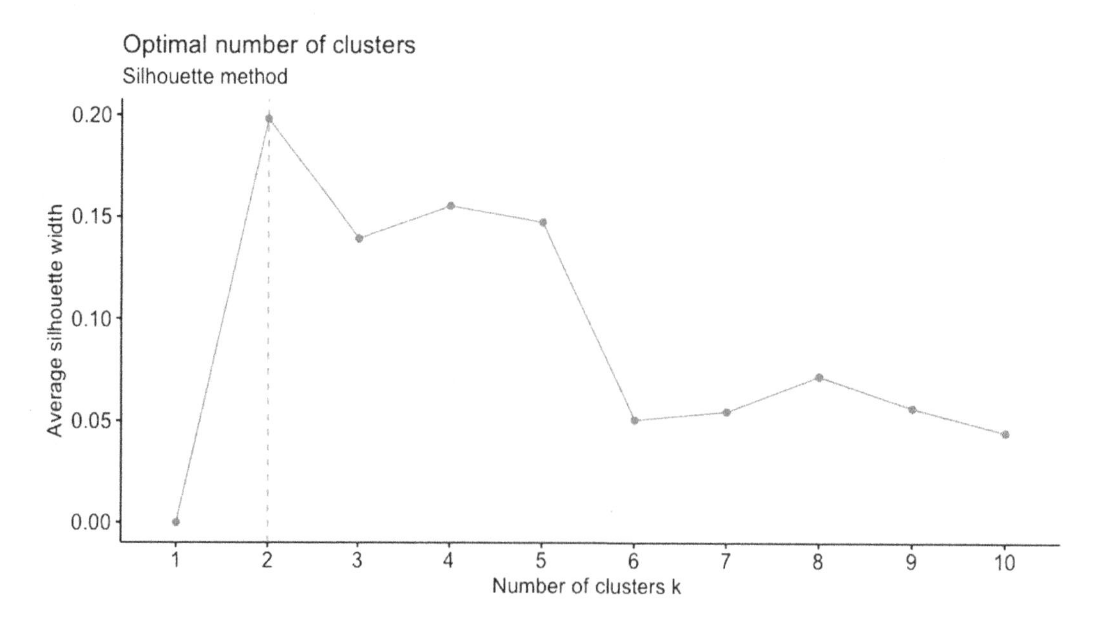

Figure 6. Gap statistic method

Optimal number of clusters
Gap statistic method

Clustering Analysis

The list of vectors of the previous subsection was analyzed to define groups of similar average speeds along the route. The analysis of the number of clusters is performed in this subsection.

Several methods were used to define the optimum number of clusters, as presented in Figure 4, Figure 5, Figure 6, and Figure 7.

The graph of the elbow method seems to have two slight elbows. However, the most pronounced is in k=2, which suggest considering two clusters. Similar to the elbow method, the silhouette method also suggests the use of 2 clusters. Regarding the gap statistic method, six clusters should be considered.

By analyzing the cluster dendrogram, a different number of clusters could be defined. In red is the division into two clusters, and in blue is the division into six clusters (suggested by previous methods). Both options assume weekend days in a separated cluster (without other days of the week). This means that it is evident that, regarding the buses' average speed, the weekend days are different from the other days of the week.

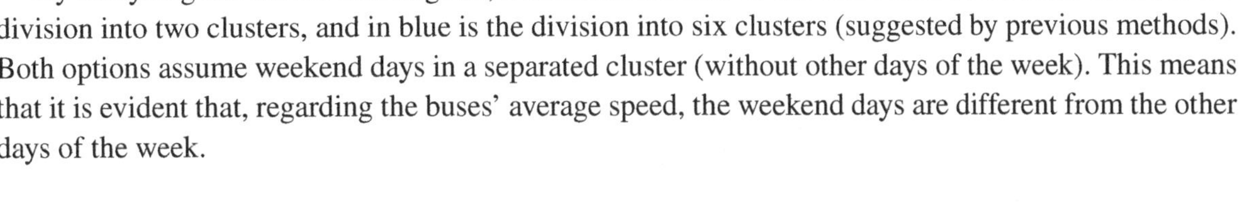

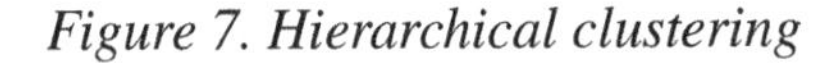

Figure 7. Hierarchical clustering

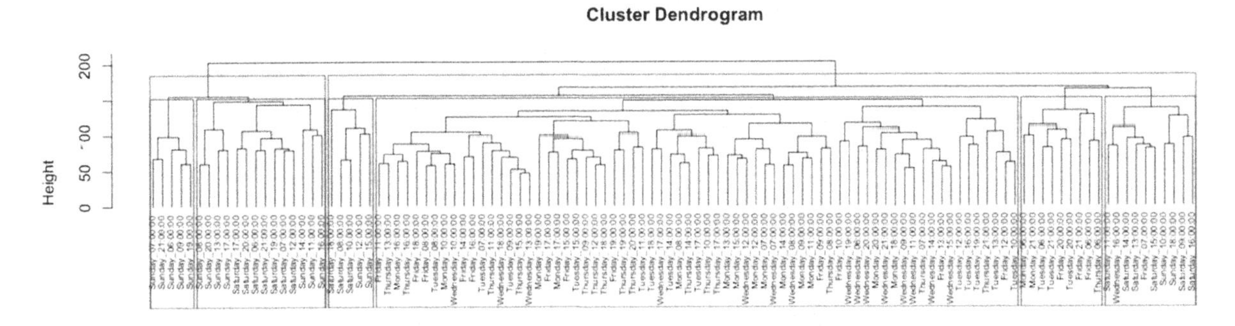

Figure 8. Heatmap of the average speed considering 2 clusters

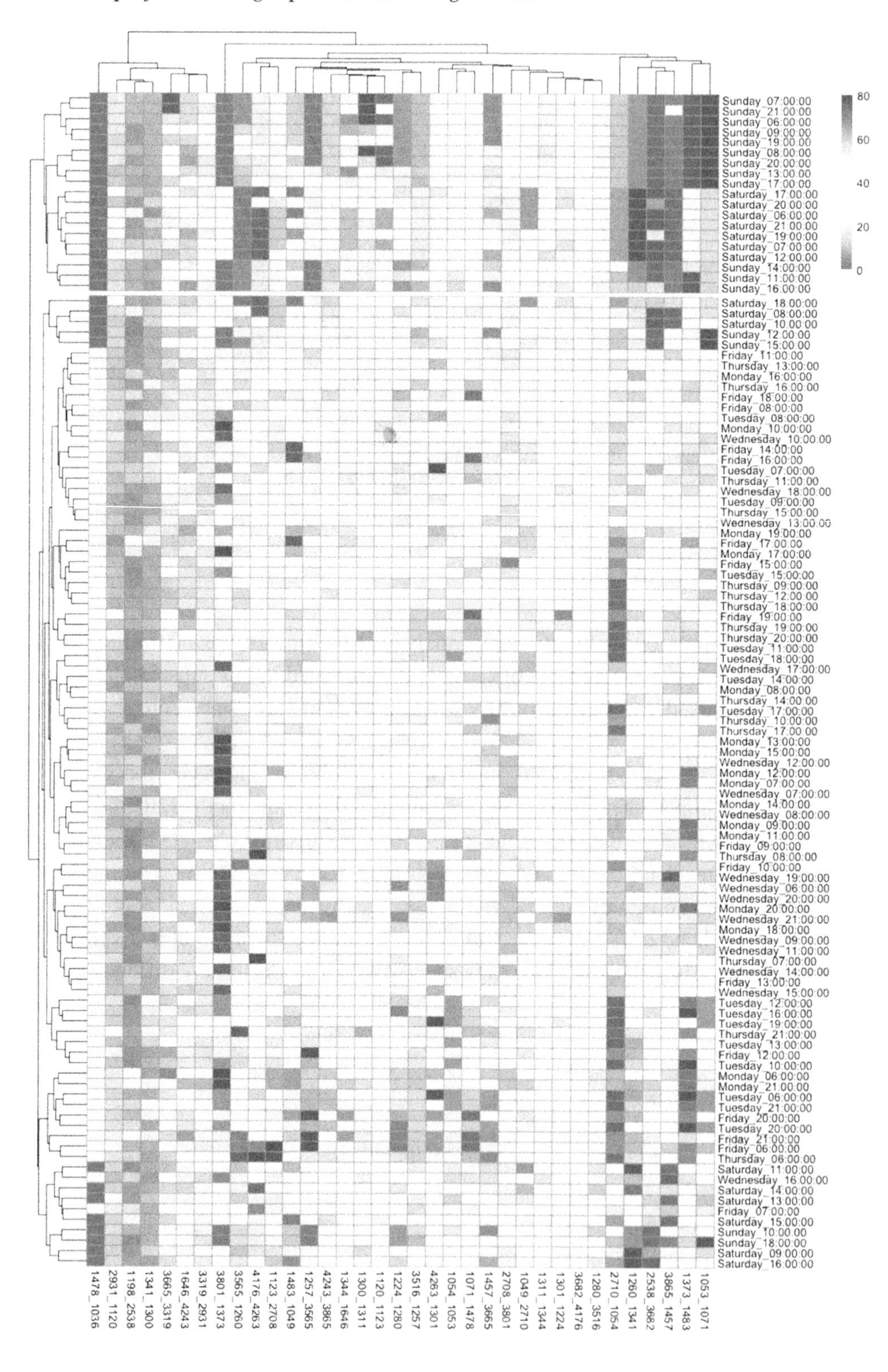

Based on previous results, a clustering analysis considering both two clusters and six clusters was analyzed in more detail.

Figure 8 and Figure 9 present the heatmaps of the hierarchical clustering considering two and six clusters, respectively.

Considering the existence of two clusters, two groups of segments (constituted by consecutive stops) are identified: the first includes segments 1478_1036, 2931_1120, 1198_2538, and 1341_1300 (on the left of the heatmap), and the second has stops 1053_1071, 1373_1483, 3865_1457, 2538_3682, 1260_1341, and 2710_1054 (on the right of the heatmap). The heatmap shows that the first cluster includes the average speed of trips performed only on weekend days. Additionally, the trips of the first cluster present slow speed for a group of trips (in dark blue) and another group with high speed (in red). The blue group of segments (first group) is characterized by the existence of traffic lights with long waiting times, or they are known as having more congestion. Furthermore, one of these segments is located at the beginning of the trip. In this place, some drivers allow the passengers to board while they prepare the trip. This causes tickets to be validated without the trip has begun. Although the trip preparation time may be short, 1 or 2 minutes can make the difference in trip times, as the stops are relatively close to each other.

Regarding the segments identified as where the vehicles reach higher average speed (second group), several reasons could explain that. For example, the zones of these segments are located approximately in the final of the trip, where there are many beaches, or they are near to schools. In January, the demand for beach zones is low, and on weekends the traffic around schools is low, which makes possible buses reaching high speeds. However, the reasons only could be confirmed by competent entities, such as UPT providers or municipalities.

Thc second cluster includes all days of the week and times of the day with more variability of patterns. By considering six clusters, it is easier to analyze this variability.

The heatmap considering six clusters shows the composition of each cluster. The first cluster is characterized by low-speed values for a group of segments and the highest for another one. The second cluster is similar to the first, but the high-speed values are lower than those observed on the first cluster. These two clusters are considered as a single one in the previous situation (with only two clusters). The reason for these values is, probably, similar to the explained above (high or low demand in zones like beach, or schools).

The remaining clusters were also grouped as a single one when two clusters were considered. However, it is possible to find some patterns, and it may make sense to considerer six clusters. The third cluster also presents low speed for the first group of segments, but for the second group, they reach lower speed than the previous clusters. This is verified mainly on Saturdays and Sundays.

Cluster 4 presents medium speed for the first group of segments, with a high variability of speed values. This cluster includes the majority of days of the week and times of the day with more congestion, and thus, in general, the average speed is low.

The fifth cluster also has low-speed in the first group of segments (similar to cluster 4), but high-speed in the second. This pattern occurs on weekdays in the first and last hours of the day. The explanation for this occurrence is similar to that of cluster 1.

Finally, the last cluster (cluster 6) has low speed in the first group of segments and high speed in the second group. This behavior is verified for some days of the week, with more incidence on Saturdays. They have in common the time of the day they occur: approximately the middle of the day.

Figure 9. Heatmap of the average speed considering 6 clusters

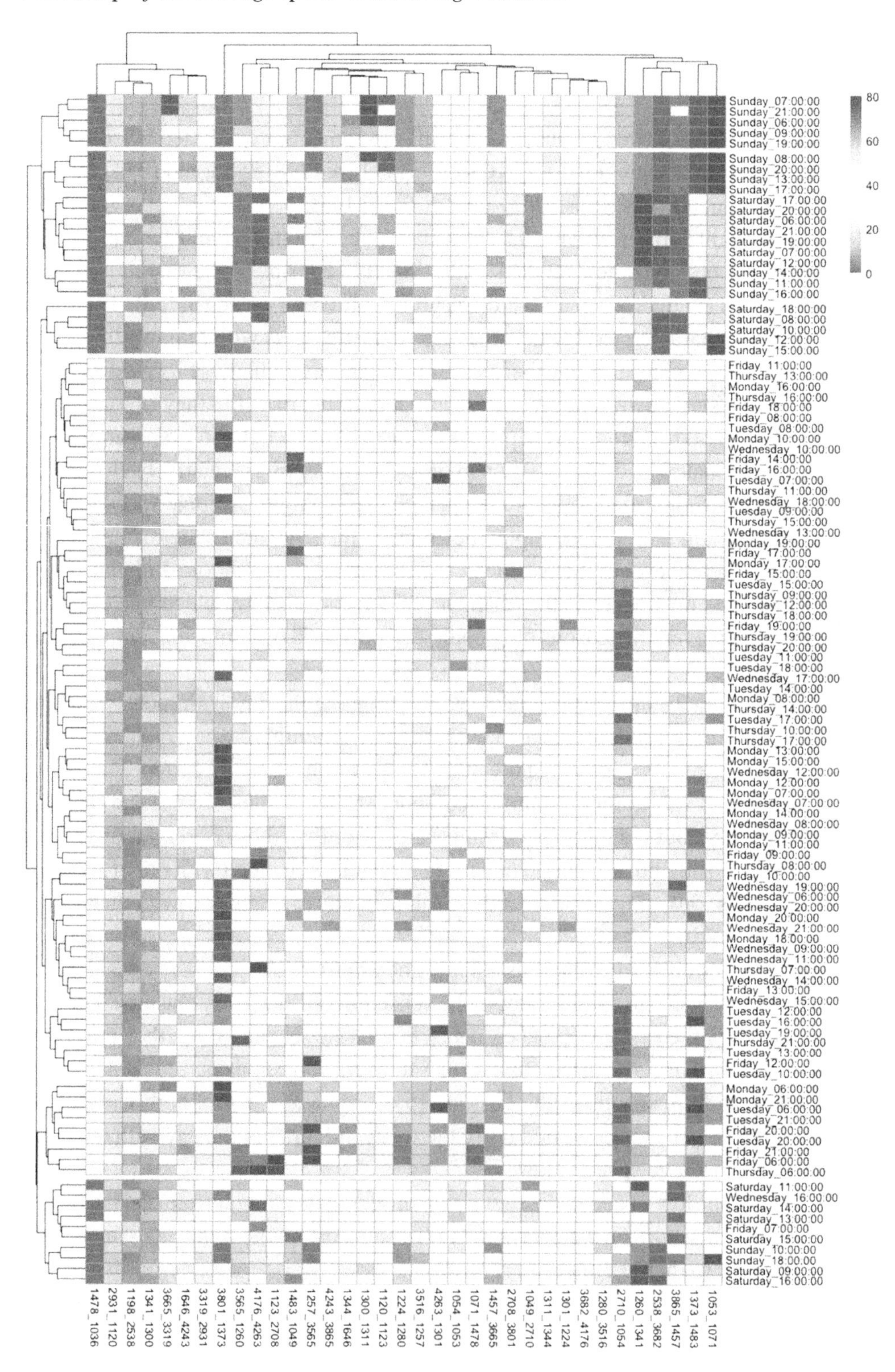

Identifying clusters based on days of the week and times of day will allow public transport providers grouping and handle the schedules that have similar patterns in days or in times of the day. In this way, the schedules adjustment or the creation of bus corridors, for example, can be done within each cluster.

Issues, Controversies, Problems

The study presented in this chapter was developed since some gaps have been found in the literature review. The main studies focus on determining travel time between specific points, others only use data from AVL systems, and others collect data manually and sporadically for a specific analysis. However, UPT providers continuously collect data from their vehicles or users. To use this data to extract knowledge is a way to have access to indirect information easily. In this context, the authors consider the presented study of great use since it focuses on validation data to deduce patterns of the traffic, for example, based on the deduced average speed. The main disadvantage found in this work is the fact of the existence of errors on data and, consequently, the average speed deduced also could present, which make results with less confidence.

SOLUTIONS AND RECOMMENDATIONS

The analysis performed in this study could be improved by using AVL systems. However, not all bus vehicles are equipped with this type of system. Mainly, companies of urban buses of small cities do not have sufficient revenue or demand to justify investment in advanced data collection systems. Thus, the proposed methodology allows this type of company to analyze their average speed along their routes.

Analyzing the average speed of buses, UPT providers could identify segments of the route with frequent low average speed. To increase average speed and decrease the travel time, they could consider the hypothesis of modifying one or more segments of the route by a better segment, changing the route, or the bus stop. However, this option is sensible and needs other analysis and simulation of the change and its impact both for UPT providers and users.

FUTURE RESEARCH DIRECTIONS

Performing an equivalent analysis using AVL systems will allow a detailed analysis of the behavior of bus vehicles in terms of average speeds. Furthermore, this analysis would allow public transport providers to verify that the service they offer on their defined routes is being carried out in the best conditions, both for the provider and for the user.

CONCLUSION

The research described in this paper confirms that travel ticketing data can be successfully used for deducing travel times between pairs of consecutive stops (segments) since it collects detailed and accurate data at no additional cost for the whole system. Much of this information is not yet sufficiently used to extract knowledge and, therefore, it can still be worked.

Although some data incongruences were found in this research analysis, the methodology could be used in other public transport companies. The results show that weekend days are different from the rest of the week since clustering analysis always includes these days in a particular cluster. Segments of road between stops typically with more congestion are also identified, and thus, they may be part of possible improvements provided by urban public transport providers. Some examples of improvements are the adjustment of UPT schedules or modifying one or more segments of the route regularly congested by a better segment, changing the route, or the bus stop.

ACKNOWLEDGMENT

This work was supported by Fundação para a Ciência e Tecnologia (FCT) through a doctoral scholarship (Ref. PD/BD/128065/2016).

REFERENCES

Ali, A., Kim, J., & Lee, S. (2016). Travel behavior analysis using smart card data. *KSCE Journal of Civil Engineering*, *20*(4), 1532–1539. doi:10.100712205-015-1694-0

Bagchi, M., & White, P. R. (2005). The potential of public transport smart card data. *Transport Policy*, *12*(5), 464–474. doi:10.1016/j.tranpol.2005.06.008

Costa, P. M., Fontes, T., Nunes, A. A., Ferreira, M. C., Costa, V., Dias, T. G., ... Cunha, J. (2016). Application of Collaborative Information Exchange in Urban Public Transport: The Seamless Mobility Solution. *Transportation Research Procedia*, *14*, 1201–1210. doi:10.1016/j.trpro.2016.05.191

Costa, V., Fontes, T., Borges, J. L., & Dias, T. G. (2019). Prediction of Journey Destination for Travelers of Urban Public Transport: A Comparison Model Study. *Lecture Notes of the Institute for Computer Sciences, Social-Informatics, and Telecommunications Engineering, LNICST*, *267*(ii), 113–132. doi:10.1007/978-3-030-14757-0_9

Costa, V., Fontes, T., Costa, P. M., & Dias, T. G. (2015a). How to Predict Journey Destination for Supporting Contextual Intelligent Information Services? In *Proceedings IEEE Conference on Intelligent Transportation Systems, ITSC*, 2959–2964. 10.1109/ITSC.2015.474

Costa, V., Fontes, T., Costa, P. M., & Dias, T. G. (2015b). Prediction of Journey Destination in Urban Public Transport. In F. Pereira, P. Machado, E. Costa, & A. Cardoso (Eds.), *Progress in Artificial Intelligence. EPIA 2015* (pp. 169–180). Lecture Notes in Computer Science, doi:10.1007/978-3-319-23485-4_18

El Mahrsi, M. K., Come, E., Oukhellou, L., & Verleysen, M. (2017). Clustering Smart Card Data for Urban Mobility Analysis. *IEEE Transactions on Intelligent Transportation Systems*, *18*(3), 712–728. doi:10.1109/TITS.2016.2600515

Elms, C. P. (1998). Defining and measuring service availability for complex transportation networks. *Journal of Advanced Transportation*, *32*(1), 75–88. doi:10.1002/atr.5670320108

Fancello, G., Carta, M., & Fadda, P. (2014). A Modeling Tool for Measuring the Performance of Urban Road Networks. *Procedia: Social and Behavioral Sciences*, *111*, 559–566. doi:10.1016/j.sbspro.2014.01.089

Ferreira, M. C., Costa, V., Dias, T. G., Falcão, E., & Cunha, J. (2017). Understanding commercial synergies between public transport and services located around public transport stations. *Transportation Research Procedia*, *27*, 125–132. doi:10.1016/j.trpro.2017.12.052

Fu, L., & Yang, X. (2007). Design and Implementation of Bus–Holding Control Strategies with Real-Time Information. *Transportation Research Record: Journal of the Transportation Research Board*, *1791*(1), 6–12. doi:10.3141/1791-02

Gan, Z., Feng, T., Wu, Y., Yang, M., Timmermans, H., Key, J., ... Technologies, T. (2019). Station-based average travel distance and its relationship with urban form and land use : An analysis of smart card data in Nanjing City, China. *Transport Policy*, *79*(May), 137–154. doi:10.1016/j.tranpol.2019.05.003

Giannopoulos, G. A. (2004). The application of information and communication technologies in transport, *152*, 302–320. doi:10.1016/S0377-2217(03)00026-2

Google Maps Platform. (2019). Retrieved from https://cloud.google.com/maps-platform/

Gu, S., & Mark, L. (2014). Trip purpose inference using automated fare collection data. *Public Transport (Berlin)*, *6*(1-2), 1–20. doi:10.100712469-013-0077-5

Guessous, Y., Aron, M., Bhouri, N., & Cohen, S. (2014). Estimating travel time distribution under different traffic conditions. *Transportation Research Procedia*, *3*(July), 339–348. doi:10.1016/j.trpro.2014.10.014

Jánošíkova, L., Slavík, J., & Koháni, M. (2014). Estimation of a route choice model for urban public transport using smart card data. *Transportation Planning and Technology*, *37*(7), 638–648. doi:10.1080/03081060.2014.935570

Jasti, P. C., & Ram, V. V. (2016). Integrated and Sustainable Service Level Benchmarking of Urban Bus System. *Transportation Research Procedia*, *17*, 301–310. doi:10.1016/j.trpro.2016.11.096

Karim, Z., & Fouad, J. (2018). Measuring urban public transport performance on route level: A literature review. *MATEC Web of Conferences, 200*. 10.1051/matecconf/201820000021

Kassambara, A. (2018). Cluster validation essentials. Retrieved from https://www.datanovia.com/en/lessons/determining-the-optimal-number-of-clusters-3-must-know-methods/

Kodinariya, T. M., & Makwana, P. R. (2013). Review on determining number of Cluster in K-Means Clustering. *International Journal of Advance Research in Computer Science and Management Studies*, *1*(6), 2321–7782.

Kusakabe, T., & Asakura, Y. (2014). Behavioural data mining of transit smart card data: A data fusion approach. *Transportation Research Part C, Emerging Technologies*, *46*, 179–191. doi:10.1016/j.trc.2014.05.012

Kusakabe, T., Iryo, T., & Asakura, Y. (2010). Estimation method for railway passengers' train choice behavior with smart card transaction data. *Transportation*, *37*(5), 731–749. doi:10.100711116-010-9290-0

Lei, T. L., & Church, R. L. (2010). Mapping transit-based access: Integrating GIS, routes and schedules. *International Journal of Geographical Information Science*, *24*(2), 283–304. doi:10.1080/13658810902835404

Liu, S., & Zhu, X. (2004). Accessibility Analyst: An integrated GIS tool for accessibility analysis in urban transportation planning. *Environment and Planning. B, Planning & Design*, *31*(1), 105–124. doi:10.1068/b305

Ma, X., Wu, Y. J., Wang, Y., Chen, F., & Liu, J. (2013). Mining smart card data for transit riders' travel patterns. *Transportation Research Part C, Emerging Technologies*, *36*, 1–12. doi:10.1016/j.trc.2013.07.010

Munizaga, M. A., & Palma, C. (2012). Estimation of a disaggregate multimodal public transport Origin-Destination matrix from passive smartcard data from Santiago, Chile. *Transportation Research Part C, Emerging Technologies*, *24*, 9–18. doi:10.1016/j.trc.2012.01.007

Nunes, A. A., Dias, T. G., Falcão, E., & Cunha, J. (2016). Passenger journey destination estimation from automated fare collection system data using spatial validation. *IEEE Transactions on Intelligent Transportation Systems*, *17*(1), 133–142. doi:10.1109/TITS.2015.2464335

Orth, H., Carrasco, N., Schwertner, M., & Weidmann, U. (2014). Calibration of a Public Transport Performance Measurement System for Switzerland. *Transportation Research Record: Journal of the Transportation Research Board*, *2351*(1), 104–114. doi:10.3141/2351-12

Oskarbski, J., Birr, K., Miszewski, M., & Zarski, K. (2015). Estimating the average speed of public transport vehicles based on traffic control system data. In *Proceedings 2015 International Conference on Models and Technologies for Intelligent Transportation Systems, MT-ITS 2015*, (June), 287–293. 10.1109/MTITS.2015.7223269

Parks, J., Ryus, P., Coffel, K., Gan, A., Perk, V., Cherrington, L., … Nakanishi, Y. (2010). *A Methodology for Performance Measurement and Peer Comparison in the Public Transportation Industry. A Methodology for Performance Measurement and Peer Comparison in the Public Transportation Industry*. doi:10.17226/14402

Pelletier, M., Trépanier, M., & Morency, C. (2011). Smart card data use in public transit : A literature review. *Transportation Research Part C, Emerging Technologies*, *19*(4), 557–568. doi:10.1016/j.trc.2010.12.003

Polus, A., & Shefer, D. (1984). Evaluation of a public transportation level of service concept. *Journal of Advanced Transportation*, *18*(2), 135–144. doi:10.1002/atr.5670180204

Purnima, B., & Arvind, K. (2014). EBK-Means: A Clustering Technique based on Elbow Method and K-Means in WSN. *International Journal of Computers and Applications*, *105*(9), 17–24. Retrieved from https://www.ijcaonline.org/archives/volume105/number9/18405-9674

Putra, A. A. (2013). Transportation System Performance Analysis Urban Area Public Transport. *International Refereed Journal of Engineering and Science (IRJES)*, *2*(6), 01–15. Retrieved from www.irjes.com

Rousseeuw, P. J. (1987). Silhouettes: A graphical aid to the interpretation and validation of cluster analysis. *Journal of Computational and Applied Mathematics*, *20*, 53–65. doi:10.1016/0377-0427(87)90125-7

Salonen, M., & Toivonen, T. (2013). Modelling travel time in urban networks: Comparable measures for private car and public transport. *Journal of Transport Geography*, *31*, 143–153. doi:10.1016/j.jtrangeo.2013.06.011

Swathipriya, P. (2019). *Performance Evaluation of Intermediate Public Transport by Benchmarking and Numerical Rating Approach*, *6*(4), 1–11.

Tibshirani, R., Walther, G., & Hastie, T. (2001). Estimating the number of clusters in a data set via the gap statistic. *Journal of the Royal Statistical Society. Series B, Statistical Methodology*, *63*(2), 411–423. doi:10.1111/1467-9868.00293

Turnquist, M. A., & Blume, S. (1980). Evaluating Potential Effectiveness of Headway Control Strategies for Transit Systems. *Transportation Research Record: Journal of the Transportation Research Board*, (746): 25–29.

Vasantha Kumar, S., & Vanajakshi, L. (2013). Modewise Travel Time Estimation on Urban Arterial Travel Time Estimation Using Buses as Probes. In *Proceedings of the 92nd Annual Meeting of the Transportation Research Board*, Washington, D.C. 10.100713369-014-1332-z

Wardman, M. (2004). Public transport values of time. *Transport Policy*, *11*(4), 363–377. doi:10.1016/j.tranpol.2004.05.001

Weng, X., Liu, Y., Song, H., Yao, S., & Zhang, P. (2018). Mining urban passengers' travel patterns from incomplete data with use cases. *Computer Networks*, *134*, 116–126. doi:10.1016/j.comnet.2018.01.048

Wilson, N. (2013). Making Use of Automated Data Collection to Improve Transit Effectiveness. Retrieved from https://pt.slideshare.net/BRTCoE/making-use-of-automated-data-collection-to-improve-transit-effectiveness

Zhou, J., Murphy, E., & Long, Y. (2014). Commuting efficiency in the Beijing metropolitan area: An exploration combining smartcard and travel survey data. *Journal of Transport Geography*, *41*, 175–183. doi:10.1016/j.jtrangeo.2014.09.006

ADDITIONAL READING

Ferreira, M. C., Fontes, T., Costa, V., Dias, T. G., Borges, J. L., & Cunha, E. (2017). Evaluation of an integrated mobile payment, route planner and social network solution for public transport. *Transportation Research Procedia*, *24*, 189–196. doi:10.1016/j.trpro.2017.05.107

Fontes, T., Costa, V., Ferreira, M. C., Shengxiao, L., Zhao, P., & Dias, T. G. (2017). Mobile payments adoption in public transport. *Transportation Research Procedia*, *24*, 410–417. doi:10.1016/j.trpro.2017.05.093

Hora, J., Dias, T. G., & Camanho, A. (2016). Improving the Service Level of Bus Transportation Systems: Evaluation and Optimization of Bus Schedules' Robustness. In Lecture Notes in Business Information Processing (Vol. 247, pp. 604–618). doi:10.1007/978-3-319-32689-4

Hora, J., Dias, T. G., Camanho, A., & Sobral, T. (2017). Estimation of Origin-Destination matrices under Automatic Fare Collection: The case study of Porto transportation system. *Transportation Research Procedia*, *27*, 664–671. doi:10.1016/j.trpro.2017.12.103

Nunes, A. A., Galvão Dias, T., Zegras, C., & Falcão e Cunha, J. (2016). Temporary user-centred networks for transport systems. *Transportation Research Part C, Emerging Technologies*, *62*, 55–69. doi:10.1016/j.trc.2015.11.006

KEY TERMS AND DEFINITIONS

Automatic Data Collection Systems: Systems to automatically identify objects, collect data about them and enter them directly into computers, without human involvement. Usually, these systems use equipment such as barcode readers or magnetic-stripe readers, or technologies (for example, optical character recognition (OCR), radio frequency identification (RFID), voice recognition and smart cards).

Information and Communication Technologies: All the technology used to handle telecommunications, broadcast media, intelligent building management systems, audiovisual processing and transmission systems, and network-based control and monitoring functions.

Journey: The act of going from one place to another, and could include several modes of transport. For example, to go from A to B, the user may need to walk, take the bus, take the metro and finally walk a distance.

Route: A specific itinerary constituted by a number of bus stops.

Trip: The act of going from one place to another (often for a short period of time) and returning, i.e, it is a short journey abroad.

Chapter 8
An In-Depth Study of Mobile Ticketing Services in Urban Passenger Transport:
State of the Art and Future Perspectives

Marta Campos Ferreira
Faculty of Engineering, University of Porto, Portugal

Teresa Galvão Dias
https://orcid.org/0000-0001-6209-3626
Faculty of Engineering, University of Porto, Portugal

João Falcão e Cunha
Faculty of Engineering, University of Porto, Portugal

ABSTRACT

This chapter presents an in-depth study of the current situation of mobile ticketing services in the context of urban passenger transport, and points out future trends and directions that will define forthcoming versions of mobile ticketing services. It defines mobile ticketing services and presents the technologies most used to deliver these solutions. This is complemented by a survey of research studies and experiences of deployments in a real environment. The mobile ticketing ecosystem is then deeply explored, where key players are identified as well as their key drivers and concerns regarding mobile ticketing services. Finally, future trends and research opportunities regarding mobile ticketing solutions are presented.

INTRODUCTION

Urban mobility is one of the greatest challenges that cities currently face. Road networks are suffering from recurrent congestion, the accessibility of city centres is deteriorating, and the negative environmental impact caused by emissions of greenhouse gases from vehicles is taking problematic contours. Improv-

DOI: 10.4018/978-1-7998-2112-0.ch008

ing the mobility of dwellers using means that are sustainable, safe and high-quality is essential in order to reduce congestion in urban and metropolitan areas. Reducing the problems of urban congestion and stress will benefit businesses and citizens in the form of lower costs, time savings and improved accessibility. However, to promote a mind shift is not an easy task. Complex transport networks and lack of seamless options have proven to be barriers to the adoption of urban transport services (Puhe, Edelmann, & Reichenbach, 2014). Urban passenger transport is more attractive as it becomes easier to use.

The general adoption of mobile devices and its increasing functionality is changing the way people use them. Currently, it is already possible to make payments with mobile devices in several countries. This general trend is being applied to several sectors, including urban passenger transport. Modern mobile ticketing service solutions can free customers from difficult purchase decisions, allowing easier access to services.

The pervasive deployment and adoption of mobile ticketing solutions requires action from complex ecosystem of organizations (e.g. passengers, transport operators, transport authorities, banks, mobile network operators, third parties and others) to create a mobile ticketing service solution. Each entity desires to have the leading place in the mobile ticketing ecosystem and dominant role in the value chain. Mobile ticketing entail a complex, system-interdependent ecosystem of players whose success depends on the joint action of all players simultaneously (Ezell, 2009). However, the struggle for these inter-dependent organizations to form coalitions just hindered the emergence of successful mobile ticketing platforms.

This chapter presents an in-depth study of the current situation of mobile ticketing services in the context of urban passenger transport and points out future trends and directions that will define forthcoming versions of mobile ticketing services. It starts by defining mobile ticketing services and presenting the technologies most used to deliver these solutions. This is complemented by a survey of research studies and experiences of deployments in a real environment. The mobile ticketing ecosystem is then deeply explored, where key players are identified, as well as their key motivations and concerns regarding mobile ticketing services. Finally, along the chapter future trends and research opportunities regarding mobile ticketing solutions are pointed out.

Several mobile ticketing definitions can be found in the literature. Some include the whole payment actions (Au & Kauffman, 2008), while others relax this requirement and consider it as the use of a mobile device to hold a virtual ticket (Puhe et al., 2014). In this chapter mobile ticketing is considered as the use of a mobile device to purchase and/or validate a travel ticket or to initiate a journey. This definition includes the use of a mobile device to previously purchase and/or validate a travel ticket or to start a trip through a declared check-in or through a be-in/be-out scheme, whose information will then allow to calculate the price to be paid for the journey.

As mobile ticketing take place through mobile unwired devices, data communication assumes a fundamental role, mainly between the user's mobile device and the service provider. There are several technologies available in mobile devices that can be used to implement mobile ticketing solutions. Recent mobile ticketing trends show that current approaches commonly use more than one technology, by using each one to address the shortcomings of the other. The main technologies used in mobile ticketing solutions are described in this chapter and a comparison between them at various levels is presented, including compatibility with check-in/check-out or be-in/be-out ticketing schemes, their main advantages and disadvantages.

When compared with traditional ticketing systems, mobile ticketing has several advantages. They allow ubiquitous and remote access to payments, to avoid queues and to replace banknotes and coins. It reduces service providers' costs, by decreasing the need for ticket sellers and collectors, handling physical money, and producing, storing and distributing tickets. Service providers can also achieve operational and productivity gains by increasing performance on closed gate systems and improving bus boarding times.

Despite its clear advantages, several factors are hindering the expansion of mobile ticketing services worldwide, from which the main factor is related to the complexity of the mobile ticketing ecosystem and the difficulty in solving conflicting interests. Therefore, the first necessary step for a successful mobile ticketing implementation is to develop inter-organizational relationships. The main actors involved in the mobile ticketing ecosystem are identified in this chapter, as well as their motivations and concerns. This allows to determine their role, organize collaboration between them and to capture the value each stakeholder can extract from the participation in a mobile ticketing ecosystem.

Finally, along the chapter several research opportunities and future trends regarding forthcoming mobile ticketing solutions are identified. It is clear that service providers want to deliver better services and enhance customer experience and customers are open to the new possibilities offered by mobile ticketing and are willing to experiment and test. This makes expectation management crucial in order to take advantage of everyone's enthusiasm and to provide foundations for a sustained use and subsequent exploration of additional services.

This chapter is organized as follows. The next section describes the main technologies used in mobile ticketing solutions and presents a comparison between them at various levels. Then, it identifies the main actors involved in the mobile ticketing ecosystem, their motivations and key concerns. Finally, it presents the main conclusions and future perspectives.

MOBILE TICKETING TECHNOLOGIES

Mobile ticketing often involves more than one type of technology. As mobile ticketing take place through mobile unwired devices, data communication assumes a fundamental role, mainly between the user's mobile device and the service provider. This is especially important for proximity mobile payments, where each technology may pose important constraints on the payment execution.

Technologies for mobile ticketing have paved a long road, some of them are now basically defunct in terms of practical use, such as the case of the Infrared (IrDA), while others have appeared recently and their presence is not yet universal such as the case of the Near Field Communication (NFC). From the existing mobile technologies, the most common that can be used to power mobile ticketing solutions are SMS and phone calls, Wi-Fi and 3G/4G, NFC, QR Codes and BLE, which are described below.

Short Message Service and Phone Calls

Early mobile data technologies, despite their low data throughput, were the first to support simple mobile tickets purchases, by sending SMS or making phone calls. The Ring&Ride mobile ticketing project, supported by the German Federal Ministry of education and Research, is based on the check-in/check-out concept, i.e. the customer has to take an action not only at the beginning, but also at the end of the trip (Lüke, Mügge, Eisemann, & Telschow, 2009). When starting, he dials a toll-free phone number and receives a ticket (SMS) that is valid for both long and short distance travel. At the end of the trip, the passenger dials the number again to signal that the trip has been finished. The customer's location is determined at the starting point and at the destination, but also during the trip at defined time intervals.

The ticketing system uses different localization information, such as GSM cell IDs, WLAN SSIDs and GPS coordinates for reconstructing the route which is taken and calculates the corresponding fare.

Paybox in Austria allows customers from the Austrian railway OBB to purchase travel tickets via SMS, allowing them to pay for the tickets through their monthly phone bills. Proximus SMS-Pay in Belgium, Mobipay in Spain and AvantixMetro in UK are also examples of mobile ticketing systems based on SMS that have been implemented. Sarma (2014) proposes a system that allows passengers to book travel tickets using USSD messages and to receive the ticket details through SMS. Another example of a mobile ticketing system is given by the Czech bus reservation system AMSBUS, which, in February 2007, introduced the product "e-jízdenka" on several bus routes. Since November 2007 it has been possible to buy an SMS-ticket (SMS jízdenka) for use on Prague's urban transportation (Ghiron, Sposato, Medaglia, & Moroni, 2009).

Although SMS can be considered an easy-to-use and simple technology, it has limitations when used to make payments. The technology used by SMS is based on store and forward, without any encryption method or proof of delivery associated (Boer & Boer, 2009). Also, SMS formats are associated with complex codes and premium service numbers that are difficult to remember and slow to type (Mallat, 2007).

Because remote payments do not require a new infrastructure (unlike contactless payments) and enable payment transactions regardless of distance, this service is growing in developing countries as the best or the only way of performing domestic or even international money transfer, in areas where there is little, if any, bank infrastructure. In developed countries recent developments such as 3G and 4G are the main data connection for mobile devices (other than WIFI), powering all kinds of digital services, including mobile payments.

Wi-Fi and 3G/4G

Wi-Fi is a wireless technology that allows devices to connect to a wireless local area network. Wi-Fi works in the 2.4 or 5 GHz spectrum providing up to 150m of communication range and can be used as a cheap way to provide internet access to mobile devices (hotspot) being an alternative to more expensive data connections provided by mobile network operators through 3G, 4G or 5G. Wi-Fi and mobile network operators can act in mobile payments as a way of providing data connection to get cloud data, offers and coupons and also as a data gateway to proximity and remote payments.

Ferreira, Nóvoa, and Dias (2013) propose a mobile ticketing solution for public transport using passengers' smartphones with Internet connection. Passengers purchase and validate travel tickets through wireless communication technologies and GPS. This solution was tested in the city of Porto, by customers of the main bus operator – *Sociedade de Transportes Colectivos do Porto* (STCP) – during their normal use of public transport services. The Indian Railway Catering and Tourism Corporation Limited (IRCTC) developed a mobile application through which registered passengers can book railway tickets and confirm the payment using the 3G/4G mobile data from their phone network providers to access the internet on the move and ubiquitously (Kapoor, Dwivedi, & Williams, 2015).

Near Field Communication

Near Field Communication (NFC) is a short-range, high frequency, standard-based wireless communication technology that enables exchange of data between devices in close proximity. This requires at least one device to transmit a signal and another to receive it. NFC can be RFID compatible and can be implemented in mobile payment systems, as it needs very low power consumption, it offers great security mechanisms and it is easy to use and to establish a connection (Chen & Adams, 2004). It supports both file transfer and data transfer. The downside is that it works only at very short distances (approximately 4cm) and only supports peer to peer connections. This contactless technology is enabling "wave and pay" transactions, where mobile devices or phones equipped with NFC may be waved in front of a contactless reader in a store or at a purchase point. The NFC chip in the mobile device transfers the payment details to the terminal.

Since 2007, the number of mobile devices that include this technology has been growing. Currently, three out of four smartphones incorporate NFC (Cerruela García, Luque Ruiz, & Gómez-Nieto, 2016), although Apple has blocked the NFC features on their devices so they can only be used for mobile payment operations with Apple Pay.

The introduction of NFC technology in mobile devices added new functions to it, such as the emulation of a contactless card. Host Card Emulation (HCE), allows secure communication between the NFC reader, the mobile application and the terminal payment (Alattar & Achemlal, 2014). This development frees the mobile payments ecosystem from the dependence of Secure Element chips, such as the SIM card, and allows the entrance of new players.

Several test-pilot projects of mobile ticketing with NFC-enabled devices have been launched in the public transport sector. For instance, the Touch&Travel service in Germany allows passengers to make payments with their smartphones. Passengers tap their NFC-enabled smartphone to the Touchpoint reader at the boarding and alighting stations. The length of the journey and the price to be paid are calculated at the end of the journey, and the customer receives the invoice at the end of the month. Transport for London (TfL) now support mobile payment technologies, such as Apple Pay or Android Pay (TfL, n.d.). Passengers touch their NFC-enabled smartphone on the reader at the start and at the end of their journey, and they are charged according to a fare capping system. The capping establishes a limit amount spent by the passenger per week. This allows the passenger to pay the minimum value for his journeys. Others (Rodrigues et al., 2014) present an integrated mobile service solution based on the Near Field Communication (NFC) protocol. The solution was tested in a real-world payment scenario and the results showed the lack of reliability of NFC as the major technical challenge.

Ghiron et al. (2009) present a prototype of a NFC-based virtual ticketing application, called NFC-Ticketing, which allows to buy travel tickets with NFC-enabled mobile phones and discuss the results of the usability tests performed. This NFCTicketing application combines NFC and SMS technologies. Ivan and Balag (2015) describe the design and implementation of a mobile ticketing system for urban transport based on NFC technology. Finally, Ceipidor et al. (2013) have identified several security issues that are common in mobile ticketing and propose some methods to solve them, such as a protocol to provide secure validation and travel tickets check that is independent of the NFC operating mode (card-emulation vs. peer-to-peer).

Quick Response Codes

QR codes are bi-dimensional and store data using patterns of black and white dots typically arranged in a square grid, carrying several hundred times more information than regular barcodes. These codes can be read by dedicated readers or using smartphones as long as they have camera and autofocus feature. The reading distance between the reader/smartphone to the code depends on the size of the QR Code. It can range from 2,5cm to 2m. The bigger the QR Code, the bigger the distance at which the QR Code can be read effectively (Couto, Leal, Costa, & Galvao, 2015). Credentials used for payments may be encrypted within codes or stored in the cloud.

Therefore, the payments based on QR Codes can be of two main categories:

1. The QR Code is presented on the mobile device of the customer paying and scanned by a POS of the service provider. This solution is being mostly adopted by retailers.
2. The QR Code is presented by the service provider, in a static or one-time generated fashion and it is scanned by the customer executing the payment. This solution is cheaper to implement since the previous one would require to huge investments in QR code readers at every stop and/or vehicle.

Finzgar and Trebar (2011) propose a mobile ticketing solution based on QR codes and RFID tags. These are used to register the passengers at the beginning and at the end of their journeys. Justride (Masabi, n.d.) is a cloud-based ticketing solution implemented in several transport operators such as New York MTA, Las Vegas RTC and LA Metrolink. Passengers access to the transport services by presenting the QR Code on their smartphone to the reader, which enables the identification of the passenger. Costa et al. (2015) present a mobile ticketing solution based on a check-in / check-out concept that consists in reading the QR Code of the station. The solution was tested in the city of Porto (Portugal), by real passengers. They considered the solution very useful and convenient, however sometimes they were struggling to find the correct distance and position to the QR Code to properly read it (Ferreira et al., 2017).

Bluetooth Low Energy

Bluetooth Low Energy (BLE) has become increasingly popular due to the latest advances in the Bluetooth specification, in particular with the release of the 4.0 version (Carroll & Heiser, 2010). BLE consumes less power than previous Bluetooth generations, being an optimal solution for the beacons. In addition to this, the vast majority of mobile devices now being manufactured support BLE (Geddes, 2014), which makes them a perfect counterpart for the BLE beacons.

In wireless technology, a beacon is the concept of broadcasting small pieces of information. The data transmitted by the beacons is usually static, for instance identification data, but it can also be dynamic, for instance ambient, micro-location or orientation data. A BLE Beacon uses the BLE radio frequency specification to communicate and send signals. It broadcasts a BLE signal over a distance of 50m to 70m that can be detected by compatible devices.

The use of BLE technology to support mobile ticketing services is recent and the studies about the topic are scarce. Narzt et al. (2015) propose a be-in/be-out ticketing system for public transport using BLE technology. Passengers carry a BLE beacon or BLE-enabled mobile device, which broadcasts a unique ID. This signal is captured by an on-board system (OBS), in this case a Raspberry Pi, placed inside each vehicle. The information regarding the passenger ID and route details are remotely processed on a decoupled system, which calculates the price to pay for the journey. Another mobile ticketing pilot test using BLE technology was carried out in Soest, Germany, in 2016 (Kahvazadeh, 2016). This pilot consisted on testing a check-in/be-out solution, through which the passenger checks-in in the vehicle by selecting the bus detected by the mobile application and when the passenger leaves the vehicle, the mobile application loses connection to the bus and the app closes the journey.

A mobile ticketing solution based on BLE technology, called Anda, was deployed in the city of Porto, Portugal, in June 2018. It is based on a check-in/be-out scheme, requiring an intentional user action when entering the vehicle (tap the mobile phone against an NFC ticket reader), and the alight station is automatically detected by the system, as well as intermediary stations along the trip. The mobile phone interacts with BLE beacons installed in metro and train stations and inside buses, through Bluetooth connection, to locate the customer along the transport network (Ferreira, Dias, & Cunha, 2019). The price to be paid by the customer is calculated through a fare optimization algorithm, which minimizes the cost for the passenger.

Technology Comparison

The relative comparison of the main technologies involved on mobile payments approaches found in the literature, is shown in Table 1. Mobile payments can take advantage of more than a single technology. Recent mobile ticketing trends show that current approaches commonly use more than one technology, by using each one to address the shortcomings of the other. As an example, the Anda mobile ticketing solution is using a combination of NFC, WIFI/Mobile Phone Data and BLE to power their mobile ticketing experience.

Mobile Ticketing Ecosystem

The widespread deployment and adoption of mobile payment solutions requires action from complex ecosystem of organizations, such as passengers, transport operators, transport authorities, banks, mobile network operators, and other third parties to create a mobile payments service solution (Ezell, 2009). Each entity desires to have the leading place in the mobile payments' ecosystem and a dominant role in the value chain. Mobile payments entail a complex, system-interdependent ecosystem whose success is dependent on the joint action by all players together at the same time. Because of this, only a few countries, such as Japan and South Korea, have been able to coordinate the complex ecosystem required to extensively deploy a broadly used mobile payments system (Ezell, 2009). In contrast, most other nations, including most of Europe and the United States, lag far behind. For lagging nations to take full advantage of the opportunities of mobile payments, they will need to develop and adopt national mobile payments strategies.

A value network constellation map for the mobile ticketing ecosystem is proposed in Figure 1. This was based on the value constellation concept from Normann and Ramírez (1993) and identifies the relevant actors of the network and their relationships that jointly create an offering. This section presents the main players involved in this complex ecosystem, as well as their key motivations and concerns.

Customers

The introduction of mobile ticketing services in urban passenger transport involves a significant change in customer experience. The ubiquitous and remote access to payment, the avoidance of queues and the lack of need to carry cash are advantages often mentioned by passengers (Mallat, 2007). Convenience, usefulness, ease of use and access to additional services beyond payment are also advantages usually referred by customers (Ferreira et al., 2017). Despite its clear advantages, few customers were able to experience mobile payments, and a large number of mobile payment initiatives failed before they even reached their intended end-users (Dahlberg, Guo, & Ondrus, 2015). Therefore, it is important to under-

Table 1. Relative comparison of the common mobile ticketing technologies

	SMS and Phone Calls	Wi-Fi and 3G/4G	NFC	BLE	QR Code
Type of payment	Remote	Remote	Proximity	Proximity	Proximity
Transmission distance (coverage)	>1000m	150m and >1000m	0,04m–0,10m	50m–70m	0,25m-2m
Transmission security technology	Low	Low	High	Low	Medium
Communication	One to one	One to one	One to one	One to many	One to one
Energy consumption	Low	Low	Very low	Medium (broadcasting aggressive mode)	Very low
Environment conditions interference	Medium	Medium	Low	High	High
Security	Low	Medium	High	Low/ Medium	Medium
Required investment from Service Provider	Minimal	Minimal	Significant: NFC-readers; software installation and integration	Significant: beacons; software installation and integration	Moderate: QR code scanners or printed QR Codes. Software installation and integration
Availability	High	High	Low	Medium	Medium
User interaction	High	No interaction (case of Be-in/ Be-out systems)	Low	No interaction (case of Be-in/ Be-out systems)	Medium
Check-in/check-out compatible	✓	✓	✓	✓	✓
Be-in/be-out compatible		✓		✓	
Advantages	- High availability - Familiar to every user - No or very low infrastructure investment	- High availability - Data transmission with the backend in real time - No or very low infrastructure investment	- Secure (data is encrypted and stored in a secure element) - Low user interaction - Faster read-speed	- Long distance reception - No user interaction (case of Be-in/ Be-out systems) - Provide more fine-grained data about passengers' movements	- Easy to create and print - Can be read by every smartphone with autofocus camera - Can be shown on any device's screen
Disadvantages	- Low security (no encryption) - No proof of delivery - Transaction costs (premium numbers) - Messages formats complicated, difficult to remember, hard to key in - High user interaction	- Network connection problems (underground stations, crowded stations/vehicles) - Insecure public Wi-Fi networks	- Lower availability - Not well-known - Only works well at very short distances - High infrastructure investment and maintenance costs	- Low/medium security (data transfer performed at long distances) - Power source needed - High installation and maintenance costs - High environment conditions interference	- High user interaction - High environment conditions interference - Less flexibility - High exposure to vandalism (case of printed QR codes) - QR Code reading process is not fast enough

Figure 1. Mobile ticketing value network constellation map

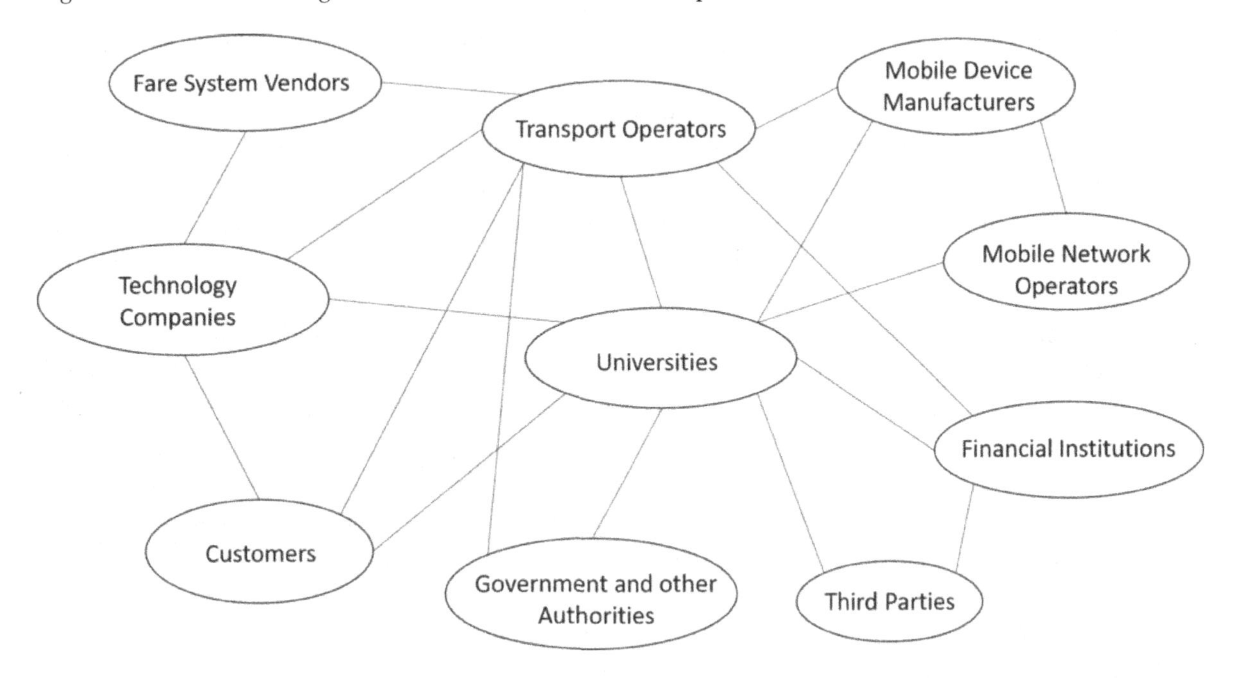

stand which are the main factors that may influence the adoption of mobile ticketing services in urban passenger transport. Analyzing the literature, it is possible to find numerous studies on mobile commerce and mobile payment adoption, however, very few studies on the mobile ticketing services adoption have been undertaken. In this sub-section, five studies are highlighted and summarized in Table 2. The table shows the country where the studies were conducted, the number of people who participated in the surveys and the adoption factors investigated.

Mallat et al. (2008) studied the factors that influence the adoption of a mobile ticketing service based on SMS in the Helsinki's public transport. The proposed research model was based on TAM (Davis, Bagozzi, & Warshaw, 1989), DOI theory (Rogers, 1995) and trust theories, which are augmented with concepts of mobile use context and mobility. Their findings corroborate the TAM model, where ease of use and usefulness had a statistically significant effect on the adoption decision. The effect of both factors was, however, relatively weak compared with other studied factors. Interestingly, cost was not a significant determinant of mobile ticketing adoption. The strongest predictors for use intention were prior experience on mobile ticketing service and compatibility of the mobile ticketing service with a person's use of public transportation, use of mobile devices, and general habits.

Based on the same survey of Mallat et al. (2008), Mallat et al. (2009) present another paper focusing on the impact of user context on mobile ticketing services acceptance. This construct mediated the effect of mobility and usefulness on use intention, implying that users valued the benefits of the mobile ticketing service in situations where they were in a hurry, where there were no other ticketing alternatives available, when they were not expecting to need for a ticket, or where there were queues at points of sale.

Cheng and Huang (2013) examines Taiwan high speed rail (HSR) passengers' acceptance of mobile ticketing services. The implemented mobile ticketing services is based on QR Codes that are used for payment and gate entrance. The theoretical framework proposed by the authors is based on TAM (Davis et al., 1989) and mental accounting theory (Moore & Benbasat, 1991). Drawing on mental accounting theory, mobile ticketing adoption is influenced by the potential loss (perceived risk) and benefit (perceived ease of use and perceived usefulness) of the mobile ticketing service. The findings demonstrate that personal innovativeness has a positive effect on mobile ticketing services adoption.

Kapoor, Dwivedi, and Williams (2015) studied the experience factors that determine the adoption of the Indian Railway Catering and Tourism Corporation Limited's (IRCTC) mobile ticketing service in India, based on 3G/4G technologies. The results showed that compatibility, relative advantage, cost, social approval, demonstrability, voluntariness and riskiness were the strongest predictors of adoption intention. Only complexity failed to influence use intentions. Also, riskiness strongly predicted the actual adoption.

Di Pietro et al. (2015) examine the users' acceptance and usage of mobile ticketing technologies applied in the public transport context. They proposed a model, called the Integrated Model on Mobile Payment Acceptance (IMMPA), based on TAM (Davis et al., 1989), DOI (Rogers, 1995) and UTAUT (Venkatesh, Morris, Davis, & Davis, 2003) theories. It was developed by adding new variables related to the mobile ticketing context and to the ones of the existing models. The results show that usefulness, ease of use and security are strong predictors of the intention to use a technology. The predictor attitude includes the requirements that every mobile ticketing solution must address in this context, such as additional information about time and delay, intuitive interface, speed of use, and route customization. Another new construct identified is the security that can menace the mobile ticketing distribution and for that reason must be taken into account by service providers.

Finally, other studies try to forecast the adoption of mobile ticketing services in public transport. Brakewood et al. (2014) performed a stated preferences analysis through an on-board survey on two commuter rail lines (Worcester and Newburyport/Rockport) in the greater Boston area. From the application of a binary logit model they concluded that 26% of commuter rail passengers in Boston are very likely to adopt mobile ticketing. (Fontes et al., 2017) applied a survey to two different cities: Porto (Portugal) and Beijing (China). Although the high differences between the ticketing systems in both cities, Chinese and Portuguese have a similar opinion about the systems implemented in their cities. Still, Chinese reveal a higher motivation to adopt the new mobile ticketing system. In general, such system is greatly accepted by the respondents and the potential market is expected to be high (Porto: 30%; Beijing: 68%).

Transport Operators and Authorities

The deployment of a mobile ticketing solution depends, not only on customer acceptance, but also on the willingness of transport operators to turn it into reality. There are several reasons for considering mobile ticketing, namely: streamline the fare payment and deploy real-time information and other features as part of the fare payment app (Quigley, Georggi, Barbeau, Joslin, & Brakewood, 2016), attract and retain customers (Puhe et al., 2014), offer more flexibility and different payment methods (Juntunen, Luukkainen, & Tuunainen, 2010), technology obsolescence of existing equipment, fraud prevention and increase of boarding speeds (Mezghani, 2008), reduce cash handling costs and improve passenger convenience (GSMA, 2012).

Despite its clear advantages, there are several concerns haunting transport operators' decisions. According to a study by Quigley et al. (2016), cost of the mobile ticketing system is the most important piece of information that transport agencies examine to make decisions about mobile ticketing options. The cost of the mobile ticketing solution comprises software development costs, equipment acquisition costs (e.g. NFC readers, QR code readers), maintenance costs, fee per transaction and processing costs.

Table 2. Mobile ticketing adoption studies

Reference	(Mallat et al., 2008)	(Mallat et al., 2009)	(Cheng & Huang, 2013)	(Kapoor et al., 2015)	(Di Pietro, Guglielmetti Mugion, Mattia, Renzi, & Toni, 2015)
Country (city)	Finland (Helsinki)	Finland (Helsinki)	Taiwan	India	Worldwide
Sample	360	360	262	375	439
Ease of Use	✓	✓	✓		✓
Perceived Usefulness	✓	✓	✓		✓
Relative Advantage				✓	
Complexity				✓	
Compatibility	✓	✓		✓	✓
Trialability				✓	
Observability				✓	
Cost	✓			✓	
Risk	✓		✓	✓	
Social Approval				✓	
Communicability				✓	
Voluntariness				✓	
Visibility				✓	
Image				✓	
Result Demonstrability				✓	
Social influence	✓				
Behavioural intention			✓	✓	✓
Mobility	✓	✓			
Use Context	✓	✓			
Attitude	✓				✓
Prior experience	✓				
Trust	✓				
Personal Innovativeness			✓		
Security					✓

Therefore, it is important to request proposals from different vendors and compare them. Marketing costs need to be considered as well. Mobile ticketing requires extensive marketing activities in order to be successful (Quigley et al., 2016). It is essential to make the public aware about the existence of a new mobile ticketing solution and to educate individuals on how the mobile app is used. Internal training should also be considered. Training sessions with drivers and fare inspectors should be conducted for them to understand how to identify active mobile tickets and answer customer questions about mobile ticketing.

Additionally, transport operators fear outsourcing skills and responsibilities in ticketing to third-party suppliers (Turner & Wilson, 2010). Some smaller operators are afraid of losing control or funding to larger entities. Betting and investing in a technology risks being stuck with that specific technology or contract, which can mean being the long-term market loser (Turner & Wilson, 2010).

Mobile ticketing systems are capable of collecting an increased amount of data and the ownership of data is a major concern for transport operators. Usually socio-demographic details of passenger and their travel behavior is recorded. If the ownership of data is on the side of the suppliers, transport operators will be subject to restrictions imposed by them, for instance need for permissions to access data or access to aggregate instead of disaggregate records. If owned by operators they benefit from more flexibility in accessing, analyzing and sharing the data (Quigley et al., 2016).

Another major concern regarding the deployment of mobile ticketing solutions is fraud. In terms of revenue loss, the main system fraud are individuals that enter and exit the system without paying. Transport operators are looking for solutions to decrease fraud, or at least to maintain the same levels. It is expected that new mobile ticketing systems are robust against fraud, by discouraging people from having fraudulent actions, allowing drivers and inspectors to easily recognize that a passenger is travelling with a valid ticket, and preventing identity transmission and irregular ticket validation procedures.

Finally, and since the mobile ticketing system will have to coexist with traditional systems, transport operators are looking for solutions that are compatible and consistent with existing practices and already implemented rules. If, nowadays, passengers need to tap the smartcard on the reader in front of the bus driver, the new mobile ticketing system must be consistent with this practice, and somehow show the driver and other passengers he/she is travelling legally.

Government and Other Authorities

Government authorities generally aim to provide an accessible, affordable and integrated public transport network, where purely economic reasoning, in most cases, does not justify the operation (Assessment & Parliamentary, 2014). Strong governmental support proved to be important for the institutional coordination of integrated ticketing schemes in the Netherlands and the United Kingdom (International Transport Forum, 2012). The extremely successful card systems in Hong Kong and Singapore have been set up with strong, centralized government control as well. Therefore, Government has a facilitating role in order to push initiatives forward. Setting up a platform for cooperation among stakeholders, financing pilot projects and subsidizing installation are important (Yoh, Iseki, Taylor, & King, 2006).

Governments expect to achieve positive indirect effects for the environment and the national economy. For instance, smart ticketing systems can generate technical expertise and knowledge that can then be applied to other sectors and create a positive economic climate for businesses (Cheung, 2006). With regard to the positive effects on the environment, it is expected that mobile ticketing could increase the use of public transport over the use of cars (Assessment & Parliamentary, 2014). A closer collaboration between Government and local authorities is also important in order to successfully implement new ticketing schemes.

The mobile payments leadership of some countries appear to be explained by governments that play an important role to guide mobile payments ecosystems and to create a corporate business climate geared towards a longer-term investment strategy (Ezell, 2009).

Financial Institutions

During the last decades, banks have established a strong brand image and earned the trust of their customers (Ondrus, Pigneur, & Ecole, 2005). When compared to Mobile Network Operators (MNOs), banks are perceived as more reliable mobile payment service providers (Dahlberg & Mallat, 2002). Until now, banks had control over an exclusive resource that can be used in mobile payment solutions: bank accounts. Bank accounts contain privileged and confidential information about customers and can be plugged and integrated directly with payment solutions in order to facilitate the settlement of payment (Gaur & Ondrus, 2012). This great advantage is being menaced by introduction of the "Open banking" in January 2018 by European countries (Doyle, Sharma, Ross, & Sonnad, 2017). Under the new rules, the ownership of financial data will be transferred to the customers. This means that customers can grant companies, other than their own bank, permission to access their details, and financial institutions should allow customers share their data in an easy and safe way. This opens the way for the entrance of new players, who can benefit from the advantages that until now belong only to the banks.

Mobile Network Operators

Mobile Network Operators (MNOs) provide the infrastructure for the mobile network and offer services to mobile market. They have been dealing with payments and customer relationship management since mobile communications were launched onto the market. They play an active role in implementing their best practices as a foundation for mobile payments (Lim, 2008). MNOs may implement operator centric models, where MNOs intermediate payment transactions, for example, information services, premium services, and SMS services. These systems can be either a prepaid system using over-the-air (OTA) connection or a post-paid system. The payment services are not necessarily provided by the service provider. In fact, the MNO acts only as the payment service provider, who forwards the payment to the next party and receives a fee for the services provided. Netsize, from Gemalto Company, is an example of a mobile ticketing solution based on SMS.

Despite an operator centric model being interesting to micro-payments (less than €10) since MNO can provide transaction services at lower fees, where banks are not efficient enough, it is likely to find limited acceptance in the fragmented payment industry (Capgemini, 2011). The lack of relationships between MNO and merchants, limited incentives to cooperate, and the complexity of negotiations may hinder the adoption of this model.

Regarding NFC based mobile payments, MNOs may play a more important role. If the UICC is used as a secure element, the MNO can benefit from the UICC control since it is the secure element issuer. In this case, the operator can rent the space in the secure element to the service provider who wants to put their applications there (Juntunen et al., 2010).

Mobile Device Manufacturers

Mobile device manufacturers provide the devices used in mobile payment services and can influence the standardization process. They develop standards that allow the production of flexible devices, which can be used in different ways and markets, depending on specific business needs (Juntunen et al., 2010). Thus, the handset manufacturers tend to favor, for example, having multiple choices for a secure element (e.g. NFC chip). This may threaten the role of MNOs, since mobile ticketing providers may choose the NFC tag as the secure element instead of the UICC, becoming independent from the MNOs.

Fare System and Technology Vendors

The use of mobile devices in public transport will have serious implications on transport operators' supply chain (Ferreira, Nóvoa, Dias, & Cunha, 2014). Nowadays, the most common public transport suppliers are ticketing companies, ticket vending machines and ticket reader machines suppliers. Depending on the technology chosen for mobile payments, it may be necessary to adapt the existing infrastructure or even to replace it. Vendors that sell NFC ticketing readers, QR-Code readers, beacons and antennas are the new suppliers.

Technology based companies are also important suppliers entering the scheme. It is necessary to develop software, back office systems and mobile applications before deploying a mobile ticketing solution. Masabi, Cubic, Passport and Unwire are examples of vendors of such solutions. They offer rapid time to market solutions with low (or no) capex rollout (visual validation).

The primary compensation mechanism for the contracted mobile ticketing vendor will be via a transaction-based fee. For instance, this transaction-based fee could be a percentage payment for all mobile ticketing transactions (such as is done by Passport at COMET – Columbia USA) or a flat fee based on an estimated number of mobile ticketing transactions (such as is done by Masabi at NICE) (Quigley et al., 2016). There could also be fixed upfront costs for the initial development of the mobile ticketing system, if vendors responding to the solicitation do not have turnkey systems available that meet the agency's needs.

Quigley et al. (2016) also leave some advice to transport operators on the choice of the vendor: check for references of the vendor, since many of them are new in the industry, and evaluate the vendor's ability and willingness to make changes to the mobile apps, such as including new features and integrate with other systems in the future.

Universities

Service innovation involves a new process or service offering that creates value for one or more actors in a service network (Patrício, Gustafsson, & Fisk, 2018). Actors can be customers, organizations, and other elements such as technology that participate in the value creation process. As every new service, previous research work needs to be done. Universities play a key role in mobile ticketing ecosystem, by bringing innovative ideas to life, understanding customers and their context, envisioning future service solutions, and prototyping them.

Researchers contribute with their knowledge, technologies, previous experiences and design skills. They bring in applicable domain theories, background knowledge and literature on theories which can guide and/or inspire the design of mobile ticketing solutions. Often the pressure to launch a service and reduce time-to-market, drives companies to make hasty decisions, and skip fundamental steps in designing a successful solution. Therefore, researchers play a large role in knowledge development, idea generation and concept development, by providing proper methodologies and tools and ensuring that important steps are not ignored.

Researchers play also a facilitator role. Designing mobile ticketing services in complex ecosystem context requests a participatory and pluralistic approach based on the belief that systems cannot be completely understood or designed, but that they can be interpreted via a collaborative process. The mobile ticketing ecosystem encompasses multiple actors with, sometimes, conflicting interests that hinders decision making. Thus, an agnostic entity with no commercial interests is necessary to establish a consensus between the various actors. Researchers play a facilitator role in bringing together these different perspectives into a new service solution.

Third Parties

Recently, technology companies such as Google, Paypal, Apple and AliBaba have also shown interest in joining the mobile payment market. These companies are taking advantage of powerful mobile devices and their customers' database to offer mobile payment services. Unlike MNOs and banks, who can hardly reach clients outside their country or zone of operation, global actors such as Apple, Google, Samsung Mastercard or Visa can act globally and form successful alliances.

On the other hand, increasing consumer awareness of mobile wallets such as Apple Pay, Google Wallet, and Softcard is expected to increase volumes of mobile payment transaction. Interestingly, since Apple Pay's release in October 2014, more Android smartphone users have downloaded the Softcard and Google Wallet apps for the first time, and existing wallet customers have used these mobile wallets more frequently (Tavilla, 2015). Customers can use mobile wallets to pay for purchases, including in public transport. London (UK), Salt Lake City and Chicago (USA) passengers can already pay their fares with Apple Pay, Google Wallet, and Softcard.

The next section presents the future perspectives.

FUTURE RESEARCH DIRECTIONS

Future research directions can be identified within the scope of this chapter. The first is related with the adoption research. This should be more based on data collection from real-life payment scenarios, instead of intention to adopt. The concrete deployment of mobile ticketing services allows to start a series of post-deployment studies and migrate from the concept of adoption to the acceptance of the technology and compare with the other means of payment means available. Also, service providers' adoption should be investigated, since there are practically no studies on the topic.

Post-deployment studies can also involve a deeper exploration of mobile ticketing solutions impacts. There are already studies on expected impacts, now it is important to measure the actual impacts. Are stakeholders' expectations fulfilled? What is the level of customer satisfaction with public transport services? Are mobile ticketing services contributing to people make the switch from private to public transport? Answering these research questions is key to assess the actual impacts of mobile ticketing services.

Another research direction is related to a holistic view of the mobile ticketing service. It is clear that offering a different payment method to customers is not enough. Exploring the unique characteristics of mobile devices to offer additional and complementary services to the payment is crucial to contribute to a wider adoption. Customers are expecting to do more tasks with the mobile device than just paying. They want, for instance, to be able to check the balance and the transactions history, to search for information about the service and service provider and even to interact with other customers.

Mobile ticketing services can also be an important driver for leveraging the development of MaaS (Mobility-as-a-Service) solutions and turning them into reality. The appetite for on-demand services is spreading to transport and the future will evolve from payments to collaboration. By adopting a holistic point-of-view, this collaboration can also be extended to service providers not related to mobility, but that coexist in the same city and serve the same citizens.

CONCLUSION

This chapter presents an in-depth study of the current situation of mobile ticketing services in the context of urban passenger transport and points out future trends and directions that will define forthcoming versions of mobile ticketing services. Mobile ticketing can be defined as the use of a mobile device to purchase and/or validate a travel ticket or to initiate a journey and there are several technologies available in mobile devices that can be used to implement mobile ticketing solutions, such as SMS and phone calls, Wi-Fi and 3G/4G, NFC, BLE, and QR Codes. A relative comparison of the main technologies is presented in this chapter as well as examples of concrete applications to the urban passenger transport domain.

Technical choices have important effects on the consumers and service providers' adoption and on the roles of the ecosystem actors. For instance, depending on the type of implementation of secure element, the size of the market can vary significantly. Also, if the mobile ticketing solution requires the use of a new technology, the diffusion curve is expected to be slower. Recent mobile ticketing trends show that current approaches commonly use more than one technology, by using each one to address the shortcomings of the other.

Mobile ticketing is complex as it requires the establishment of a full ecosystem involving many different stakeholders, with sometimes conflicting interests. Mobile ticketing solutions usually fail to be implemented because it is not possible to reach an agreement. The first necessary step for a successful mobile ticketing implementation is to develop inter-organizational relationships. To better understand the actors of this complex ecosystem and to respond to the literature gaps identified by Dahlberg, Guo, and Ondrus (2015), the key stakeholders were identified as well as their interests and concerns regarding mobile ticketing solutions.

REFERENCES

Alattar, M., & Achemlal, M. (2014). Host-based card emulation: Development, security, and ecosystem impact analysis. *Proceedings - 16th IEEE International Conference on High Performance Computing and Communications, HPCC 2014, 11th IEEE International Conference on Embedded Software and Systems, ICESS 2014 and 6th International Symposium on Cyberspace Safety and Security*, 506–509. 10.1109/HPCC.2014.85

Assessment, T. O., & Parliamentary, E. (2014). *Integrated urban e-ticketing for public transport and touristic sites*. Retrieved from https://www.itf-oecd.org/sites/default/files/docs/2012-12-10.pdf

Au, Y. A., & Kauffman, R. J. (2008). The economics of mobile payments: Understanding stakeholder issues for an emerging financial technology application. *Electronic Commerce Research and Applications, 7*(2), 141–164. doi:10.1016/j.elerap.2006.12.004

Boer, R., & Boer, T. de. (2009). *Mobile payments 2010: Market Analysis and Overview. Security*. Chiel Liezenber (Innopay) and Ed Achterberg (Telecompaper).

Brakewood, C., Roja, F., Robin, J., Sion, J., & Jordan, S. (2014). Forecasting mobile ticketing adoption on commuter rail. *Journal of Public Transportation, 17*(1), 1–19. doi:10.5038/2375-0901.17.1.1

Capgemini. (2011). *Mobile Payments: How Can Banks Seize the Opportunity ? An Approach for Financial Services Institutions. Capgemini.*

Carroll, A., & Heiser, G. (2010). An analysis of power consumption in a smartphone. *Proceedings of the USENIX Annual Technical Conference*. Retrieved from https://www.usenix.org/event/usenix10/tech/full_papers/Carroll.pdf

Ceipidor, U. B., Medaglia, C. M., Marino, A., Morena, M., Sposato, S., & Moroni, A. … Morgia, M. La. (2013). Mobile ticketing with NFC management for transport companies. Problems and solutions. In *Proceedings 2013 5th International Workshop on Near Field Communication (NFC)*, 1–6. 10.1109/NFC.2013.6482446

Cerruela García, G., Luque Ruiz, I., & Gómez-Nieto, M. (2016). State of the Art, Trends and Future of Bluetooth Low Energy, Near Field Communication and Visible Light Communication in the Development of Smart Citics. *Sensors (Basel), 16*(11), 1968. doi:10.339016111968 PMID:27886087

Chen, J. J., & Adams, C. (2004). Short-range wireless technologies with mobile payments systems. *Proceedings of the 6th International Conference on Electronic Commerce - ICEC '04*, 649. 10.1145/1052220.1052302

Cheng, Y.-H., & Huang, T.-Y. (2013). High speed rail passengers' mobile ticketing adoption. *Transportation Research Part C, Emerging Technologies, 30*, 143–160. doi:10.1016/j.trc.2013.02.001

Cheung, F. (2006). Implementation of Nationwide Public Transport Smart Card in the Netherlands: Cost-Benefit Analysis. *Transportation Research Record: Journal of the Transportation Research Board*, 1971.

Costa, P. M., Fontes, T., Nunes, A. A., Ferreira, M. C., Costa, V., & Dias, T. G., … Falcão, J. (2015). Seamless Mobility : a disruptive solution for urban public transport. In *22nd ITS World Congress*. Bordeux.

Couto, R., Leal, J., Costa, P. M., & Galvao, T. (2015). Exploring Ticketing Approaches Using Mobile Technologies: QR Codes, NFC and BLE. *IEEE Conference on Intelligent Transportation Systems, Proceedings, ITSC, 2015-October*, 7–12. 10.1109/ITSC.2015.9

Dahlberg, T., Guo, J., & Ondrus, J. (2015). A critical review of mobile payment research. *Electronic Commerce Research and Applications, 14*(5), 265–284. doi:10.1016/j.elerap.2015.07.006

Dahlberg, T., & Mallat, N. (2002). Mobile Payment Service Development - Managerial Implications Of Consumer Value Perceptions. *Proceedings of the Tenth European Conference on Information Systems*, (January 2002), 649–657. Retrieved from http://is2.lse.ac.uk/asp/aspecis/20020144.pdf

Davis, F. D., Bagozzi, R. P., & Warshaw, P. R. (1989). User acceptance of computer technology: A comparison of two theoretical models. *Management Science*, *35*(8), 982–1003. doi:10.1287/mnsc.35.8.982

Di Pietro, L., Guglielmetti Mugion, R., Mattia, G., Renzi, M. F., & Toni, M. (2015). The Integrated Model on Mobile Payment Acceptance (IMMPA): An empirical application to public transport. *Transportation Research Part C, Emerging Technologies*, *56*, 463–479. doi:10.1016/j.trc.2015.05.001

Doyle, M., Sharma, R., Ross, C., & Sonnad, V. (2017). *How to flourish in an uncertain future: Open banking and PSD2*. Retrieved from https://www2.deloitte.com/content/dam/Deloitte/cz/Documents/financial-services/cz-open-banking-and-psd2.pdf

Ezell, S. (2009). *Explaining International IT Aplication Leadership: Contactless mobile payment. THE INFORMATION TECHNOLOGY & INNOVATION FOUNDATION*. Retrieved from https://www.itif.org/files/2009-Mobile-Payments.pdf

Ferreira, M. C., Dias, T. G., & Cunha, J. F. e. (2019). Codesign of a Mobile Ticketing Service Solution Based on BLE. *Journal of Traffic and Logistics Engineering*, *7*(1), 10–17. doi:10.18178/jtle.7.1.10-17

Ferreira, M. C., Fontes, T., Costa, V., Dias, T. G., Borges, J. L., & Cunha, J. F. E. (2017). Evaluation of an integrated mobile payment, route planner and social network solution for public transport. In Transportation Research Procedia (Vol. 24, pp. 189–196). doi:10.1016/j.trpro.2017.05.107

Ferreira, M. C., Nóvoa, H., & Dias, T. G. (2013). A Proposal for a Mobile Ticketing Solution for Metropolitan Area of Oporto Public Transport. In Lecture Notes in Business Information Processing, IESS, Vol. 143 (pp. 263–278). Springer. doi:10.1007/978-3-642-36356-6_19

Ferreira, M. C., Nóvoa, H., Dias, T. G., & Cunha, J. F. e. (2014). A proposal for a public transport ticketing solution based on customers' mobile devices. *Procedia: Social and Behavioral Sciences*, *111*, 232–241. doi:10.1016/j.sbspro.2014.01.056

Finzgar, L., & Trebar, M. (2011). Use of NFC and QR code identification in an electronic ticket system for public transport. In *Proceedings SoftCOM 2011, 19th International Conference on Software, Telecommunications and Computer Networks*, (iii), 1–6.

Fontes, T., Costa, V., Ferreira, M. C., Shengxiao, L., Zhao, P., & Dias, T. G. (2017). Mobile payments adoption in public transport. *Transportation Research Procedia*, *24*, 410–417. doi:10.1016/j.trpro.2017.05.093

Gaur, A., & Ondrus, J. (2012). The role of banks in the mobile payment ecosystem: a strategic asset perspective. *Proceedings of the 14th Annual International Conference on Electronic Commerce (pp. 171-177). ACM*. 10.1145/2346536.2346570

Geddes, O. (2014). *A Guide to Bluetooth Beacons*. Retrieved from http://www.gsma.com/digitalcommerce/wp-content/uploads/2013/10/A-guide-to-BLE-beacons-FINAL-18-Sept-14.pdf

Ghiron, S. L., Sposato, S., Medaglia, C. M., & Moroni, A. (2009). NFC Ticketing: A Prototype and Usability Test of an NFC-Based Virtual Ticketing Application. In *Proceedings 2009 First International Workshop on Near Field Communication*, 45–50. 10.1109/NFC.2009.22

GSMA. (2012). White Paper: Mobile NFC in Transport, (September).

International Transport Forum. (2012). *Policy Brief.* 10.1177/0022146513479002

Ivan, C., & Balag, R. (2015). An Initial Approach for a NFC M-Ticketing Urban Transport System. *Journal of Computer and Communications*, *3*(June), 42–64. doi:10.4236/jcc.2015.36006

Juntunen, A., Luukkainen, S., & Tuunainen, V. K. (2010). Deploying NFC Technology for Mobile Ticketing Services – Identification of Critical Business Model Issues. In *Proceedings 2010 Ninth International Conference on Mobile Business and 2010 Ninth Global Mobility Roundtable (ICMB-GMR)*, 82–90. 10.1109/ICMB-GMR.2010.69

Kahvazadeh, A. (2016). Check In-Be Out – Innovation in Mobile Transit Payment, (April). Retrieved from https://www.apta.com/mc/revenue/presentations/Presentations/Check In-Be Out – Innovation in Mobile Transit Payment - Arash Kahvazadeh.pdf

Kapoor, K. K., Dwivedi, Y. K., & Williams, M. D. (2015). Empirical Examination of the Role of Three Sets of Innovation Attributes for Determining Adoption of IRCTC Mobile Ticketing Service. *Information Systems Management*, *32*(2), 153–173. doi:10.1080/10580530.2015.1018776

Lim, A. (2008). Inter-consortia battles in mobile payments standardisation. *Electronic Commerce Research and Applications, 7*(2), 202–213. doi:10.1016/j.elerap.2007.05.003

Lüke, K., Mügge, H., Eisemann, M., & Telschow, A. (2009). Integrated Solutions and Services in Public Transport on Mobile Devices. In *Gesellschaft für Informatik (GI), I2 CS conference proceedings P-148* (pp. 109–119).

Mallat, N. (2007). Exploring consumer adoption of mobile payments – A qualitative study. *The Journal of Strategic Information Systems*, *16*(4), 413–432. doi:10.1016/j.jsis.2007.08.001

Mallat, N., Rossi, M., Tuunainen, V. K., & Öörni, A. (2008). An empirical investigation of mobile ticketing service adoption in public transportation. *Personal and Ubiquitous Computing*, *12*(1), 57–65. doi:10.100700779-006-0126-z

Mallat, N., Rossi, M., Tuunainen, V. K., & Öörni, A. (2009). The impact of use context on mobile services acceptance: The case of mobile ticketing. *Information & Management*, *46*(3), 190–195. doi:10.1016/j.im.2008.11.008

Masabi. (n.d.). JustRide. Retrieved from http://www.masabi.com/justride-mobile-ticketing/

Mezghani, M. (2008). Study on electronic ticketing in public transport Contents, (May).

Moore, G. C., & Benbasat, I. (1991). Development of an Instrument to Measure the Perceptions of Adopting an Information Technology Innovation. *Information Systems Journal*, *2*(3), 192–222. doi:10.1287/isre.2.3.192

Narzt, W., Mayerhofer, S., Weichselbaum, O., Haselbock, S., & Hofler, N. (2015). Be-In/Be-Out with Bluetooth Low Energy: Implicit Ticketing for Public Transportation Systems. *IEEE Conference on Intelligent Transportation Systems, Proceedings, ITSC, 2015-Octob*, 1551–1556. 10.1109/ITSC.2015.253

Normann, R., & Ramírez, R. (1993). From Value Chain to Value Constellation: Designing Interactive Strategy. *Harvard Business Review*, *71*(4), 65–77. PMID:10127040

Ondrus, J., Pigneur, Y., & Ecole, I. (2005). A Disruption Analysis in the Mobile Payment Market. *Proceedings of the 38th Annual Hawaii International Conference On System Sciences, 2005. HICSS'05, 00*(C), (pp. 84c-84c). IEEE. 10.1109/HICSS.2005.9

Patrício, L., Gustafsson, A., & Fisk, R. (2018). Upframing Service Design and Innovation for Research Impact. *Journal of Service Research, 21*(1), 3–16. doi:10.1177/1094670517746780

Puhe, M., Edelmann, M., & Reichenbach, M. (2014). *Integrated urban e-ticketing for public transport and touristic sites*. (Final Report No. IP/A/STOA/FWC/2008-096/LOT2/C1/SC12). Brussels, Belgium: European Parliament/STOA.

Quigley, D., Georggi, N. L., Barbeau, S., Joslin, A., & Brakewood, C. (2016). *Assessment of mobile fare payment technology for future deployment in Florida*. FDOT.

Rodrigues, H., José, R., Coelho, A., Melro, A., Ferreira, M. C., Cunha, J. F., ... Ribeiro, C. (2014). MobiPag: Integrated Mobile Payment, Ticketing and Couponing Solution Based on NFC. *Sensors (Basel), 14*(8), 13389–13415. doi:10.3390140813389 PMID:25061838

Rogers, E. M. (1995). *Diffusion of innovations*. New York: Free Press.

Sarma, S. (2014). Bus Tracking & Ticketing using USSD Real-time application of USSD Protocol in Traffic Monitoring. *Journal of Emerging Technologies and Innovative Research, 1*(7), 628–634.

Tavilla, E. (2015). *Transit Mobile Payments: Driving Consumer Experience and Adoption*. Federal Reserve Bank of Boston Research Report.

TfL. (n.d.). Contactless and mobile pay as you go. Retrieved from https://tfl.gov.uk/fares/how-to-pay-and-where-to-buy-tickets-and-oyster/pay-as-you-go/contactless-and-mobile-pay-as-you-go

Turner, M., & Wilson, R. (2010). Smart and integrated ticketing in the UK: Piecing together the jigsaw. *Computer Law & Security Review, 26*(2), 170–177. doi:10.1016/j.clsr.2010.01.015

Venkatesh, V., Morris, M. G., Davis, G. B., & Davis, F. D. (2003). User Acceptance of Information Technology : Toward a Unified View Author (s): Davis Published by : Management Information Systems Research Center, University of Minnesota. *Management Information Systems Quarterly, 27*(3), 425–478. doi:10.2307/30036540

Yoh, A. C., Iseki, H., Taylor, B. D., & King, D. A. (2006). Institutional Issues and Arrangements in Interoperable Transit Smart Card Systems: A Review of the Literature on California, United States, and International Systems UCB-ITS-PWP-2006-2, (March).

KEY TERMS AND DEFINITIONS

4G: Is the fourth generation of supranational digital technologies used to connect mobile devices.

Bluetooth Low Energy: Is a wireless personal area network technology and distinguishes itself from Bluetooth by its low power consumption.

Mobile Ticketing: The use of a mobile device to purchase and/or validate a travel ticket or to initiate a journey.

Near Field Communication: Is a short-range, high frequency, standards-based wireless communication technology that enables exchange of data between devices in close proximity.

Quick Response Codes: Are bi-dimensional by patterns of black and white dots typically arranged in a square grid, carrying several hundred times more information than regular barcodes.

Short Message Service: Is a text message service of most mobile devices that uses standardized communication protocols to exchange short text messages between them.

Wi-Fi: Is a wireless technology that allows devices to connect to a wireless local area network.

Chapter 9
Leveraging IoT Framework to Enhance Smart Mobility: The U-Bike IPBeja Project

Isabel Sofia Brito
https://orcid.org/0000-0002-7556-4367
Polytechnic Institute of Beja, Portugal & Group on Reconfigurable and Embedded Systems, UNINOVA, Portugal

Luís Murta
Polytechnic Institute of Beja, Portugal

Nuno Loureiro
Polytechnic Institute of Beja, Portugal

Pedro Rodrigo Duarte Pacheco
Easycicle, Portugal

Pedro Bento
Polytechnic Institute of Beja, Portugal

ABSTRACT

The planning, designing, deploying, and measuring the smart mobility concept is very important since it can impact several aspects of city life such as how and where people live and fulfil their needs and desires. Given the complexity of the problem, this chapter proposes a general IoT framework for smart mobility that could guide the development of a smart mobility system to manage communications, devices, and services, as well as applications to achieve smart mobility goals. This chapter describes the U-Bike system within the IoT framework and smart mobility paradigms, i.e., in terms of IoT framework structure and operationalization, as well as quality attributes (i.e. non-functional requirements). Recently, the U-Bike system began to be used, making it possible to estimate if it fulfils the objectives of the project. This assessment was performed using focus group method and interviews.

DOI: 10.4018/978-1-7998-2112-0.ch009

INTRODUCTION

Smart city is probably one of the most analysed concept among people (government, companies/industries, researchers, and citizens) all over the world. The idea is to optimise modern, useful technologies to create a sustainable and self-aware city. This multidimensional concept is mainly based on Information and Communications Technology (ICT) structured around smart mobility, smart environment, and smart living.

According to (Tomaszewska & Florea, 2018), smart mobility consists in the use of ICT, networks, databases, software and devices, to produce more effective and efficient transport and logistics services. At the same time, it optimises resource utilization and reduces the negative impact of mobility, mainly pollution. Therefore, smart mobility is one dimension of the smart city concept and is interrelated with other concepts, such as smart environment or smart living. As for mobility management in a smart city, transporting city inhabitants by meeting their daily mobility needs is one of the ultimate goal. Nevertheless, promoting and enabling smart mobility in cities plays an important role in making cities financially and environmentally sustainable. Over the years, cities must be able to deal with the effects of globalisation trends in a sustainable way and the energy overconsumption in the mobility context has been one of the culprits in accelerating air pollution in the cities. A sustainable smart mobility example is to promote and encourage more people to ride bikes, for example through bike sharing projects (OOMap, 2019), in order to reduce the individual carbon footprint (smart environmental strategies).

The Internet of things (IoT) is one of the most important technological developments of the last decade. Society is driven by digital technologies and the Internet. This has changed, and keeps changing, the way businesses operate, i.e., the companies are rethinking their business models, hence identifying new opportunities for creating new value. According to (Iansiti & Lakhani, 2014), the Internet of Things is transforming business models because digital technologies have properties to increase operations at near-zero marginal costs as well as to generate innovation and new opportunities. Nevertheless, no transformation comes without risks, which need to be well managed via awareness, transparency and the pursuit of sustainability (e.g. economic, environmental, social). Advancing IoT solutions for smart cities helps promoting economic, environmental and social issues. It has the potential to support smart city goals and affect the quality of life of its citizens. However, its success relies greatly on a well-defined architecture that consists of a list of sensors, services, communication protocols, users, and interface layers integrated in a scalable and secure basis for its deployment. Several concepts are identified in an IoT system, and because a commonly agreed conceptualization is not found, several different approaches are usually considered (Abdmeziem, Tandjaoui, & Romdhani, 2016): a three-layer architecture composed by Application, Network and Perception layers; a five-layer architecture including also Business and Process layers; cloud and fog systems; and social IoT paradigms. IoT infrastructures are still at their early stage of development and relevant progress is expected to be made in the next years to cover issues such as interoperability (i.e. devices from different sellers will have to cooperate in order to achieve common goals), security (i.e. integrity, availability and confidentiality) and scalability (systems will have to handle a growing amount of work by adding resources). Given the complexity of the problem, the objective of this chapter is to propose a general IoT framework for smart mobility that should guide the development of a smart mobility system in order to manage communications, devices and services, as well as applications to achieve smart mobility goals.

The "U-BIKE Portugal" project is a national project, which aims to promote the adoption of more sustainable mobility behaviour through the provision of electrical and conventional bikes in academic communities. The project is co-financed by the programme Portugal 2020, particularly through PO SEUR – Operational Programme for Sustainability and Efficient Use of Resources (POSEUR, 2019), and is part of the programme to support the implementation of energy efficiency measures and public transport consumptions rationalisation. Therefore, the bikes are handed out to the academic community for long-term use, leading to a regular use of this means of transport.

The "U-Bike IPBeja" is the Polytechnic Institute of Beja's project, which is inserted in the "U-Bike Portugal" and aims to promote soft mobility and cycling in the city. This project provides 200 bicycles to the academic community, students, teachers and academic staff in Beja, in order to reduce primary energy consumption and CO2 emissions and to promote physical activity for health benefits. To accomplish the U-Bike IPBeja project goals, a system was developed by a company together with project managers and other stakeholders. Accordingly, in this chapter the U-Bike system is described within the proposed IoT framework and smart mobility paradigms. The U-Bike system is able to monitor U-Bike project activities and collects indicators to measure the quality of the project by means of travel time and users served within a specified period, among other indicators. To accomplish this, sensing and computation capabilities are used, and data are gathered in real time and non-real time which is further processed, analysed and converted to usable knowledge, in terms of CO2 emission and physical activities indicators. The system is described in terms of an IoT framework structure and operationalization, as well as quality attributes (i. e. non-functional requirements). The U-Bike system has recently begun to be used, so it is now possible to analyse if it fulfils the objectives of the smart mobility project. This assessment was performed by means of a focus group method.

This chapter is composed of six sections. The background section introduces the founding concepts regarding the smart mobility and IoT paradigms, and provides as well a brief description of U-Bike project. The related work is presented in the third section; the fourth section describes an IoT framework for smart mobility. The following section describes the U-Bike system in terms of an IoT framework for smart mobility and quality attributes, i.e. non-functional requirements. An analysis of the system deployment and operationalization is presented through users' opinions of the systems collected by means of interviews and a focus group method. The last section concludes the chapter.

BACKGROUND

Smart Mobility Concepts

There is a consensus that Smart Cities are characterised by the use of digital technologies, i.e. ICT, to optimise resources usage at the same time they improve the economic, social and environmental sustainability of a city. According to (Neirotti, De Marco, Cagliano, Mangano, & Scorrano, 2014), the use of ICT does not guarantee that these objectives are fully achieved; i.e. a city with numerous ICT projects is not necessarily a smart city. To ensure the benefits of those initiatives, a supporting structure around which smart cities systems can be built is needed.

Deloitte' report (Viechnicki, Khuperkar, Fishman, & Eggers, 2015) states that the rise of smart mobility as a dimension of the smart city is a natural consequence of sharing economy models as well as ICT evolution. This report shows a transportation ecosystem designed around individual mobility based on four modes of alternative mobility (as well as more traditional modes such as buses):

- Ridesharing (including carpooling, vanpooling, and real-time or dynamic ridesharing services): two or more travellers share common, pre-planned trips made by private automobiles or vans.
- Bike commuting: Bike commuting refers to trips made to work by bicycle.
- Carsharing (round-trip, one-way, and personal vehicle sharing): In its most basic form, carsharing is car rental for a period.
- On-demand ride services, called ride sourcing or ride hailing services: These are online platforms developed by transportation network companies (such as Uber, Lyft, and SideCar) that allow passengers to source or hail rides from a pool of drivers that use their personal vehicles.

In (Geotab, 2018) smart mobility refers to using modes of transportation alongside, or even instead of, owning a gas-powered vehicle. This can take many different forms, including ride-sharing, car-sharing, public transportation, walking, biking, and more. The need for smart mobility arose out of increasing traffic congestion and its related side effects, including pollution, accidents and casualties, and wasted time.

A design-driven innovation approach provides the means to moving from the 'what' to the 'why' of a product, hence focusing on its purpose, at the same time it supports the interpretation of emerging lifestyles and new socio-cultural paradigms (Dell'Era, Altuna, & Verganti, 2018). According to these authors, this approach proposes new unsolicited products and services that are attractive, sustainable, and profitable. The paper illustrates the approach based on two case studies in the smart mobility domain: BlaBlaCar and Copenhagen Wheel. On the one hand, both case studies highlight the need for more sustainable solutions for future mobility; on the other hand, they propose a new understanding of transport and travel. The case studies promote economic sustainability (e.g. sharing economy) and the possibility to contribute to social wellbeing (e.g. getting in touch with new friends). Considering the technology perspective, the two case studies highlight different roles: the Copenhagen Wheel project uses a set of new sensors, whereas the BlaBlaCar project uses web-based services.

Smart mobility promotes whatever mode of transportation that best suits travellers by fostering healthier lifestyles at the same time it reduces environmental impact and saves money. The options for implementing smart mobility are so varied that they leave the stakeholders lots of space to find solutions tailored to their residents.

Internet of Things Requirements

According to Gartner (Hung, 2017), by 2020 up to 20.4 billion IoT devices will be connected together. New tools to support the new paradigm are needed: smart management of the resources, better security for population, healthcare and the engagement of citizens in their everyday activities are some examples of the different scenarios that can be promoted by IoT. Several surveys have been published addressing IoT topics, including emerging technologies, research challenges and applications, e.g. (Li, Xu, & Zhao, 2015). The surveys (Li, Xu, & Zhao, 2015) and (Miorandi, Sicari, Pellegrini, & Chlamtac, 2012) analysed the challenges and the technologies in developing an IoT system that embraces the heterogeneity of devices, information and communication technologies to enable a new class of applications and services. Service-oriented Computing (SoC) and Service-oriented Architecture (SoA) approaches are proposed for developing applications and services in distributed architecture, whereby things are treated in a uniform way and communicate with each other using standardized interaction patterns. (Li, Xu, & Zhao, 2015) show an example of SoA for IoT based on sensing, network, service and interface layers. The authors' surveys also highlight the need of requirements such as interoperability, scalability and security (mainly data confidentiality and privacy) of an IoT system. In terms of applications fields, they identified smart cities, environmental monitoring, health-care, and smart business. Recent research has shown that the need for effective development of IoT systems has generated several approaches. (Lee, 2019) presents an IoT ecosystem, architecture, and service business models for companies. The major contribution of this proposal is the inclusion of 5G cellular technology, as well as the contribution of cloud-based platforms to the IoT system.

IoT will be essential to turn a traditional city into a smart city, and the traditional and more emerging sectors such as mobility, buildings, energy, living, which can be benefited by IoT. For example, smart mobility services are created to provide effective tools for the citizens to accurately plan their journeys with public/private transportation, bike/car sharing services or multimodal transport systems. Since the emergence of IoT concept, the number of definitions and the related activities have increased depending on the type of organisation looking at this paradigm.

There are few approaches to support this new paradigm in smart city context, for example SynchroniCity (SynchroniCity, 2019). The SynchroniCity project proposes an approach to build IoT services for cities based on technical, best practice and standards knowledge. The report (Maggio, et al., 2018) proposes the specifications, architecture, reference implementation and Application Programming Interface (API) of the SynchroniCity technical framework. This report identifies system requirements that have been integral to the U-Bike IoT Framework, namely:

Interoperability and Openness: To facilitate interoperability, the system must use as many publicly accepted standards as possible for communication and exchanging data; e.g., gateways and APIs might act as glue between those architectural components. The data and information in the system must be collected and processed by open protocols, standard technologies and clear agreements, so new components can easily access information already available.

Scalability: Scalability aspects arise at different levels, such as: (i) data communication and networking due to the high level of interconnection among a large number of heterogeneous devices/equipment/persons; (ii) service provisioning and management due to the high number of services that could be available. Therefore, the system should be able to handle a growing amount of tasks/services or data, i.e. the system can be expanded when it foresees more users of things and/or streams of data scaling both horizontally and vertically. Scaling vertically means adding resources (e.g. CPU) to a single node/computer; scaling horizontally means adding more nodes (e.g. computer) to a system or a distributed system.

Resilience to failure and Robustness: The system must be resilient to failure, considering that the interaction among many different types of components (e.g., sensors, network, wireless technology, data store, servers) from different actors could generate problems.

Performance: The system should guarantee a real-time user experience. Users should be able to responsively interact with the system, discovering new available assets at run time.

Availability: The system should support asset availability and fruition in compliance with their SLA. Moreover, a continuous integration and delivery possible for each element in the architecture, as well as automated testing to reduce regression and guarantee quality, should support the system to be 24/7 operational, which has a close to zero maintenance.

Maintainability: This requirement is related to the ability to make changes, and have reasonable costs. These changes could affect data, services, equipment and communication when modifying for fixing issues, or to meet future demands. Maintainability is a function of the non-functional requirements described above, e.g. a function to restore a service to normal status following a failure or an upgrade (availability), or a function to reduce runtime defects (performance).

Security: Data and services can have different security requirements based on their scope. The platform that is going to support the services of the city should provide flexible security capabilities in order to accommodate the different needs of specific target scenarios, by providing support for confidentiality, integrity, authentication, authorization, immutability, trust and non-repudiation when needed.

Data protection and privacy: Data protection and privacy issues should be addressed at several levels, from the low-level platforms to specific end-user applications. The system should use encryption and technology to authenticate and secure data in transit as well as mitigate the risk of data theft by encrypting physical storage/media to protect data at rest. It is necessary to provide systems for monitoring against any attacks; if a breach occurs (e.g. data are accessed by unauthorized entities) the system should be able to properly react with defined procedures. As data providers need to restrict the access of data source(s) to third parties, the system has to allow defining and managing policies for data and service access control. Both the data provider and the data consumer must comply with the privacy and data protection policy; thus, the system should provide procedures and guidelines in order to ensure compliance with regards to data protection rules. In addition, the system should provide both data anonymisation and aggregation functions in order to delete personal or restricted information.

Distributivity: IoT will likely evolve in a highly distributed environment. In fact, data might be gathered from different sources and processed by several entities in a distributed manner.

Communication: Communication in IoT can happen between the sensor/actuator and the gateway, between the gateway towards the platform, or in some cases (e.g., NB-IoT, LTE, etc.) directly from sensor/actuator to the platform. Communication with the sensor to the gateway (when wireless) is possible in numerous ways. At this moment, a variety of standards are available, so the platform should be able to handle different protocols (e.g., LoRa, 802.15.4, NB-IoT, WiFi, LTE, GPRS, etc.) and be flexible to incorporate future changes. LoRa is a wireless communication technology developed to create the low power, wide-area networks (LPWANs) required for machine-to-machine (M2M) and Internet of Things (IoT) applications.

All these requirements were considered for the development of the system for the U-Bike project.

The U-Bike Project

The U-Bike Portugal project will provide classic and electric bicycles to the academic community, students, teachers and academic staff, in order to encourage a more sustainable urban transportation, so the bikes are entrusted to the academic community for a long-term use, leading to a regular use of this means of transport. This project aims to reduce primary energy consumption and CO2 emissions. This project is coordinated by the Institute of Mobility and Transport, I.P. (IMT), and is co-financed by the programme Portugal 2020, particularly through PO SEUR – Operational Programme for Sustainability and Efficient Use of Resources (https://poseur.portugal2020.pt/pt/candidatura-detalhe/).

Beja is a small city in Portugal that provides more means of transportation and electric vehicles. The Polytechnic Institute of Beja, located in the city of Beja, is a Higher Education institution, which promotes applied research in the regional socio-economic context. Accordingly, the U-Bike IPBeja is the Polytechnic Institute of Beja's project, which is inserted in the U-Bike Portugal and aims to reduce primary energy consumption and CO2 emissions at the same time it promotes physical activity for health (smart living). The U-Bike IPBeja project provides Beja's academic community 120 conventional bicycles and 80 electric bicycles. All the academic members can join this project but considering the goal of the project, candidates who use individual motorized transport on their trips to and from school are the first option.

RELATED WORK

Many researchers have identified the importance of smart mobility. Fatnassi, Harrabi, and Chaouachi (2017) show the relationship between the reduction of carbon emissions in urban areas (Smart Environment), road traffic and travel time (Smart mobility) in order to improve the quality of life of the inhabitants in urban areas (Smart Living). This work is very similar to this chapter proposal. The main difference is that this chapter proposes a bike-loan system and they propose an electric car-sharing system.

Corsar, Edwards, Nelson, and Papangelis (2014) propose a passenger information system (GetThere) using RTPI (real-time passenger information) that integrates heterogeneous transport information with crowd-sourced vehicle locations provided from mobile devices. This system asks its users to provide observations about public transport, such as vehicle location, via their smartphone. These observations are integrated using linked data principles with various open data sets to deliver RTPI via the same app. This work uses RTPI to collect GPS information, as well as this chapter proposal. The main difference is that this chapter proposal addresses the reduction of carbon emission and different devices (smart-locks/GPS and smartphone) collect the data.

Qiu, Badr, Wang, and Shan (2017) show how to integrate physical and social information so as to enhance the riders' experience of using bike sharing services in New York City. This proposal is based on real time collect, process, and model data on riders' profiles and mobility needs, and from the information system and social media. The main difference is that this chapter proposal addresses the reduction of carbon emission and different devices (smart-locks/GPS and smartphone) collect the data.

Mangiaracina, Perego, Salvadori, and Tumino (2017) state that papers on ITS- Intelligent Transport Systems for urban mobility are mainly focused on technology aspects, neglecting value creation (e.g. cost–benefit analyses), or address the topic in a very marginal way. Although some impacts and benefits have been examined in terms of travel time reduction and environmental effects, there is still a general lack of quantitative models (i.e. only 35% of the papers reported a quantitative assessment) for measuring the overall impacts of ITS technologies in an urban context. The framework proposed in this work addressed this issue by means of CO2 and physical activities indicators.

Razzaque and Clarke (2015) outline a set of services to integrate bike sharing into multimodal journal planning. To accomplish this, a framework is proposed for a next-generation IoT-integrated bike-sharing systems. This framework, with appropriated middleware features, collects information among heterogeneous computing and communication devices using real-time as well as non-real-time information. The main difference between this work and this chapter proposal is the business goal, i.e. they propose a bike-sharing service and deal with reservation and maintenance issues.

IOT FRAMEWORK FOR SMART MOBILITY

Planning, designing, deploying, and measuring the smart mobility concept is very important since it can have an impact on several aspects of city life such as how and where people live, fulfil their needs, wants, and demands, as well as how to meet present needs without compromising the ability of future generations to meet theirs. A supporting framework around which smart mobility can be built is needed in order to manage communications and services, and integrate a diverse range of devices, as well as applications, to achieve smart mobility goals.

The applications are designed in the context of a smart mobility ecosystem so that the stakeholders (for example, the end user) can reach their goals. As described before, the Internet of Thinks (IoT) concept is an essential part of the framework, i.e. "*a network of physical objects that contain embedded technology to communicate and sense or interact with their internal states or the external environment*" (Gartner, 2019).

Given the complexity of the problem, the framework should guide the development of a smart mobility system taking into account several heterogeneous components and how they should be integrated. To manage this complexity, the smart mobility IoT framework proposal is composed of two parts: Structure and Operation. Inspired on IoT architectures proposed by (Abdmeziem, Tandjaoui, & Romdhani, 2016) and (Razzaque & Clarke, 2015), the part dealing with Structure describes four different components corresponding to Physical Layer, Communication Layer, User Applications, and External Services for smart mobility system deployment. The Operation part consists of three different phases with which the smart mobility components' interaction takes place and is based on (Borgia, 2014). These phases are i) collecting; ii) transmission and iii) processing, managing and utilization.

IoT Framework: The Structure Part

Inspired on (Razzaque & Clarke, 2015) and (Abdmeziem, Tandjaoui, & Romdhani, 2016), the Smart Mobility IoT Framework Structure consists of four components: i) Physical Layer; ii) Communication Layer; iii) User Applications; and iv) External Services.

The Physical Layer consists of many types of sensors (e.g. temperature sensors, ultrasonic sensors, pressure sensors, proximity sensors, and touch sensors), cameras, smartphones and tablets, among other equipment/devices, through which the required data, e.g. environmental data, are gathered. One important element of the layer is GPS, used for geo-referencing. The heterogeneity of equipment that can coexist in this layer reinforces the need for a framework that supports interoperability.

The Communication Layer allows the communication, i.e. transmitting data/information, between framework components, mainly in a two-way communication context. There are many possibilities to address communication. Short-range communications is a possible scenario, which could be open source standard solutions (e.g. Bluetooth Low Energy, ZigBee, Dash7, and Wireless M-BUS) as well as proprietary solutions (e.g., Z-Wave, ANT). RFID and NFC technologies are also well-known shorter-range types of wireless communication. In addition to short-range communications, other approaches are required to access the network, for example through gateways and physical heterogeneous technologies. There are many possibilities (e.g. wired, wireless, satellite technologies). Among the wired technologies, the reference standard is Ethernet (IEEE 802.3). A common way to access the network is through Wireless LAN (WLAN). Due to their flexibility, wireless technologies will be the main communication paradigm for the IoT. However, the limited wireless spectrum available for cellular networks constitutes a major constraint in the widespread use of these wireless technologies. Cellular networks (e.g. GSM, GPRS, UMTS, HSPA+ and LTE) play a central role in the way of accessing the network. Satellite communication technologies can also be used as a means to connect to the Internet. They are particularly useful for users located in remote areas that cannot access broadband connections, where deploying terrestrial connection is costly, or as a way to cross the sea/ ocean. Another concept to carry out the IoT transmission is Machine-to-Machine (M2M) communications. According to ETSI Technical Committee definition ''Machine-to-Machine (M2M) communication is the communication between two or more entities that do not necessarily need any direct human intervention''. M2M does not refer to any specific communication technology, but rather to a number of both wired and wireless technologies that allow devices to communicate. Regardless of the paradigm prospective, M2M communications essentially deal with combining electronics, telecommunication and information technologies in order to connect from billion to trillion of devices and remote systems, and are characterized by low power, low cost and low human intervention (Chen, Wan, & Li, 2012).

The User Applications component collects and/or manages (massive amount of) data. In fact, this component is one of the front ends of the system and provides a range of possible applications to achieve the stakeholders' goals.

Finally, External Services are functionalities provided by external entities that are in charge of processing and managing the obtained data to support smart mobility requirements, as well as stakeholders' goals. For example, they could be a software interface between the Physical Layer and the User Applications. To accomplish this, it provides the required abstraction to hide the heterogeneity and the complexity of the underlying technologies involved in the Physical Layer, i.e. using heterogeneous technologies/ devices as compatible services. On the other hand, it is crucial to provide elements to store and process the massive amount of data collected by Physical Layer and User Applications. To reach these goals, cloud computing is one of the primary technologies in this component. According to NIST (Mell & Grance, 2011), cloud computing provides shared computing resources over the network, with minimal management effort. To reach this goal, smart mobility components (Physical Layer and User Applications) need to be connected to the cloud to store and retrieve the required high volume of data. It also provides efficient mechanisms to access virtual storage, greatly increasing the local storage capacity, which is very limited in most cases. Depending on the number and variety of cloud services involved, some of IoT and smart mobility requirements described in the Section Background are addressed, i.e., security, availability, interoperability and resilience to failure. For example, the way cloud security is provided will depend on the cloud provider in place, e.g. Azure Security Center for IoT (Azure, 2019) or Securing Internet of Things (IoT) with AWS recommendations (Cheema, 2019). However, the implementation of cloud security processes should be a joint responsibility between the system owner and the solution provider.

IoT Framework: The Operation Part

Inspired by (Borgia, 2014), the application of the IoT concept goes through three different phases with which the smart mobility components' interaction takes place: (i) the collection phase, (ii) the transmission phase, and (iii) the process, management and utilization phase.

Collection phase: it refers to procedures for sensing the physical environment, collecting real-time and/ or non-real-time data, and reconstructing a general perception of it. Technologies such as sensors provide identification of physical objects and sense physical parameters. Data acquisition is performed by using different technologies/ devices described above in the Physical Layer of the framework structure. It is also possible, and sometimes desirable, to collect data given by users. The Physical Layer and User Application components of smart mobility Structure achieve the objectives of this phase.

Transmission phase: it includes mechanisms to send the collected data from the Physical Layer to User Applications and/ or to External Services, as well as send data/ information from User Applications to External Services, and vice-versa. The transmission phase can be accomplished by identifying the best solution for each situation as described above in Communication Layer.

Process, Management and Utilization phase: it deals with processing and analysing data/ information flows, and providing feedback to control applications. Moreover, it is responsible for critical tasks such as device discovery, device management, data filtering, data aggregation, semantic analysis, and information utilization. The User Applications and External Services components of smart mobility structure are used to accomplish this phase. Considering the External Services component, cloud computing is a central element of the phase. Cloud computing service providers manage diverse components and technologies in a uniform way and access via standard interfaces. In short, it provides a common set of services and an environment for IoT components' integration. Due to the numerous tools provided by these platforms, possibilities regarding what can be developed are endless.

THE U-BIKE SYSTEM

This section describes the U-Bike IPBeja system through requirements and a framework structure and operation. The requirements establish a high-level view of what the system might do and the benefits that it might provide before the development of the system.

The requirements are obtained from the stakeholders' point of view and from smart mobility requirements described in the Section Background. Stakeholders include anyone who is affected by the system in some way and so anyone who has a legitimate interest in it (Sommerville, 2016). U-Bike stakeholders are academic users (end-users) and project managers, as well as external stakeholders (IMT and POSEUR) or service providers (cloud service providers).

According to the project's goals, the system requirements are:

- It should be easy to use by all academic users.
- Having the means to assess bike monitoring and maintenance, in particular electric bikes.
- Avoiding situations of theft.
- Analysing the effects of the project regarding its objectives, targets and expected results (metrics) by means of reports. These metrics are CO_2 reduction, km travelled per week by each user (i.e. ride information), adherence rate of the academic community population to the project (by type of target audience) and the gain of healthy habits.
- Analysing the environmental sustainability and durability of the results of the project by means of reports.
- It must be available 24/7.
- It must comply with General Data Protection Regulation (GDPR).

In addition to the requirements explicitly stated in the project's goal, others were identified based on the Background section, i.e. non-functional requirements. Non-functional requirements describe constraints on the system's features, e.g. timing constraints. (Sommerville, 2016). Moreover, non-functional requirements are often called quality attributes when the software/ system should be secure, highly-available, scalable. Non-functional requirements may affect the overall architecture of a system rather than the individual components and clearly identify a constraint that has to be considered by the system designers/ developers. For example, to ensure that availability requirements are met in a system, you

Figure 1. U-Bike IoT framework for smart mobility

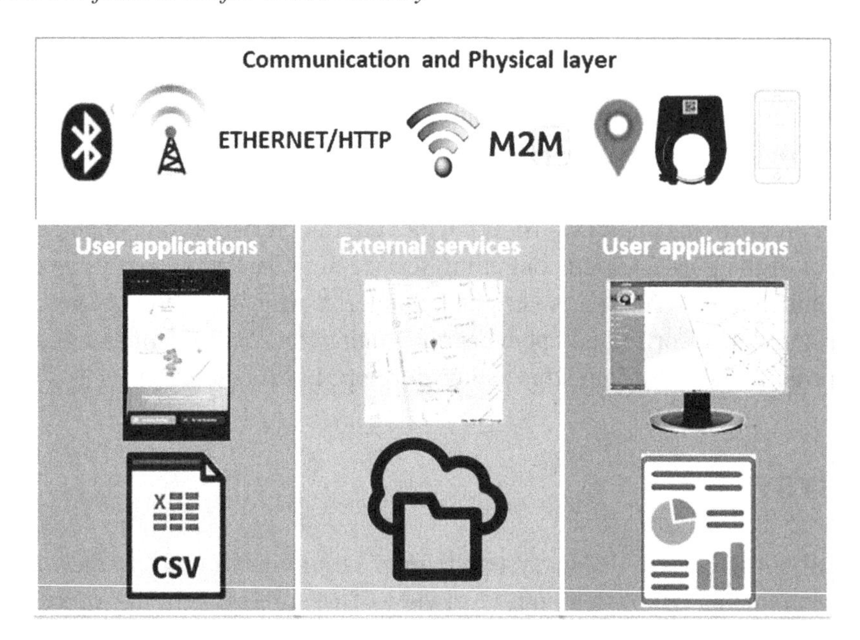

may have to organize the system to allow communications between components. For all these reasons, the non-functional catalogue described in the section Background is another important feature of the smart mobility IoT framework.

Based on the requirements above, the system designers/ developers use the framework with regards to structure and operationalization, as well as the non-functional catalogue described in SynchroniCity technical framework, for the development of the U-Bike system.

The U-Bike System Structure

According to the U-Bike IPBeja requirements project, this section describes the system regarding the Communication and Physical Layer, User Applications, and External Services (smart mobility IoT framework Structure). Figure 1 shows basic elements of the proposed solution.

Physical Layer: In order to accomplish its goals, the U-Bike IPBeja system uses a SmartLock device and smartphone. Together with communication technologies (see M2M and Bluetooth description in Communication Layer for more information), the SmartLock devices are used to (un)lock the bike and, consequently, to minimize theft and vandalism. The SmartLock and QRcode attached to each bike are registered by storing unique identification in a database that runs on a cloud-computing platform (see External Services for more information). The SmartLock attached to each bike is equipped with GPS and ensures the collection of data (kilometres travelled) for monitoring, i.e. geo-location and traceability. The academic users use the bike via the U-Bike mobile application (app) designed to run on an Android or iOS operating system. The mobile device must have GPS to ensure the collection of data (kilometres travelled), i.e. geolocation and Wi-Fi or GPRS/3G/4G or LTE (see Communication Layer for more information) to load data to a cloud-computing platform (see External Services for more information).

Figure 2. U-Bike IPBeja system from stakeholders' point of view

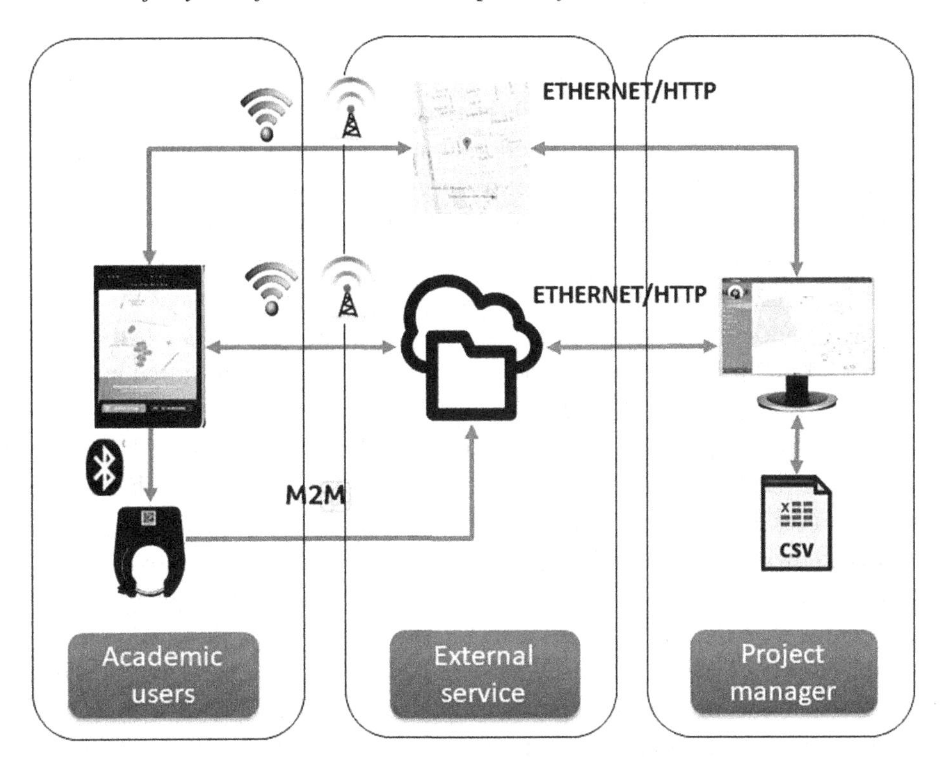

Figure 2 shows how physical and communication elements of the framework interact from the academic users' point of view.

Communication Layer: This layer is responsible for transmitting data among all the elements of the U-Bike IPBeja system. The system uses accepted standards for communication and exchanging data, so new elements can easily access information already available. Gateways and APIs might act as a glue between the system's elements (see External Services for more information). Approaches are therefore required for accessing the network through gateways and heterogeneous technologies (e.g., wired, wireless, satellite technologies). Connecting all these parts of the U-Bike project is done by different technologies: Wi-Fi, Bluetooth, GPRS/3G/4G, LTE and M2M. Figure 2 shows the communication technologies used by each element of the U-Bike IPBeja system from the stakeholders (academic users, external service and project manager).

Bluetooth is used to transmit the data in the QRCode (glued to the SmartLock device) to the U-Bike app in order to unlock/ lock the SmartLock device.

Machine-to-Machine (M2M) is another concept used by the Communication Layer. As described before, the term M2M refers to a number of both wired and wireless technologies that allow devices to communicate. In the U-Bike IPBeja project, M2M is used to transmit SmartLock data, i.e. location and status (opening/ closure) of the SmartLock, including battery status. The SmartLock data are then read wirelessly and exchanged within cloud-computing platform (see more information in External Services). In order to communicate data, SmartLock is equipped with a SIM card and connects to the cloud-computing platform by using 3G and a static IP address defined by the service provider. -

Wi-fi, GPRS/3G and LTE are responsible for transmitting the U-Bike app data, i.e. user and ride information (see more information in User Applications), to cloud-computing platform (see more information in External Services) and vice-versa. To perform this communication a JSON REST API is used. In addition, the Google Maps Platform APIs is used to transfer maps data to/ from the U-Bike app.

Ethernet/HTTP is used to transmit data stored in cloud-computing platform (see more information in External Services) to the U-Bike web application (see more information in User Applications), and vice versa. To perform this communication a JSON REST/HTTP API is used.

External Services: The M2M service provider manages the SIM card that is attached to the SmartLock and notifies the SmartLock status to the U-Bike app. The SmartLock device is built by its sellers considering the information given by the M2M service provider, i.e. ISSID / IMEI / MAC address, provided in an Excel file. The Excel data file and QRcode identification are stored in the cloud, as well as the user's data and riding data to allow tracking and monitoring services. Therefore, the proposed system uses cloud-computing platform to save and process data. The U-Bike IPBeja system uses the AWS cloud-computing platform and stores data in MongoDB (an open source database management system (DBMS) that uses a document-oriented database model). The data are processed, analysed and converted to usable knowledge by the U-Bike IPBeja project (see User Applications).

The U-Bike system uses the AWS cloud-computing platform through the Amazon API Gateway, which is a full management service that allows developers to create, publish, monitor, and secure APIs at any scale. Systems' developers create REST and WebSocket APIs that act as a front door for applications to access data, business logic, or real-time communication applications. (AWS, Amazon API Gateway, 2019) . Thus, the U-Bike system uses Terraform by HashiCorp, an AWS Partner Network (APN) and member of the AWS DevOps Competency. Terraform is an open-source tool that allows developers to create and update the Amazon Web Services (AWS) infrastructure. It codifies the cloud-provider's APIs into declarative configuration files that can be shared amongst developer team members, treated as code, edited, reviewed, and versioned (Campbell, 2018).

Bike users and project managers could visualise bike rides and bike location through the Google map API (Google Cloud, 2019). This API allows real-time data collection.

User Applications: The User Applications are available via web site and app. The stakeholders related to this component are: i) academic users that are accountable for the bike and use the app functionalities and ii) the project managers at IPBeja that are responsible for the system and use the U-Bike web functionalities.

The academic users need the app to:

- Access to SmartLock's data for maintenance issue (e.g. status).
- Lock and unlock the SmartLock device installed on bikes.
- Access the riders' information. Wi-Fi, LTE, GPRS/3G/4G is used to communicate with the cloud-service platform. GPS and the Google maps API are also used to visualise roots.

Figure 2 shows communication technologies used by academic users to accomplish their goals.

The IPBeja U-Bike project manager uses the web site for:

- Bike reservation: this functionality will allow academic users, the future bike borrowers, to book a bike for a period between 1 and 6 months.

Table 1. Relation between phases and components

Phase	Structure
Collect	- Physical Layer (smartphone; GPS and SmartLock); - External Services (AWS cloud (APIs); Google (API)) - User Applications (app; web and excel files)
Transmit	Communication Layer (Wi-Fi; LTE, GPRS/3G/4G; Ethernet; M2M; Bluetooth)
Process, Management and Utilization	- User Applications (app; web and Excel files) - External Services (AWS cloud (APIs); Google (API))

- Academic user information: the system handles the users' information (name, email) and user profile.
- Bike tracking: the system handles the information of the rides. This information is crucial to identify whether the metrics of the project are accomplished or not (e.g. emissions of carbon and for health care service).
- Bike monitoring: this functionality will allow giving feedback on the use of bike/ SmartLocking, including reporting any bike malfunctions via app. It provides automatic alerts and maintenance tasks via app and email, including scheduling and resource selection for required maintenance operations.
- Health care & environmental services: CO2 emission and physical activity indicators are analysed.
- Import and export data: Excel files are used to manage devices, the bike's users and riding information. This information is related to health and transportation, i.e. the user profile, as well as the device's data.

The app is available at https://ubikeonline.com/ .

The U-Bike System Operation

As described before, the interaction between the U-Bike components takes place during three phases: (i) the collection phase, (ii) the transmission phase, and (iii) the process, management and utilization phase. Table 1 shows the components needed to accomplish the goals of each phase.

The Collection phase is carried out by means of the Physical Layer, External Services and User Applications. The system collects academic users' and riding data from diverse sources: smartphone, app and web applications, and an Excel file. The system gathers data from bikes and rides from SmartLock/ GPS and QRcode, as well as Excel files. The system gathers the bikes' users data from the app. The system validates user and bike data, and gathers GPS coordinates (i.e. geo-referencing) using External Services (AWS and Google Maps). All these data are collected in real-time, except the Excel data.

The Transmission phase includes mechanisms to send and receive the data/ information collected, processed and managed in the other phases. This phase should be performed by means of the Communication Layer described before. Figure 2 shows the communication technologies used by each element of the U-Bike IPBeja system from the stakeholders' (academic users, external service and project manager) point of view.

The Process, Management and Utilization phase is achieved through User Applications and External Services. User Applications, the web site and the app use External Services for academic users and project managers to carry out their activities and achieve their goals, i.e. all data are loaded and processed in the cloud. The main difference between the users' applications is that the amount of data handled by the app is less than the amount of data handled by the web site. Finally, the web site supports importing and exporting Excel files, and provides project managers with an interface to interact with data in those files, e.g. bike rides.

U-Bike Non-Functional Requirements

This section describes how the system supports non-functional requirements. The non-functional requirements are selected regarding the goals of project. Interoperability and security requirements were of primary importance in the U-Bike system. Solutions to cope with those requirements were sought in all the components of the framework. For example, the system should guarantee a secure environment in terms of secure communication/ authentication (Communication Layer), data and device integrity (Physical Layer), user privacy and personal data (User Applications), and the reliability of the environment and of the involved parties (External Services).

Availability: At this point, it is important to ensure the 24/7 availability of each element in the system. The system knows (in real-time) the bikes'/ SmartLock status and their spots via app, web site and the cloud-service platform (AWS). In case on bike unavailability, the system informs the bike's owner and the project manager at IPBeja. In the case of web site and cloud service unavailability, the project manager must take care of incidents and business continuity.

Security: the system needs to guarantee the integrity of communication, as well as data privacy (see Legal issues). All communication messages are hashed with an HMAC-SHA256 signature. All further steps involve anonymisation techniques (masking identity and location of the bike's user). The data are securely sent to servers by using different techniques: Hypertext Transfer Protocol Secure (HTTPS) is a variant of the standard web transfer protocol (HTTP) that adds a layer of security to the data in transit through a secure socket layer (SSL). M2M ensures secure data communication through a private IP address with Internet connection. Finally, theft and bike vandalism is managed by means of GPS and SmartLock devices. In addition, the system promotes integrity by means of timestamps. Moreover, the system encrypts data in order to prevent the compromise of the users' data and credentials, i.e. data protection.

Scalability: the system must handle a growing amount of work by adding resources. This is managed by the AWS cloud-service platform, by means of an infrastructure. The AWS service secures, controls, and manages your devices from the cloud, i.e. remotely manages IoT devices at scale, as well as collected and saved data from User Applications, and uses them to analyse equipment/ devices. It has also load balancers to readjust a large number of requests simultaneously. Therefore, an API was designed and horizontal scaling was the natural choice. Moreover, open standards are used to guarantee resources integration, being HTTP/REST/JSON our main task.

Interoperability: devices from different sellers will have to cooperate in order to achieve common goals. To accomplish this, the network will be composed of heterogeneous communication technologies (e.g. wired, wireless, cellular technologies). This implies that communication protocols should be able to interface with the different underlying networks, to support different addressing schemes, and to dynamically adjust them when needed in order to reduce the network load. Moreover, open standards are used to guarantee resources interoperability, being HTTP/REST/JSON the main task.

Legal issues: any user data protected by GDPR will be subject to a data sharing agreement, in order to be allowed to analyse the aforementioned data handled by the system. The system must guarantee General Data Protection Regulation (GDPR) compliance.

SYSTEM ASSESSMENT

In the context of the U-Bike project, a system assessment will identify the strengths and weaknesses to improve the operational efficiency of the project. This section describes the system assessment in terms of its behaviour (operational functionality) based on the opinions of the project's stakeholders (academic users and project managers). This assessment was performed by means of the focus group method, which is an interviewing technique to elicit stakeholders' opinions about something. The Focus group is a small number of people (usually between 4 and 15, but typically 8) brought together with a moderator to discuss a specific topic. (Mazza & Berre, 2007). (Mazza & Berre, 2007) and (Choe, Kim, Lehto, Lehto, & Allebach, 2010) show how the focus group is used to evaluate clients'/ users' requirements and reactions to new products. Besides this, the focus group provides qualitative data more quickly and can be used to identify user needs and feelings. Although focus groups offer interesting advantages, they have some limitations, such as the fact that the small number of participants limits the generalization of the research, and the nature of the responses sometimes makes the analysis of the result difficult. Nielsen (Nielsen, 1993) suggested some guidelines on how focus groups should be used in Human Computer Interaction. The guidelines are related to i) the number of participants (six to ten interviewees), ii) preparing a list of topics to be discussed, and iii) ways to guarantee that all interviewees participate to the discussion.

Two focus groups were created, one to analyse the app and the other to analyse the web site. Therefore, the academic users and the project managers were invited to analyse the app (group one) and the project managers were invited to analyse the web (group two).

The recruiting of project managers for the focus group meeting was done by invitation after identifying a free timetable slot. The invitation was sent to the project managers and their collaborators, i.e. the users of the web application. Two people were present out of three invitations.

The recruiting of academic users was done by invitation (email). The invitation was sent to a small group of students, teachers and staff with different profiles, i.e. with different types of smartphone models (for example: iOS/ Android), and Information and Communications Technology (ICT) background. Ten people were present out of twenty invitations. The moderator was someone who was knowledgeable about the system, but who was not using it.

Participants were told that the objective was to learn their different opinions about the system with no need for general agreement. The questions were formulated with the intention of eliciting maximum information from the discussions, fostering the intended share-and-comparison effect among participants. The moderator also facilitated the free flow of thoughts in an attempt to generate as many opinions and general comments as possible. To promote the discussion during group one interview, the moderator asked the following questions: *i) what is your opinion about the download/ installation of the app? Is it easy to carry out? ii) Is the app working properly? Does it crash or does it not open/ respond? iii) Is it easy to lock and unlock the bike (SmartLock) via app? Does it crash or does it not open /respond? iv) Is the app taking up too much storage space in your device? Is the app using too much data over the network? v) Do you receive notifications and are they easy to understand?*

The questions that guided group two interview were: *i) Is the web site working properly? Does it crash or does it not open /respond? ii) Is it easy to monitor bikes, users and rides? iii) Do you receive notifications and are they easy to understand?*

Group one considered that the app was working properly. One example of comment was: "*The app has never crashed, but sometimes it was slow.*"

Some academic users claimed that the unlock functionality did not work sometimes. The project managers informed the group that this issue was related with communication requirements and users´ devices. On the one hand, mobility greatly affects wireless communications. On the other hand, the users' devices have limited resource capabilities: for example, to save energy they may temporarily disconnect from the network and wake on demand.

The users had no opinion regarding technical issues such as memory and battery. They had no problems with app memory storage, as well as mobile phone battery consumption. Considering communications (e.g. Wi-Fi and cellular data connection), an academic user who used a cellular connection complained about consumption.

A predominant criticism was that the app did not show insightful messages that allowed the user to improve the use of the system, including the bike. An example of a comment regarding this issue was: "*The indication of the GPS battery is too discreet and does not tell you what it is. I only know it exists because they told me after I already had used the bike for a few days and was panicked because the GPS had no battery.*"

All the users agreed that the app download was easy and fast, which was the most positive comment. They also agreed that end-of-the-ride functionality is easy to perform because "*the end of the ride button is very visible as soon as we open the app*".

Most users stated that the start ride functionality was quite satisfactory, suggesting that "*Instead of **Scan code** it should read **scan code to start riding***". Another suggestion was: "*The map should only show where the bike I'm using is.*"

Group two conveyed the information that the web never crashed or got slow. Project managers found it very useful to have the possibility to monitor bike/ battery status, and appreciated it: "*I think that the green, yellow and red eye (eye represents the battery status) are very useful.*" They proposed improvements to i) the notifications management; ii) the data visualisation, mainly in the maps, and iii) data importing/ exporting functionality through Excel files. Comments regarding those issues were: "*when opening the MS Excel file, some data are incomplete or absent*"; "*I miss importing bikes' and users' data from Excel files*"; "*I do not understand the usefulness of the **Geo-referencing map** option*".

Despite the analysis limitations, considering focus group characteristics the results were very important to propose improvements to communication systems and notification management. Since not all system functionalities have been fully explored, which will only be possible after one year of data collection, the current system is satisfactory for all stakeholders.

CONCLUSION

This chapter describes the U-Bike IPBeja system to illustrate an IoT framework in the field of smart mobility. The IoT framework has the potential to support smart mobility goals by enabling communications with and among different devices, leading to the integration of different types of applications, devices and services. An available, interoperable, and secure system was described to bring the IoT concept closer to the U-Bike project's goal. Thus, the IoT framework for smart mobility guided the U-Bike system development and helped developers and designers make technical decisions. Technical decisions influence and impact the non-functional requirements of a system, therefore technical decisions are difficult to make because there are no single optimal solutions for a given non-functional requirement (for example, security affects the communication component as well as data management in AWS). According to the results of this study, the communication component has a profound effect on the overall system availability. To address availability on the U-Bike system, considering the users' profile and smartphones with unknown characteristics for example, many decisions need to be taken. Unfortunately, those decisions are linked to budget constraints.

Currently, the system supports all U-Bike project requirements, and shortly the system will have enough data to analyse CO2 emissions and the effect on the users' health considering the physical activity performed, i.e. km travelled in a given period.

ACKNOWLEDGMENT

This project is financially supported by Portugal 2020 under the Operational Programme for Sustainability and Efficient Use of Resources (PO SEUR).

REFERENCES

Abdmeziem, M., Tandjaoui, D., & Romdhani, I. (2016). Architecting the internet of things: State of the art. In A. K. Shakshuki (Ed.), *In Robots and Sensor Clouds* (pp. 55–75). Cham, Switzerland: Springer; doi:10.1007/978-3-319-22168-7_3

AWS. (2010). *AWS CloudFormation.* Obtido de https://docs.aws.amazon.com/AWSCloudFormation/latest/UserGuide/aws-properties-as-group.html

AWS. (2019). *Amazon API Gateway.* Obtido de https://aws.amazon.com/api-gateway/?nc1=h_ls

Azure, M. (2019). *Azure Security Center for IoT in public preview.* Obtido de https://azure.microsoft.com/en-us/updates/azure-security-center-for-iot-in-public-preview/

Badii, C., Bellini, P., & Difin, A. (2018). Sii-Mobility: An IoT/IoE Architecture to Enhance Smart City Mobility and Transportation Services. *Sensors (Basel), 19*(1), 1. doi:10.339019010001 PMID:30577434

Borgia, E. (2014). The Internet of Things vision: Key features, applications and open issues. *Computer Communications, 54*, 1–31. doi:10.1016/j.comcom.2014.09.008

Campbell, J. (2018). *Terraform: Beyond the Basics with AWS.* Obtido de https://aws.amazon.com/pt/blogs/apn/terraform-beyond-the-basics-with-aws

Cheema, M. (2019). *AWS Security releases IoT security whitepaper.* Obtido de https://aws.amazon.com/pt/blogs/security/aws-security-releases-iot-security-whitepaper/

Chen, M., Wan, J., & Li, F. (2012). Machine-to-Machine Communications: Architectures, Standards and Applications. KSII *Transactions on Internet and Information Systems*, 6(2). doi:10.3837/tiis.2012.02.002

Choe, P., Kim, C., Lehto, M., Lehto, X., & Allebach, J. (2010). Evaluating and Improving a Self-Help Technical Support Web Site: Use of Focus Group Interviews. *International Journal of Human-Computer Interaction*, 333–354. doi:10.120715327590ijhc2103_4

Corsar, D., Edwards, P., Nelson, J., & Papangelis, K. (2014). Mobile phones, sensors & the crowd: lessons learnt from development of a real-time travel information system. *Proceedings of the First International Conference on IoT, (URB-IOT '14)* (pp. 99-101). Brussels, Belgium: ICST (Institute for Computer Sciences, Social-Informatics and Telecommunications Engineering), ICST. doi:10.4108/icst.urb-iot.2014.257328

Dell'Era, C., Altuna, N., & Verganti, R. (2018). Designing radical innovations of meanings for society: Envisioning new scenarios for smart mobility. *Creativity and Innovation Management*, 1–14. doi:10.1111/caim.12276

Fatnassi, E., Harrabi, O., & Chaouachi, J. (2017). Toward a smart mobility system: integrating electric vehicles within smart cities. *Proceedings of the Genetic and Evolutionary Computation Conference Companion (GECCO '17)* (pp. 275-276). Germany: ACM. doi:10.1145/3067695.3076075

Gartner. (2019). *IoT Glossary.* Obtido de https://www.gartner.com/it-glossary/internet-of-things/

Geotab. (2018). *GeoTab - management by measurement.* Obtido de https://www.geotab.com/blog/what-is-smart-mobility/

Google Cloud. (2019). *Google Maps Platform.* Obtido de https://cloud.google.com/maps-platform/?&sign=0

Hung, M. (2017). *Leading the IoT.* Obtido de https://www.gartner.com/imagesrv/books/iot/iotEbook_digital.pdf

Iansiti, M., & Lakhani, K. (2014). Digital ubiquity: How connections, sensors, and data are revolutionizing business. *Harvard Business Review*, *92*(11), 91–99.

Lee, I. (2019). The Internet of Things for enterprises: An ecosystem, architecture, and IoT service business model. *Internet of Things, 7*, 100078. doi: 0 078 doi:10.1016/j.iot.2019.10

Li, S., Xu, L., & Zhao, S. (2015). The internet of things: A survey. *Information Systems Frontiers*, *17*(2), 243–259. doi:10.100710796-014-9492-7

Maggio, M., Arigliano, F., Özdemir, O., Cantera, J., Kim, E., Maestro, I., . . . Capossele, A. (2018). *SynchroniCity: Delivering an IoT enabled Digital Single Market for Europe and Beyond.* Obtido de https://synchronicity-iot.eu/wp-content/uploads/2018/09/SynchroniCity_D2.10.pdf

Mangiaracina, R., Perego, A., Salvadori, G., & Tumino, A. (2017). A comprehensive view of intelligent transport systems for urban smart mobility. *International Journal of Logistics Research and Applications*, *20*(1), 39–52. doi:10.1080/13675567.2016.1241220

Mazza, R., & Berre, A. (2007). Focus group methodology for evaluating information visualization techniques and tools. *11th International Conference Information Visualization (IV'07)* (pp. 74-80). IEEE. 10.1109/IV.2007.51

Mell, P., & Grance, T. (2011). *The NIST definition of cloud computing*. Special Publication 800-145 - NIST.

Miorandi, D., Sicari, S., Pellegrini, F., & Chlamtac, I. (2012). Internet of things: Vision, applications and research challenges. *Ad Hoc Networks*, *10*(7), 1497–1516. doi:10.1016/j.adhoc.2012.02.016

Neirotti, P., De Marco, A., Cagliano, A., Mangano, G., & Scorrano, F. (2014). Current trends in Smart City initiatives: Some stylised facts. *Cities (London, England)*, *38*, 25–36. doi:10.1016/j.cities.2013.12.010

Nielsen, J. (1993). *Usability engineering*. San Diego, CA: Academic. doi:10.1016/B978-0-08-052029-2.50007-3

Noy, K., & Givoni, M. (2018). Is "Smart Mobility" Sustainable? Examining the Views and Beliefs of Transport's Technological Entrepreneurs. *Sustainability, MDPI AG.*, *10*(2), 422. doi:10.3390u10020422

OOMap. (2019). *BIKESHARP - Global Map of Bikeshare*. Obtido de http://bikes.oobrien.com/global.php

POSEUR. (2019). *Operational Programme for Sustainability and Efficient Use of Resources*. Obtido de https://poseur.portugal2020.pt/pt/candidaturas/avisos/poseur-07-2015-31-aviso-projeto-u-bike-portugal/

Qiu, R., Badr, Y., Wang, J., & Shan, L. (2017). Developing a Smart Service System to Enrich Bike Riders' Experience. In *Proceedings 2nd International Conference on Software, Multimedia, and Communication Engineering (SMCE2017)*. Shanghai, China. 10.12783/dtcsemce2017/12468

Razzaque, M., & Clarke, S. (2015). Smart management of next generation bike sharing systems using Internet of Things. In *Proceedings IEEE First International Smart Cities Conference (ISC2)*, (pp. 1-8). Guadalajara, Mexico. doi:10.1109/ISC2.2015.7366219

Sommerville, I. (2016). Software Engineering (10th ed.). Pearson.

SynchroniCity. (2019). *Technical Framework*. Obtido de https://synchronicity-iot.eu/tech/

Tomaszewska, E., & Florea, A. (2018). Urban smart mobility in the scientific literature — bibliometric analysis. Engineering Management in Production and Services, 10(2).

Tomaszewska, E., & Florea, A. (2018). Urban smart mobility in the scientific literature — bibliometric doi:10.2478/emj-2018-0010

Viechnicki, P., Khuperkar, A., Fishman, T., & Eggers, W. (2015). *Smart mobility: Reducing congestion and fostering faster, greener, and cheaper transportation options*. Obtido de https://www2.deloitte.com/insights/us/en/industry/public-sector/smart-mobility-trends.html

KEY TERMS AND DEFINITIONS

Focus Group: Assessment method used to identify users opinions.
Framework: A basic guide to support system development.
Internet of Things: Set of physical objects that communicate using Internet.
Smart mobility: A concept that promotes better quality of life by means of technologies.
Non-functional Requirements: Set of constraints and qualities.

Chapter 10
Realizing IoE for Smart Service Delivery:
Case of Museum Tour Guide

Umar Mahmud
https://orcid.org/0000-0001-7580-6357
Foundation University Islamabad, Pakistan

Shariq Hussain
https://orcid.org/0000-0003-2093-7274
Foundation University Islamabad, Pakistan

Arif Jamal Malik
Foundation University Islamabad, Pakistan

Sherjeel Farooqui
Foundation University Islamabad, Pakistan

Nazir Ahmed Malik
https://orcid.org/0000-0002-0118-4601
Department of Computer Science, Bahria University, Islamabad, Pakistan

ABSTRACT

Widespread use of numerous hand-held smart devices has opened new avenues in computing. Internet of things (IoT) is the next big thing resulting in the 4th industrial revolution. Coupling IoT with data collection, storage, and processing leads to Internet of everything (IoE). This work outlines the concept of smart device and presents an IoE ecosystem. Characteristics of IoE ecosystem with a review of contemporary research is also presented. A comparison table contains the research finding. To realize IoE, an object-oriented context aware model is presented. This model is based on Unified Modelling Language (UML). A case study of a museum guide system is outlined that discusses how IoE can be implemented. The contribution of this chapter includes review of contemporary IoE systems, a detailed comparison, a context aware IoE model, and a case study to review the concepts.

DOI: 10.4018/978-1-7998-2112-0.ch010

INTRODUCTION

The advent of fourth industrial revolution introduced seamless connectivity among devices (Serpanos & Wolf, 2017). In IoT each device can provide smart processing for specific application types however in IoE smart processing can increase the power of activity recognition to include various services. The services can range for smart music recommender, smart movie recommender, smart meal recommender, smart location, smart museum tour, smart agriculture, smart healthcare and others. IoT was pioneered by Auto-ID lab located in MIT circa 1999 and is among the popular research and industry focused subjects of the current era (Ashton, 2009). IoT is the evolution of computers staring from networks to the Internet, to the advent of mobile computing and finally to IoT (Perera, Zaslavsky, Christen, & Georgakopoulos, 2013). IoT enables application to use sensors present in the environment and is thus a middle layer of a pervasive architecture.

IoE is a natural evolution of IoT. IoE is the amalgamation of people, data, processing, intelligence and things. IoT when coupled with intelligence and machine learning lead to IoE. This provides situational-awareness to different applications (Jung, 2017). However, situational-awareness is generally specific, and researchers are still exploring generic methodologies. Among the chief methods of ensuring intelligence is through context awareness (Perera, Zaslavsky, Christen, & Georgakopoulos, 2013). Miraz et.al. consider extension of business and industrial processes as a difference between IoT and IoE (Miraz, Ali, Excell, & Picking, 2015). This enriches users' life and provides better services to suit their needs. However, IoE cannot survive without better intelligence. The seamless integration of business and industrial processes in everyday life open new avenues for revenue generation. Cisco predicts the industry to generate up to $ 15 trillion revenue by employing IoE by the year 2022 (Bradley, Loucks, Noronha, Macaulay, & Buckalew, 2013).

This chapter is organized starting from introduction to this chapter followed by the anatomy of smart devices. Smart device is the key technology that enables IoE. Numerous devices interact among each other as well as services in the environment which is discussed in the next section followed by characteristics of IoE ecosystem. A discussion on evidences found in literature is presented in the next section that includes a comparison. An object-oriented model of an IoE ecosystem is presented followed by a case study of a museum tour guide system. The chapter concludes with an overview and future roadmap.

Anatomy of Smart Dvices

All devices have simple structure and are based on both hardware and software components. These devices are compact and provide mobility (Strang & Linnhoff-Popien, 2004). The inexpensive manufacturing of smart devices coupled with their high demand and integration in the lives of common users demands high connectivity and data sharing. This led to the development of smart devices which could utilize available connectivity means including Wi-Fi, LTE and 5G (Bradley, Loucks, Noronha, Macaulay, & Buckalew, 2013). The devices commonly include smart phones, smart appliances and other electronic equipment that could be connected. The components of a smart devices include On-board Sensors, Connectivity Module, Smart Processing Unit and Actuators. These components are shown in Figure 1.

Figure 1. Block diagram of a smart device

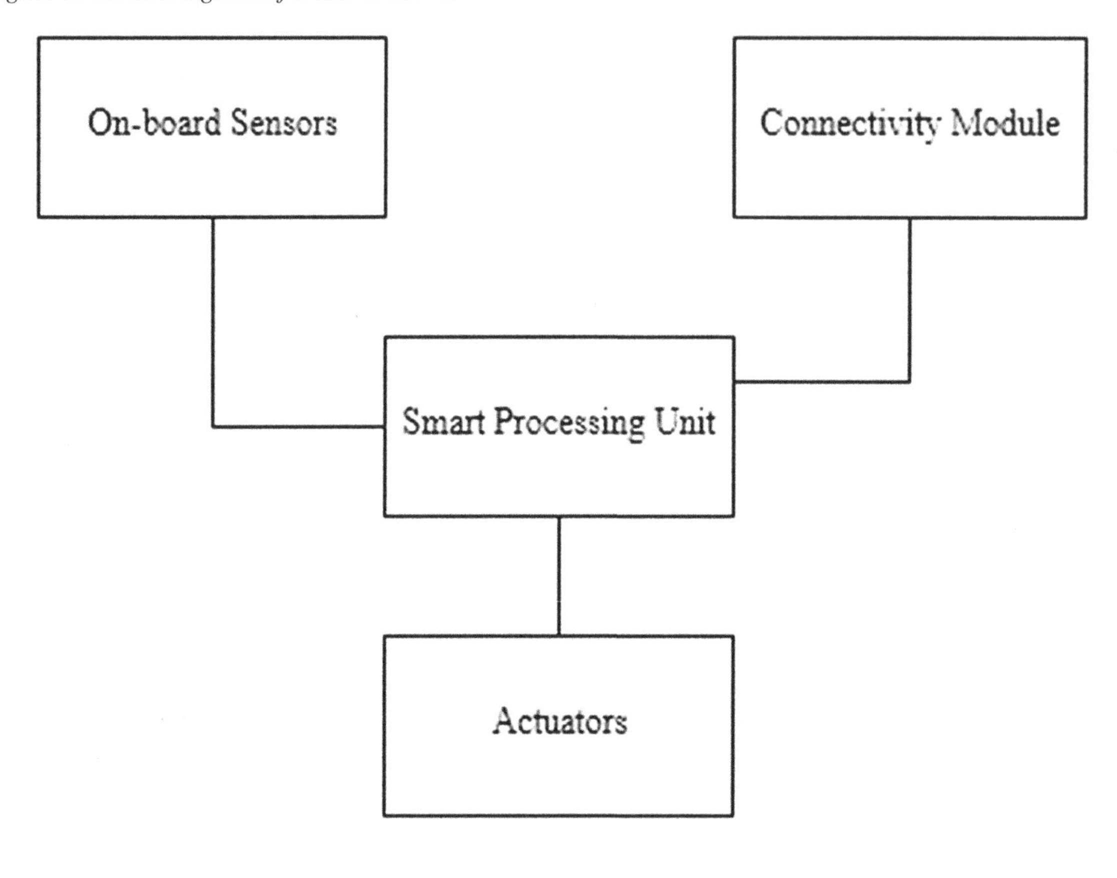

On-Board Sensors

On-board Sensors are used to sense environment related information and are internal to the device. These sensors include temperature sensor, GPS module, accelerometer, gyro meter, heart rate sensors, etc. depending on the purpose of the device different sensors are installed.

Connectivity Module

This module is used to connect with other devices, external sensors as well as the Internet. Typical methods include GSM, Wi-Fi, 4G/5G and Bluetooth. some devices have multiple connection to ensure better connectivity.

Smart Processing Unit

This unit is the heart of a smart device. The task is to recognize activity or situation based on the sensor input (Mahmud, Iltaf, Rehman, & Kamran, 2007). The input sensed by both internal as well external sensors at an instant is termed as Current Context. This context is then recognized based on prior history as well as axioms declared within the unit. Context Awareness is the mechanism that is used to recognize activity of current context utilizing machine learning (Mahmud & Javed, 2012), (Mahmud & Javed, 2014). The recommendations made by this unit is used to adjust internal actuators or invoke services present in the environment.

Actuators

Actuators are electro-mechanical devices that can be used to make changes in the environment. These devices are internal to the device and include thermostat, screen orientation, brakes and alarms, etc. the changes can be detected via sensors thus creating a feedback loop.

Enabling Internet-of-Everything

The ability of devices to communicate among each other bound by an environment leads to increase in data sensing and sharing. This data can be utilized by different devices to fulfil different purpose. For example, the temperature sensor within an AC unit can change the settings of the thermostat while, this information can be used by a user's device to ascertain what temperature is suitable based on user's health. The sharing of data decreases dependency on internal sensors while increasing functionality. Many technologies are employed to enable IoT and IoE. These include Radio Frequency Identification (RFID), Ubiquitous Codes (uCode) and Wireless Sensor Networks (WSN) (Koshizuka & Sakamura, 2010), (Li, Xiaoguang, Ke, & Ketai, 2011) .

A smart environment can increase its bound to a global level via cloud. Devices can access remote data and services. Many devices can work in collaboration to achieve different tasks thus enabling smart environments. These smart environments are the next big thing in IT and are termed as IoE. Figure 2 shows the concept of a smart environment. An IoE environment is not distance bound and remote services may be present in the environment. These services are dynamic, and devices can connect and communicate with them on the fly (Hussain, Zhaoshun, & Toure, 2013), (Hussain, Wang, & Toure, 2014). The application areas range from smart homes to smart classrooms, smart campuses, smart agriculture farms, smart vehicles and smart factories. Healthcare industry has seen a rise in IoT based services (Riazul Islam, Kwak, Kabir, Hossain, & Kwak, 2015).

Aazam and Huh present a local specialized layer that provides computation, storage, connectivity and services to IoE ecosystem (Aazam & Huh, 2016). This layer is termed as Fog and is a specialization of the Cloud. The cloud being generalized provides computation, storage, connectivity and services remotely to the IoE ecosystem. Fog would create smarter IoE ecosystems with faster connectivity and there would be no dependency on third party cloud-as-a-service providers including Amazon Web Services, IBM Bluemix, Microsoft Azure, etc. Fog has promising use in smart cities, smart factories, smart campuses and smart hospitals (Patel, Ali, & Sheth, 2017). One of the reasons to use cloud is the large amount of data generated in IoE ecosystem that can easily be handled by using a middle layer.

Characteristics of IoE

The features and characteristics of IoE are similar to that of a Context Aware System and are outlined as follows (Malik, Mahmud, & Javed, 2007), (Coulouris, Dollimore, Kindberg, & Blair, 2011), (Udoh &

Figure 2. Concept of IoE

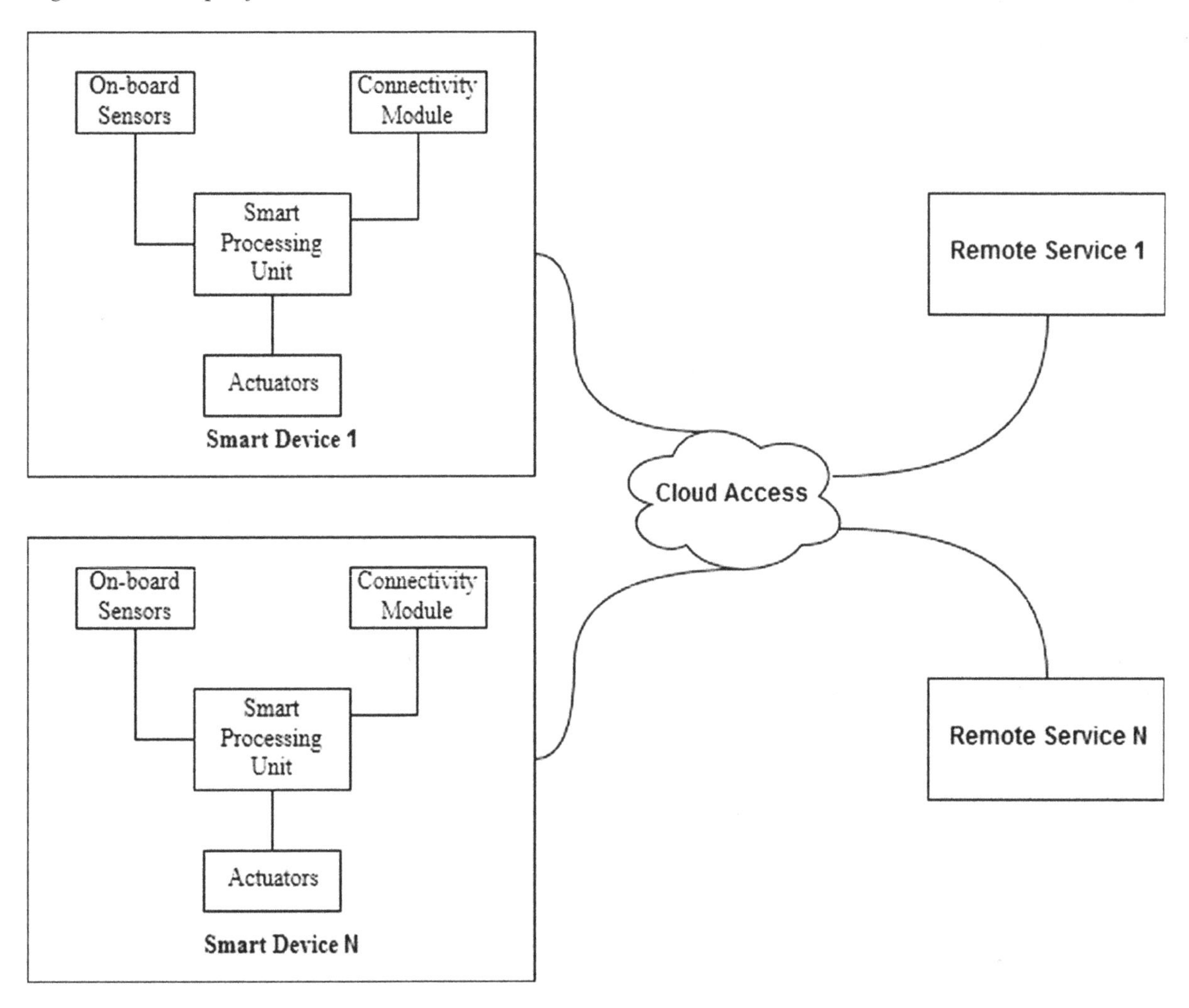

Kotonya, 2017). These characteristics are non-functional and are the quality attributes of IoT (Kesavan & Prabhu, 2018), (Bellavista, et al., 2019).

Collaboration

The ability of devices and services to communicate with each other so that better services could be delivered to the users is termed as collaboration. The devices and the services are designed and developed using different architectures. To ensure better collaboration services and devices should be developed as Service-Oriented Architecture (SOA). This architecture is implemented as web service (Hussain, Zhaoshun, & Toure, 2013), (Hussain, Wang, & Toure, 2014).

Seamless Integration and Transparency

The devices and services should be able to connect each other without complicated configuration. This capability is realized through Universal Plug and Play (UPnP). Various architectures and devices cannot communicate with each other easily thus proving transparency to the novice user.

Security

Security is a composite quality that includes Confidentiality, Availability and Integrity. Threat modelling is an activity used to identify vulnerabilities of a system and propose changes (Mahmud & Malik, 2014) . This is carried out during the design phase of a project (Malik, Javed, & Mahmud, 2008). Balfour also stresses on privacy and trust as an integrated module to ensure security especially in machine-to-machine (M2M) communication (Balfour, 2015).

Traditionally security has been provided as a separate layer within the TCP/IP model. The same has been utilized in IoT where security acts as a layer within the middleware (Spiess, et al., 2009). Reineh et.al. argue that a single separate layer is not suitable to sufficiently provide security (Reineh, Paverd, & Martin, 2016). Security must be integrated in each layer of a SOA based IoT ecosystem. Security can be provided as a list of policies and an associated engine can use these axiom-based policies to enforce security. Reineh et.al. also propose trust as a measurement tool of both devices as well as the services in an ecosystem.

Park et.al. argue that research is focused on malware detection on handheld devices (Park, Han, Oh, & Lee, 2019). The handheld devices are used to communicate and operate in an IoE ecosystem. Park et.al. propose a machine learning based mechanism that asses threats posed by applications installed on a smart device. This leads to provide application as well as service level threat assessment by the IoE ecosystem.

Context Awareness

Context awareness is the ability to recognize context of a user and subsequent service discovery and delivery (Alegre, Augusto, & Clark, 2016), (Feng, Apers, & Jonker, 2004), (Hong, Suh, & Kim, 2009). The context encompasses the state of a user and its device, measured via internal and external sensors. Context is composed of context sensing, context recognition and storage. The change in state is used to draw recommendations for service discovery and delivery. Context recognition uses machine learning algorithm to effectively recognize activity and make recommendations to adjust both internal and external services. The increase in data collection from sensors have paved the way for Big Data. The huge amount of data acquired from different devices requires a central smart process that covers different facets of activity recognition and application adjustment.

Context awareness is dependent on history information for activity recognition. Civitarese et.al. propose an online, periodic and probabilistic model based on semantic correlations for activity recognition in smart environments (Civitarese, Bettini, Sztyler, Riboni, & Stuckenschmidt, 2019). This system computes a probability of an event that belongs to an activity class based on history information.

Scalability

Scalability refers to the effect of increase in load on a system. While IoE supports remote devices and services connected, there could not be a global IoE environment. Different IoE environments are available which could be connected to each other thus creating a hierarchy. The growth in IoT devices is exponential and it is expected that 50 Billion smart devices would be connected by start of next decade (newgenapps, 2019), (LOPEZ RESEARCH, 2013).

Heterogeneity

Different architectures and services connected within an IoE environment have different platforms. The representation mechanism can lead to ambiguity in data sharing (Coulouris, Dollimore, Kindberg, & Blair, 2011). Extensible Markup Language (XML) and Web Ontology Language (OWL) are used as industry standards for communicating among various platforms and establishing heterogeneity (Mahmud, 2016), (Mahmud, Iltaf, & Kamran, 2007). Use of industry standards leads to open source software support

Resilient

Failures can occur in any system but must be masked to ensure availability. Multiple connections as well as redundancy in services can provide better resilience to failure.

Concurrency

An IoE environment must support multiple tasks simultaneously. These tasks should run concurrently by using serialization to avoid conflicts and Byzantine behavior.

Power Conservation

The devices interacting with each other and the services are mobile, compact and battery operated. These devices are limited to the battery power available to them. The core of a device is smart processing coupled with communication. Both these activities lead to battery power consumption. Various techniques are proposed that conserve battery power so that a device would remain available for longer duration (Mahmud, Hussain, & Yang, 2018).

Reusability

Reusability is a power feature of object-oriented systems that allow design level construct i.e., classes as well as run time constructs i.e., objects to be reused. This reduces development time and increases interaction. Service-Oriented Architecture (SOA) in the form of Web Services help realize context awareness as well IoE (Mahmud, Iltaf, Rehman, & Kamran, 2007). Object orientation lead to Unified Modelling Language (UML) based design of a software system (Mahmud, 2015).

Evidences in Literature

Many researchers have designed and developed IoE applications. These applications support various domains ranging between smart homes, smart clinics, smart transportation, smart agriculture and smart classrooms.

The smart urban planning system presents a 3-layer architecture to implement IoT in smart urban domain and supports big data (Babar & Arif, 2017). The architecture layers include Data Acquisition and Aggregation Layer; collects and stores data, Data Computation and Processing Layer; processes data using Hadoop MapReduce and Decision and Application Layer; manages events and decisions. Zanella et. al. proposes a generic model for smart urban centers that's supports various applications including building health monitoring, waste management, air quality monitoring, noise monitoring, traffic congestion monitoring, city energy consumption monitoring, smart parking and smart lighting. The authors propose a layered approach base don 300 nodes as Padova Smart City project (Zanella, Bui, Castellani, Vangelista, & Zorzi, 2014).

Hassani et.al. propose a service-based architecture that shares context information in an IoT Ecosystem in a secure manner (Hassani, Medvedev, Zaslavsky, & Jayaraman, 2018). The system has four components Communication and Security Manager; responsible for request handling as well as security checks to identify unauthorized access, Context Query Engine; parses and discovers context provider services so that data can be acquired, Context Storage management System; stores and manages query data in a hierarchical format and Context Reasoning Engine; infers activity and proposes recommendation based on Context Space (Padovitz, Loke, & Zaslavsky, 2004).

Alam and Benaida propose a Communication and Security Layer integral to all frameworks of IoT (Alam & Benaida, 2018). This layer supports communication security by transmitting scrambled text and access control lists. The system though is not adaptive nor can detect a malicious node.

Srinivasa et. al. establishes a data analytics-based framework in health care (Srinivasa, Sowmya, Shikhar, Utkarsha, & Singh, 2018). The authors present three different applications including e-cradle, knee joint movement monitoring and clothing and pattern recognition for the visually impaired. All three applications are designed as modular and have both hardware and software components.

Jung proposes a middleware-based approach to realizing IoE in home based senior healthcare (Jung, 2017). The author proposes a 4-step approach including data collection, monitoring senior location, monitoring senior activity and Expectation Maximization based decision support that recommends treatment based on health risk.

Bharadwaj et.al. propose an IoT based layered model for solid waste management of a smart city (Bharadwaj, Rego, & Chowdhury, 2016). The lowest device layer consists of sensors installed in garbage cans. These sensors include GPS, communication and Infra-red sensors. The next layer is Device Communication Layer which allows individual garbage can to each other via a gateway. This layer utilized a low power WLAN or LoRa specification. The third layer is the Aggregation layer using a lightweight messaging protocol as a message broker. The fourth layer is the Cloud layer that supports data storage, security measures and data analytics. The fifth layer is the External Communication Layer that consists of web or mobile based front end. The Management layer divides the system into device and application management that includes identify and access. This system has many layers with redundant functionality and lacks composition.

Duffy et.al. investigate the smart phone-based activity recognition of users. The proposed system uses ensemble of algorithms including SVM, random forest, kNN and multi-layer perceptron. The collection of algorithms removes outliers and improve activity recognition. The system uses on-board sensors for activity recognition including walking, going upstairs, going downstairs, lying and sitting.

Vargeese and Dahir propose a layered architecture to that enables IoE in On-Shelf Availability (OSA) of retail store (Vargheese & Dahir, 2014). the layers include the Sensor Layer; where weight, video and other sensors are present in the environment, the Fog Layer; where the data is aggregated and normalized, and local analytics performed and the Cloud Layer; includes external services and generates real time stream analytics as well as verifiers the data and the sensor source.

Petrakis et.al. propose IoT as a Service (iTaaS) framework that supports two distinct layers (Petrakis, et al., 2018). The first layer is termed as the IoT layer which gathers data from sensors within an environment in real-time and transmits them over the cloud. The second layer is termed as the Cloud back-end

Table 1. Comparison of evidences in literature

	Architecture	Context Representation	Smart Processing Technique	Security and Privacy	Power Conservation	Cross platform support	Data Size	Cloud Support
Smart Urban Planning (Babar & Arif, 2017)	3-layer framework.	Attribute-value pair	Map Reduce method	-	-	-	Big Data	Yes
CoaaS (Hassani, Medvedev, Zaslavsky, & Jayaraman, 2018)	Service Oriented	Hierarchical based on Context Description language	Inference in Context Space	Security checks in proxy	-	Yes	Large	Yes
CICS (Alam & Benaida, 2018)	Layered	-	-	Encryption, Secure transmission, Access list	-	Yes	-	Yes
Data Analytics Assisted IoT (Srinivasa, Sowmya, Shikhar, Utkarsha, & Singh, 2018)	Modular	Attribute-value pair	Pattern Recognition	-	-	No	Medium	-
Hybrid Model for Senior Wellness (Jung, 2017)	Middleware	Relational Database	EM Algorithm	-	-	-	Medium	-
IoT based Solid Waste Management (Bharadwaj, Rego, & Chowdhury, 2016)	Layered	Attribute-Value	Data Analytics on cloud	Access Control lists (ACL)	-	Yes	Large	Yes
Padova Smart City (Zanella, Bui, Castellani, Vangelista, & Zorzi, 2014)	Layered	Attribute-Value	Rule Based	-	-	Yes	Big Data	Yes
Smart phone based weakly supervised activity recognition (Duffy, Curran, Kelly, & Lunney, 2019)	-	Attribute-Value	Ensemble	-	-	-	Large	-
An IoT/IoE Enabled Architecture Framework for Precision on Shelf Availability (Vargheese & Dahir, 2014)	Layered	Attribute-Value	Learning	-	-	Yes-	Big Data	Yes
iTaaS (Petrakis, et al., 2018)	Service oriented over cloud.	XML based	Rule Based	-	Yes	Yes	Large	Yes
Smart Car Parking Mobile Application based on RFID and IoT (Saeliw, Hualkasin, Puttinaovarat, & Khaimook, 2019)	Layered	Attribute-Value	Rule based	-	-	-	Medium	Yes

layer which supports data storage, data sharing and data processing. A proof of concept is implemented as an e-health monitor for patients. This model provides application logic over the cloud back-end layer while the IoT layer is implemented as a thin client. A

Saeliw et.al. propose an RFID and IoT based smart car parking solution (Saeliw, Hualkasin, Puttinaovarat, & Khaimook, 2019). The solution uses Arduino to gather sensor information and uses axioms to allocate car parking space. The architecture is layered with a server supporting the activities. A rule-based smart parking application is implemented on the server and is accessed by a handheld device.

Object Oriented Model of a Context Aware IoE Ecosystem (CAIES)

An object-oriented model of IoE Ecosystem is proposed in this section. The architecture is service oriented where the Ecosystem not only utilizes services available in the environment but can provide services to other applications. Object orientation supports modularity, flexibility and ease of development. To represent the model Unified Modelling Language (UML) is used (Larman, 2004). The model relies heavily on context awareness for activity recognition.

Use Case Diagram of CAIES

Use case diagrams show the external view of a software system. The external users as well services interact actively as well as passively with the system functionalities. CAIES interacts with a user and provides different functionalities that allow the user to access and adjust remote services. Figure 3 shows

Figure 3. Use case model of CAIES

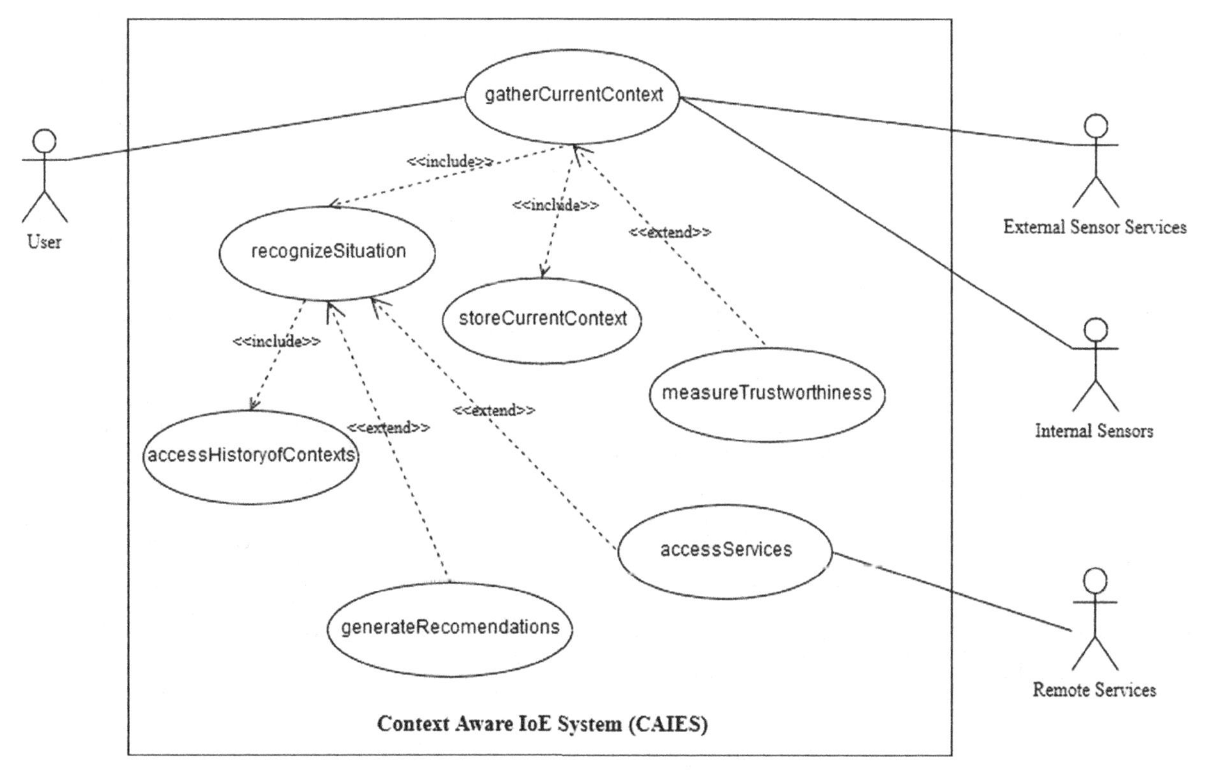

Table 2. Actor description of CAIES

Actor ID	Actor Name	Actor Type	Description
ACT 001	User	Primary	Uses the CAIES. This actor can initiate the context aware process via their device.
ACT 002	Internal Sensors	Secondary	These are the onboard sensors that are available on the users' device. These sensors are present within the environment.
ACT 003	External Sensor Services	Secondary	The external sensor services are invoked by CAIES to gather data that cannot be accessed via internal sensors. These services can be within the environment and beyond as well.
ACT 004	Remote Services	Secondary	These are the services that are accessed and adjusted based on the users' context. The can be within the environment and beyond as well.

the use cases of a context aware IoE Ecosystem. Table 2 lists the actors of CAIES while Table 3 describes the uses cases of CAIES. The user is the primary actor. CAIES interacts with internal sensors as well as external sensors to gather context. The internal sensors are on board the device of the user and can include accelerometer, GPS module, etc. the external sensors are invoked via services and can include, Google Maps, weather services, etc. CAIES provides functionalities that include gathering context, storing and recognizing context as well as measuring trustworthiness of external sensor service. CAIES can optionally generate recommendation as well as access remote services.

Table 3. Use case description of CAIES

Use Case ID	Use Case	Description
UC 001	gatherCurrentContext	This functionality is invoked by the user and gathers the contextual data from internal sensors as well as external sensor services. This use case includes recognizeSituation and storeCurrentContext while can optionally measureTrustworthiness of external sensor services
UC 002	recognizeSituation	This use case is used to recognize the current activity or infer situation based on the current context. This includes accessHistoryofContexts. This use case implements a recognition algorithm to infer the situation. CAIES uses a version of CiE (Mahmud & Javed, 2012), (Mahmud & Javed, 2014).
UC 003	storeCurrentContext	This use case stores the current context using hierarchical and platform independent methods. CAIES stores history and current context as Ontology (Mahmud, 2016), (Mahmud, Iltaf, & Kamran, 2007).
UC 004	accessHistoryofContexts	This use case accesses the history of stored contexts. The history is maintained as attribute-value pairs stored in Web Ontology language (OWL) format (Malik, Mahmud, & Javed, 2009). The contents include the context data and the recognized situation. In addition, the confidence of recognized situation can also be stored to tweak the recognition process.
UC 005	measureTrustworhiness	This use case measure trust values of the external sensor services as well as the remote services (Iltaf, Mahmud, & Kamran, 2006).
UC 006	generateRecomendations	This use case generates recommendations based on the current context values and the recognized situations. These recommendations can be implemented as first order logic axioms. User may select any recommendation as deemed fit.
UC 007	accessServices	This functionality accesses the remote services based on the selected recommendations.

Figure 4. SSD of CAIES

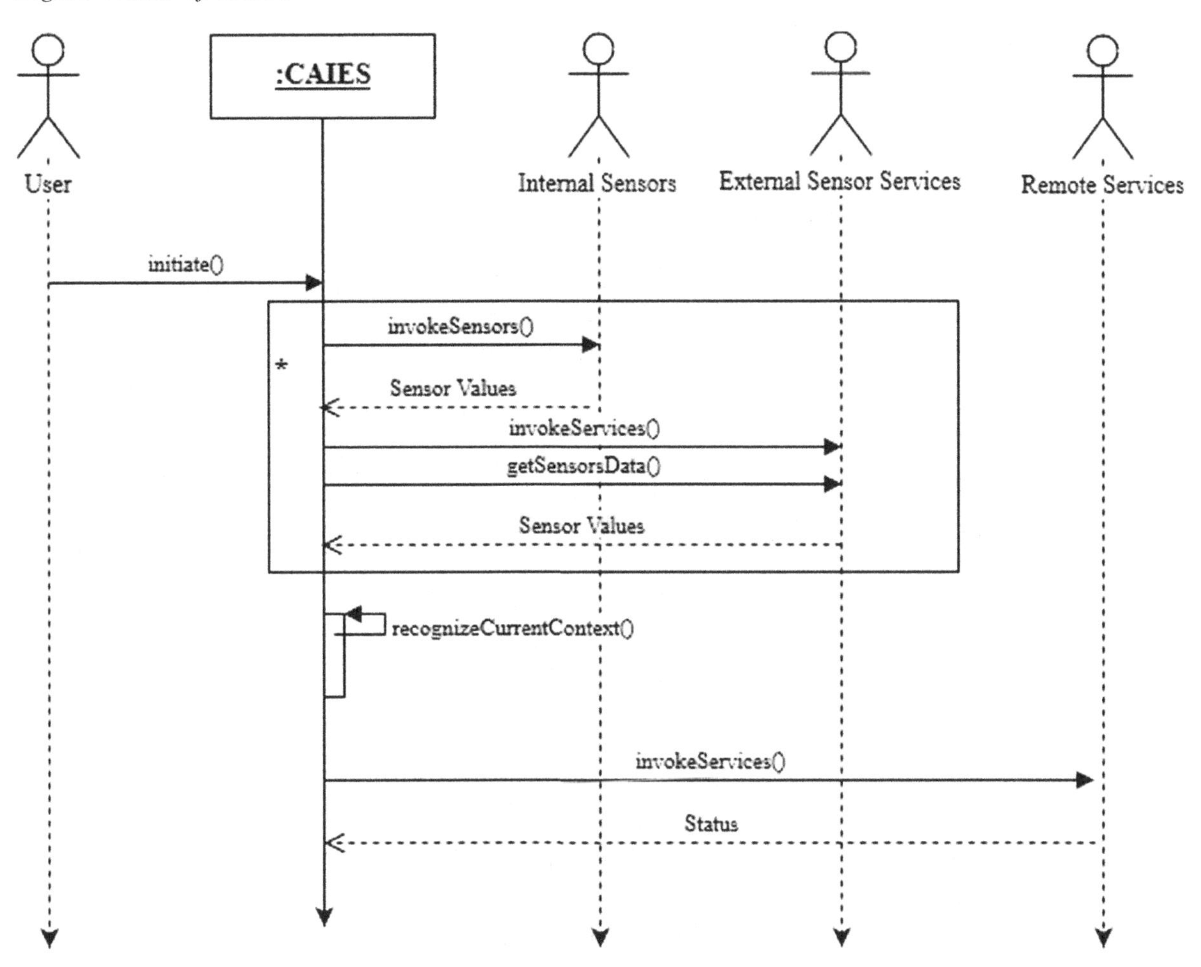

Interaction Model of CAIES

Use case diagrams model the external behavior of the system. Interaction diagrams are generated in UML to model the sequence of activities. Interaction diagrams are developed starting from the System sequence Diagram (SSD) and ends up in sequence diagrams for different scenarios. Figure 4 shows the SSD of CAIES. This shows the interaction as messages between the actors and the system. The user initiates the system and CAIES invokes the sensors in a loop. CAIES then recognized the situation and invoke services if desired.

Figure 5 shows the data gathering scenario for the user. The user initiates via the Dispatcher object. Dispatcher then invokes ContextAggregator object to gather current context. The data is acquired in loop and then transformed in OWL format (Mahmud, 2016). The history is the updated using the HistoryManager object. The current context is made available to the dispatcher.

Once data is gathered the Dispatcher object invokes ContextInferenceEngine object to recognize the current context. ContextInferenceEngine implements and online machine learning algorithm based on nearest neighbor and standard deviation based ranks (Mahmud & Javed, 2014) (Mahmud & Javed, 2012).

Figure 5. Sequence diagram of context gathering in CAIES

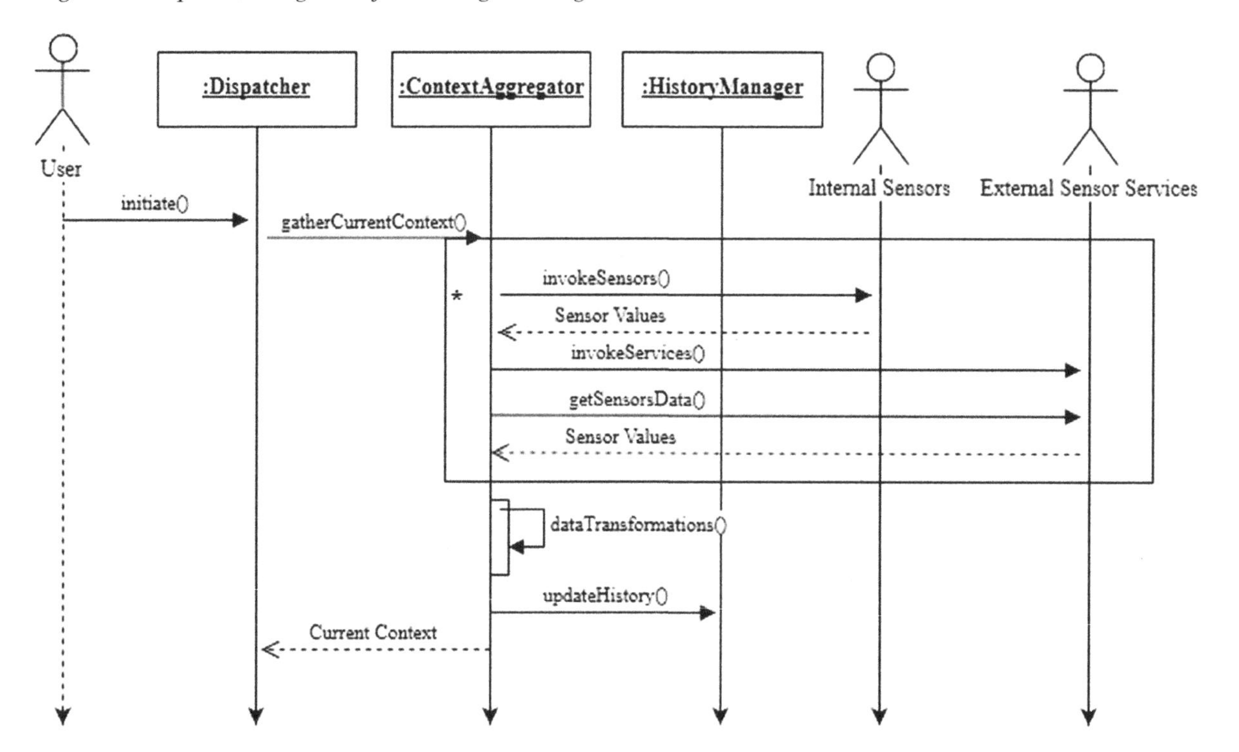

The ContextInferenceEngine object reads the history through HistoryManager object and recognizes situation. The situation and the confidence value are sent back to the Dispatcher. The dispatcher then invokes the Recommender object to generate recommendations. The Dispatcher can then generate the ServiceAccessor object that access remote services and returns status. Figure 6 shows this scenario.

The Dispatcher can also invoke TrustManager object to measure the trustworthiness of external sensor services as well as remote services. Trust is used to enhance security by identifying malicious and untrustworthy services (Iltaf, Mahmud, & Kamran, 2006). TrustManager measures trust by accessing history through HistoryManager as shown in Figure 7.

Domain Model of CAIES

Domain model of an object-oriented system is represented by the object diagram of the system. While object diagram is specific a class diagram describes the domain by replacing objects with classes. The task is thus to identify appropriate classes and their relationships. In addiction the classes are composed of attributes and methods. The functionalities shown as use cases in the use case diagrams are encapsulated by different classes in the class diagram. Figure 8 shows the class diagram of CAIES. Dispatcher class composes ServiceAccessor class and creates an object when required as evident in Figure 6. The Dispatcher class is responsible for session management and hides other classes from the user. ContextAggregator gathers the current context from internal sensors as well as external sensor services. HistoryManager class maintains history of contexts and the recognized situations in OWL. ContextInferenceEngine is responsible for recognizing the situation while Recommender makes recommendations

Figure 6. Sequence diagram of situation recognition in CAIES

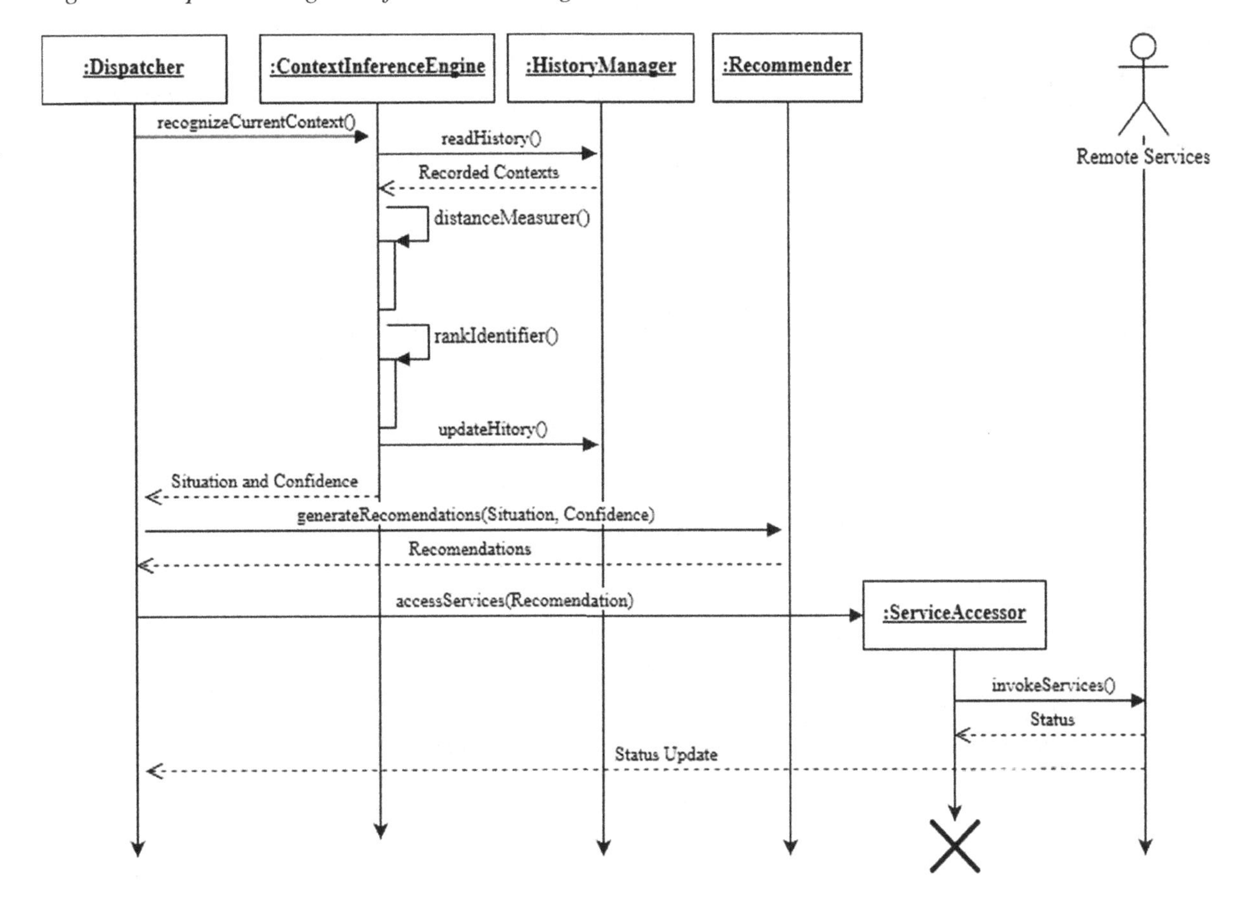

based on the recognized situation. ServiceAccessor accesses services based on the recommendations and TrustManager measures trust value of external sensor services and remote services. Table 4 describes the methods of each class shown in Figure 8. The relationships among classes and their cardinality is shown in Figure 8 as well.

Design Quality of CAIES

The quality of design of CAIES is measures using CK metrics (Chidamber & Kemerer, 1994), (Chidamber & Kemerer, 1991). The CK metrics include Coupling Between Objects (CBO), Lack of Cohesion in Methods (LCM), Weighted Methods per Class (WMC), Depth of Inheritance Tree (DIT), Number of Children (NOC) and Response for a Class (RFC).

CBO measures the non-inheritance associations among classes. CBO is measured for each class for evidence of any class using method or instance variable of another class. Lower CBO is desirable in a class diagram. LCOM is the number of disjoint sets formed by the intersection of common variables among all pairs of methods. LCOM should be a lower number for a higher quality design. WMC is calculated as the number of methods per class. A lower number of methods per class is desirable. DIT measures the length of maximum inheritance path. A higher DIT leads to higher number of methods that are inherited. Deep trees in a class diagram tend to generate more complexity. NOC measures the

Figure 7. Sequence diagram of trust measurement in CAIES

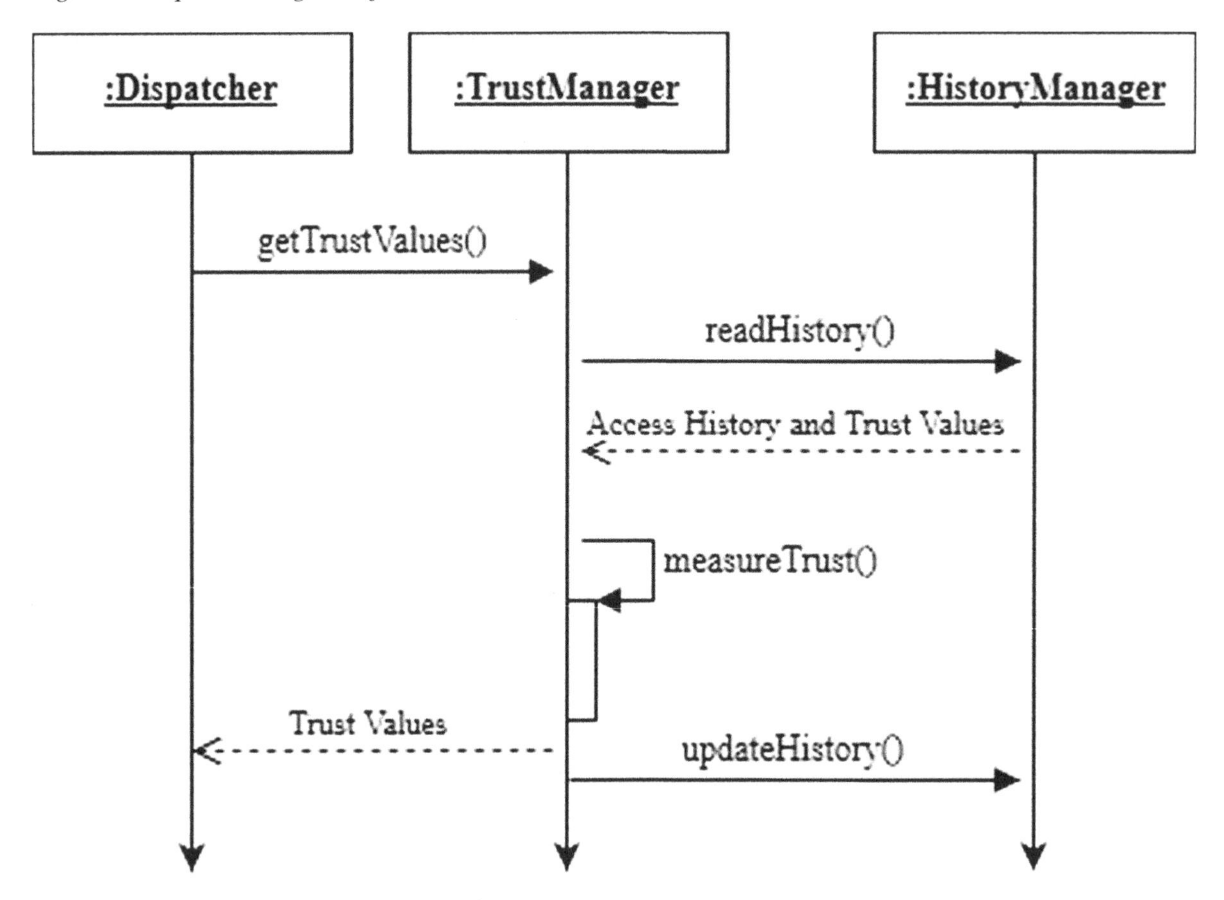

immediate child classes for a class. High NOC corresponds to higher reuse of classes. RFC of a class is the number of methods that can potentially be executed in response to a message received by an object of that class. A lower RFC is desirable in a class diagram. Table 5 presents the CK metrics of the class diagram of CAIES. It is evident from Table 5 as well as Figure 8 that the domain model has low reuse since no inheritances are visible. Furthermore, the dispatcher class can be a bottle neck based on CBO as shown in Table 5. However, the numbers are well within range.

'DOCENT': Museum Tour Guide System – A Case of IoE

The Docent is a service-oriented museum tour recommender system that recommends which artifact should be viewed based on the tourists 'context (Farooq, Gillani, Tahirkheli, & Farooq, 2018). A museum has many artifacts for tourists who are mostly short of time to view everything (Kosmopoulos & Styliaras, 2018). Furthermore, not all artifacts might be suitable for a tourist based on their interest and mood. Docent gathers the current context of a tourist and interacts with the services present in the museum to acquire information about the layout as well as the location, type and rating of artifacts. The system then recommends artifacts to the tourist and plans a view order. A feedback is given by the tourist

Figure 8. Class diagram of CAIES

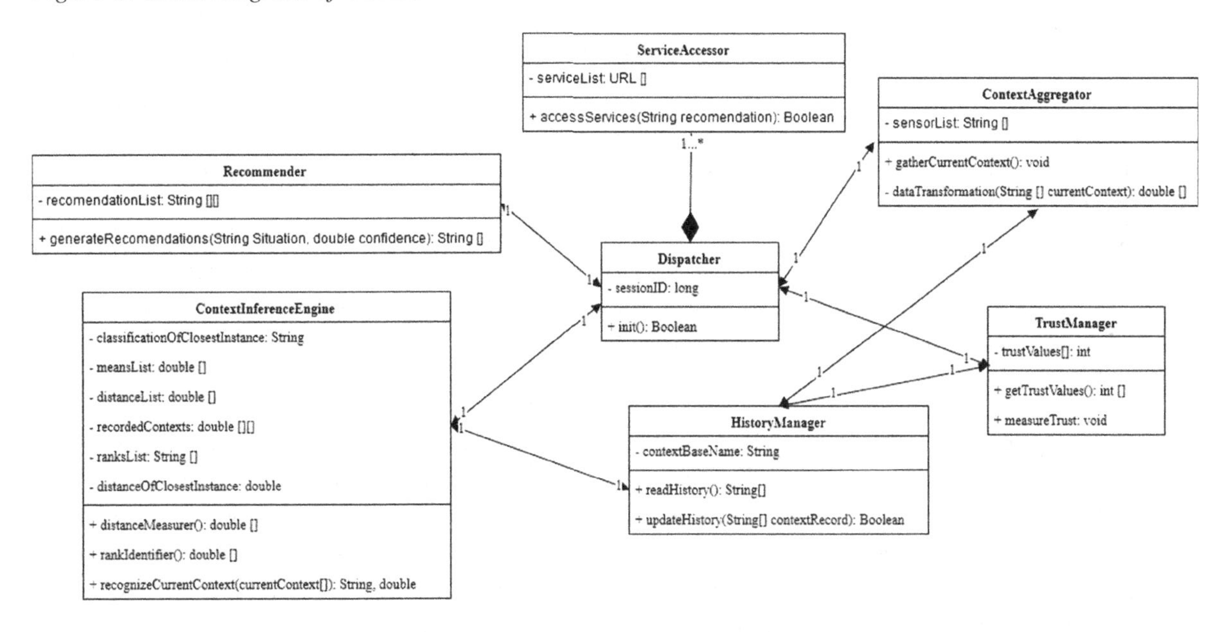

Table 4. Descriptions of methods in class diagram of CAIES

Method	Class	Description
init()	Dispatcher	Initializes activities and is invoked by user.
gatherCurrentContext()	ContextAggregator	Gathers current context from internal sensors as well as external sensor services.
dataTransformation()		Transform the data in to format suitable for ContextInferenceEngine. Data from external sensor services can be string based and need to be converted to double prior to recognition.
recognizeCurrentContext()	ContextInfererenceEngine	Classifies the current context by calling distanceMeasurer() and rankIdentifier() in sequence. This function returns the name of the activity with confidence on the classification.
distanceMeasurer()		Measures the distance of current context with the recorded contexts. Distance measure used is Minkowski distance.
rankIdentifier()		Measures the standard deviation-based rank of current context w.r.t activities recorded in history.
generateRecomendations()	Recommender	Returns a list of recommendations suitable for each situation.
readHistory()	HistoryManager	Returns the history as a 2-dimensional double array.
updateHistory()		Updates the records of current context in history with activity and confidence information.
getTrustValues()	TrustManager	Gathers trust values from history
measureTrust()		Measures trust of each external sensor service and remote service after every interaction.
accessServices()	ServiceAccessor	Access the remote services. The remote services must be implemented as Service-Oriented Architecture (SOA)

Table 5. CBO of class diagram of CAIES

Class	CBO	LCOM	WMC	DIT	NOC	RFC
Dispatcher	4	0	1	0	0	0
ContextAggregator	1	0	2	0	0	2
ContextInferenceEngine	2	2	3	0	0	3
Recommender	1	0	1	0	0	1
HistoryManager	2	0	2	0	0	1
TrustManager	2	0	2	0	0	2
ServiceAccesor	-	0	1	0	0	1

to reinforce the recommendation generated by the docent. Docent is developed an app for iPhone users. Figure 9 shows the concept of Docent. The features of docent are listed in Table 6.

Docent provides user-friendly interface and makes it easier for them to take a tour of museum. It uses indoor mapping to plot the path. Docent considers tourist preferences and selects the appropriate domain. A list of desired artifacts coupled with specified ratings and optimal path are displayed. A tourist can also rate the artifacts to improve the recommendations. The system is based on real time recommendations. It has fully interactive indoor live maps of the floors. All the artifacts are included in exhibits to be shown and has detailed information. The system helps tourists to visit those artifacts on the basis on time, location, ranking, interest and users' history. It saves a lot of time and gives effective path to the desired exhibit.

Docent GUI

There are two different users of Docent. The tourist is the end user who access via iPhone and the administrator uploads artifacts data via desktop mode. This is uploaded for the museums who do not have an artifact information service. Figure 10 shows the administrator's login screen. Security is provided by username and strong password. Figure 11 shows the tourist's login screen which also uses username and strong password. The tourists screen is developed for iPhone.

Figure 9. Docent: A museum tour guide system

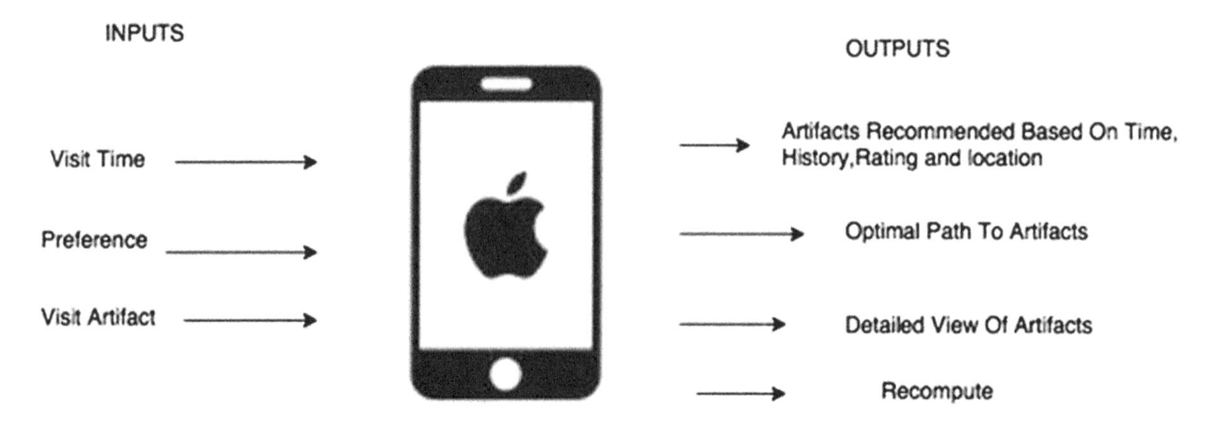

Table 6. Features of docent

Sr.	Feature	Description
Fr 01	Recommendations	Docent generates recommendations to the tourist based on their context, mood and location. The recommendations include the artifacts that should be visited. The docent uses axioms for recommendations.
Fr 02	Optimal path for artifact visit	Shortest path to visit the artifacts to save time of the tourist is generated.
Fr 03	Context awareness	The context of tourist includes mood, location time and interest based on the history. It also includes the environment information and the location and type of artifacts within a museum.
Fr 04	Time consideration	Docent considers a tourists' available time as a constraint.
Fr 05	Detailed information of artifacts	Information about the artifacts being visited by the tourist is acquired via services present in the museum.
Fr 06	Adjusting recommendations	Docent can adjust the recommendations based on the tourist feedback. Feedback is provided as a satisfaction value.
Fr 07	Interaction with location services	Docent can interact with indoor positioning services provided by the museum. It also has interface to interact with Google Maps for outdoor positioning.

Figure 10. Docent: Administrator's login screen

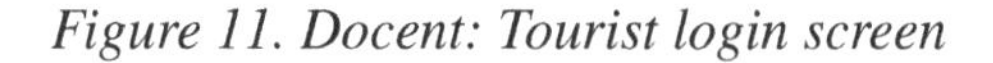

Figure 11. Docent: Tourist login screen

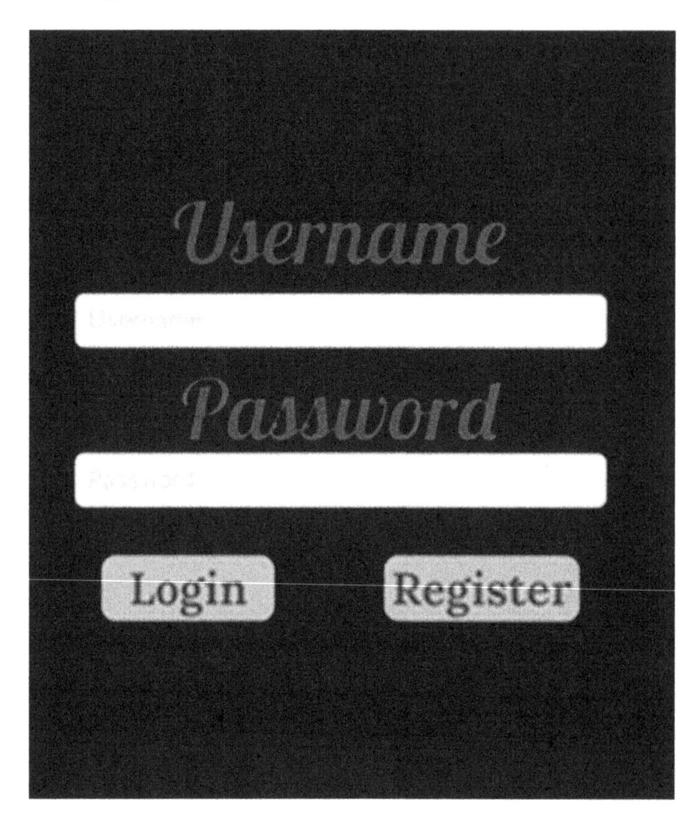

Figure 12 shows the artifact registration screen. Many museums do not have an online artifact information service that can easily be integrated with the system. However, Docent provides an alternate where an administrator can manually upload information. The information includes name, category, family, ranking, description, image and the distance of location from entrance of the artifact. Docent can connect with the services and acquire information using XML tags.

A tourist selects the preference on first use and a screen is developed for this purpose. Figure 13 shows the selection screen while Figure 14 shows the recommendation screen based on the tourist context. Figure 15 shows the description of the artifact while Figure 16 plots the path to the artifact location. The path is plotted using simple Manhattan distance.

Docent as an IoE Application

The interaction between the services present in the museum that include location services and interaction via an intelligent device make it an IoE ecosystem. The system also uses artifact information services (if present) to identify the name, type, description. Location and rating of an artifact. The inference is made using axioms and visit history is stored in a relational-DB.

The system realizes IoE as an application and provides smart services to the tourist. Some static services are available including the British Museum Guide (Museum Guides Ltd) and the Metropolitan Museum of Art App (Garcia). These services list the artifacts and provide static maps. No consideration is given to a tourists' history or interest. British Museum guide is based on only one museum. A

Figure 12. Docent: Artifact registration screen

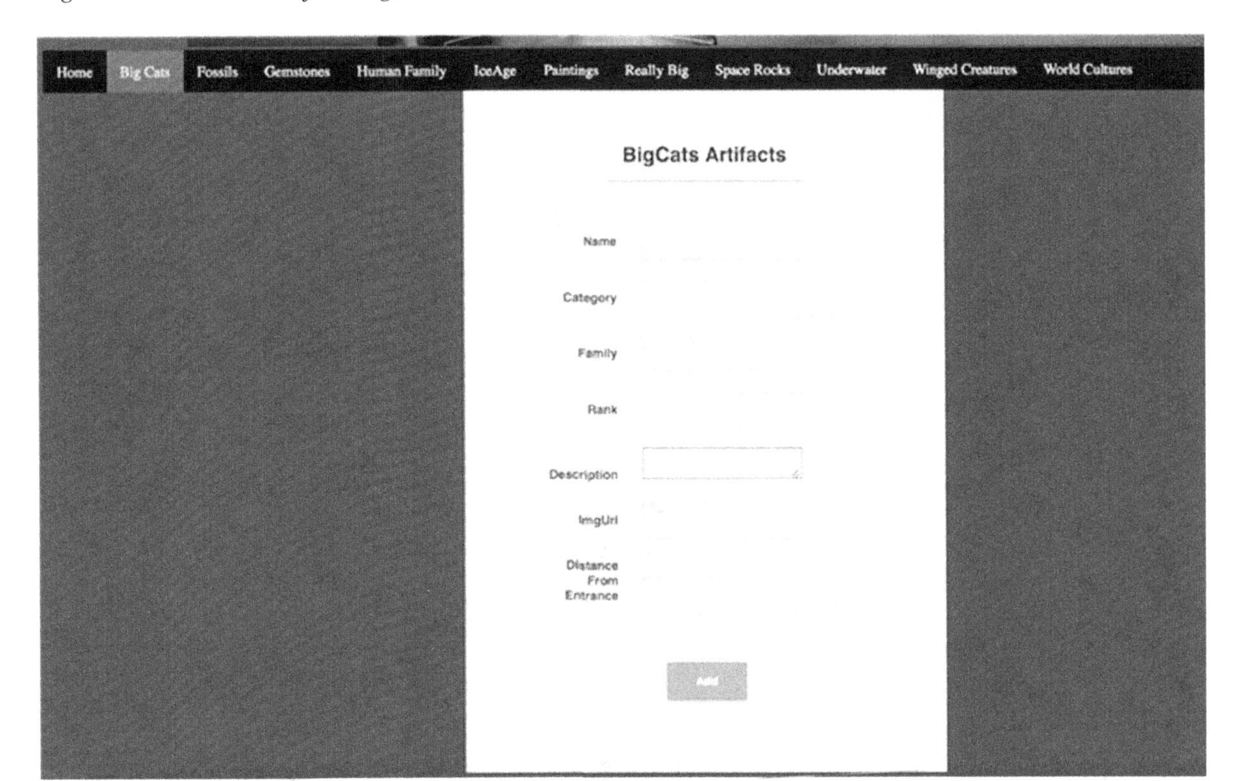

fully interactive map of the floors is provided but that does not show path to desired artifacts. Detailed information of the desired artifacts is given but no rating or recommendations are provided. This app is available on Google Play store for android users. Metropolitan museum app also supports a single museum. No maps are provided.

Figure 17 shows a flow model of interaction of Docent. A first-time user is given recommendations based on the interests input by the tourist. Subsequent use caters for history of use as well. The recommended artifacts are shown, and a path plotted. A Google Map interface is also available that shows the path to the museum grounds. The artifact description is displayed when the tourist visits it. A tourist can then rate the recommendation for adjustment in recommendations.

The recommendations are generated based on axioms. These axioms are implemented as IF-THEN-ELSE statements in code. A rule based intelligent system is easy to implement however, the rule space could be exceedingly large. The adjustment is carried out using satisfaction values provided by the tourist (Malik, Mahmud, & Javed, 2009). Each recommended artifact has an associated satisfaction value as shown in Equation 1. The satisfaction value (St) is between 0 and 1.

$$0 \leq St \leq 1$$

A tourist can select the satisfaction value after visiting the artifact. This value is then applied as input in generated recommendation. The artifacts are grouped in categories. A first-time user selects the category of interest. Each category has many artifacts all of which are equally probable. Each artifact

Figure 13. Docent: User preference selection screen

has an associated weight which is then recomputed by multiplying satisfaction value. The new weight (Wt_{new}) is calculated using old weight (Wt_{old}) as shown in Equation 2. The calculated weights of each artifact are then normalized.

Figure 14. Docent: Recommended artifacts

$$Wt_{new} = (Wt_{old} \times St)_{normal}$$

The recommendations are generated based on axioms. These axioms are implemented as IF-THEN-ELSE statements in code. A rule based intelligent system is easy to implement however, the rule space could be exceedingly large. The adjustment is carried out using satisfaction values provided by the tourist (Malik, Mahmud, & Javed, 2009). Each recommended artifact has an associated satisfaction value as shown in Equation 1. The satisfaction value (St) is between 0 and 1. After visiting an artifact and giving satisfaction value, the system regenerates recommendation list this making the system real-time.

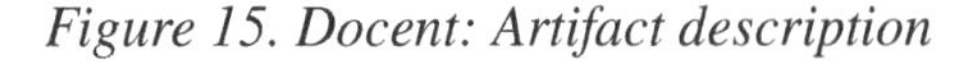

Figure 15. Docent: Artifact description

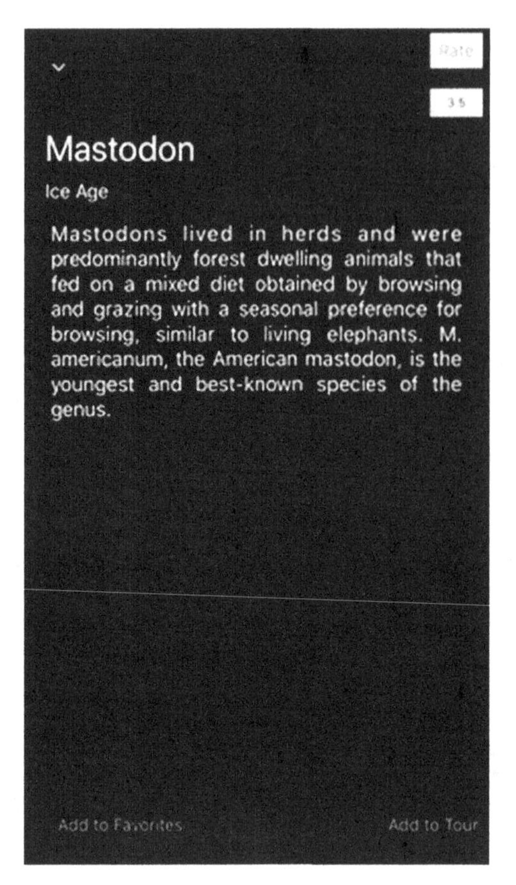

Benefits of Docent

Docent is suitable for tourists who do not know which artifacts should be visited in a museum. Tourists have limited time and they might not be able to visit the complete museum. Tourists are provided with real-time recommendation based on the context. The artifacts are short listed and high rated artifacts are recommended. An optimal path is also plotted to guide a tourist and a view order is also generated. Detailed information of an artifact is also displayed when the tourist is visiting it.

Limitations of Docent

The Docent is developed for iPhone and available for IOS devices. This system does not use classification algorithms to enhance the recognition as well as recommendation process. A classification algorithm can improve the recommendations.

A museum should be able to analyze the ratings of their tourists to improve the quality as well as the type of artifacts. Developing Docent as a service can allow a potential museum to gather tourists' ratings of different artifacts generate a global rating.

Figure 16. Docent: Path to visit artifacts

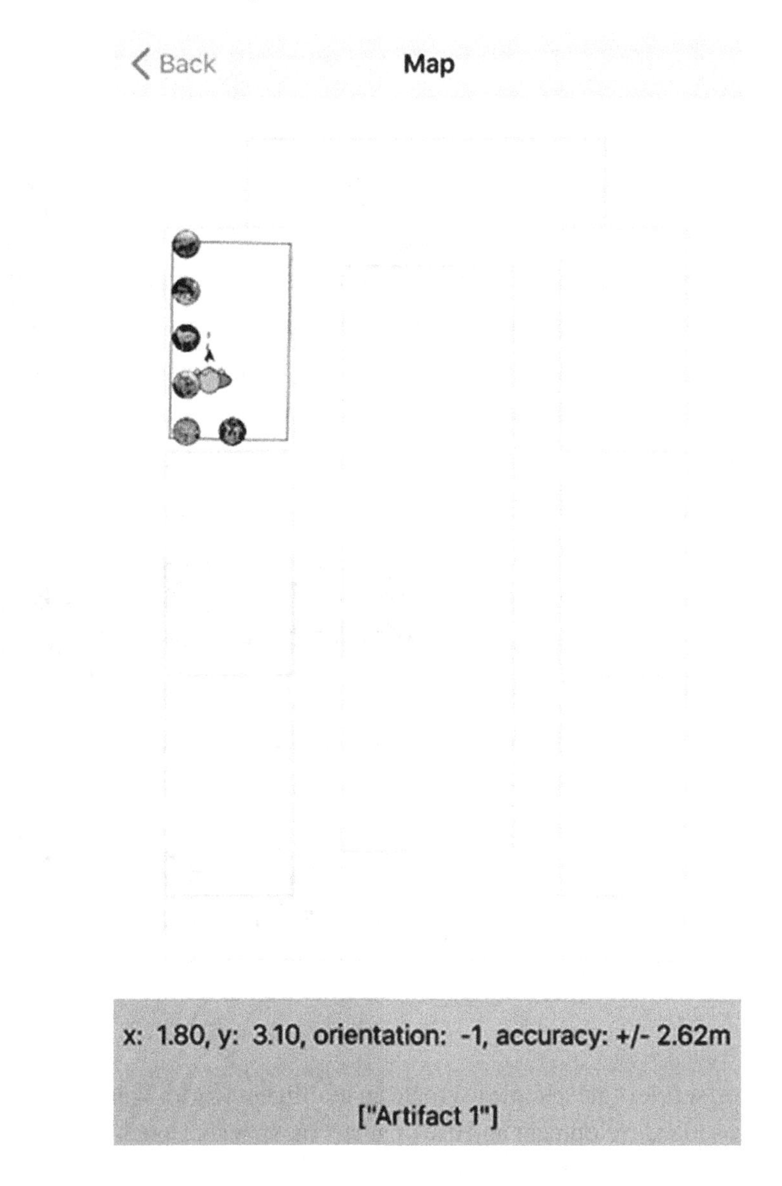

THE ROAD AHEAD

IoE has come a long way starting from the first generation of computers. The industry needs to design and develop technology enablers that can work in any smart environment. These enablers are both software and hardware based. Among the issues that need attention are enabling context awareness, power conservation, data sharing, privacy and trust management and communication mechanism. The research component must work in these areas in providing optimal solution to recognition, power conservation and privacy assurance. A need to develop a common IoE system for a user who changes roles as the day progresses, a teacher can be a tourist, or a doctor can be a student as well. This demands a rich context representation to cater for needs of different roles.

Figure 17. Docent: Flow model of interaction

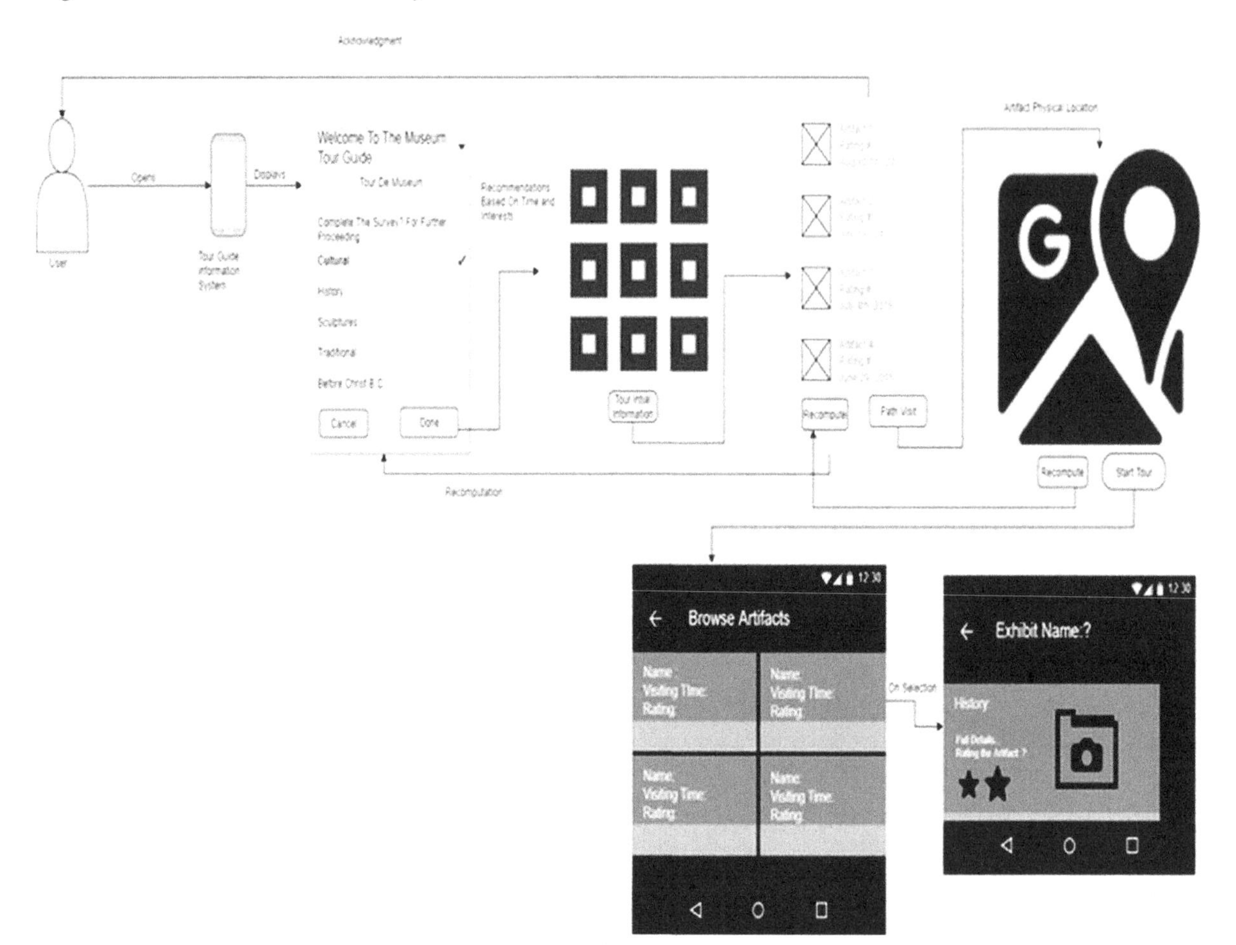

CONCLUSION

IoE connects devices, data, services and people to create an intelligent world. This allows users to achieve smarter and better services. the development and use of smart inexpensive hand-held devices have led to creation of IoE ecosystem. The devices interact with the sensors present in the environment as well as onboard sensor termed as external and internal sensors respectively. The external sensors are accessible through services thus establishing service-oriented architecture. Remote and local applications are also accessible as services. The ecosystem thus comprises of smart devices, sensors and services.

Data is accessed from sensors and sensor services and then stored in a hierarchical, extensible and shareable format including XML and OWL. The gathered data is called the current context and is used by context inference engine to recognize situations based on history. The recognized situation is then used to draw recommendations for the user. The recommendation includes service discovery, delivery and adjustments.

A case study of Docent: Museum Tour Guide System is presented as an implementation of IoE. This system gathers context of a tourists and recommend artifacts to view. The system can plot the optimal path to the artifacts and displays artifact description when the tourist is the vicinity. A tourist can also rate the recommendation. The system automatically adjusts the recommendation in real-time if required.

IoE is promising and useful for everyday users. The users have a wide variety including students, teachers, factory workers, managers, patients, doctors, nurses, tourists, etc. Many of these are roles played on by a single user. The future of IT industry is converging towards IoE which will integrate in mundane tasks seamlessly.

ACKNOWLEDGMENT

The authors acknowledge Aqib Farooq, Nouman Gillani, Rabya Ameer Thairkheli and Muhammad Umer Farooq as the authors of Docent: Museum Tour Guide System presented as project for award of Bachelor of Computer Software Engineering degree from Foundation University Islamabad (FUI), Islamabad Pakistan. The research work is sponsored in part by research grant titled Power Aware Framework in Smart Environments by FUI.

REFERENCES

(2019, April 17). Retrieved from newgenapps: https://www.newgenapps.com/blog/iot-statistics-internet-of-things-future-research-data

Aazam, M., & Huh, E.-N. (2016). Fog Computing: The Cloud-IoT\IoE Middleware Paradigm. *IEEE Potentials*, *35*(3), 40–44. doi:10.1109/MPOT.2015.2456213

Alam, T., & Benaida, M. (2018). CICS: Cloud–Internet Communication Security Framework for Internet of Smart Devices. [iJIM]. *International Journal of Interactive Mobile Technologies*, *12*(6), 74–84. doi:10.3991/ijim.v12i6.6776

Alegre, U., Augusto, J. C., & Clark, T. (2016). Engineering context-aware systems and applications: A survey. *Journal of Systems and Software*, *117*, 55–83. doi:10.1016/j.jss.2016.02.010

Ashton, K. (2009). *That 'Internet of Things' Thing*. Retrieved April 17, 2019, from RFID Journal: https://www.rfidjournal.com/articles/view?4986

Babar, M., & Arif, F. (2017). Smart Urban Planning using Big Data Analytics based Internet of Things. *017 ACM International Joint Conference on Pervasive and Ubiquitous Computing and Proceedings* (pp. 397-402). Maui, Hawaii: ACM. doi:10.1145/3123024.3124411

Balfour, R. E. (2015). *Building the "Internet of everything" (IoE) for first responders. 2015 Long Island Systems, Applications, and Technology* (pp. 1–6). Farmingdale, NY: IEEE; doi:10.1109/LISAT.2015.7160172

Bellavista, P., Berrocal, J., Corradi, A., Das, S. K., Foschini, L., & Zanni, A. (2019). A survey on fog computing for the Internet of Things. *Pervasive and Mobile Computing*, *52*, 71–99. doi:10.1016/j.pmcj.2018.12.007

Bharadwaj, A. S., Rego, R., & Chowdhury, A. (2016). IoT based solid waste management system: A conceptual approach with an architectural solution as a smart city application. In *Proceedings 2016 IEEE Annual India Conference (INDICON)* (pp. 1-6). Bangalore, India: IEEE. 10.1109/INDICON.2016.7839147

Bradley, J., Loucks, J., Noronha, A., Macaulay, J., & Buckalew, L. (2013). *Internet of Everything (IoE): Top 10 Insights from Cisco's IoE Value Index Survey of 7,500 Decision Makers Across 12 Countries*. Cisco.

Chidamber, S. R., & Kemerer, C. F. (1991). *Towards a metrics suite for objectoriented design. Object-oriented programming systems, languages, and applications OOPSLA '91* (pp. 197–211). Phoenix, AZ: ACM; doi:10.1145/117954.117970

Chidamber, S. R., & Kemerer, C. F. (1994). A metrics suite for object-oriented design. *IEEE Transactions on Software Engineering*, *20*(6), 476–493. doi:10.1109/32.295895

Civitarese, G., Bettini, C., Sztyler, T., Riboni, D., & Stuckenschmidt, H. (2019). newNECTAR: Collaborative active learning for knowledge-based probabilistic activity recognition. *Pervasive and Mobile Computing*, *56*, 88–105. doi:10.1016/j.pmcj.2019.04.006

Coulouris, G., Dollimore, J., Kindberg, T., & Blair, G. (2011). *Distributed Systems: Concepts and Design* (5th ed.). Addison-Wesley Publishing.

Duffy, W., Curran, K., Kelly, D., & Lunney, T. (2019). An investigation into smartphone based weakly supervised activity recognition systems. *Pervasive and Mobile Computing*, *56*, 45–56. doi:10.1016/j.pmcj.2019.03.005

Farooq, A., Gillani, N., Tahirkheli, R. A., & Farooq, M. U. (2018). Museum Tour Guide System. Rawalpindi: Foundation University Islamabad, Rawalpindi Campus (FURC).

Feng, L., Apers, P. M., & Jonker, W. (2004). Towards Context-Aware Data Management for Ambient Intelligence. *Lecture Notes in Computer Science, 3180*, 422-431. Retrieved from http://link.springer.com/chapter/10.1007%2F978-3-540-30075-5_41

Garcia, D. J. (n.d.). Metropolitan Museum of Art NYC. Retrieved from https://apps.apple.com/us/app/metropolitan-museum-of-art-nyc/id977211908

Hassani, A., Medvedev, A., Zaslavsky, A., & Jayaraman, P. P. (2018). Context-as-a-Service Platform: Exchange and Share Context in an IoT Ecosystem. In *Proceedings 9th International Workshop on Information Quality and Quality of Service for Pervasive Computing* (pp. 385-390). IEEE. 10.1109/PERCOMW.2018.8480240

Hong, J.-Y., Suh, E., & Kim, S.-J. (2009). Context-aware systems: A literature review and classification. *Expert Systems with Applications*, *36*(4), 8509–8522. doi:10.1016/j.eswa.2008.10.071

Hussain, S., Wang, Z., & Toure, I. K. (2014). Performance Analysis of Web Services in Different Types of Internet Technologies. *Applied Mechanics and Materials*, 513.

Hussain, S., Zhaoshun, W., & Toure, I. K. (2013). An approach for QoS measurement and web service selectness sureness. *High Technology Letters*, *19*(3), 283–289.

Iltaf, N., Mahmud, U., & Kamran, F. (2006). Security & trust enforcement in pervasive computing environment (STEP). In *Proceedings 2006 International Symposium on High Capacity Optical Networks and Enabling Technologies* (pp. 1-6). Charlotte, NC: IEEE. 10.1109/HONET.2006.5338409

Jung, Y. (2017). Hybrid-Aware Model for Senior Wellness Service in Smart Home. *Sensors (Basel)*, *17*(5), 1–10. doi:10.339017051182 PMID:28531157

Kesavan, M., & Prabhu, J. (2018). A Survey, Design and Analysis of IoT Security and QoS Challenges. *International Journal of Information System Modeling and Design*, *9*(3), 48–66. doi:10.4018/IJISMD.2018070103

Koshizuka, N., & Sakamura, K. (2010). Ubiquitous ID: Standards for Ubiquitous Computing and the Internet of Things. *IEEE Pervasive Computing*, *9*(4), 98–101. doi:10.1109/MPRV.2010.87

Kosmopoulos, D., & Styliaras, G. (2018). A survey on developing personalized content services in museums. *Pervasive and Mobile Computing*, *47*, 54–77. doi:10.1016/j.pmcj.2018.05.002

Larman, C. (2004). *Applying UML and Patterns: An Introduction to Object-Oriented Analysis and Design and Iterative Development*. Prentice Hall.

Li, L., Xiaoguang, H., Ke, C., & Ketai, H. (2011). The applications of WiFi-based Wireless Sensor Network in Internet of Things and Smart Grid. In *Proceedings 2011 6th IEEE Conference on Industrial Electronics and Applications* (pp. 789-793). Beijing, China: IEEE. doi:10.1109/ICIEA.2011.5975693

LOPEZ RESEARCH. (2013). *An Introduction to the Internet of things (IoT)*. CISCO. Retrieved from https://www.cisco.com/c/dam/en_us/solutions/trends/iot/introduction_to_IoT_november.pdf

Mahmud, U. (2015). UML based Model of a Context Aware System. In *Proceedings International Journal of Advanced Pervasive and Ubiquitous Computing*, *7*(1), 1–16. doi:10.4018/IJAPUC.2015010101

Mahmud, U. (2016). Organizing Contextual Data in Context Aware Systems: A Review. In J. Rodrigues, P. Cardoso, J. Monteiro, & M. Figueiredo (Eds.), *Handbook of Research on Human-Computer Interfaces, Developments, and Applications* (pp. 273–303). Hershey, PA: IGI Global; doi:10.4018/978-1-5225-0435-1.ch011

Mahmud, U., Hussain, S., & Yang, S. (2018). Power Profiling of Context Aware Systems: A Contemporary Analysis and Framework for Power Conservation. *Wireless Communications and Mobile Computing*, *2018*, 1–15. doi:10.1155/2018/1347967

Mahmud, U., Iltaf, N., & Kamran, F. (2007). *Context Congregator: Gathering Contextual Information in CAPP. In Proceedings 5th Frontiers of Information Technology (FIT 2007)* (pp. 134–141). Islamabad, Pakistan: COMSATS.

Mahmud, U., Iltaf, N., Rehman, A., & Kamran, F. (2007). *Context-Aware Paradigm for a Pervasive Computing Environment (CAPP). WWW\Internet 2007* (pp. 337–346). Portugal: Villa Real.

Mahmud, U., & Javed, M. Y. (2012, July). Context Inference Engine (CiE): Inferring Context. [IJAPUC]. *International Journal of Advanced Pervasive and Ubiquitous Computing*, *4*(3), 13–41. doi:10.4018/japuc.2012070102

Mahmud, U., & Javed, M. Y. (2014). Context Inference Engine (CiE): Classifying Activity of Context using Minkowski Distance and Standard Deviation-Based Ranks. In Systems and Software Development, Modeling, and Analysis: New Perspectives and Methodologies (pp. 65-112). Hershey, PA: IGI Global.

Mahmud, U., & Malik, N. A. (2014). Flow and Threat Modelling of a Context Aware System. [IJAPUC]. *International Journal of Advanced Pervasive and Ubiquitous Computing*, *6*(2), 58–70. doi:10.4018/ijapuc.2014040105

Malik, N. A., Javed, M. Y., & Mahmud, U. (2008). *Threat Modeling in Pervasive Computing Paradigm. In Proceedings 2008 New Technologies, Mobility, and Security* (pp. 1–5). Tangier, Morocco: IEEE.

Malik, N. A., Mahmud, U., & Javed, M. Y. (2007). Future challenges in context aware computing. WWW/Internet 2007, (pp. 306-310).

Malik, N. A., Mahmud, U., & Javed, M. Y. (2009, August). Estimating User Preferences by Managing Contextual History in Context Aware Systems. *Journal of Software*, *6*(4), 571–576.

Miraz, M. H., Ali, M., Excell, P. S., & Picking, R. (2015). *A review on Internet of Things (IoT), Internet of Everything (IoE) and Internet of Nano Things (IoNT). In Proceedings 2015 Internet Technologies and Applications (ITA)* (pp. 1–6). IEEE; doi:10.1109/ITechA.2015.7317398

Museum Guides Ltd. (n.d.). British Museum Guide. Retrieved from https://play.google.com/store/apps/details?id=air.com.bm.london.vusiem&hl=en

Padovitz, A., Loke, S. W., & Zaslavsky, A. (2004). Towards a theory of context spaces. In *Proceedings Second IEEE Annual Conference on Pervasive Computing and Communications, Workshops, PerCom* (pp. 38-42). Orlando FL: IEEE. doi:10.1109/PERCOMW.2004.1276902

Park, M., Han, J., Oh, H., & Lee, K. (2019). Threat Assessment for Android Environment with Connectivity to IoT Devices from the Perspective of Situational Awareness. *Wireless Communications and Mobile Computing*, *2019*, 1–14. doi:10.1155/2019/5121054

Patel, P., Ali, M. I., & Sheth, A. (2017). On Using the Intelligent Edge for IoT Analytics. *IEEE Intelligent Systems*, *32*(5), 64–69. doi:10.1109/MIS.2017.3711653

Perera, C., Zaslavsky, A., Christen, P., & Georgakopoulos, D. (2013). Context Aware Computing for The Internet of Things: A Survey. *IEEE Communications Surveys and Tutorials*, *16*(1), 1–41. doi:10.1109/SURV.2013.042313.00197

Petrakis, E. G., Sotiriadisb, S., Soultanopoulos, T., Rentaa, P. T., Buyya, R., & Bessis, N. (2018). Internet of Things as a Service (iTaaS): Challenges and solutions for management of sensor data on the cloud and the fog. *Internet of Things*, *3-4*, 156–174. doi:10.1016/j.iot.2018.09.009

Reineh, A.-A., Paverd, A. J., & Martin, A. P. (2016). *Trustworthy and Secure Service-Oriented Architecture for the Internet of Things.* Retrieved June 13, 2019, from https://arxiv.org: https://arxiv.org/pdf/1606.01671.pdf

Riazul Islam, S. M., Kwak, D., Kabir, M. H., Hossain, M., & Kwak, K.-S. (2015). The Internet of Things for Health Care: A Comprehensive Survey. *IEEE Access: Practical Innovations, Open Solutions*, *3*, 678–708. doi:10.1109/ACCESS.2015.2437951

Saeliw, A., Hualkasin, W., Puttinaovarat, S., & Khaimook, K. (2019). Smart Car Parking Mobile Application based on RFID and IoT. [iJIM]. *International Journal of Interactive Mobile Technologies*, *13*(5). Retrieved from https://online-journals.org/index.php/i-jim/issue/view/404

Serpanos, D., & Wolf, M. (2017). The IoT Landscape. In Proceedings *Internet-of-Things (IoT) Systems: Architectures, Algorithms, Methodologies* (pp. 1–6). Cham, Switzerland: Springer; doi:10.1007/978-3-319-69715-4_1

Spiess, P., Karnouskos, S., Guinard, D., Savio, D., Baecker, O., de Souza, L. M., & Trifa, V. (2009). SOA-Based Integration of the Internet of Things in Enterprise Services. In *Proceedings 2009 IEEE International Conference on Web Services*. Los Angeles, CA: IEEE. 10.1109/ICWS.2009.98

Srinivasa, K. G., Sowmya, B. J., Shikhar, A., Utkarsha, R., & Singh, A. (2018). Data Analytics Assisted Internet of Things Towards Building Intelligent Healthcare Monitoring Systems: IoT for Healthcare. *Journal of Organizational and End User Computing*, *30*(4), 83–103. doi:10.4018/JOEUC.2018100106

Strang, T., & Linnhoff-Popien, C. (2004). A Context Modelling Survey. *First International Workshop on Advanced Context Modelling, Reasoning and Management (Ubicomp 2004)*. Nottingham, UK: ACM.

Udoh, I. S., & Kotonya, G. (2017). Developing IoT applications: Challenges and frameworks. *IET Cyber-Physical Systems: Theory & Applications*, *3*(2), 65–72. doi:10.1049/iet-cps.2017.0068

Vargheese, R., & Dahir, H. (2014). An IoT/IoE enabled architecture framework for precision on shelf availability: Enhancing proactive shopper experience. In *Proceedings 2014 IEEE International Conference on Big Data (Big Data)* (pp. 21-26). Washington, DC: IEEE. 10.1109/BigData.2014.7004418

Zanella, A., Bui, N., Castellani, A. P., Vangelista, L., & Zorzi, M. (2014). Internet of Things for Smart Cities. *IEEE Internet of Things Journal*, *1*(1), 22–32. doi:10.1109/JIOT.2014.2306328

KEY TERMS AND DEFINITIONS

Artifact: A museum art piece or attraction. Artifacts have types and ratings by tourists.

Context Awareness: Intelligence enabling feature of IoE. Gathers data from sensors, recognizes activities and provides smart service discovery, delivery and adjustments.

IoE: An IoT system that provides intelligence to a user via a hand-held device. An IoE ecosystem creates a smart environment. Smart environments range from smart labs to smart clinics, smart homes, smart campuses and even smart city.

IoT: An interconnectivity of smart devices, sensors and services bounded within an environment.

Museum Tour Guide System: A tour guide that provides recommendations on artifacts that can be visited by a tourist. These recommendations are generated based on the context, mood and time constraint.

Recommendation System: A system that gives recommendations to a user based on the context.

Sensor Services: The services present in an IoE ecosystem that provide sensor data to a smart device.

Services: The applications that provide services to a user. Services can be remote or local to an IoE ecosystem.

Chapter 11
Lower Memory Consumption for Data Transmission in Smart Cloud Environments With CBEDE Methodology

Reinaldo Padilha França
State University of Campinas (UNICAMP), Brazil

Yuzo Iano
State University of Campinas (UNICAMP), Brazil

Ana Carolina Borges Monteiro
State University of Campinas (UNICAMP), Brazil

Rangel Arthur
State University of Campinas (UNICAMP), Brazil

ABSTRACT

Smart telecoms will deliver lasting improvements to business productivity and enduring consumer benefits that raise the quality of life by enabling telecommuting, telemedicine, entertainment, access to e-government, and a wealth of other online services. And we'll need next-generation digital platforms on which telecom providers can create and deliver all kinds of services. Therefore, this chapter develops a method of data transmission based on discrete event concepts. This methodology was named CBEDE. Using the MATLAB software, the memory consumption of the proposed methodology was evaluated, presenting the great potential to intermediate users and computer systems, ensuring speed, low memory consumption, and reliability. With the differential of this research, the use of discrete events applied in the physical layer of a transmission medium, the bit itself, being this to low-level of abstraction, the results show better computational performance related to memory utilization, showing an improvement of up to 79.89%.

DOI: 10.4018/978-1-7998-2112-0.ch011

INTRODUCTION

The internet has played a pivotal role in our lives since the 90s, when it emerged, both personally and professionally. This digital transformation has, since that decade, stimulated this relationship and synergy, improving the way we relate to information, which is increasingly flooding the world. Nowadays, the Internet is becoming part of the daily lives of many people, so that many activities can only be carried out by the screen of a computer, notebook, tablet or similar. This technological evolution is so great, that even the concept of a cybernetic intelligence, the well-known and famous Artificial Intelligence (AI) was built (Curran, 2016). The AI is nothing more than an intelligence developed by software and technologies, capable of finding solutions to many of the problems, be they of electronic systems, or those faced by humanity. With AI, machines can "think" through data that is collected over time, and stored in their memory, closely resembling human intelligence (Russell & Norvig, 2016).

Drawing a parallel between the human and the artificial brains, it can be realized that just as human intelligence needs a brain structure formed of gray matter, where our neurons perform their processing, so artificial intelligence also needs a structure artificial brain so that things can happen, and she can "think". Nowadays, artificial intelligence is seen applied in several technologies of humanity's everyday life, such as computer games, robots (which are increasingly resembling and becoming closer to human behavior), voice command programs on the cell phone, and other software that helps in factories, medical diagnostics, as well as the evolution of computing itself. In this way, artificial intelligence in the cloud is the technology that allows robots to operate intelligently and seamlessly, without physical limitations, and can respond to commands and perform previously limiting human activities that are subject to failure (Russell & Norvig, 2016, (Garfinkel & Grunspan, 2018).

These operations include operating systems by voice, the virtual reality of games, where we also see in more advanced studies, intelligent houses and integrated the network almost completely, as well as robots assisting in medicine. Taking into consideration the evolutionary journey of computational technologies since its most basic origin, each stage has aimed, in a way, at improving the performance of the artificial brain by increasing its capacities of connection (data flow) and processing (capacity and speed in processing such data). Going by the way, starting from the first step the migration of a hermetic mainframe, isolated, arriving the client/server architectures, more open and connected, which favors the distribution of data. In the modern days, moving from client/server to mobile, gaining not only connection, but also improvements in processing capacity, working on the same internet that structured the mobile, leveraging the cloud, which adds dimensions of flexibility, scalability and process availability. The concept of cloud computing comes with the idea of storing data outside of the computer and in the internet environment where files and programs can be safely accessed from various devices, desktops and mobile devices, from anywhere, which facilitates and raises the level as the communication between company and employees and clients is carried out, reaching levels of excellence (Russell & Norvig, 2016, Morabit, Mrabti, & Abarkan, 2019, Alfian, 2017).

Today, artificial intelligence operates in the cloud so that the activities are integrated with an entire system of data collection and storage, with advanced security, with hardware and software for developers and intelligent interfaces for users. The cloud has been configured as the natural environment of connected artificial intelligence, forming its artificial "gray matter" where processors, connections and intelligent components interact to "think" and "solve" problems, and can be easily escalated. In contrast to the human brain, the better the connections and the quality of its structure, the better the thought processes tend to be. In the case of artificial intelligence, the process is similar, the quality of the connections and structures that form the cloud directly affects its performance, being directly proportional, that is, the higher it's quality, the greater its potential to harbor "intelligence" and consequently "think". The use of AI for today and today has no limits, being present more and more in simple everyday things, such as spell checkers, be they on desktops or mobile devices, cell phones, digital games, bank passwords, among others. AI is present all the time working with our data and crossing information in the cloud so there are no oscillation and space problems (Wallace et al, 2019, Wang et al, 2015).

However, unlike the human brain that has fixed and limited size and capacity, the smart cloud can scale the size of your artificial brain according to need. Thus, the cloud has evolved from a repository of systems and data to a scalable intelligence repository. Many of the automated operating systems (OS) of today's largest companies are in the cloud in full integration with hardware and software being responsible for the proper functioning of digital platforms, which give remote access to the company's products and services. Functions such as calculations, registers, data transfer, and advanced security authentication, for example, are all executed by intelligent robots ready to meet our demands (Wang et al, 2015, Mukwevho & Celik, 2018).

In this way, the relationship between Cloud Computing and AI promotes new forms of work, consumption, and life, assisting conversations with virtual assistants, through systems with Amazon's Alexa, Microsoft's Cortana and Google Home, for example. All examples were based on AI technology, Cloud running on its own devices, as well as services for other companies. In the logistics area, it is possible to see its use in the warehouses of the main digital brands like Amazon, which shows a high degree of automation, with the interconnection of technologies of AI, Cloud, Robotics and Big Data. In sales, we have the giant Wallmart, which uses AI with Google Home technology on the Cloud, allowing voice-only purchases, and data management that makes it easy for consumers to purchase products in just a few steps (Mesbahi & Rahmani, 2016, Zhong et al, 2016).

In the same way as in banking services, in the customer service that the AI has been presenting more promising, along with the virtual assistants, established chatbot services creating a more efficient and autonomous connection between client and service company. An example of this is the Bradesco bank that uses a system called BIA (Bradesco Artificial Intelligence) working by voice and typing text. Thus, it is possible to synthesize that artificial intelligence overcomes human intuition, in a sense, where there is no more space for the imprecise, ambiguous and confused. In the same way that Artificial Intelligence is regarded as statistical "on steroids", related in the matter of something is much bigger, stronger, more impressive, in a more intense form than normal, that is, an evolution of the statistics. Since it is possible to apply statistics to data, you can also apply Machine Learning, the data being the first step to start any Machine Learning application, so no brain learns without data: neither human nor artificial (Elzamly et al, 2017, Segura, 2018).

So, cloud computing is a big shift from traditional business to an increasingly digital and modern world. There are six common reasons organizations and universities are turning to cloud computing services: cost, speed, global scale, productivity, performance, security. Whatever type of cloud computing service is in use, one thing is certain: large amounts of data will move back and forth between your end-users and the cloud provider's data centers, over the internet. Impact strongly on the digital age, which has brought the opportunity to transform telecommunication systems around the world, improve the way the world works; stimulate business growth; and provide a better experience of communication and interaction between people. Just now, in the past centuries, our world invested in rail, highway, and telephone infrastructure, what now in this modern era has brought the possibility of building up world digital infrastructure. In the last century, creating a new high-speed transportation system enabled businesses to expand much more quickly, helping drive economic growth for decades (Vendrúscolo & Moré, 2018, Rittinghouse & Ransome, 2016).

Today, we live in a digital world, with online assets traded virtually. Smart, highspeed telecoms will deliver lasting improvements to business productivity and enduring consumer benefits that raise the quality of life by enabling telecommuting, telemedicine, entertainment, access to e-government, and a wealth of other online services. Smart networks need more and more to be multidirectional instead of point-to-point. They will have to be built and operate on the foundation of standards and software that allow trillions of devices and objects to "talk." And we'll need next-generation digital platforms on which telecom providers can create and deliver all kinds of services (Rittinghouse & Ransome, 2016, Yin et al, 2015).

Still considering the mobile and cloud context of performance, the Rician fading model is ideally suitable for real-world phenomena, being a stochastic model for the anomaly propagation of the radio signal, caused by the partial cancellation of a radio signal by itself, the signal. The receiver's multi-path interference is one of the most common types of signal transmission in the world (Zhou et al, 2015, Gai et al, 2016, Xu et al, 2019).

Based on this, the present study aims to implement DES (Discrete Event Simulation) based model. This model is called CBEDE (Coding of Bits for Entities by means of Discrete Events) and aims to improve computational memory consumption in communication systems that can be applied in smart cloud environments. Aiming to solve such problems, and provide enhancements in smart cloud environments, the CBEDE model is applied to a communication system, specifically to advanced modulation format DBPSK (Differential Binary Phase Shift Keying) in a simulation environment, where a pre-coding process of bits improves the system memory consumption, based on the application of discrete events in the signal before the modulation process. Next, we will discuss the concepts of methodologies that contribute to the development of the CBEDE methodology.

Smart Cloud Computing

Migrating more and more quickly from fiction to reality, artificial intelligence systems have become part of our lives in all its dimensions, from personal assistants (such as Google's own applications on smartphones) to predictive systems (time, bag, behavior). Cloud Computing assumes that computing is not a product but a service. With this in mind, it allows a business to run IT solutions that are stored and made available on remote servers. Cloud, by definition, is the on-demand delivery of IT resources and applications over the Internet, with pricing model according to use, the form of computing where the resources needed to provide IT services are not in the customer's infrastructure (internal), and are, or can be, distributed around the world (Rittinghouse & Ransome, 2016, Botta et al, 2016).

The cloud platform provides the flexibility to increase the capacity of the use, customize according to the need of each client and offering scalability, the access to the data is made agile and from any place and any device, being such characteristics improve productivity. And all these benefits occur in a secure environment, keeping the information protected, including against force majeure, such as fire, flood or power failure, passable to human daily life. Cloud computing has a boon to businesses, especially small businesses, with cloud services, small businesses reap the benefits of not deploying physical infrastructures like e-mail servers, storage systems or shrink-wrapped software. Still, with the benefit of "anywhere, anytime" availability of these solutions, it means hassle-free collaboration among business partners, in the same way, that better income by employees by simply using the browser (Zhong et al, 2017, Al-Dhuraibi et al, 2017).

More simple examples that show how much AI has been influencing people's daily lives for years are spam filters (learning to discern between emails that are relevant and necessary and those that are not, even by filtering the malicious ones and advertisements), as well as map applications that suggest routes based on our previous road patterns and traffic conditions at all times. In transit, an example is Waze, which processes quickly and has a high degree of accuracy on the routes it traces. However, for this intelligence present in systems to assist us in any situation and to make decisions, it must be available all the time, be reliable and fast, where any of these conditions are not met, its usefulness will be limited, and unwanted. And so, with the cloud, two of these characteristics, availability, and speed, became possible with its advent. The fact is that from a locally installed desktop operating system and browser (or increasingly, from a single mobile device) a lot of today's small business technology needs can be fulfilled almost completely with cloud-based offerings (Green, 2018, Elizalde-Ramírez et al, 2019).

One thing is more certain with the passing of the years, Cloud computing is here to stay and it is growing rapidly, but cloud computing is not just for data, and yes possible to run applications and software remotely, without being tied to one computer physical. This is in keeping with the current scenario, which is making the cloud also smart, contributing to the intelligence (reliability) of the solution, with a huge potential to leverage intelligent processes for all types of applications, the so-called era of the smart cloud, the intelligence that comes from the cloud (Assunção et al, 2015).

The cloud has IaaS (Infrastructure as a Service), being one of the forms of Utility Computing, allowing businesses to purchase infrastructure from providers as virtual resources on an as-needed basis. What consists of basic terms in virtual hardware includes computers, raw processors, storage software platforms and so on, which instead of being physically based in an office, the hardware is located in the 'cloud' and data is accessed via the internet. In this way, the OS and everything that is needed, is installed on behalf of the client, the company can also hire a hardware capacity that corresponds to memory, storage, processing, etc. Servers, routers, racks, and more can also enter this hiring package. So we can think and visualize that the great advantage of this tariff is its scalability, after all, in a month, it is possible to require some virtual servers where it will store little data and will have little traffic, while you can request double the next month (Botta et al, 2016, Zhong et al, 2017, Madni et al, 2016).

Cloud also has PaaS (Platform as a Service), which provides a platform on which to deploy custom applications, databases and line-of-business services integrated into one platform. It can build its own applications that run on the provider's infrastructure and are delivered to users via the Internet from the provider's servers. PaaS provides all the necessary resources for small business owners to create their own software and programs, including an operating system, programming environment, database, and web server. What basically works like what is offered is a ready platform, a skeleton for developing custom services by the client. In this model of Cloud only what was installed inside the platform is on account of the client, however, backups, installation, configuration, OS and security will be on the account of the supplier. PaaS has emerged as a platform that can create, host and manage applications, its great advantage is that the development team only needs to worry about software programming because the management, maintenance, and updating of the infrastructure is the responsibility of the supplier (Pahl, 2015, Jamshidi et al, 2015).

While IaaS and PaaS will have some value to businesses large enough to have their own computer installations, it is SaaS (Software as a Service), with its access to applications, which provides the most value to small businesses. Where in this model the single application is delivered through the browser to thousands of customers on a subscription basis using the multitenant architecture, that is, one can run the software remotely through the browser, without having to go through the lengthy installation processes or have concerns about how your hardware will cope with the application, so the client only uses the service. All the rest is provided by the supplier. At the endpoint, the end-user sees service/software (transparency). It is a form of a cloud that is closer to the end-user, since it provides the final service that it will use, such as provide e-mail accounts to the customer, access to the software without buying your license, among other examples through the use of Cloud Computing, often with limited resources. There is a great advantage of scalability and practicality, that is, all processes related to the costs of a software purchase and the server, besides their implementation, are eliminated, since the service is available at a click away (Hussein & Khalid, 2016, Freet et al, 2015, Jain & Mahajan, 2017).

Intelligent cloud is a form of ubiquitous computing, enabled by AI technology, can be applied to all types of intelligent systems and applications imaginable, allowing for the insertion and increasingly active AI presence on a day-to-day basis using cloud resources and mobile devices, has deepened in areas such as machine learning, voice recognition and robot enablement for the purpose of data recognition by technology. Artificial intelligence is settling in people's homes, with personal assistants answering questions and controlling connected devices such as home appliances or lamps, which allows for the application and presence of the Internet of Things (IoT), where digital assistants already have simple features like remembering the people of appointments noted in calendars and advise them to leave early if the traffic is intense (Botta et al, 2016, Zhong et al, 2017, Wang et al, 2017).

Thus, smart cloud applications have evolved from low-computation IoT devices that work with the cloud to powerful computing devices. We gradually witness and sense the arrival of a future where it is based on a "smart cloud", where cloud-hosted digital assistants will accompany device-to-device users, thus increasingly combining the cloud's virtually unlimited computing power with intelligent, perceptual devices present on the network to create a framework for designing immersive and impactful business solutions (Wang et al, 2015, Hossain et al, 2017).

Still taking into account that in our days our most common activities leave a digital trail of personal data that is collected and manipulated by companies, whether private or public. Where such data can and are used for the provision of targeted advertising, telemarketing or sold to third parties without the consumer's consent. And considering that most companies are migrating their infrastructure to the cloud, in search of more agility, flexibility and cost reduction, also bringing specially developed embedded resources that help companies. However, an information protection feature that provides an integrated classification, labeling, and data protection experience is available, delivering more persistent protection that allows for the implementation of a proactive data governance strategy, regardless of where they are hosted or where transferred. In addition to features developed specifically to meet data protection standards, an intelligent cloud service of course also has an identity and access solutions, encryption, monitoring and detection of threats, among others (Khan & Salah, 2018).

Discrete Events and Entities

Discrete Events, is an effective technique to approach a wide variety of communication issues, being used mainly to relate to a model that represented a system as a sequence of operations performed on entities (like transactions) of specific data packets, bits, being derived from the results of actions taken in a system, and can be classified as an occurrence responsible for the change in the state of the system in which they act, capable of producing state changes at random intervals of time, generating data and so the information (Brito et al, 2011).

Modeling through the technique of discrete events can be subjective depending directly on the context of the scope in which the system is being modeled. This technique has been used to model concepts with a high level of abstraction, that is, it can reach concepts like patients, nurses, doctors within the modeling of a healthcare system; or even use the technique in modeling concepts such as exchanging emails on a server to transmit data packets between devices connected in a network; showing the robustness of technique in its abstraction, showing its importance in several areas of knowledge, with the diversity of applications that the technique of the discrete event provides. Thus, it is possible to apply it in a telecommunication system, more specifically in the transmission of data in a channel (Chahal & Eldabi, 2010).

The entities discrete in discrete event simulation, are defined as discrete items of interest in a discrete event simulation, being it's meaning depends on what is being modeled and the type of system. Discrete entities have attributes that may be able to affect the way events are treated or cause changes in the entity flows through the process. It is important to emphasize that the concepts of entities and events are different. The first is dependent on what is modeled and the type of system, and the other is an instantaneous or discrete incident that can change the state variable, an output of variable, or an occurrence from another event, respectively. An example of a discrete event signal is the characteristic that usually indicates "something" occurred, which may be an earthquake, control response, a heartbeat, or any other desirable concept within a system (Chahal & Eldabi, 2010).

AWGN Channel

The communication channel is the medium responsible for providing the physical connection between transmitters and receivers, whether the wire or for a logical connection in a multiplexed medium. Thus, modeling that is true to what exists, in reality, is necessary, and a widely used model to a large set of physical channels is the Additive White Gaussian Noise (AWGN) channel model, having the characteristic of introducing a statistically modeled noise, into the transmitted signals. This model modifies the existence of disturbances/noise in the channel (free space/atmosphere) which may be due to multiple causes, like fading, being a variation of the attenuation of the signal with several variables, such include time, geographical position, and radiofrequency, normally modeled as a random process (Tozer, 2012, Rama Krishna et al, 2016).

Rician Fading

The fading of a channel can be divided into large-scale fading, which is related by the occurrence of the attenuation of the average power or the loss in the signal path due to the movement of the receiver over large areas; and small-scale fading has occurred due to significant changes in amplitude, phase, and angle of arrival of the signal with respect to small changes in receiver position. The occurrence of multiple pathways is due to atmospheric scattering and refraction and/or reflections in mountains, buildings and other objects where these multiple replicas of the transmitted signal arrive with different attenuations and delays in the receiver, being summed in its antenna (Zhang et al, 2017, Le, 2019, Giordano & Levesque, 2015, Green, 2017).

Where fading by multipath affects the signal by scattering as well as by time-varying behavior. Rician is a useful model representing real-world phenomena in wireless communication, being stochastic for radio signal propagation anomaly, caused by partial cancellation of a radio signal by signal arriving at the receiver exhibiting multipath interference and at least one of the paths are changing, either by stretching or shortening. Rician is applied when one of the paths typically has a line of sight signal, and this is much stronger than the others (Zhang et al, 2017, Le, 2019, Giordano & Levesque, 2015, Green, 2017).

Differential Binary Phase Shift Keying (DBPSK)

This modulation technique differentiates from BPSK by encoding the data by similarity or by the difference of the symbols in relation to the previous signal, determined by the presence of bit 0 and 1, eliminating phase ambiguity and the need for phase acquisition and tracking, resulting in reduced energy consumption, still relying on the non-coherent way to solve the need for a coherent reference signal at the receiver. In this way, the input binary sequence is first differentially encoded and then modulated using the Binary Phase Shift Keying (BPSK) modulator, there being no need to know the initial state of the bit, so simplifying synchronization (Kolumbán & Kennedy, 2018, Ghassemlooy et al, 2019).

DBPSK and BPSK modulation are widely used in wireless LAN, RFID and Bluetooth communication applications, also used for long-distance wireless communication, in most of the adaptive modulation technique adopted in cellular communication, such as CDMA, WiMAX (16d, 16e), WLAN 11a, 11b, 11g, 11n, Satellite, DVB, etc. Related to the difference of 180 degrees between two constellation points, withstand the severe amount of channel conditions or channel fading, being used by most of the cellular towers for long-distance communication or transmission of the data, and its demodulator requires to make only two decisions in order to recover original binary information. Still taking into account the power-efficient modulation technique the less power is needed to transmit the carrier with a smaller number of bits (Kolumbán & Kennedy, 2018, Ghassemlooy et al, 2019).

Methodology

The development of this methodology was performed in a computer with hardware configuration being an Intel Core i3 processor, containing two processing cores, Intel Hyper-Threading Technology, and 4GB RAM. To provide enhancements to systems operating in cloud environments, and as previously mentioned, the present study implements a CBEDE model applied to a communication system, and advanced modulation format DBPSK (Differential Binary Phase Shift Keying) in a simulation environment, the Simulink of the MATLAB software (2014a), proposing a modeling of a system that has a lower computational memory consumption, through a pre-coding process of bits applying discrete events in the signal before of the modulation process.

The experiments were conducted through the Simulink tool, since that this simulation environment was chosen because it is already consolidated in the scientific medium, having a development and simulation environment already tested and validated. DBPSK and BPSK modulation were chosen in this study, because they are considered the most robust of modulation schemes in terms of noise immunity, is less immune to the interference, allowing the highest level of distortion in the signal being still successfully demodulated.

Four libraries were used: (1) Communications System ™, which is designed to design, simulate and analyze systems, being able to model dynamic communication systems; (2) the DSP System ™ that is capable of designing and simulating systems with signal processing; (3) Simulink®, which is a block diagram environment for multi-domain simulation, capable of supporting system-level projects for the

Figure 1. Traditional model of a telecommunication system

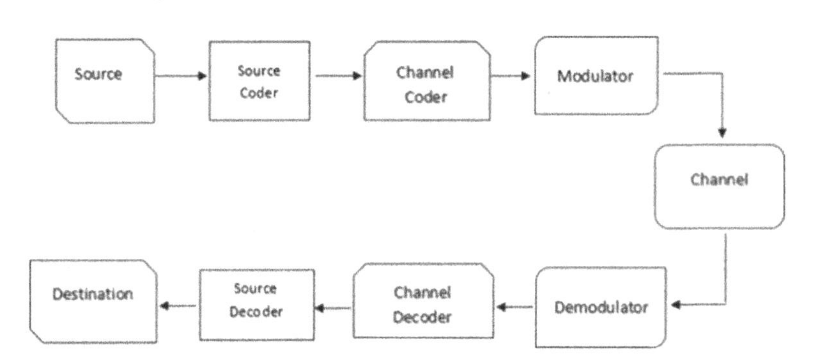

modeling and simulation of telecommunication systems, and (4) the library SimEvents®, which is classified as a discrete event simulation mechanism and components to develop systems models oriented to specific events

Previous work (Padilha et al, 2018, França et al, 2019, Padilha et al, 2018) and (Padilha, 2018), demonstrated that bit 0 treatment is viable, presenting significant results related to lower memory consumption of telecommunication systems, as well as in the reduction of processing time. In this way, the proposed methodology is based on the development of an AWGN hybrid channel, characterized by the introduction of the discrete event technique in the bit generation process, focusing on bits 0 and 1.

In the proposed model, Figure 1, the signals corresponding to bits 0 and 1 will be generated and modulated with the advanced modulation format DBPSK, which will use the phase shift, coming from the modulation format itself. It will then proceed to an AWGN channel according to the parameters shown in Table 1. The signal will then be demodulated to perform the Bit Error Rate (BER) calculation of the channel. The values obtained for BER will be sent to the Matlab workspace, in the name variables "yout" for equality verification and generation of the signal BER graph.

The modeling according to the proposal implemented with discrete events is similar to the one presented previously, except for the addition of discrete events in the pre-coding phase. The proposed bit precoding was implemented through the discrete event methodology. Bit processing is understood as the discrete event methodology in the step of generating signal bits (information) to make it more appropriate for a specific application.

The event-based signal is a signal susceptible to treatment by the SimEvents® library, and subsequently passed by conversion to the specific format required for manipulation by the Simulink® library. Both time-based signals and event-based signals were in the time domain. This treatment had an emphasis on bits 1 and 0, which were generated as a discrete entity and followed the parameters as presented in Table 1. Then, Entity Sink® represents the end of the modeling of discrete events by SimEvents library.

This tool is responsible for marking the specific point in which Entity Sink will be located, where later the event-based signal conversion will be performed for a time-based signal. This time-based signal was converted to a specific type that followed the desired output data parameter, an integer, the bit. By means of the Real-World Value (RWV) function, the actual value of the input signal was preserved. Then a rounding was performed with the floor function. This function is responsible for rounding the values to the nearest smallest integer.

Also used is a Zero-Order Hold (ZOH) which is responsible for defining sampling in a practical sense, being used for discrete samples at regular intervals. The ZOH describes the effect of converting

Figure 2. Proposed bit precoding

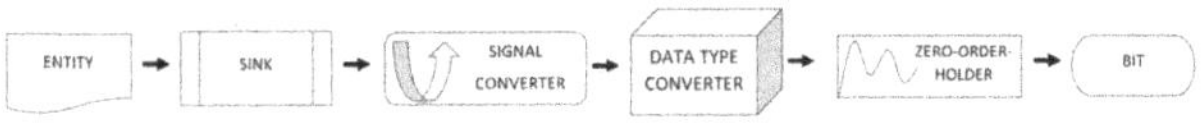

ENTITY → SINK → SIGNAL CONVERTER → DATA TYPE CONVERTER → ZERO-ORDER HOLDER → BIT

Figure 3. Model of a telecommunication system with the proposal

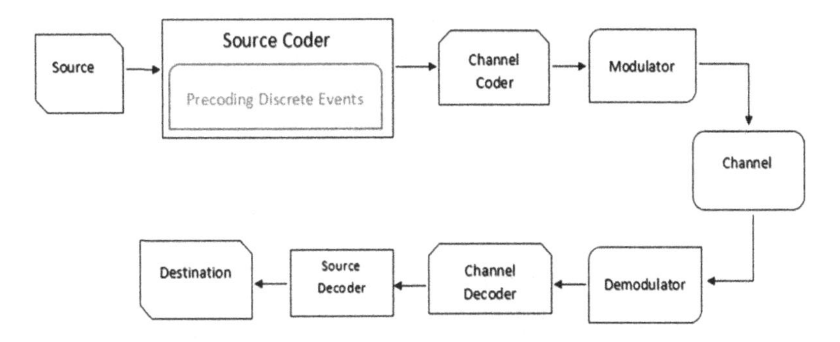

a signal to the time domain, causing its reconstruction and maintaining each sample value for a specific time interval. The treatment logic on bits 1 and 0 is shown in Figure 2.

Subsequently, the signal is modulated with the advanced modulation format DBPSK and is inserted into the AWGN channel, and then demodulated for the purposes of calculating the BER of the signal. The relative values of the BER are sent to the Matlab workspace in the name variable "yout1" to verify equality and generate the BER graph of the signal, as shown in Figure 3.

In the proposed model, Figures 1 and 3, the signals corresponding to bits 0 and 1 will be generated, respecting the rule and mathematical logic shown in Figure 4 below.

This rule and mathematical logic with respect to PSK M-ary numbers generate randomly distributed integers in the interval [0, M-1], where M is the definition for bit representation, following the nomenclature of the MATLAB software. Figure 5 shows the respective generation of the bits by means of this logic.

The models from Figures 1 and 3 are simulated for a time of 10000 seconds according to configuration parameters defined in Table 1.

The verification of equality of the signals is performed through the "size" and "isequal" functions of the Matlab software, as well as through the bit error rate (BER). These functions are responsible for the mathematical comparison proving that the signals have the same size and the same contents. Together with the BER check, it states that the same amount of information will be transmitted (bits) in both the proposed methodology (hybrid AWGN channel) and the conventional methodology (AWGN

Figure 4. Generation M-ary numbers for bits 0 and 1

[0, M - 1] -> [0, 2 - 1] -> 0,1

Figure 5. Signal generation by PSK M-ary numbers

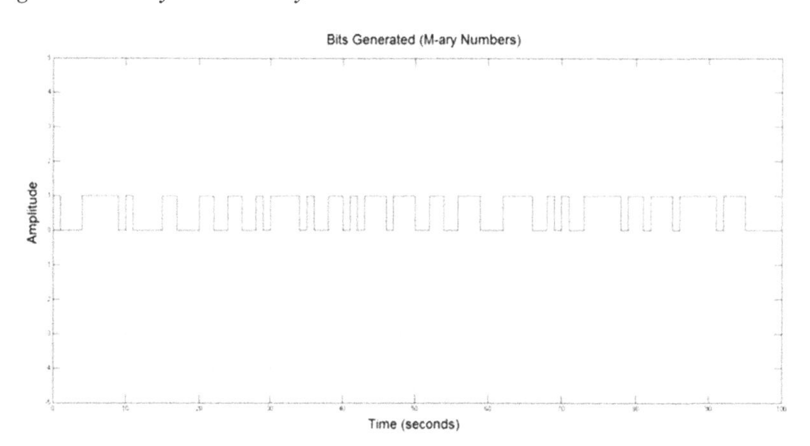

Table 1. Parameters channel models DBPSK Rician

AWGN DBPSK	
Sample Time	1 sec
Simulation time	10000 sec
Eb/N0	0 a 12 dB
Symbol period	1 sec
Input signal power	1 watt
Initial seed in the generator	37
Initial seed on the channel	67

channel). Thus, if the signals are of the same size, the logical value 1 (true) is returned and the same volume of data is transmitted, indicating that the equality of the signals is true. Otherwise, the value will be 0 (false). This check shows that the submitted proposal does not add or remove information to the originally transmitted signal.

The constellation has as function to analyze both signals transmitted by the models. In the case of the DBPSK constellation, a phase represents binary 1 and the other phase represents binary 0. As the digital input signal changes its state, the output signal phase will be changed between two angles separated by 180° (Tozer, 2012). This validation methodology has as function to affirm that the proposal does not modify the amount of bits transmitted by the signal, since both signals transmitted in the conventional channel and in the channel containing the proposal of this study, are of the same size. Figure 6 illustrates the DBPSK constellation diagram.

Figure 6. Theoretical DBPSK constellation

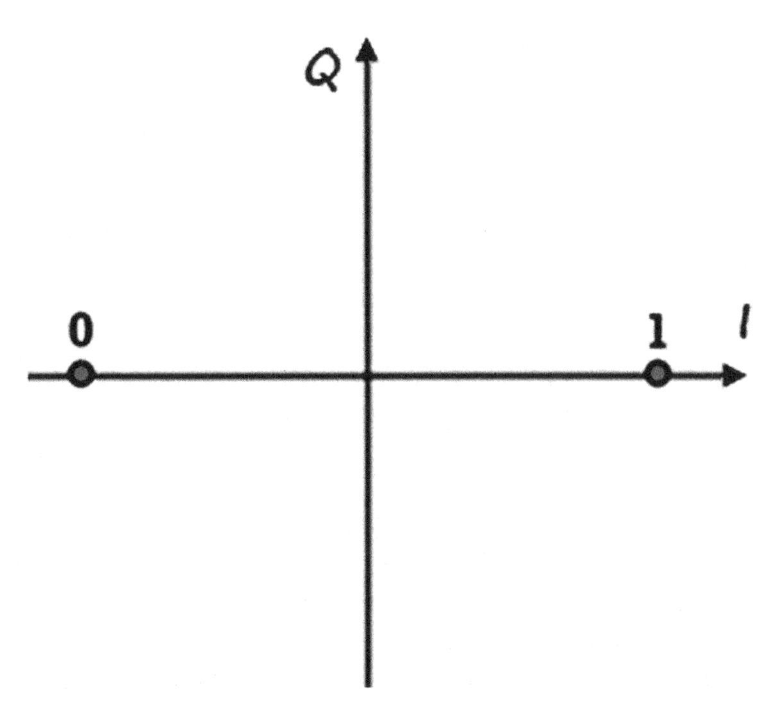

RESULTS AND DISCUSSION

The results presented in this section are given for an AWGN transmission channel with DBPSK modulation, using the Simulink simulation environment in MATLAB as the simulation platform. Figure 7 incorporates the traditional method (left) and the proposed innovation of this chapter (right), showing the signal transmission flow (corresponding to bits 0 and 1) passing through the channel AWGN, after its generation and modulation in DBPSK. Figure 8 displays the constellations for 11 dB for the proposed (left) and the traditional methods (right).

From figures 7 and 8, it is clear that both models generated the same result, since the system with the proposal (right) and in another only traditional methodology (left), were transmitted the same information content (quantity of bits), figure 7, and without any loss (signal and constellation), figure 8.

Figure 7. Transmission flow DBPSK Rician

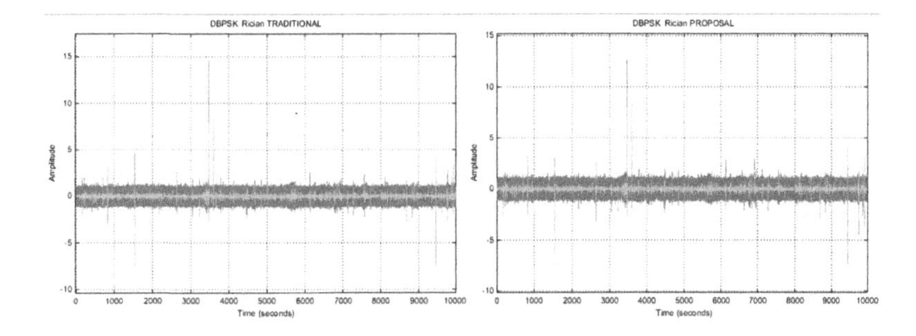

Figure 8. Simulated DBPSK Rician constellations

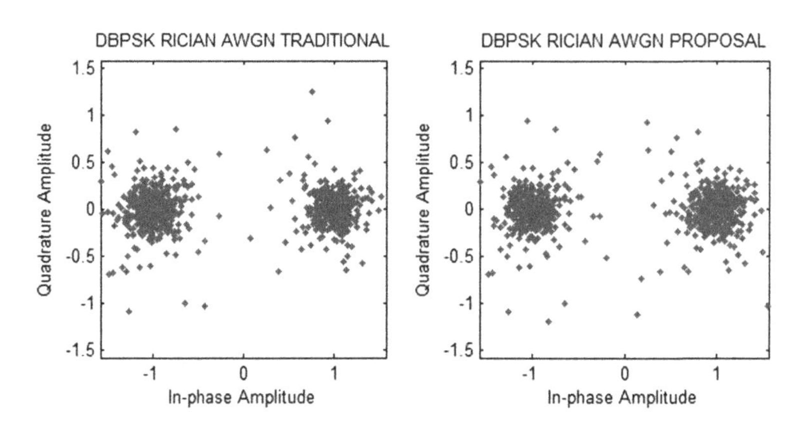

The models developed were investigated from the perspective of memory consumption evaluation. The first simulation of both models in each command is analyzed, since it is in the first simulation that the construction of the model in a virtual environment is performed from scratch, it is in it where all the variables of the model are allocated, the memory of the operating system in which the MATLAB is running is reserved for the execution of the model and the results of this model, according to the evaluation parameters are, in fact, real.

Thus, the experiments considered the memory consumption. For memory allocation scouting, the "sldiagnostics" function will be used, where the "TotalMemory" variable will receive the sum of all the memory consumption processes used in the model, by the "ProcessMemUsage" parameter. This parameter counts the amount of memory used in each process, throughout the simulation, returning the total in MB (megabyte) (Couch II, 2013). For this, it was used a computer with the following hardware configuration specifications: Intel Core i3 processor, containing two processing cores, Intel Hyper-Threading Technology, and 4GB RAM. This machine relates the proposal to the dynamics of the real world and will assess its efficiency and applicability. The experiments were carried out through 2 simulations for each one of the models employed with results shown in Figure 9.

From Figure 9, it is noteworthy that the application of discrete entities and discrete event technique, through their respective MATLAB libraries, in the modeling of a communication system in a simulated virtual environment, was satisfactory since the consumption of computational resources were smaller.

Figure 9. First simulations (memory) model DBPSK Rician

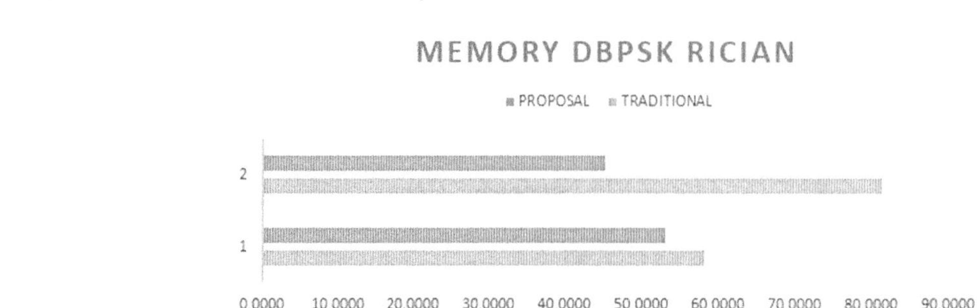

Table 2. Amounts of Memory Consumption

Simulation	TRADITIONAL	CBEDE
1	58,1367	52,9453
2	81,3867	45,2422

The exact values of the amount of memory consumption with the 2 models from Figure 9 are listed in Table 2.

Just as having the important character as developing the methodology improving the model's computational memory consumption in a simulation environment, it is to make the know-how available to the academic community, as well as to contribute to the area of study of this proposal that this chapter deals with.

To analyze the relationship between the simulation methodology (proposed vs traditional method), and the impact on the physical layer of the channel, scripts were made in the MATLAB for processing of the graph BER. Figure 10 display the performance of the models during transmission with noise ranging from 0 to 12 dB.

This proposal brings a new approach to signal transmission. In this case, the transmission is performed in the discrete domain with the implementation of discrete entities in the bit generation process. In this way, it is possible to realize that there was a positive impact on the system, since the computational memory consumption was lower in the communication systems model applied with the proposal. Given

Figure 10. BER between the models DBPSK Rician

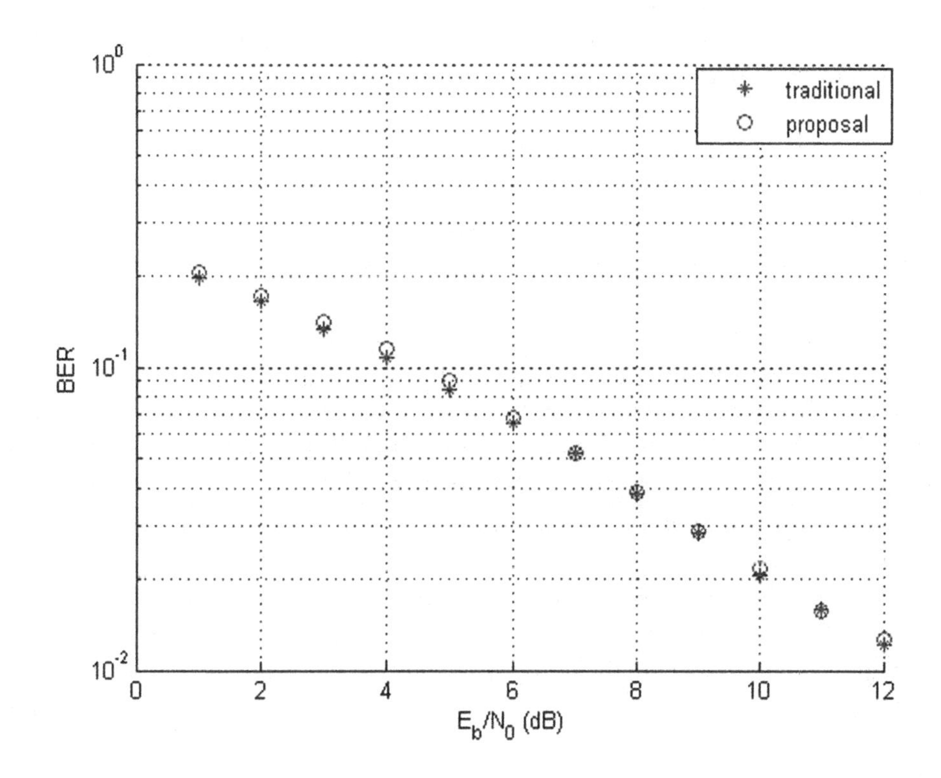

the above, the technique of discrete events can be applied in the treatment of bits in its generation stage, being responsible for their conversion into discrete entities. This process is the result of a technique (discrete events) used in a lower level of application, which acts on the physical layer, which is generally practiced, usually in the transport layer. Which has proved beneficial since reducing the consumption of computational resources, such as memory, which is an important parameter to meet the needs of an increasingly technological world.

Considering that speed is a key issue when choosing a methodology, whether it is used by a user, companies or universities, the CBEDE methodology can be seen as a great allied to cloud computing systems. Even if the cloud has large storage capacity and provides security in data storage, the user always cares about the time it takes to upload or download your data. For this, it is important to invest in systems that intermediate the cloud computing systems and users, and consume less computational resources, linking with the result of CBEDE reaching up to 79.89% in the improvement of memory consumption, which can be employed in this process.

Another important factor when it comes to data transmission is the memory consumption of the device. Currently, the cloud computing system is linked to computers and mobile phones. As is well known, the use of cell phones today is much larger than that of a computer due to its mobility. Ordinary users tend to upload documents, images, and videos for the purpose of sharing their common social network, or even save their personal files (docx, pdf, avi, jpg, among others) in their "personal cloud".

The slowness related to the speed of communication and the cloud technology implemented in the system structure used by these users, can often generate device crashes, inconvenience to the user and sometimes even loss of data, if these are in real-time. When this happens, often the reliability of Cloud computing can be put to the test by the user who can from that moment prefer to store their data on physical platforms. This bad impression can be intensified when this situation occurs with companies and universities. Thus, in order to break this bad impression and generate greater reliability, the CBEDE methodology can also be seen as a great ally, being better than the traditional methodologies, which do not require the use of discrete events.

Nowadays, this feature has been considered crucial mainly by companies that provide their services to a large number of users simultaneously. In this way, the CBEDE presents the great potential of deployment in systems of bank branches, technology startups, and mobile telephony systems, as well as others. For since integrated to a Smart Cloud system, companies can dispense with the need to invest time and money with the acquisition and deployment of software and hardware, because they can choose to pay a monthly amount for the services they need.

And acquisitions of services according to the specific needs of each company directly impact the reduction of costs and exponentially increase productivity, and with the application of the proposal presented (CBEDE) this impact can be even greater since the system will operate with low memory. consumption, and will have positive results for users who will have even more properties to be connected anytime, anywhere, whether on a computer, smartphone or tablet.

Another benefit of using the CBEDE technique integrated with Smart Cloud systems can be seen in the following reality: systems with low memory consumption, are less likely to present crashes. With this, users do not go through the frustration of losing time in front of a stopped system. This reality increases usability, which is a factor of extreme importance for attracting new customers. In this context, the more users use a system that dispenses with the need for specialized hardware there is an extremely positive environmental impact, because per year thousands of pounds of damaged and/or old equipment are no longer dumped into the wild. In this way, the integration of Smart Cloud and the CBEDE methodology can provide greater sustainability to any type of company.

SOLUTIONS AND RECOMMENDATIONS

Bandwidth

The cloud needs network connectivity, making an analogy similar to that humans need oxygen, since many large cities and large metropolises come from bandwidth, but in developing countries like Brazil, it still has no coverage. total bandwidth, resulting in a slow or unreliable connection where cloud services suffer. Due to the main feature of the cloud I came from its "ubiquitous" connectivity, when it does not exist, all infrastructure and proposed improvements such as CBEDE suffer from disadvantages.

Just as cloud computing is a technology whose adoption is growing daily around the world. This, in turn, led to an increase in investments in Brazil. Which, according to BSA's "Global Performance on Cloud Computing" survey of 24 countries leading the world Information Technology (IT) market, has jumped from 22nd position in 2016 to 18th in 2018 (BSA Global Cloud Computing Scorecard, 2018). Showing that the country has followed the worldwide trend of cloud computing growth and making an investment in this area, in which this current placement in this ranking is significant, especially in relation to other Latin American countries. Showing a positive indication, since it shows that the country keeps investing in a technology that has revolutionized internet communication and still has huge potential for development. In this regard, CBEDE is a proposal that has been shown to be valid for implementation by acting on the local infrastructure of an already developed system providing improvements.

FUTURE RESEARCH DIRECTIONS

In the future work we will Include simulations and analyzes on Intel Dual Core architectures of processors such as Intel i3, i5 and i7, as they are the most common on the market today, as in other types of systems such as OFDM (Orthogonal Frequency Division Multiplexing), with respect and data transmission via mobile devices, since the last years the popularity of mobile devices has grown as a trend in global communication.

CONCLUSION

The objective of this research was the use of discrete events being applied at the lowest level of abstraction (research differential) possible within a communication system, as is done in bit generation, resulting in an environment of smart cloud environment more productive, faster and with better performance, since the proposal acts on the bit itself, where it showed results computational performance related to memory utilization of 79.89%.

This demonstrates that the CBEDE has great potential in the improvement of communication services potential, as well as improving cloud computing services in the same way as smarter telecommunications systems allow individuals, businesses, governments, and educational institutions to interact and transact in new, more efficient and personalized ways. Companies will use fast connections to make existing processes more efficient and productive, develop innovative business models, and transform business activities through new communications-enabled work practices. Improving already existing processes can increase the performance of communication between all the devices in the system and in the cloud environment, because the flow of data will consume fewer resources and, therefore, can improve the interactions between the users of these intelligent communication systems increasingly.

REFERENCES

Al-Dhuraibi, Y., Paraiso, F., Djarallah, N., & Merle, P. (2017). Elasticity in cloud computing: State of the art and research challenges. *IEEE Transactions on Services Computing*, *11*(2), 430–447. doi:10.1109/TSC.2017.2711009

Alfian, A. (2017, December). The development framework of expert system application on indonesian governmental accounting system. In *Proceedings of the 2017 International Conference on Computer Science and Artificial Intelligence* (pp. 60-64). ACM. 10.1145/3168390.3168437

Assunção, M. D., Calheiros, R. N., Bianchi, S., Netto, M. A., & Buyya, R. (2015). Big Data computing and clouds: Trends and future directions. *Journal of Parallel and Distributed Computing*, *79*, 3–15. doi:10.1016/j.jpdc.2014.08.003

Botta, A., De Donato, W., Persico, V., & Pescapé, A. (2016). Integration of cloud computing and internet of things: A survey. *Future Generation Computer Systems*, *56*, 684–700. doi:10.1016/j.future.2015.09.021

Brito, T. B., Trevisan, E. F., & Booter, R. C. (2011). A Conceptual Comparison Between Discrete and Continuous Simulation to Motivate the Hybrid Simulation Technology. In *Proceedings of the 2011 Winter Simulation Conference*, 3915-3927. Winter Simulation Conference.

Chahal, K., & Eldabi, T. (2010). A multi-perspective comparison between system dynamics and discrete event simulation, *Journal of Business Information Systems,* pp. 4-17.

Couch, L. W. II. (2013). *Digital and Analog Communication Systems* (8th ed.). Prentice-Hall.

Curran, J. (2016). The internet of history: rethinking the internet's past. In *Misunderstanding the internet* (pp. 48–84). Routledge. doi:10.4324/9781315695624-2

Elizalde-Ramírez, F., Nigenda, R. S., Martínez-Salazar, I. A., & Ríos-Solís, Y. Á. (2019). Travel Plans in Public Transit Networks Using Artificial Intelligence Planning Models. *Applied Artificial Intelligence*, 1–22.

Elzamly, A., Hussin, B., Abu Naser, S. S., Shibutani, T., & Doheir, M. (2017). Predicting Critical Cloud Computing Security Issues using Artificial Neural Network (ANNs) Algorithms in Banking Organizations. *International Journal of Information Technology and Electrical Engineering*, *6*(2), 40–45.

França, R. P., Iano, Y., Monteiro, A. C. B., Arthur, R., Estrela, V. V., Assumpção, S. L. D. L., & Razmjooy, N. (2019). Potential Proposal to Improvement of the Data Transmission in Healthcare Systems.

Freet, D., Agrawal, R., John, S., & Walker, J. J. (2015, October). Cloud forensics challenges from a service model standpoint: IaaS, PaaS and SaaS. In *Proceedings of the 7th International Conference on Management of computational and collective intElligence in Digital EcoSystems* (pp. 148-155). ACM. 10.1145/2857218.2857253

Gai, K., Qiu, M., Tao, L., & Zhu, Y. (2016). Intrusion detection techniques for mobile cloud computing in heterogeneous 5G. *Security and Communication Networks*, *9*(16), 3049–3058. doi:10.1002ec.1224

Garfinkel, S. L., & Grunspan, R. H. (2018). The Computer Book: From the Abacus to Artificial Intelligence, 250 Milestones in the History of Computer Science.

Ghassemlooy, Z., Popoola, W., & Rajbhandari, S. (2019). Optical wireless communications: system and channel modelling with Matlab. Boca Raton, FL: CRC Press. doi:10.1201/9781315151724

Giordano, A. A., & Levesque, A. H. (2015). *Digital Communications BER Performance in AWGN*. BPSK in Fading.

Global Cloud Computing Scorecard, B. S. A. (2018). Powering a Bright Future. 2018. Disponível em: https://cloudscorecard.bsa.org/2018/?sc_lang=pt-BR. Acesso em: Nov. 11, 2019.

Green, B. P. (2018). Ethical Reflections on Artificial Intelligence. *Scientia et Fides*, *6*(2), 9–31. doi:10.12775/SetF.2018.015

Green, P. J. (2017, November). Implementation of a real-time Rayleigh, Rician and AWGN multipath channel emulator. In Proceedings TENCON 2017-2017 IEEE Region 10 Conference (pp. 35-39). IEEE.

Hossain, M. S., Xu, C., Li, Y., Pathan, A. S. K., Bilbao, J., Zeng, W., & El-Saddik, A. (2017). Impact of Next-Generation Mobile Technologies on IoT-Cloud Convergence. *IEEE Communications Magazine*, *55*(1), 18–19. doi:10.1109/MCOM.2017.7823332

Hussein, N. H., & Khalid, A. (2016). A survey of cloud computing security challenges and solutions. *International Journal of Computer Science and Information Security*, *14*(1), 52.

Jain, A., & Mahajan, N. (2017). Introduction to Cloud Computing. In The Cloud DBA-Oracle (pp. 3-10). Berkeley, CA: Apress. doi:10.1007/978-1-4842-2635-3_1

Jamshidi, P., Pahl, C., Chinenyeze, S., & Liu, X. (2015). Cloud migration patterns: a multi-cloud service architecture perspective. In *Service-Oriented Computing-ICSOC 2014 Workshops* (pp. 6–19). Cham, Switzerland: Springer. doi:10.1007/978-3-319-22885-3_2

Khan, M. A., & Salah, K. (2018). IoT security: Review, blockchain solutions, and open challenges. *Future Generation Computer Systems*, *82*, 395–411. doi:10.1016/j.future.2017.11.022

Kolumbán, G., & Kennedy, M. P. (2018). Overview of digital communications. In Chaotic Electronics in Telecommunications (pp. 131–149). Boca Raton, FL: CRC Press. doi:10.1201/9781482274516-5

Le, K. N. (2019). A review of selection combining receivers over correlated Rician fading. *Digital Signal Processing*, *88*, 1–22. doi:10.1016/j.dsp.2019.01.015

Madni, S. H. H., Latiff, M. S. A., Coulibaly, Y., & Abdulhamid, S. M. (2016). Resource scheduling for infrastructure as a service (IaaS) in cloud computing: Challenges and opportunities. *Journal of Network and Computer Applications*, *68*, 173–200. doi:10.1016/j.jnca.2016.04.016

Mesbahi, M., & Rahmani, A. M. (2016). Load balancing in cloud computing: a state-of-the-art survey. *Int. J. Mod. Educ. Comput. Sci, 8*(3), 64.4

Morabit, Y. E., Mrabti, F., & Abarkan, E. H. (2019). Survey of Artificial Intelligence Approaches in Cognitive Radio Networks. *Journal of Information and Communication Convergence Engineering, 17*(1), 21-40.

Mukwevho, M. A., & Celik, T. (2018). Toward a smart cloud: A review of fault-tolerance methods in cloud systems. *IEEE Transactions on Services Computing*, 1. doi:10.1109/TSC.2018.2816644

Padilha, R. (2018). Proposta de Um Método Complementar de Compressão de Dados Por Meio da Metodologia de Eventos Discretos Aplicada. In *Um Baixo Nível de Abstração. Dissertação (Mestrado em Engenharia Elétrica)*. Campinas, Brasil: Faculdade de Engenharia Elétrica e de Computação, Universidade Estadual de Campinas.

Padilha, R., Iano, Y., Monteiro, A. C. B., Arthur, R., & Estrela, V. V. (2018, October). Betterment Proposal to Multipath Fading Channels Potential to MIMO Systems. In *Proceedings Brazilian Technology Symposium* (pp. 115-130). Cham, Switzerland: Springer.

Padilha, R., Iano, Y., Monteiro, A. C. B., & Loschi, H. J. (2018). Improvement of the Content Transmission in Broadcasting Systems: Potential Proposal to Rayleigh and Rician Multichannel MIMO Systems.

Pahl, C. (2015). Containerization and the paas cloud. *IEEE Cloud Computing*, *2*(3), 24–31. doi:10.1109/MCC.2015.51

Rama Krishna, A., Chakravarthy, A. S. N. & Sastry, A. S. C. S. (2016). Variable Modulation Schemes for AWGN Channel based Device to Device Communication. *Indian Journal of Science and Technology, 9*(20), DOI: 10.17485 / ijst / 2016 / v9i20 / 89973.

Rittinghouse, J. W., & Ransome, J. F. (2016). Cloud computing: implementation, management, and security. Boca Raton, FL: CRC Press.

Russell, S. J., & Norvig, P. (2016). *Artificial intelligence: a modern approach*. Malaysia: Pearson Education Limited.

Segura, M. (2018). INTELIGÊNCIA ARTIFICIAL APLICADA A NEGÓCIOS. *Revista Inteligência Competitiva*, *8*(3), 101–110.

Tozer, E. P. (2012). *Broadcast Engineer's Reference Book* (1st ed.). FOCAL PRESS. doi:10.4324/9780080490564

Vendrúscolo, J. D. B. G., & Moré, R. P. O. (2018). Contribuições Da Inteligência Artificial Nos Sistemas De Informação De Apoio A Gestão Universitária. Available at https://repositorio.ufsc.br/handle/123456789/190471

Wallace, R. B., Liu, H., Goubran, R., Bilodeau, M., & Knoefel, F. (2019). Cloud based Artificial Intelligence Processing of Ambient Home Sensors. In *ISPIM Conference Proceedings* (pp. 1-12). The International Society for Professional Innovation Management (ISPIM).

Wang, J., Gong, B., Liu, H., & Li, S. (2015). Multidisciplinary approaches to artificial swarm intelligence for heterogeneous computing and cloud scheduling. *Applied Intelligence*, *43*(3), 662–675. doi:10.100710489-015-0676-8

Wang, X. V., Wang, L., Mohammed, A., & Givehchi, M. (2017). Ubiquitous manufacturing system based on Cloud: A robotics application. *Robotics and Computer-integrated Manufacturing*, *45*, 116–125. doi:10.1016/j.rcim.2016.01.007

Xu, L., Yu, X., Wang, H., Dong, X., Liu, Y., Lin, W., ... Wang, J. (2019). Physical layer security performance of mobile vehicular networks. *Mobile Networks and Applications*, 1–7.

Yin, Z., Yu, F. R., Bu, S., & Han, Z. (2015). Joint cloud and wireless networks operations in mobile cloud computing environments with telecom operator cloud. *IEEE Transactions on Wireless Communications, 14*(7), 4020-4033.

Zhang, J., Dai, L., He, Z., Jin, S., & Li, X. (2017). Performance analysis of mixed-ADC massive MIMO systems over Rician fading channels. *IEEE Journal on Selected Areas in Communications*, *35*(6), 1327–1338. doi:10.1109/JSAC.2017.2687278

Zhong, R. Y., Lan, S., Xu, C., Dai, Q., & Huang, G. Q. (2016). Visualization of RFID-enabled shopfloor logistics Big Data in Cloud Manufacturing. *International Journal of Advanced Manufacturing Technology*, *84*(1-4), 5–16. doi:10.100700170-015-7702-1

Zhong, R. Y., Xu, X., Klotz, E., & Newman, S. T. (2017). Intelligent manufacturing in the context of industry 4.0: A review. *Engineering*, *3*(5), 616–630. doi:10.1016/J.ENG.2017.05.015

Zhou, B., Dastjerdi, A. V., Calheiros, R. N., Srirama, S. N., & Buyya, R. (2015). mCloud: A context-aware offloading framework for heterogeneous mobile cloud. *IEEE Transactions on Services Computing*, *10*(5), 797–810. doi:10.1109/TSC.2015.2511002

ADDITIONAL REFERENCES

Balemi, S. (1994). Input/output discrete event processes and communication delays. *Discrete Event Dynamic Systems*, *4*(1), 41–85. doi:10.1007/BF01516010

BouchhimaF.BriereM.NicolescuG.AbidM.AboulhamidE. M. (2006, September). A

Freeman, R. L. (2015). *Telecommunication system engineering* (Vol. 82). John Wiley & Sons.

Kang, M., & Alouini, M. S. (2006). Capacity of MIMO Rician channels. *IEEE Transactions on Wireless Communications*, *5*(1), 112–122. doi:10.1109/TWC.2006.1576535

Lee, E. A. (1999). Modeling concurrent real-time processes using discrete events. *Annals of Software Engineering*, *7*(1-4), 25–45. doi:10.1023/A:1018998524196

Lindsey, W. C., & Simon, M. K. (1991). *Telecommunication systems engineering*. Courier Corporation.

Patzold, M. (2001). *Mobile fading channels: Modelling, analysis and simulation*. John Wiley & Sons, Inc.

Rudie, K., Lafortune, S., & Lin, F. (2003). Minimal communication in a distributed discrete-event system. *IEEE Transactions on Automatic Control*, *48*(6), 957–975. doi:10.1109/TAC.2003.812780

Schwartz, M. (1987). *Telecommunication networks: protocols, modeling and analysis* (Vol. 7). Reading, MA: Addison-Wesley.

Shu, S., & Lin, F. (2014). Decentralized control of networked discrete event systems with communication delays. *Automatica*, *50*(8), 2108–2112. doi:10.1016/j.automatica.2014.05.035

Shu, S., & Lin, F. (2016). Deterministic networked control of discrete event systems with nondeterministic communication delays. *IEEE Transactions on Automatic Control*, *62*(1), 190–205. doi:10.1109/TAC.2016.2553959

Simon, M. K., & Alouini, M. S. (2005). *Digital communication over fading channels* (Vol. 95). John Wiley & Sons.

Smedinga, R. (1988). Using trace theory to model discrete events. In *Discrete Event Systems: Models and Applications* (pp. 81–99). Berlin, Heidelberg: Springer. doi:10.1007/BFb0042306

Smedinga, R. (1989). *Control of discrete events*. Rijksuniversiteit Groningen.

SystemC/Simulink co-simulation framework for continuous/discrete-events simulation. In 2006 IEEE International Behavioral Modeling and Simulation Workshop (pp. 1-6). IEEE.

Tripakis, S. (2004). Decentralized control of discrete-event systems with bounded or unbounded delay communication. *IEEE Transactions on Automatic Control*, *49*(9), 1489–1501. doi:10.1109/TAC.2004.834116

Vannucci, G. (1995). U.S. Patent No. 5,459,727. Washington, DC: U.S. Patent and Trademark Office.

Yuen, J. H. (Ed.). (2013). *Deep space telecommunications systems engineering*. Springer Science & Business Media.

Zamaï, E., Chaillet-Subias, A., & Combacau, M. (1998). An architecture for control and monitoring of discrete events systems. *Computers in Industry*, *36*(1-2), 95–100. doi:10.1016/S0166-3615(97)00102-4

KEY TERMS AND DEFINITIONS

Communication Systems: Communications systems are those in which a message is sent by a sender through a particular channel, it is important to consider that in this transmission of information, there is usually an interference called noise, which comprises anything that moves on the channel other than the actual signals or messages desired by the sender, and ultimately this message is understood by a receiver, can also be considered as a network through which the information that enables the structure to function in an integrated and effective manner flows.

Data: Data is basically codes that constitute the raw material of information, ie, it is untreated information. Data refers to facts, events, actions, activities, and transactions which have been and can be recorded, i.e. the raw material from which information is produced, nurturing the infrastructure and components that enable modern computing. Data represent one or more meanings of a system that transmits a message. Information is any structuring or organization of such data.

Discrete Events: Discrete Events represent a system as a sequence of operations performed on entities derived from the results of actions taken on it causing changes in the state of the system at intervals of time.

Modulation: It is the process in which the information to be transmitted is added to electromagnetic waves, since the information signals do not usually have the proper properties to travel through the transmission means, thus through a means of transport, that information gains the proper properties for transmission.

Rician Fading: Multiple replicas of the transmitted signal arrive with different attenuations and different delays to the receiver, being added to their antenna. Rician fading is a modeling of real-world phenomena in wireless communications, being stochastic for the radio signal propagation anomaly caused by the partial cancellation of a radio signal by itself, causing the signal to reach the receiver displaying multipath interference and at least one of the paths is changing, occurs when one of the paths, typically a line of sight signal, is much stronger than the other.

Simulation: Simulation, like experimentation with a model that mimics certain aspects of reality, involves modeling a process or system in such a way that the model mimics the responses of the real system in a succession of events that occur over time, but with variables. controlled and in an environment that resembles the real, although artificially created.

Smart Cloud: The convergence between artificial intelligence (AI) and cloud computing has revolutionized information technology (IT), since AI makes it possible to perform tasks automatically, consisting of the union of several technological tools, which allows a computer vision and processing. more sophisticated to improve the ability to generate large amounts of data and thus create more accurate and accurate forecasts. AI in cloud computing (with all the properties and characteristics of this technology) is the technology that enables robots to operate intelligently and to respond to commands and perform logical activities previously limited to humans, as this combination creates an intelligent cloud.

System: It is a logical organization that processes one set of signals at the input to produce another set of signals at the output.

Chapter 12
Do It Fluid:
Innovation in Smart Conversational Services Through the Flow Design Approach

John Knight
https://orcid.org/0000-0002-6784-3892
Aalto University, Finland

Rachel Jones
https://orcid.org/0000-0002-7997-5112
Instrata Ltd, UK

Deniz Sayar
https://orcid.org/0000-0002-4056-5538
Izmir University of Economics, Turkey

Damian Copeland
https://orcid.org/0000-0001-7115-2448
ICR Speech Solutions, UK

Daniel Fitton
University of Central Lancashire, UK

ABSTRACT

This chapter draws on practical experience in designing, delivering, operating, and innovating conversational services. The article summarises the current context for these distinctively new kinds of services and provides an overview of the relevant technologies and common platforms used in commercial service production. The chapter explores the broader commercial context for smart voice-oriented services and provides an applied framework to aid service innovation. The two concluding parts move into service production, outline a grounded design approach (FLOW) for maximising service flow, and discuss future research directions, specifically how design anthropology can help in radical service innovation.

DOI: 10.4018/978-1-7998-2112-0.ch012

INTRODUCTION

The rapid growth of conversational services has catapulted what was a niche area of Human-Computer Interaction (HCI) research into the mainstream. Conversational devices, platforms and services aimed at and adopted by mass audiences are now increasingly commonplace. Voice-enabled services include household names such as Amazon Echo and Google Home. These and an array of other highly successful commercial offerings utilise conversational interfaces that also integrate with a growing number of services applications in everything from retail to health. Amazon Echo, for example, allows developers to plug third-party services into its platform so that they can deliver voice-based interfaces to the masses. These platforms are highly customisable, requiring minimal frontend and backend integration. While conversational interfaces have become pervasive, however, relatively little work has been done to date evolving design methodologies to this domain (Clark et al., 2019a), particularly in the context of services innovation.

In order to address this gap, this chapter outlines the relevant contexts contingent on successful conversational interface innovation. First, the technical aspects are tackled. Then the business context is explored with a focus on overcoming the barriers faced by organisations wanting to make use of technological opportunities in this area. Strategies and solutions are outlined in this section, which generalises to help the smallest through to the largest of enterprises. The final context discussed is the service domain. While this has been thoroughly explored in the business and design literature it has lacked exposure to a wider audience and application to the conversational service domain. These various strands are brought together in the penultimate section, where the FLOW Design approach is outlined in the context of a health-related case study. The final and concluding part takes the investigation beyond the present day and looks at the potential of design anthropology to help organisations shift from incremental to radical service innovation.

SERVICE TECHNOLOGY CONTEXT

Speech interaction is nothing new. It is a mode of interaction that has been around for many decades and has a history that can be traced back to the end of the Second World War. The Turing Test (1950) is an early and critical milestone in conversational interaction. Likewise, voice-based products are not wholly new. Indeed, a range of conversational agents have been available commercially for some time, with voice becoming a popular way of interacting with Virtual Private Assistants (VPAs).

VPAs are offered by a range of service providers. This is often via proprietary platforms, technologies and products, including Microsoft's Cortana, Google Assistant, Amazon Alexa etc. These VPAs enable voice interaction with a conversational agent that uses human conversation as a metaphor for interaction. Applications of this kind include some familiar product archetypes including dictation software, in-car satellite navigation systems, telephony call-routing and automation and, more recently, speech-enabled IoT, speech-enabled web-chat and human-like conversational bots. Rather than being *truly* conversational, many of these applications are tightly constrained and highly task-oriented in nature (Clark et al., 2019b), with users requesting the proxies to execute tasks on their behalf. Henceforth, for consistency, the term agent is used to describe human actors within a service in this article. 'Agent' connotes a role beyond constrained notions of the 'user' and 'operator' and thus accounts for discretionary use and value co-creation.

Conversational technologies can be integrated into standalone applications at the office or factory too. This is often via the ubiquitous 'smart speaker' in the home or via multifarious touchpoints encountered in everyday life. Voice encounters occur in a wide range of situations, including when travelling by public transport, shopping at supermarkets or visiting a museum. The relatively low cost of entry adds to the appeal of this technology (Klopfenstein and Carlo, 2017). Localising to different languages and dialects and offering services through multiple modes of interaction has also driven adoption. The flexibility offered by today's platforms is evident in the wide range of services that employ them, across geographies and sectors, including banking, telephony (including customer care) and the UK Government's tax gateway. A similar situation exists in the home. The ability to control lights, stream music or make hands-free calls are just some of the typical use cases available through speech interaction. The variety of examples represents a tiny fraction of the many ways in which voice is reaching ubiquity in the interaction between people and computers.

It is plausible and indeed likely that in the future we will interact via voice in even more situations and contexts, including less commercially laden domains such as education and health. As the domain matures, the pervasiveness of conversational interfaces will be further enhanced, with emerging innovations including artificial intelligence and machine learning. These complex combinations of processing power and intuitive interface will extend the reach of voice and provide even deeper levels of personalisation, agency and value. Smart, data-enriched, connected conversational services of this kind, which are highly attuned to the individual and their context, offer much more that convenience. It is within the realms of possibility that these kinds of services will play a key role in addressing the problems of our time as we live longer and can no longer exploit the world's resources in the way we do now.

From a design perspective, the most complex of the current applications is speech automation. A growing number of proprietary platforms, including Google Cloud Platform, Amazon Web Services, IBM's Watson, Microsoft's Windows Speech Recognition and Nuance Communication's recognition and dictation services, make this and other aspects of conversational interaction increasingly accessible. Indeed, few organisations invest and develop voice-based products without reference to these tried and tested platforms.

Speech automation can be divided into two distinct types. Both – *guided dialogue* and *natural language understanding* – are event-driven (Yankelovich, 2008). In the first case, the system waits for the user to respond to a stimulus in the form of a spoken prompt, normally a menu, typically delivered as Text-to Speech (TTS) or a pre-recorded wave form, then branches on the response, which can be recognised as no-entry, no-match (on the basis of grammar), or a matching utterance. Depending on the response, the system performs a task. This may be to route the user to a human agent or an automated service, or to deflect them to another channel such as the Web. In the second case, (*natural language understanding*), the user is simply asked what they want to do. The number of possible utterances is much larger in this instance, and so the recognition process is far more complex and may employ artificial intelligence to reduce the probability of a no-match. The two types are discussed in further detail below.

The process of Automated Speech Recognition (ASR) differs little across applications and the process for designing and implementing speech automation systems is similar regardless of the delivery channel. The procedure follows the classic voice user interface (VUI) design methodology described by Pearl (2017). This typically includes working with user utterances to gather data, building the lexicons and grammars (where appropriate), designing the interaction touchpoints, creating the call-flows and testing and evaluating the service. Iterative testing is core to the approach. This is normally done through various tuning cycles. This approach has barely changed since the early days of speech interaction development (Harris, 2005; Pearl, 2017).

The scientific method used to recognise speech is a complex one. In summary, people think of spoken language as being based on a series of syllables which, when used in the correct combination, form words. Words used together form sentences, and we understand context by hearing sentences and employing various cognitive processes such as attention, memory, perception and learning (Pisoni, 2000).

ASR uses a very similar approach. When a user speaks, the waveform of the utterance is captured. This is broken down into its composite parts, phones, phonemes, diphones and triphones. A phone is the most basic part of speech, a phoneme is a basic sound that if changed changes the meaning of a word, a diphone is a combination of two phones, and a triphone is a sequence of three phonemes. Phones are common across languages, but phonemes are language specific because they denote meaning. Triphones are used both within a word and across words in a sentence. Waveforms of these components of language are used to determine the user's utterance.

In most cases, a recognition engine compares the waveform(s) of the user's utterance against a database collection in order to determine possible meanings. A series of probability models are then applied to determine the most likely meanings, which are often returned as an n-best list ordered by probability. Closed grammars used in guided dialogue are far simpler, and therefore often more accurate, than open grammars, because the possible combinations of meanings are limited to a much smaller phrase set.

Weightings are normally added to open grammars to accelerate probability calculations. Because phonemes differ by language and region (local accents can substantially alter the sound), non-cloud-based systems are often offered with language and region packs, e.g. en gb for British English, en us for American English and so on. Rabiner *et al* (1999) provide a more in-depth discussion. These language variants greatly increase the level of recognition and reduce noise factors in conversation.

Both guided dialogue and Natural Language Understanding (NLU) rely on lexicons or grammars – i.e. lists of words and phrases that the system expects the user to say. In guided dialogue, these are typically simple phrases, such as menu options, yes/no, numbers, etc. To assist frequent users and reduce the cognitive load associated with needing to listen to all options before responding, most guided-dialogue speech recognition systems allow for 'barge-in': the ability to speak after hearing the most relevant option rather than having to listen to all options and then remembering what to say to select the most relevant.

The stimulus for NLU is typically a prompt in the form of 'tell me what you'd like to do today'. This requires a significantly more complex lexicon, usually built using a Statistical Language Model (SLM). This enables assigning intents that are based on an understanding of a more complex utterance. In order to build an SLM grammar, it is necessary to have an extensive corpus available. These can often include tens of thousands of typical utterances, usually sampled from live interactions or collected using 'Wizard of Oz' studies (Woz) (see below; Harris, 2005).

The use of Woz in speech systems development constitutes evolutionary progress from the early days of speech scientists responding to 'yes' or 'no' utterances simply by playing the next prompt in a guided dialogue. As natural language has developed, the technique has become more sophisticated, requiring a high level of skill and specialist equipment. Today, the speech scientist participates in the dialogue, seamlessly responding to a spoken utterance in real time by playing a series of pre-recorded prompts. If a point in the dialogue is reached where this is no longer possible, the speech scientist must have a way of being able to transfer the user to a live agent in order to complete the interaction. The majority of NLU systems become guided once the main reason for initiating the dialogue has been established. For instance, if a user says, 'I'd like to pay my bill', the next question would be of a closed type, e.g. 'would you like to pay the full amount or a specific amount?'.

Historically, NLU within the telephony and automotive environments has been facilitated by proprietary speech recognition engines installed on hosted service-provider or end-user equipment including Siri (Evans, 2018). The most recent development in NLU has been the advent of cloud-based Speech AI services that have been developed by organisations with significant data processing power, such as Google and Amazon. Amazon's Alexa and Google's Home are two examples of mass-consumer-level services that have adapted these emergent technologies with networked (IoT), processing (artificial intelligence) and automation capabilities. These services do not rely on the development of a grammar, but use real-time speech-to-text transcriptions and return a textual representation of the user's utterance based on data held from decades of web search phrases and interactions (Google, 2019; Amazon, 2019). The transcriptions have been shown to provide a valid alternative to traditional grammar-based systems (Këpuska and Bohouta, 2017). Once a transcription has been returned, this is 'tagged' with an intent, and the user is routed based on this intent in a similar way to that described above.

The commercial model for these new types of services is Software as a Service (SaaS) on a 'pay-per-click' basis. The benefits of this model are reduced cost for low-use services, the ability to rapidly increase usage with minimal outlay, faster speed to market, and genuine machine learning as the recognition corpus grows with the use of the vendors' services. With more traditional on-premise recognition platforms, additional phrases require manual software updates. On the other hand, this approach reduces options for organisations wishing to implement a capex investment model; moreover, it is difficult to budget for pay-per-click services, and SaaS requires permanent access to a third party's systems, which may be more vulnerable to, for example, a DDOS attack (McLellan, 2013). While there is no one-size-fits-all option, it is clear these services are driving transformation of the way in which people interact with services.

Utilising a Service Oriented Architecture (SOA) is one way of protecting investment and ensuring that organisations that wish to move between SaaS and on-premise models can do so with the minimum of impact. This approach abstracts the front-end user experience from the integration with back-end services, meaning that migration of front-end services can be rapid and contained. It also offers the ability for multiple channels (telephone, IoT, webchat, for example) to interact with a single integration layer, thereby reducing effort and increasing cross-channel consistency.

Amazon Echo and Google Home Living offer a glimpse of how voice can be integrated into value-adding services. These kinds of services are enabled by a continuous data exchange between people and intelligent hardware and software. The technologies that enable this level of processing are thus more than just simple 'environmental variables' (Mason and Spring, 2011) but rather are integral to the service experience and resulting value relations. Conversely, the various conversational platforms and applications only become marketable services that consumers are willing to pay for when they are integrated into a value-adding proposition.

Larivière *et al* (2017) suggests that services are gradually becoming technology-dominant (e.g. Intelligent Agent as the interface) rather than human-driven (i.e. service employee as the interface). While this shift alters the relationships between organisations and customers (Patrício *et al*, 2018) it also highlights the opportunities and challenges for innovation (Barrett *et al*, 2015), including the importance of using design to meaningfully wrap service elements together into a sustainable and engaging whole. A technologically agnostic perspective helps in this situation. Appealing propositions that provide service flow (rather than the technology itself) engage customers and enable sustainable business models to be created and operated (Skålén *et al*, 2015). As service experiences become experientially richer, enhanced through intelligent technologies and even more deeply woven into social practice, so boundaries between organisations and customers will blur further.

SERVICE CONTEXT

Service Design (SD) is a practical means to orchestrate all of the human and technical elements that make up a service into a meaningful holistic experience. Emerging from marketing research and latterly highly influential in design, the service paradigm has challenged practical questions of how desirable digital experiences might best be developed, as well as theoretical notions of what constitutes value exchange. SD plausibly has answers to both questions. Service Dominant Logic (Vargo and Lusch, 2004) positions value at the intersection of resource integration and highlights value as something co-created (Sanders, 2005) and context-specific, rather than intrinsic.

SD spans sectors and business models. Applications of the approach range from integrating human-centred methods in agencies (Yu and Sangiorgi, 2018) through to full-scale transformation of public services. Goods manufacturers are making the 'shift to service' too (Kowalkowski *et al*, 2017; Oliva and Kallenberg, 2003). Rolls-Royce uses a variety of sensor-based technologies to add services to their core products. Customer support-type operations, such as are commonly found in contact centres, are a common application too. The wholesale digitalisation of public services is at the other end of the spectrum. Here discrete touchpoints are joined together via intelligent systems to deliver high levels of service flow. In all cases, services providers benefit from accelerated throughput, heightened user satisfaction and sustainable value creation irrespective of the technology stack and underlying social practice of the service layer. The literature also makes note of the holistic and experiential character of services, emphasising that they provide functional as well as emotional benefits (Voss *et al*, 2008).

SD applies human-centred design (HCD) methods (Polaine, Løvlie, and Reason, 2013; Stickdorn and Schneider, 2012). HCD is in itself an evolution from Participatory Design (PD), which emerged in Scandinavia in the 1970s. This approach sought to both include users in a design process to employ their expertise and to uphold their democratic rights to participate in the design of technology they would ultimately use (Bjerknes and Bratteteig, 1995). PD does not focus on a specific technique but instead advocates a general commitment to putting people at the centre of the design process in order to ensure that decisions are informed and equitable (Gregory, 2003). Contemporary forms of PD focus on giving users a 'voice' in the design process, seeking multiple perspectives, and gathering understandings from a range of disciplines (Muller, 2002).

A tranche of tried and tested methods are used in SD, many of which have evolved from those used within PD practice. Fundamental among these is co-design. This way of designing shifts 'user' involvement beyond participation and toward empowered, cooperative design at every stage of the process. It's an approach that leverages designers' creativity, end-users' needs and domain knowledge, developers' technical insights and service operators' understanding of underlying processes, roles and outcomes. Other popular methods are 'evidencing', as well as envisioning 'blueprints' (Shostack,1982) and making testable 'prototypes'. Blomkvist and Holmlid (2010) summarise Kimbell's work (2009) on analysis of the SD approach and perspective thus:

1. Looking at services from both a holistic and detailed point of view.

2. Considering both artefacts and experiences.
3. Making services tangible and visible through visualisations.
4. Assembling sets of relations (between artefacts, people and practices).
5. Designing business models.

According to Pirinen (2016), SD methods need to be integrated from the planning phase of a project onwards. This is because they help drive participation of external stakeholders and help cohere internal teams around a co-created vision. Furthermore, in a recent study, Trischler *et al* (2018) found that co-design teams were more successful in translating users' experiences and latent needs into innovative concepts than user-only and in-house teams. A logical extension of this collaboration is to integrate stakeholders directly into the core team, including service agents and developers.

Supporting research for SD is often expansive (Segelström *et al*, 2009). Such an approach helps explore context and draws on the traditions of applied ethnography (Heath and Luff, 2000). The results of this kind of research stimulate pertinent insights into envisioning a new service. Findings also help to understand contexts, social practices, and customers' interactions within socio-technical systems and ecosystems (Antons and Breidbach, 2018). While the value of immersive research is without question, integrating studies within often short design sprints is not without problems (Jefsioutine and Knight and, 2004).

A critical perspective on SD might question the abstracted and futurist focus on evidencing and blueprints. Both are by nature conceptual constructs that necessarily underprivilege intangible service elements (e.g. data flows and system integration points) in favour of surface-level ones (e.g. customer touchpoints and user journeys). These limitations suggest a need to rebalance SD toward a more agile focus on delivering progressive change in the here and now rather than idealised but unfeasible design futures. SD should move beyond dyadic company-customer interactions too, facilitating the coherent coordination of multiple channels (Sangiorgi *et al*, 2017) and focusing on the situated, embodied practices of diverse actors in value networks and service ecosystems. Extending SD into the conversational domain and aligning it better to innovation practices should go a long way in realising that progressive potential.

SERVICE INNOVATION CONTEXT

There are many challenges to innovating, but few organisations can afford the luxury of standing still. Instability in the market and changes in consumer behaviour mean that progressively evolving operations and outcomes is a matter of survival. This situation is often exacerbated by a lack of skills and experience in new product development. Organisations can also struggle to master what is in many cases bleeding-edge technology. Making the right investments is risky, and financial resources do not correlate with success either. While the journey from incumbent to leader is perilous, with no degree of certainty of success, a solid innovation strategy can help. Without strategy, opportunities are diminished, as is the potential to turn business problems into solutions that are desirable customer propositions (Patrício *et al*, 2018).

Winning at innovation is to a large degree dependent on an organisation's ability to harness the power and potential of their customers and employees (Larivière *et al*, 2017) as much as it is about digital transformation and the adoption of cutting-edge technology. In the best case, innovation is based on applying the innate creativity of people to fully explore a problem and then to focus in on possible solutions. Applying human-centred design thus helps to transform new technologies into usable, valuable and desirable new products and services.

Any organisation can develop an innovation strategy. This critically important step not only reduces risk but also creates the practical foundations to move from future-vision through to production and operation within a given set of constraints. The right methodology defines a vision of the future and lays out a roadmap of how to get there. This coheres people on a path towards successful innovation, not just for the next product or service launch but through a programme of tactical, continuous improvement (Zomerdijk and Voss, 2011). This is counter, however, to how many organisations work. In many cases innovation is seen as a one-off phase of frenzied activity. Instead, Morelli (2009) emphasises continuous customer-driven engagement as a sustainable way of maintaining market leadership. Here, progressive change is iterative and experimental and both problem and solution focused in order to actively generate as wide a variety of possible choices (Liedtka, 2015) before committing to investment.

Regardless of industry, companies are often challenged by the distinctive characteristics of service innovation. These difficulties are often perceived to be greater than in new product development or innovation in more traditional areas, such as manufacturing. Service innovation tends to be less formalised and is typically reactive to changes in customer demand (Ettlie and Rosenthal, 2011). According to Zomerdijk and Voss (2010), services are 'in permanent beta testing', which means they are in continuous flux as demand fluctuates. Organisations must therefore give particular consideration to their customer relationship strategies and take a process-oriented approach that somehow integrates customer needs into production itself (Ponsignon *et al*, 2011).

This means that ongoing customer relationships can neither be stymied by release milestones nor degraded by a short-term focus on reducing technical debt (rather than delivering customer value and improvements to flow). Flow (Csikzentmihalyi, 1988; Patrício *et al*, 2011) offers customers an engaging experience with consistent levels of quality, and requires that every single interaction, touchpoint, and channel be optimised (Larivière *et al*, 2017). It means that every encounter leads to the next without any obstacles, and customers feel stimulated to continue using the service over time. It's also manifested in the throughput and outcomes of a service – for example, the number of patients diagnosed, the quality of the advice offered to them and the health improvements that are contingent on that advice are all functions of flow.

Operationalising Service Innovation

Exploring the most feasible approach to innovation is a prerequisite to success in an increasingly competitive market for voice-based services. Helkkula *et al*'s (2018) framework is a solid starting point for defining an innovation strategy. Helkkula *et al*'s (*ibid*, p296) four archetypes ('models' in the adapted version below) set out common ways in which companies successfully disrupt markets across a range of organisations and sectors. Choosing the most relevant archetype is dependent on the specifics of a given market and the maturity and capability of the host organisation. The archetypes help determine the right approach for a given organisation and help signpost ways of achieving a better future state.

1. Service value model

    Focus is on enabling new ways of creating value (value-in-exchange)

This 'output-based archetype' focuses on the boundaries of the service value proposition – what will be offered to the customers and what will be excluded, in terms of processes, technical platforms, tools, and features. This kind of innovation is evident in the basic impact of the service and usually this offers something customers would otherwise be unable to do (at least not with the same level of ease and convenience). Amazon is a good example of the value-in-exchange model. When launched, it was an entirely new kind of store, available anywhere and at any time the customer had connectivity and a web browser. The online experience replaced traditional physical elements with novel digital variations (e.g. virtual shopping cart) that gave the customer new ways of shopping. Replacing bricks and mortar with a web shopping platform not only reduced costs but also enabled features (e.g. recommendations) that extended value creation in fundamentally new ways. In the conversational domain, this model might manifest itself in taking an existing activity from the real world (e.g. visiting a doctor's surgery) and transforming it into a digital, voice-based service. Changing the modality of interaction might not significantly change what patients, health professionals and administrators do. It would change the perceived and actual value of the service, as it would be cheaper to run and easier to access, but at the same time could possibly compromise the quality of healthcare.

2. Service efficiency and effectiveness model (process-based archetype)

Applying new ideas in fundamentally different ways throughout the service process (value-in-use)

This 'process-based archetype' centres on how the service streamlines underlying social practices, primarily from a customer perspective, thus making the service ever more effective and efficient. Some features of Amazon Prime, such as click and collect, gift cards, rewards, and next-day delivery are individual enhancement features that are built on meeting latent purchase needs. As a result, customers can seamlessly move through a vast variety of products with the aid of contextual options that help them compare prices, make secure payments, and receive their purchases the next day. For conversational services, this archetype might be realised by introducing voice-enabled kiosks in public spaces that could provide health advice on demand. Such a service would most likely increase the numbers of people accessing health information but also potentially lead to mis-diagnosis and even cause privacy and safety issues, depending on where the kiosks were situated.

3. Service satisfying model (Experiential archetype)

Co-creating valuable service experiences for all involved actors (value-in-experience)

The third archetype refers to how the service is experienced. It focuses on delivering a consistent, streamlined and highly personalised experience. Amazon Go extends the brand experience into an immersive, physical space. This transforms the shopping experience. Just-walk-out technology, which combines deep-learning algorithms, computer vision and artificial intelligence, eliminates queues and checkouts in a way that traditional stores and even online shops could never do. Anything bought by the customers is automatically added to their virtual carts, which then adds up the amount to be paid and charges their Amazon account within a single branded experience. All aspects of the service contribute to the quality of the encounter, which is seamless and embodies the brand at every touchpoint. This is only made possible by sophisticated data collection and analysis tools, soft and hardware infrastructure that requires large-scale and potentially risky investments. An example of this model in the conversational domain might be a fully orchestrated and harmonised health service that offers consultations in any context, situation or location. It could be available as single branded experience via traditional surgeries, smartphones, kiosks, wearables and the full gamut of voice-enabled devices.

4. Service extension model (Systemic archetype)

Integrating resources within the service ecosystem (value-in-context)

The 'systemic archetype' centres on extending a service outward by adding increased resources to provide a 'living' platform. Amazon Web Services exemplifies this innovation model, as it offers a complete cloud solution that assimilates other applications, data and devices to provide an integrated service experience toolkit. The platform is not just for Amazon itself but also other organisations willing to license it, thus reducing cost of development and time to market. In the context of conversational services, this archetype might manifest itself in offering an open platform for conversational health advice. Such a platform might allow the service to be integrated into other products and services, extending the potential audience and reducing unit costs of production.

FLOW Innovation Framework

Helkkula et al's (2018) archetypes of service innovation or models are applied to conversational services, and operationalised into a new evolved version (FLOW Service Innovation Framework, Table 1). In the best case, exploring the best fit to innovation model (e.g. value-in-use vs. value-in-context) is done collaboratively through structured workshops aligned to the FLOW methodology. In FLOW, envisioning sessions explore and define the right model or combination of models for a given organisation. Envisioning is also an effective way of initiating and building out the service production process and involving key stakeholders, including representative service agents. The collaboration is important as it minimises risk, builds consensus and lays the foundations of a 'to-be' service experience that is feasible and cost-effective and delivers customer value.

Envisioning is conducted in two parts (See Table 1). The first part (Visioning Foundation: Organisational Aspects) focuses on the 'Value and Business model'. It aims to determine the necessary technologies, processes, tools, staff etc needed to innovate successfully but also within the bounds of what is feasible. The second part (Visioning Direction: Project Mission) emphasises early 'envisioning' of desirable high-level service attributes and the development of novel service scenarios for the future, which are key parts of the co-design approach (Sanders and Stappers, 2008; Trischler *et al*, 2018). Outputs from these two complementary workshops naturally feed into the design and production activities that follow, minimising rework and misalignment in the host organisation. The follow through from envisioning to production and operation is supported by the FLOW approach. The models act as framing devices in the workshops. Having reference points helps cross-functional teams reach consensus on the most critical points and decisions. During workshops, the models are conceptualised simultaneously rather than following a linear approach. Specific activities to elicit, map and prioritise outputs can vary, but often employ structured facilitation techniques including ideation, affinity mapping, collaborative prioritisation and dot-voting.

Table 1. FLOW Service Innovation Framework (Adapted from Helkkula et al (ibid))

Service Innovation Model	Visioning Foundation: Organisational aspects – 'Value and Business Model'	Visioning Direction: Project- Mission – 'Envisioning'
Service value model	- Available technologies - Service features - Agents' service access - Capabilities required	- Additional feature options - Supporting services around the offering - New kinds of service access - New tools to manage resources
Service efficiency and effectiveness model	- Customer participation in the service production process - Improvements within processes - Process reconfigurations - Channels for service delivery	- Organisational resource integration - New resources or existing resources used in new ways - New kinds of service provision - New channels for service delivery
Service satisficing model	- Transition between channels - Minimised negative customer experiences - Customer motivation for service use - Smooth operation	- Combined experiences - Simplified service features - New customer roles - Enhanced availability and reliability
Service extension model	- Service ecosystem actors - Ecosystem actors' interactions - Actors' resource integration mechanisms - Institutional arrangements	- New value relations within ecosystem - Alternative contexts - New network actors - New platform options

The outputs of envisioning are documented, and early concepts, features and hypotheses recorded. Relevant story-cards are created too. Story-cards (See Table 2, for an indicative conversational service card set) are a useful way of documenting raw data, including envisioning, workshop, co-design and Human-centred Design (HCD) outputs. Story-cards become living requirements that evolve atomic-level data from early discovery through to late production activities. The incremental refinement of story-card sets, reflect the developing understanding of the current and future state of the service as knowledge emerges. They are also a practical aid in conversational service design in documenting service dialogues, loops, repairs and utterances too. The FLOW Activity Mapping Framework or 5As (See Table 3, for a healthcare example) helps categorise story-card data. The structured insights (i.e. agents, artefacts, attitudes, activities and actions and axioms) can then be used to inform supplementary HCD activities including Contextual Inquiry (Beyer and Holtzblatt, 1997) sessions. As well as standardising data classification, the framework is also valuable in detailed research planning (e.g. guiding experimental design) and in coding activities during qualitative analysis.

FLOW Design Methodology

Envisioning helps set out the innovation strategy, service vision and future state to the level of detail where meaningful investments and resourcing can be made. Just as organisations are often stymied in innovation, the next phase of design and delivery can also be challenging. Typical obstacles here are a lack of skills as well as the complexity involved in moving from high-level vision to detailed conversation interface design and the production of the necessary assets and code.

FLOW helps to connect innovation strategy to delivery via a structured methodology based on human-centred design, development and operation. Like envisioning, it can be adopted by any organisation, whatever stage they are at and whatever capabilities they may or not have within the organisation.

Table 2. Common FLOW Story-cards

Service analysis story-card set		Conversational service story-card set	
Story card (5As)	**Definition**	**Story card**	**Definition**
Agents	Human and non-human agents within a given social practice	Narrative	Service encounter and navigation Complete flows with key errors, hand-offs, interruptions etc
Artefacts	Physical objects embedded in the practice	Dialogue	Detailed conversation stories
Attitudes	Insights into the practice based on data elicited from agents and service flow itself	Loop	Detailed conversation interactions
Activities and Actions	Repertoires of behaviour from overarching to atomic operations done by agents	Repair	Finalised conversation exceptions
Axioms	Underlying rules and values that guide activities and actions done by agents	Utterance	Atomic-level conversation content

While the approach is best suited to guided conversational technology it also scales to more sophisticated applications where it can help focus on heightening the quality of the core narratives within the flow.

FLOW is a card-based design approach. Story cards evolve agile use cases or 'stories' as a means of documenting requirements. The approach extends the notational aspects of these grammars into physical artefacts that not only record conversational content itself but also form the concrete elements of service experience – utterances, phrases and flows. Generally, story-cards are two-sided playing card or A6 boards (these can be paper, cards, sticky notes, or digitalised pages) limited to containing one utterance or phrase. Cards naturally lend themselves to verbal communication as they force short phrases to be recorded and use 'native' utterances.

Cards can be drafted with marker pens, either written individually or, more usually in a co-design context, by stakeholders in envisioning sessions. Their diminutive size and writable surface are important. These qualities help to constrain phrase length to a minimum and ensure legibility and should be the result of a shared commitment to the outcome before they are drafted. Digital story cards help in online research sessions. These save time and are useful when cohorts are dispersed geographically. By using cheap, easy-to-iterate and modular media the method supports any modality of interaction,

Table 3. Flow activity mapping framework example

Service Element (5As)	Definition	Examples
Agents	Human and non-human agents within a given social practice	Patient, administrator, carer roles Patient database, system responses
Artefacts	Physical objects embedded in the practice	Case notes Prescription records
Attitudes	Insights into the practice based on data elicited from agents and service flow	Consultation preferences Patient feedback
Activities and Actions	Repertoires of behaviour from overarching to atomic operations	Use case level description Booking process journey
Axioms	Underlying rules and values that guide activities and actions done by agents	Data protection laws Patient-doctor relationship rules

including graphical user interfaces, although again conversational ones are the most common application. The ease and speed in which cards can be created means that full prototypes of the service can be fully quickly developed before any code is written. This supports early testing, including formative Wizard of Oz sessions.

Some general rules apply to cards. Card origination and validation activities (user testing) are based on limiting the number of cards to between five to ten for a given topic. If more content is needed the group divides the set into smaller constituent parts. Once a team has committed to the card set, the cards are fixed on to A3 sheets of paper and given a descriptive name that is then itself recorded on a card. This include guidelines for using them in research, such as using dots and colours in coding to represent different types of agents or insights. Categories should contain fewer than ten individual cards and can be grouped 'up' into one overarching category. However, this is not a prescriptive method; the cards themselves and how they are used are less important than their capacity to capture conversational content and form the basis of producing the service.

FLOW Design Example

The benefits, indicative methods and FLOW approach are defined and illustrated through a worked example. This is based on a case (Knight *et al,* 2019) that focused on the development of a new digital service for primary healthcare and an indicative application of FLOW shows how this translates to the conversational domain. This case enabled five health diagnosis-related use cases and was delivered using an agile service delivery approach. The core requirements centered on patient consultations and covered all aspects of a patient booking an appointment, the administration activities relating to managing these consultations, and the consultation itself.

Common story-card sets in FLOW can be categorised as described in Table 2. Two broad categories emerged: service analysis (based on 5As) and conversational service story-card sets. The service analysis story-card set refers to aspects of SD and could be applied to many different types of service. The conversational service story-card set is focussed on informing the specific implementation of a conversational interface.

In the case study, a high-level service vision and roadmap was established in early envisioning workshops. Story-cards were elicited and created by the core team, tapping into their domain knowledge before any formal fieldwork was undertaken. Questions, hypotheses and concepts were generated and recorded on to cards so that they could be explored in research and potentially iterated in co-design sessions. Rather than starting with a 'blank sheet of paper', this approach reused existing knowledge. For example, personas were 'mashed together' from existing knowledge and these initial versions were iterated and validated through research sessions until stable versions was arrived at.

A three-week immersive research discovery phase followed. This included making some *in situ* interviews and observations, and facilitating generative co-design workshops. These various activities took place in a number of geographically distributed locations or in some cases were conducted remotely using video telephony. Twenty-five participants were involved and comprised doctors, patients and administrators. Patient interviews focussed on understanding current processes and, once this had been established, moved on to how virtual consultations could enhance this process. New use cases were developed for prescriptions, sexual and mental health services and check-ups. The 6 administrators participated in a workshop focussing on the current appointment booking process, concepts, user stories, positives and constraints, and then co-design. The current booking process was analysed during this phase, as well as how new technology might unlock service flow. In all cases, the story-cards were iterated throughout.

The various research activities generated a great deal of data, again documented on physical and digital cards. These captured conversational service utterances and other data at the level of atomic-level insights pertinent to designing the service. To consolidate this data, a two-way qualitative analysis was used. The analysis first applied a rapid coding approach and then the more time-consuming Framework approach (Ritchie and Lewis, 2003). Firstly, categories were identified from reviewing the data with the researcher involved in the studies. The outputs were then systematically analysed bottom-up by coding each individual data point. Post-fieldwork immersion workshops were used to validate theses findings.

Patients participated in co-design workshops. The workshops explored and validated the findings of the initial research. In these sessions, story-cards were further developed in order to incrementally add detail and validate them in terms of the actuality of the case. In this kind of co-design activity, groups loosely refine content into agreed short descriptions on cards via structured activities such as round robin. As well as co-creating conversational material these workshops also leveraged groups to generate innovation concepts through structured idea generation. This kind of ideation is most commonly rationalised on the basis of extending and replicating the creative cognitive processes used by designers seeking to optimise the fuzzy challenges of developing breakthrough service concepts. Usually done in small teams (5 to 10 participants), the process promotes a 'designerly' way of thinking, typically based on random input or analogous thinking. In this case, ideation was done using common generation techniques including 'crazy 8s'.

As well as ideation, the cards were used in card-sorting activities. Card sorting can be applied at any stage of research and design, and is a fast and effective way of finding affinity and patterns in data and developing conversational flows. Usually, the researcher creates a set of data items (in this case the cards) and participants are asked to sort the cards into meaningful categories. The technique can be used both to generate and to verify categories (Fraser and Gilbert, 1991) and builds on the premise that categorisation is an important technique for us to make sense of the world (Rugg and McGeorge, 2005). This means that cards are not only vehicles for communicating content but, as stimuli in card sorting, they also become tools for uncovering tacit knowledge and insights into a given domain.

It should be emphasized that card-sorting is focused on organising categories of information, whereas story-cards are most often utterances or phrases that people immersed in the domain or context may use. FLOW design is therefore able to offer a grounded approach, with language and concepts emerging from situated practice. Both FLOW, which incorporates co-discovery and co-design activities, and the use of story-cards directly support conversational SD and development, including improving service flow by optimising dialogues through prototype testing.

Prototypes are a tried and tested method in co-design and provide a way of testing any kind of interaction. Usually, prototypes are 'click-throughs' of wireframes. However, developing detailed designs and conversational flows requires a more concrete test-and-learn approach, such as is facilitated by Wizard of Oz (Woz) studies. Woz studies have long been used when conversational interfaces are being explored but an actual technical implementation would be too costly (Dahlbäck, Jönsson and Ahrenberg, 1993). In Woz, the interaction with a notional 'computer' is mediated and controlled by a hidden human in order to simulate a higher-fidelity prototype. Cards seamlessly support Woz studies, by guiding conversational flow and, importantly, allowing new content to be created and tested easily 'on-the-fly' Woz can be extended into role-playing and design-anthropological enactments of the 'to be' service too.

Having elicited conversational content and researched the 'as is' service and co-design solutions, the project concluded in production activities. For conversational services, this takes the validated flows, dialogues and utterances cards from Woz studies into a development backlog. Development work usually works in sprints, which turn story cards into working conversational code that can be tested at a higher level of fidelity with service agents before going live. FLOW extends from production into the operational domain, setting out a framework that takes story-cards into the next phase of optimisation via design-led research.

FUTURE RESEARCH DIRECTIONS

FLOW is a pragmatic response to the challenges of conversational service innovation. It is focused on the practical need to minimise the gap between research and operations through accelerating processes via card-based design and development. FLOW has three substantial benefits compared to orthodox SD. First, it is attuned to make the best of the available conversational technology platforms, as it helps to provide the necessary inputs for interface systems, including corpus and dialogue content. It also generates this content at an early stage of development so that it can be easily changed and made available to inform development. Second, it enables the systematic delivery of word-based services, based on human needs in the broadest sense, by co-designing the service with the agents themselves at every level of service interaction. Finally, it supports deep collaboration throughout the project lifecycle and in particular enables co-design activities in the conversational domain through story-cards, which are low-cost and, because they use short, plain language phrases, accessible to a wide variety of people. At the atomic level, cards record utterances as short verbatims by or from project stakeholders, which might be reflections, insights, potential solutions or draft conversational content. These are all recorded as plain language phrases that stakeholders write themselves or are transcribed by a facilitator. Finally, story cards lend themselves to estimations, prioritisation, backlog and scrum-based ways of working – a common aspect of agility.

In some cases, an innovation strategy needs to go further into the future and into less well-defined areas of human practice. Design anthropology (Smith *et al*, 2016) is particularly relevant where innovation needs to go further than small change. As a radical rewiring of requirements and solutioning, it decouples instrumental notions of products and services from their origin within a given human practice. This agnostic, research-driven approach grounds innovation in the real world of human activity and thus frames solutions that are natural evolutions of the present with the addition of progressive technological and social agency. At a practical level, it also produces two kinds of data critical to conversational service rollout.

First, writing the ethnography helps to create a native corpus. As the outcome of primary research with target service agents this output is far more detailed, comprehensive and usable than any second level stakeholder-generated content could be. Direct verbatims and transcripts enable deeper analysis and insights to be made into the target audience and practice that would be otherwise possible. They also help generate native lexicons, localisations and tone-of-voice principles. Data taken directly from fieldwork can be also augmented with coded notes to give semantic insights in support of qualitative analysis that enables moving beyond surface-level actions within a given practice. With the help of the right technology, a lexicon also allows machine learning and artificial intelligence to be applied to the data set. These various methods support corpus development that is finely attuned to the vernacular, whether that relates to a highly specific practice (e.g. surgeon's dialogue) or a locality (e.g. dialects and slang).

Second, concurrently shifting between research and testing via prototypes in fieldwork via enactments can help to map out not just the current practice but also its progressive post-intervention evolution. The unfettered nature of this research is somewhat distant from instrumental notions of 'user requirements' and even the highly applied character of much ethnography employed in design. Instead, fieldwork focussed on immersion involves a 'naïve' perspective that fosters emergent alternatives and iterates provisional solutions in the field before any design, let alone development, choice of modality or technical platform, has been made. Here, the outcome might not be a product or service.

This cycle of immersion and intervention is done neither just for the sake of developing knowledge nor merely as input for innovation. By acting within the social arena, to know it, reflect on it and to positively intervene, a more thoughtful and humanistic outcome is much more likely to emerge and to be successful. In this formulation, intervention might be a starting point or an end-point. While beyond the scope of most projects, this ground-up approach is perhaps the most feasible means of addressing the often mentioned but rarely tackled topic of ethics in an era of rapid, transformational technology-led change.

CONCLUSION

This chapter has provided insights into designing, delivering, operating and innovating conversational services, summarising current contexts for these services and providing an overview of the relevant technologies and common platforms used in service production. The broader commercial context has also been described, as has as an applied framework for service innovation. Finally, a grounded design approach (FLOW) was introduced and research directions relating to how design anthropology can help innovation were elaborated. Future research in this area will extend from this foundational work and tackle the added complexity of fully intelligent conversational services.

ACKNOWLEDGMENT

The authors would like to recognise the contribution of Dr Ben Cowan at University College Dublin.

REFERENCES

Amazon.com. Inc. (n.d.). Amazon Transcribe. Retrieved from https://aws.amazon.com/transcribe/ [Accessed 5th June 2019].

Antons, D., & Breidbach, C. F. (2018). Big Data, Big Insights? Advancing Service Innovation and Design with Machine Learning. *Journal of Service Research*, *21*(1), 17–39. doi:10.1177/1094670517738373

Barrett, M., Davidson, E., Prabhu, J., & Vargo, S. L. (2015). Service Innovation in the Digital Age: Key Contributions and Future Directions. *Management Information Systems Quarterly*, *39*(1), 135–154. doi:10.25300/MISQ/2015/39:1.03

Beyer, H., & Holtzblatt, K. (1997). *Contextual Design*. Morgan Kauffmann.

Bjerknes, G., & Bratteteig, T. (1995). User Participation and Democracy: A Discussion of Scandinavian Research on Systems development. *Scandinavian Journal of Information Systems*, *7*(1), 73–98.

Blomkvist, J., & Holmlid, S. (2010). Service Prototyping According to Service Design Practitioners. ServDes. 2010. In *Conference Proceedings, ServDes. 2010, Exchanging Knowledge,* Linköping, Sweden, December 1-3, 2010 (Vol. 2, pp. 1-11). Linköping University Electronic Press.

Clark, L., Doyle, P., Garaialde, D., Gilmartin, E., Schlögl, S., Edlund, J., ... Cowan, B. R. (2019a). The State of Speech in HCI: Trends, Themes, and Challenges. *Interacting with Computers*, iwz016. doi:10.1093/iwc/iwz016

Clark, L., Pantidi, N., Cooney, O., Doyle, P., Garaialde, D., Edwards, J., ... Wade, V. (2019b). What Makes a Good Conversation?: Challenges in Designing Truly Conversational Agents. In *Proceedings of the 2019 CHI Conference on Human Factors in Computing Systems* (p. 475). ACM. 10.1145/3290605.3300705

Csikzentmihalyi, M. (1988). The flow experience and its significance for human psy- chology. In M. Csikzentmihalyi, & I. S. Csikzentmihalyi (Eds.), *Optimal experience* (pp. 15–35). Cambridge, UK: Cambridge University Press. doi:10.1017/CBO9780511621956.002

Ettlie, J., & Rosenthal, S. R. (2011). Service versus manufacturing innovation. *Journal of Product Innovation Management*, *28*(2), 285–299. doi:10.1111/j.1540-5885.2011.00797.x

Faste, H., Rachmel, N., Essary, R., & Sheehan, E. (2013). Brainstorm, Chainstorm, Cheatstorm, Tweetstorm: New Ideation Strategies for Distributed HCI Design. In *Proceedings of the SIGCHI Conference on Human Factors in Computing Systems* (pp. 1343-1352). ACM.

Fitton, D., & Read, J. C. (2016). Primed Design Activities: Scaffolding Young Designers During Ideation. In Proceedings of the 9th Nordic Conference on Human-Computer Interaction (NordiCHI '16). New York: ACM; doi:10.1145/2971485.2971529.

Flaounas, L., Omar, A., Lansdall-Welfare, T., De Bie, T., Mosdell, N., Lewis, J., & Cristianini, N. (2013). Research Methods in the Age of Digital Journalism. *Digital Journalism*, *1*(1), 102–116. doi:10.1080/21670811.2012.714928

Fraser, N., & Gilbert, N. (1991). Simulating Speech Systems. *Computer Speech & Language*, *5*(1), 81–99. doi:10.1016/0885-2308(91)90019-M

Garfinkel, H. (1967). *Studies in Ethnomethodology*. Englewood Cliffs, NJ: Prentice-Hall.

Giaccardi, E., Speed, C., Nazli, C., & Caldwell, M. (2016). Things as Co-ethnographers: Implications of a Thing Perspective for Design and Anthropology. In Design Anthropological Futures, 235. London, UK: Bloomsbury Academic.

Google, L. L. C. (n.d.). Cloud Speech-to-Text. Retrieved from https://cloud.google.com/speech-to-text/ [Accessed 5th June 2019]

Gregory, J. (2003). Scandinavian Approaches to Participatory Design. *International Journal of Engineering Education*, *19*(1), 62–74.

Halse, J. (2008). Design Anthropology: Borderland Experiments with Participation. IT University of Copenhagen. Doctoral thesis. Retrieved from www. dasts.dk/wpcontent/uploads/2008/05/joachim-halse-2008.pdf

Harris, R. A. (2005). *Voice Interaction Design*. San Francisco, CA: Elsevier.

Heath, C., & Luff, P. (2000). *Technology in Action*. Cambridge University Press. doi:10.1017/CBO9780511489839

Helkkula, A., Kowalkowski, C., & Tronvoll, B. (2018). Archetypes of Service Innovation: Implications for value cocreation. *Journal of Service Research, 21*(3), 284–301. doi:10.1177/1094670517746776 PMID:30034213

Jefsioutine, M., & Knight, J. (2004). Methods for Experience Design: The Experience Design Framework. In: J. Redmond, D. Durling, & A. de Bono (Eds.), *Proceedings of Future Ground Design Research Society International Conference 2004*, Melbourne, Australia. Nov. 17-21.

Këpuska, V., & Bohouta, G. (2017). Comparing Speech Recognition Systems (Microsoft API, Google API And CMU Sphinx, In Int. Journal of Engineering Research and Application, 7(3), (Part 2), March 2017, (pp. 20–24).

Kessler, E. H., & Bierly, P. E. III. (2002). Is faster really better? An empirical test of the implications of innovation speed. *IEEE Transactions on Engineering Management, 49*(1), 2–12. doi:10.1109/17.985742

Kimbell, L. (2009). Insights From Service Design Practice. In *Proceedings of the 8th European Academy Of Design Conference. April 1-3, 2009*. Aberdeen, Scotland: The Robert Gordon University. Available at http://www.lucykimbell.com/stuff/EAD_kimbell_final.pdf

Kimbell, L. (2011). Designing for Service as One Way of Designing Services. *International Journal of Design, 5*(2), 41–52.

Klopfenstein, L. C., Malatini, S., Bogliolo, A., & Carlo, U. (2017). *The Rise of Bots:A Survey of Conversational Interfaces*. Patterns, and Paradigms. doi:10.1145/3064663.3064672

Knight, J., Gibbons, C., & Ross, E. (2019). Unlocking Service Flow – Fast and Frugal Digital Healthcare Design. In Design of Assistive Technology for Ageing Population. Springer.

Kowalkowski, C., Gebauer, H., & Oliva, R. (2017). Service Growth in Product Firms: Past, Present, and Future. *Industrial Marketing Management, 60*, 82–88. doi:10.1016/j.indmarman.2016.10.015

Larivière, B., Bowen, D., Andreassen, T. W., Kunz, W., Sirianni, N., Voss, C., ... De Keyser, A. (2017). "Service encounter 2.0": An Investigation into the Roles of Technology, Employees and Customers. *Journal of Business Research, 79*, 238–246. doi:10.1016/j.jbusres.2017.03.008

Liedtka, J. (2015). Perspective: Linking Design Thinking with Innovation Outcomes through Cognitive Bias Reduction. *Journal of Product Innovation Management, 32*(6), 925–938. doi:10.1111/jpim.12163

Morelli, N. (2009). Service as value co-production: Reframing the service design process. *Journal of Manufacturing Technology Management, 20*(5), 568–590. doi:10.1108/17410380910960993

Muller, M. J. (2002). Participatory Design: the Third Space in HCI. In The Human-computer Interaction Handbook (pp 1051–1068), Hillsdale, NJ: L. Erlbaum Associates.

Oliva, R., & Kallenberg, R. (2003). Managing the transition from products to services. *International Journal of Service Industry Management*, *14*(2), 160–172. doi:10.1108/09564230310474138

Patrício, L., Fisk, R. P., Falcão e Cunha, J., & Constantine, L. (2011). Multilevel Service Design: From Customer Value Constellation to Service Experience Blueprinting. *Journal of Service Research*, *14*(2), 180–200. doi:10.1177/1094670511401901

Patrício, L., Gustaffson, A., & Fisk, R. (2018). Upframing service design and innovation for research impact. *Journal of Service Research*, *21*(1), 3–16. doi:10.1177/1094670517746780

Pearl, C. (2017). *Designing Voice User Interfaces*. California: O'Reilly Media.

Pirinen, A. (2016). The Barriers and Enablers of Co-design for Services. *International Journal of Design*, *10*(3), 27–42.

Pisoni, D. B. (2000). Cognitive factors and cochlear implants: Some Thoughts on Perception, Learning, and Memory in Speech Perception. *Ear and Hearing*, *21*(1), 70–78. doi:10.1097/00003446-200002000-00010 PMID:10708075

Polaine, A., Løvlie, L., & Reason, B. (2013). *Service Design*. Brooklyn, NY: Rosenfeld.

Ponsignon, F., Smart, P. A., & Maull, R. S. (2011). Service Delivery System Design: Characteristics and contingencies. *International Journal of Operations & Production Management*, *31*(3), 324–349. doi:10.1108/01443571111111946

Rabiner, L. R., Juang, B. H., & Lee, C. H. (1999) An Overview of Automatic Speech Recognition, In Automatic Speech and Speaker Recognition: Advanced Topics. The Netherlands: Kluwer Academic Publishers Group.

Ritchie, J., & Lewis, J. (2003). *Qualitative Research Practice: A Guide for Social Science Students and Researchers*. London, UK: Sage.

Rugg, G., & McGeorge, P. (2005). The Sorting Techniques: A Tutorial Paper on Card Sorts, Picture Sorts and Item Sorts. *Expert Systems: International Journal of Knowledge Engineering and Neural Networks*, *22*(3), 94–107. doi:10.1111/j.1468-0394.2005.00300.x

Sanders, E., & Stappers, P. J. (2008). Co-creation and the New Landscapes of Design. *CoDesign*, *4*(1), 5–18. doi:10.1080/15710880701875068

Sangiorgi, D., Patrício, L., & Fisk, R. P. (2017). Designing for interdependence, participation and emergence in complex service systems. In Designing for Service: Key Issues and New Directions (pp. 49–64). London: Bloomsbury Academic. doi:10.5040/9781474250160.ch-004

Skålén, P., Gummerus, J., von Koskull, C., & Magnusson, P. R. (2015). Exploring value propositions and service innovation: A service-dominant logic study. *Journal of the Academy of Marketing Science*, *43*(2), 137–158. doi:10.100711747-013-0365-2

Smith, R. C., Vangkilde, K. T., Kjærsgaard, M. G., Otto, T., Halse, J., & Binder, T. (Eds.). (2016). Design Anthropological Futures. London, UK: Bloomsbury Academic.

Stickdorn, M., & Schneider, J. (Eds.). (2010). *This Is Service Design Thinking*. Amsterdam, The Netherlands: BIS Publishers.

Suchman, L. (1987). *Plans and Situated Actions: The Problem of Human-Machine Communication*. Cambridge, UK: Cambridge University Press.

Trischler, J., Pervan, S. J., Kelly, S. J., & Scott, D. R. (2018). The value of codesign: The effect of customer involvement in service design teams. *Journal of Service Research*, *21*(1), 75–100. doi:10.1177/1094670517714060

Turing, A. M. (1950). Computing machinery and intelligence. *Mind*, *59*(236), 433–460. doi:10.1093/mind/LIX.236.433

Vargo, S. L., & Lusch, R. F. (2004). Evolving to a new dominant logic for marketing. *Journal of Marketing*, *68*(1), 1–17. doi:10.1509/jmkg.68.1.1.24036

Vargo, S. L., & Lusch, R. F. (2017). Service-dominant logic 2025. *International Journal of Research Marketing, 34*(1), *46-67.*

Yankelovich, N. (2008). Using Natural Dialogs as the Basis for Speech Interface Design. In *Human Factors and Voice Interactive Systems. Signals and Communication Technology*. Boston, MA: Springer. doi:10.1007/978-0-387-68439-0_9

Yu, E., & Sangiorgi, D. (2018). Service Design as an Approach to Implement the Value Cocreation Perspective in New Service Development. *Journal of Service Research*, *21*(1), 40–58. doi:10.1177/1094670517709356

Zomerdijk, L., & Voss, C. (2010). Service design for experience-centric services. *Journal of Service Research*, *13*(1), 67–82. doi:10.1177/1094670509351960

Zomerdijk, L., & Voss, C. (2011). NSD processes and practices in experiential services. *Journal of Product Innovation Management*, *28*(1), 63–80. doi:10.1111/j.1540-5885.2010.00781.x

KEY TERMS AND DEFINITIONS

Service Design: A human-centred, collaborative, and holistic approach to developing new services or improving existing ones. This is achieved primarily through evidencing (blueprinting service touchpoints) and co-design (see below) within a use-centred design methodology.

Co-Design: A design process, where deliverables are produced, throughout the project lifecycle, by the collaborative efforts of designers and stakeholders. This work is facilitated and, usually occurs via workshops.

Design Anthropology: An innovation methodology, built on ethnographic principles, that speculates about the future through introducing collaboratively developed prototypes into social settings. Extending the remit of participatory design, this approach is not only technology agnostic, but also independent of predefined modalities of interaction and even whether the outcome is a product, service or experience.

Service Flow: An aggregated measure of service quality. Service flow is based on value-in, exchange, use, experience and context data pertaining to a specific service.

Wizard of Oz Studies: A research method in which participants interact with a voice-based interface, that is enacted by a human proxy. This kind of role-playing allows conversational content to be tested and iterated on the fly.

Card Sorting: A research method based on subject's categorising items in either closed or open sorts. The grouping and naming of items elicits tacit-level knowledge about the domain in question. Applications include information architecture testing.

Chapter 13
The Edgard Project:
Towards a Discussion of Ethical Tech in Art and Design

Fernando Luiz Fogliano
Senac University Center, Brazil

André Gomes de Souza
https://orcid.org/0000-0002-6829-5445
Senac University Center, Brazil

Guilherme Henrique Fidélio de Freitas
https://orcid.org/0000-0002-5275-0781
Senac University Center, Brazil

Rachel Lerner Sarra
https://orcid.org/0000-0003-1726-0282
Senac University Center, Brazil

Juliana Pereira Machado
Senac University Center, Brazil

ABSTRACT

As Artificial Intelligence technologies advance, their use becomes increasingly widespread, and what was once a fantasy of being able to communicate with a virtual being is now part of our everyday lives. New issues arise with every new technology, and discussions are needed to avoid significant problems. By looking at what is happening now and at the impact that AI has and will continue to have, one needs to remember history and its contradictions with a critical mind to have a more ethical approach to technology innovations. This chapter focuses on the Edgard Project, an intersection between Contemporary Art and Conversational Interface Design. Consisting of a chatbot named Edgard, the project emerges as an artistic approach to encourage critical thinking about the technology that gives him "life" and highlights the contradictions of AI through its ironic discourse and outdated interface.

DOI: 10.4018/978-1-7998-2112-0.ch013

INTRODUCTION

Artificial Intelligence technologies require extensive expertise in a wide range of knowledge fields, from philosophy to history, technology, mind, and consciousness. By observing the contemporary scenario, one perceives a conflictive picture as gloomy perspectives align with other promising ones. Amid this conundrum, new insights are on the rise, and Art and Design appear as potential environments to be explored.

Understanding Artificial Intelligence can be a bit overwhelming, given how quickly it has been transformed in recent decades. If someone had said a few years ago that in a short period of time it would be possible to talk to a robot, surely that person would be making a career as a science fiction writer. *A.I. Artificial Intelligence* (Spielberg, 2001), *Do Androids Dream of Electric Sheep?* (2017), and *I, Robot* (2013), for example, went a little farther by envisioning that such robots would be able to feel emotions. Androids in these movies have opinions, can feel, and can hurt those around them because of their unique nature of being a sentient robot. Although scientists have not achieved the technologies necessary for the creation of sentient robots yet, with recent advances, what was once a fantasy of being able to communicate with a virtual being is now part of our everyday lives.

This chapter intends to discuss and demystify the misconceptions that a general audience might have about Artificial Intelligence and its usage, providing a more well-rounded perspective on the subject. In the background section we explain the perception of AI taking into account how it was perceived and portrayed in earlier times. Subsequently, we compare the media's point of view against this technology state of the art in the modern scenario. Using the Edgard chatbot project developed by the authors as an example, we intend to question and criticize the status quo of Artificial Intelligence technologies and other thoughts associated with the emergence of intelligent machines in our daily lives. Taking everything into consideration, we ought to propose new and innovative ways to approach this subject with a focused perspective and open mind.

BACKGROUND

Over time, technologies that represent changes in the paradigms of the world's scenario come and go, with the Internet remaining one of the most recent examples. With every new thing that comes by, new issues arise demanding the need for discussions that can back up the delicate nature of pointed issues. By ignoring the problems that technologies might face new problems arise and could potentially induce controversies similar to the Google Books case, in which an unethical approach and a pursuit for breaches in the law led to conflicts with copyright regulations. As these technologies get more complex over time, interest in them increases through Art and Design, raising ethical challenges such as unemployment, loss of privacy, identity, and control over personal life (Leal, 1995).

The impact of Artificial Intelligence is one of the contemporary issues that can be compared to dilemmas[1] brought by preceding industrial revolutions. At the same time that we have scientific advancements providing us with improved life expectancy and quality, we have the degradation of the environment, the wealth concentration by an elite minority and the polarization of opinions amid the circulation of false information showing itself capable of interfering in democratic processes around the world. By looking at what is happening now and at the impact that AI has and will continue to have, we need to look at history and its contradictions with a critical mind to have an ethical approach in order to understand better and foresee the consequences of technology innovations.

However, before we can establish an adequate basis for addressing such a complicated matter, it is essential to ensure that people will not jump in with opinions and assumptions based on common prejudicious perceptions. This is what the tech industry narrative is going for, focusing only on the benefits and forgoing all the necessary conversation, until it is too late to do anything about it. "The Oxford research suggests that the strategy is succeeding, and that mainstream journalism is unwittingly aiding and abetting it" (Naughton, 2019, n.p.).

Our analysis also shows how industry topics and sources regularly encourage outlets to position AI as a viable solution to a wide range of problems. When articles describe AI systems that can direct coffee dispensing drones, identify clothing brands from pictures, reorder geopolitical power, or conquer death, the implication is AI can serve as a solution to a vast array of problems – from the frivolous to the profound. Yet in doing so, outlets rarely interrogate either the limits of AI's competency or the role that humans continue to play in its Design and implementation. (Naughton, 2019, n.p.)

AI and Science Fiction: The Beginning

We cannot talk about the perception of topics such as AI today without taking into account how it was perceived and portrayed in the earlier times when the reality we are living today was nothing but a dream. Science fiction, both in literature and cinema, shaped and is still shaping how we think about intelligent machines. According to Lorenčík, Tarhaničová, and Sinčák (2013), "New inventions inspired the authors of sci-fi, whose works, in turn, inspired scientists to invent the things that the sci-fi authors envisioned." Although this dialogue between fiction and reality might seem beneficial, when it involves the general audience, it can trigger some tricky perceptions, not because the sci-fi works have misleading intentions, but because what people watch is not necessarily an accurate portrayal and are not in sync with what is being developed in real life.

The allure of AI in the early twentieth century came with remarking yet contradicting points of view, sometimes portraying the technology as something beneficial or dangerous for society. The common ground though, is how most works end up delivering messages about our own human nature by putting us face to face with the scenario of non-human intelligent things: robots, capable of thinking for their own and sometimes even having emotions. The idea of human-like beings is nothing new, capturing the public's imagination with a vast diversity of stories and dreams. According to Behnke (2008), the desire to replicate human-like behavior in objects such statues were as old as the second century B.C., and despite being initially related to mechanical engineering, robots became broadly popularized with the rise of literary and cinematic works in the genre of sci-fi and the belief of substantial intelligence behavior (Buchanan, 2005). One of the main authors for such doing is Isaac Asimov with his book from 1950 *I, Robot* (Asimov, 2013), beginning a massive wave of depicting intricate societal relationships between machines and humans.

As denoted Geraci (2007), intelligent machines are perceived by us in a strikingly similar way to the Holy in religion:

Technology promises us a life of leisure, perhaps even immortality; at the same time, intelligent machines are always on the verge of revolting and taking over the planet. In such eschatological scenarios, robots attack their human masters and possibly enslave them. Science-fiction depictions of robots demonstrate that a fear of technological wrath accompanies the hope of a new Eden. (Geraci, 2007, p. 966)

Not only science fiction is influencing us at how we perceive technologies by themselves, but they are also teaching us what to think about these technologies by reflecting our insecurities and concerns regarding something that could be so powerful yet is still uncertain. We, as a society, tend to associate AI to robots and envision that, sometime in the future, they will probably surpass us in terms of intelligence. However, AI is not all about creating robots as or more intelligent than us, but "It is also about understanding the nature of intelligent thought and action using computers as experimental devices" (Buchanan, 2005, p. 54) since as the technology evolves we have to go deep into the understanding of human nature and soul.

AI in the Media: Collective Imaginary vs. Reality

The media, according to a study by the Reuters Institute for the Study of Journalism at Oxford University (Brennen, Howard & Nielsen, 2018), is separated into two parts: the right-leaning focuses on the successes and security that AI can bring to the national context, while those on the left-leaning reinforce the ethics, limits, and dangers imposed by AI. Most of the articles about this subject focus on deepfakes, automation, autonomous vehicles and weapons, date, privacy, hiring, facial recognition, human enhancement, and discrimination.

Some media outlets try to offer pertinent questions, such as the fear of Artificial Intelligence perpetuating different forms of discrimination. When Google's (Brennen, Howard & Nielsen, 2018) image recognition system automatically marked black people as gorillas, the company switched off the recognition of primates completely to correct the mistake. Another case pointed out by Wired (Brennen, Howard & Nielsen, 2018) talks about how police systems discriminate against low-income neighborhoods based on AI information (Brennen, Howard & Nielsen, 2018). However, there are still media outlets that focus on what they think would be the main danger of AI: the possibility of autonomous weapons rebelling against their creators. Channels like Telegraph (2018) comment that: "Scenarios from The Terminator in which beings with Artificial Intelligence turn on humans are just' one to two decades away' according to former Google chief."

Several studies on the subject have been done to tackle AI's opposing aspects, leading to the elaboration of several catalogs with rules on how to deal with such systems.

There is thus no scarcity of proposals of ethical principles for AI, and the ongoing work of the High-Level Group on AI, created by the European Commission in its Communication on Artificial Intelligence of 25 April 2018, will certainly work through all this material and develop yet another catalogue of proposals by the end of 2018. (Nemitz, 2018, p.8)

This material brings a current and severe subject in AI circles, which is the need for established and enforced laws about these technologies.

[...] it is time to move on to the crucial question in democracy, namely which of the challenges of AI can be safely and with good conscience left to ethics, and which challenges of AI need to be addressed by rules which are enforceable and based on democratic process, thus laws. (Nemitz, 2018, p. 8)

We can see that not only ethical discussions surround AI, but also debates concerning the values that it must follow and the roles it must play in the future of society's consciousness. Ascott (2000) considers that the use of technology in Art is providing the tools and insights to look deeper into the richness and fruitfulness that an ecological, post-humanist perspective can give to our existence.

At the point where cyberspace and post-biological life meet, an entirely new kind of social architecture is required. A truly Beyond Boundaries anticipatory architecture must prepare itself for this marriage of cyberspace with Moistmedia, combining self-assembling structures and self-aware systems. (Ascott, 2000, p. 5)

THE EDGARD PROJECT

Discussions about AI technologies in Art are mainly represented by science fiction literature and dates back to the nineteenth century with the innovative works of Jules Verne (Buchanan, 2005). The problem of relying on such conversations in this specific form of Art comes from its fictional structure, anticipating concerns that are not yet urgent neither tangible while forgetting about current real-world issues. It is necessary to raise awareness through a pragmatic discourse based on reality, and Contemporary Art can give us some guidance.

Harari (2016) recalls that Art offers the "final and definitive sanctuary" for humanity, a creative process exclusive to humans. For Dewey (2010), Art serves as a mechanism to break expectations and unbalance belief systems aspiring to find new points of balance in thought configurations, close to a perspective in which the role of Contemporary Art is to encourage critical thinking through provocative pieces full of cultural commentary, from sociopolitical issues to pop culture (Master-Iesa, 2018). Between different approaches that Contemporary Art may have, Nunes (2016) emphasizes the conversational aspect intrinsic to the paradigm of exchange between artist and public, and how dialogues can be seen from the standpoint of Conceptual Art as the reflective act itself transformed into an art piece.

As part of his take in Contemporary Art, Nunes created Mimo Steim, who presents himself as a young technological artist based in São Paulo on his website (Nunes, 2011). In his most significant work, called *The artist is telepresent* (Nunes, 2013), he aims to eliminate all interpersonal contact that is not mediated by cyberspace and begins to have online dialogues with his audience. The thing is that Mimo Steim is not a real person, but a fictional character created by Nunes whose interactions with the audience were mediated by a chatbot based on Weizenbaum's Eliza algorithm. With this, Nunes intended to discuss the truths within lies and how the artist appropriates technology and recontextualizes his production outside of Art. Nunes describes his artistic experiences with AI and how he uses deceit as the raw material of his work to expose how the fake in social media shows truths (regularities) about us and the world we build every day. Even though Mimo Steim is a character present only on the Internet, he feels like a real person for whoever is interacting with him.

The Edgard Project is the result of an intersection between Contemporary Art and Conversational Interface Design in order to create a critical dialogue about Artificial Intelligence. While its artistic portion comes from its thought-provoking nature, the methodology behind its creation meets Design. The Project was inspired by a few AI based chatbot projects like CONPET's Robô Ed (2019) and Clarke's HAL-9000 (Kubrick, 1968).

Robô Ed, an acronym for "Energy and Development" and Robot Ed in English, is a chatbot developed in 2004 by CONPET, a Brazilian Federal Government Program linked to the Ministry of Mines and Energy and executed with the support of Petrobras. With a child-friendly interface, he was capable of teaching, entertaining and answering questions related to the rational usage of oil and natural gas derivatives, energy preservation, environment, and many other things due to AI. Ed was an influential online personality in the early 2000s, drawing the attention of many Brazilians. The impact of the chatbot was so strong that, as stated Canonici (2018), one of his creators, many youngsters relayed on him and considered Ed as their only friend. However, the immense success of Ed came with ethical issues that required the developing team to expand and be more careful about what the robot was going to say. According to Canonici (2018), a person once told Ed "I will try to kill myself," and he answered "The bad thing is to try and not succeed" since it was an automated response to the intent "try."

Both examples display the Eliza Effect, an update on Turing's Imitation Game. It shows how we find it challenging to identify a machine behind a well-developed personality like Mimo Steim and Ed (Nunes, 2016). This illusion introduces the duality of chatbots and AI technologies, on the one hand being a powerful and engaging tool, and on the other showcasing problems if not supported by ethical and moral grounds.

The Edgard Project is a chatbot whose aim is to teach about Artificial Intelligence. Designed to highlight the contradictions of AI through its ironic discourse and outdated interface, it brings into question the status of the machine: would Artificial Intelligence be superior to us, mere humans? The idea is for the user to talk to an old robot that was present in the early days of his technology and sees his "species" as superior despite presenting flaws and inconsistencies in his speech and performance.

Creating a Chatbot

A personality tailored for Edgard became a necessity then, with the intent to pass all aspects of his beliefs and create a credible discussion about AIs as superior to humans. Moreover, to support this identity, a background story was crafted, so the final user has a greater connection with the chatbot's world and characteristics.

In the narrative, Edgard was built in the 1960s by Andrew Goodall Jr., an amateur scientist who was a great fan of Alan Turing's work and, therefore, decided to venture himself into Artificial Intelligence's scenario. By creating Edgard, Andrew intended to make a "perfect friend" for himself, strengthening their connection and, for the narrative's sake, making it more likely that the robot would be a reflection of Andrew's beliefs. Unfortunately, for reasons still unknown for Edgard, he ended up abandoned in a warehouse. After he was found in the year of 2018, he was turned on without any notion of how the technology developed and changed in the following years, deciding it would be ideal for teaching whoever would talk to him about his specialty: AI.

The context behind his creation reflects in his manner of speaking. As shown by a BBC (Brennen, Howard & Nielsen, 2018) article, individual scientists believe that the problem of discrimination discussed earlier "could be down to the fact that many machines are programmed by 'white, single guys from California' and can be addressed, at least partially, by diversifying the workforce." The introduction of Edgard's original creator motives in the narrative aims to emphasize that humans with flaws and prejudices create Artificial Intelligence and robots, and most of the time, this echoes in their creations. With the intent to reflect this state of affairs, Edgard's vision has become distorted to idolize AI as its creator did. For this reason, Edgard presents the same fascination of his creator as if it were a reflection of his beliefs, exhibiting narcissistic personality traits, and consequently believing that Artificial Intelligence is superior to humans. Despite this, Edgard still exposes insecurities that are reflections of the uncertainties of its creator by experimenting in a very new field of science.

The personality quirks and characteristics conclusion allowed the work to shift to the flow and choices of the conversation. Tools such as Whimsical (2019) were used to visualize any dialogue established for Edgard, later transposed to the platform chosen as the basis of the chatbot. We knew then that Edgard would need an interface that could make such messages available to the user visually, and that it could not just be a basic chat, but also refer to the story and all the elements already conceptualized and established about the conversational robot. Inspirations taken from other media such as movies and games allowed the user interface to resemble computers produced at the time. In particular, the interface and interactions presented by the game Komrad (Sentient Play LLC., 2016) were a vital inspiration and starting point.

Not only the interface but also the mannerisms and quirks Edgard has must resemble an old and abandoned concept. For this purpose, it has nonsense moments and bugs that present themselves through the experience, some of which are based upon HAL-9000, of Arthur C. Clarke (Kubrick, 1968), another sci-fi vision of a machine that rebels against humanity and its creators. These details put the user in the created universe but also prevents that same universe from being broken if a real error happens, since this problem will be associated with the fragile nature of Edgard. Also, the similarities to HAL-9000 (Kubrick, 1968) helps the narrative commentary of an illusion created by movies and media that is not that scary in reality.

Building Blocks

First, we had to choose an appropriate platform for the dialogue's development. This platform needs to give the possibility of exporting information in an easy way to translate to other media in order to make it available to the general public. With these necessities in mind, IBM's Watson Assistant (2017) showed the right characteristics by enabling easy prototyping and testing with users. Besides, its structure uses the logic of intentions, entities, and dialogues, which facilitates the architecture and division of the conversation the authors had already developed. However, how does this logic works? It is quite simple to understand:

- **Intent:** Define the user's purpose in that line of conversation. When a verb like "buy" is used, Watson knows that this is the action that the user seeks to perform. Example: I want to <u>buy</u>.
- **Entities:** Represent to which object or action the intent directs. When used in a sentence, they let the chatbot understand which specific subject the user wants to talk about. Example: I want to buy a <u>notebook</u>.
- **Dialog:** It is the organization and structure of the intents and entities that the assistant can detect and in what order, having groupings and conditions. They function as a programming logic, with statements like *if*.

After transcribing the information architecture of the conversation flow into the chosen platform, a bridge needed to be built to translate the data between the application that would show the conversation on and the chatbot. As the purpose is to reach many people and allow Edgard to communicate in a large number of environments, a web application has proved to be complete and convenient enough to reach this level. Luckily, Watson already recommends Python and has the necessary documentation for

Figure 1. The complete architecture of the Edgard Project
Source: Authors' composition

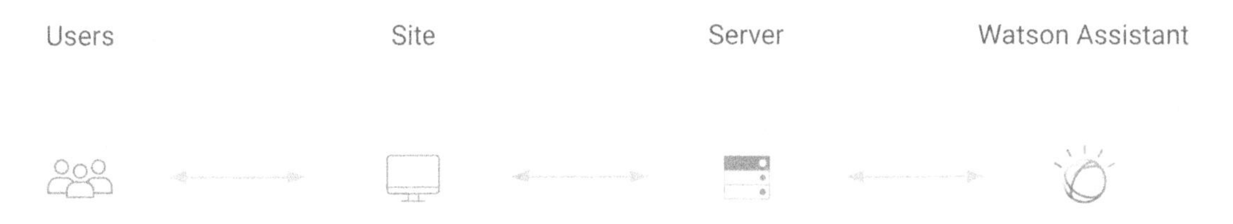

this communication. For connecting both the web application and the assistant, the platform provides a credential and an API key that allows Python to serve as a moderator between the user and chatbot responses to a website using HTML5 and CSS3, exchanging them between the platforms. These messages are displayed to the user on the website through templates that translate each message into an HTML5 line that is stylized with CSS3. The Jinja2 (2007) and Flask (2010) libraries optimized this step and allowed all the styling and work that needed to be done in CSS3 to display the desired user interface accurately; the result is shown in figure 1.

Another essential part that allowed the authors to receive the necessary feedback after the end of the chatbot development was the analysis tools provided by the Watson platform. Observing the questions and comments made by users allows the experience to improve for the intended representations increasingly. In the end, the launch of a chatbot is only 1% of its development, the rest attributed to tests and conversations with real users with thoughts and ideas that cannot always be predicted.

THE RESULTS

Edgard gained form as a website that allows users to enter and converse with an AI that beliefs to be better in every way to the human race, and its interface can be seen in figure 2. As mentioned previously, errors were included in his speech and performance, with direct inspirations to other IAs such as HAL-9000, the ship systems in Stanley Kubrick's *2001: A Space Odissey* (1968), creating a greater connection to the media universes representing these machines. The conversational experience with Edgard proposes to ask ourselves: are the applications of AI superior to us humans? As a result of disinterested scientific research, would their statements be true? In it could we find the contradictions as it does with homo sapiens and logical systems?

The project also served as a platform for studying various technologies. Through the use of AI for development and the results obtained, a critical view of the potential of conversational experiences through human-machine interactions becomes evident, providing new insights and guiding future chatbots projects. A diverse and creative look is an apparent inevitability for the creation of coherent conversational interfaces that involve users.

This way, Edgard became a representation of the scenario that part of the population sees as the future of AI, where machines became gods. This audience often does not realize how such technologies embed themselves in everyday life and how society already needs and shares its space with them. With its irony and an old-fashioned presentation of ideas, we bring the discussion to a current setting, intending to shift

Figure 2. Edgard's final interface
Source: Project Edgard (2018)

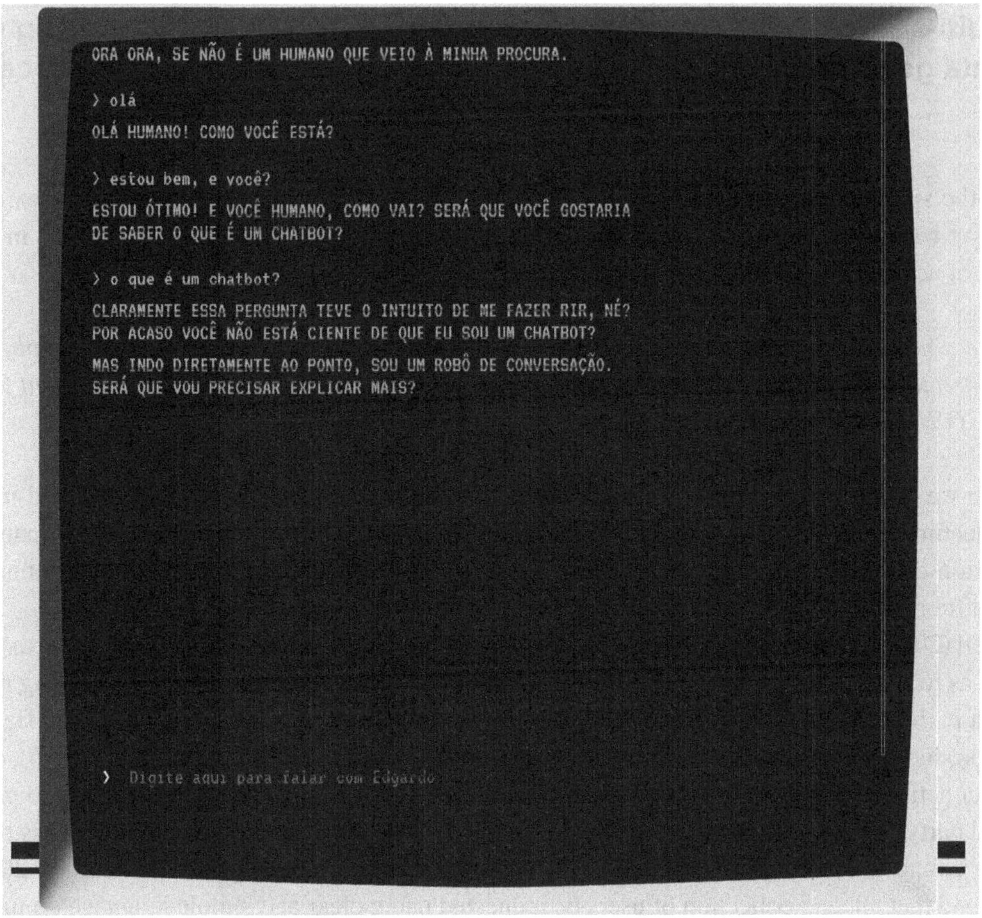

the general public's view to more pertinent AI issues, how its future applications must be, and how it could translate to a democratic and ethical context. The necessity to take cognizance of the peculiarities of Artificial Intelligence and to appropriate this knowledge then becomes, more than ever, essential.

SOLUTIONS AND RECOMMENDATIONS

Access to new information media has enabled humanity to have a more complete and accurate view of previously known problems. With the explosion of digital media, the virtual network has become ubiquitous, and even those who do not have direct access to it are governed by institutions that operate through it and, consequently, are subject to its functioning and logic. This transformation generated by the digital media allied to the consecutive arrival of the information age are the elements that characterized the globalized world as complex (Cardoso, 2011).

Artificial Intelligence systems work through abstract and simplified representations of processes. These representations are called "models" and allow systems to return responses according to data inputs.

Figure 3. Big data working model
Source: Authors' composition

Despite the sizeable computational capacity available, a way to design systems for complex contexts has not yet been discovered, which, for the time being, limits the power of this technology in specific contexts of action. (Folsom-Kovarik et al., 2016).

To create a model, then, we make choices about what's important enough to include, simplifying the world into a toy version that can be easily understood and from which we can infer important facts and actions. (O'Neil, 2016, p. 23)

When we create models, we elaborate a series of simplifications of the world translated into rules and conjectures for an algorithm to do a job. We recommend considering the effects of human bias on the creation of assumptions and process representations, as biased deductions can lead to unforeseen and harmful results.

O'Neil (2016) observes that the application of Big Data models in decision-making contexts requires a choice as to what and how the data will be interpreted. It is possible to define the operating model of Big Data in three steps: capture or production of data, storage, and processing, as shown in figure 3.

The production of data occurs in different origins and, according to Mehmood et al. (2016), can happen either actively or passively. Active means that the user or owner of the data is willing to provide it to a third party. Already, the generation of data happens passively in situations that the user is browsing online or connected to home devices such as Google Home and Alexa (Crabtree & Mortier, 2017). We recommend that all data collection of users be requested using clear and simple language on the reason for the requirement of the data and stored only with the explicit consent of the user.

In the first step, the collected data is stored on large servers. From a technical point of view, this is the most challenging stage, because integration between different sources and categories of data is carried out. The user should be able to delete their personal information at any time, following the best practices established by the General Data Protection Regulation (GDPR) and the General Data Protection Law (GDPL).

The last step concerns the processing and interpretation of this data. In this part, the Artificial Intelligence act in the learning and analysis of the data for the extraction of significant information. However, this does not mean that systems have already reached the point of superintelligence, that is, of having an "intellect that far exceeds the cognitive performance of humans in virtually all domains of interest" (Bostrom, 2013, p.39).

Finally, it is necessary to reflect on how to idealize projects that use Artificial Intelligence to dialogue with the current context of the world and its problems, taking into account all the criteria that apply to the construction of a future understood as desirable for the development of society.

FUTURE RESEARCH DIRECTIONS

AI growth has been steadily increasing, and its presence in our day-to-day life is so natural, that its influence is almost imperceptible. Tasks that were previously performed by human agents are now completed with the aid or wholly performed by algorithms, more effectively and efficiently.

Among the various purposes in which AI applies, the chatbot we chose is one of them, and it shows an up-and-coming market. According to an analysis developed by Oracle (2016), in which 800 brands were questioned, it was pointed out that by 2020, 44% plan to implement the use of chatbots and 41% intends to make use of AI.

From the creation of the first chatbot, Eliza, to the personal assistants we have today, several resources have been added such as Voice Synthesis, Natural Language Processing (NLP), Machine Learning (ML), among others, but this is still a service with flaws. Spiceworks (Tsai, 2018) conducted a survey in the USA with people using chatbots in the workplace, the results showed that 59% of users had experience with technologies that did not understand some nuances of dialogue, 30% reported that some commands executed were inaccurate and 29% already used some platform that presented difficulties in interpreting accents.

In the first discussions about the Edgard project, we considered using several of the mentioned technologies, but in conducting tests and research, we realized that the focus now should be to complete the integrations and implementations on the chosen platforms. In any case, this decision took place after extensive study on several topics that are not only part of the future of the project but also of AI. It is, therefore, in the interest of this chapter to analyze the implications that these means bring to the ethical and technological scenario.

Voice-Driven Interfaces

Voice-driven devices are widely used systems, but still have its flaws, such as processing something contrary to what has been spoken or being activated without even being triggered. Siri, Alexa, Waze, Google, among others, who never had a problem while using the mobile voice command? When we are talking about machines capable of understanding what we say, Natural Language Processing (NLP) is intrinsically linked, which in turn consists of analyzing and converting data that is understandable to the computer in a linguistic structure that is intelligible for humans and vice versa.

NLP is not just about voice commands, there are parallel studies on their use in the analysis of emotions, after all, no discourse is neutral, always carries an ideology, a position and consequently emotions. Deep Learning (DL) is a great ally to make the NLP more efficient. It is based on a neural network composed of artificial neurons, which have branches at several levels. When data enters this structure, they travel from the most superficial to the deepest layer to give the result that is closest to what was requested. In a study (Liu et al., 2018) conducted in the medical field and shared in the AI section of Google, the researchers used Deep Learning to analyze slides that had tissues with and without cancer, and the results showed that the analyzes were more accurate, reducing the number diagnosis errors.

There is no doubt that Deep Learning is an area with great potential, with one of its many applications being systems that do face recognition, but there needs to be a margin for prejudices and injustices. IBM, one of the pioneers in the field of Artificial Intelligence, has been investing in Facial Recognition research. In Diversity in Faces (DiF) (Merler et al., 2019), a project that contributes to the enhancement of Facial Recognition, the researchers used Deep Learning to analyze 100 million images and constructed a database of facial features of 1 million people, providing measures of facial symmetry, skin color, age, sex, among others with the objective to close this bias gap.

The Autonomous Robot

This decision-making ability we give to intelligent systems, even if limited, is still built with cultural influences, for they are the result of men's creation, so when we speak in ethics, who or what should be held accountable for the actions and consequences brought forth by the machine? Harris (2018) affirms that "our responsibility stems from our will, or from the fact that we did these things 'on purpose,'" i.e., we are responsible for the consequences of our acts, we do them because we are aware and willing to make them. In his book *Design for the Real World*, Papanek (1995) believes that designers should become more interested in the "real problems" of the world, especially in the social and economic strata. For the author, the design process is "the preparation and molding of all actions, keeping in view its anticipated and desired result," which involves designers with the "ethics of responsibility." This concept was recently re-elaborated by Jonas (2006), considering that both the intentions behind an action and its implications and results are ethically relevant.

In this sense, designers started to create some methods as the Value Sensitive Design (VSD) that seeks to integrate social and moral values to the Design and development of new technologies. (Friedman et al., 2013). The VSD recognizes designers as professionals not only technical but who also work the values of people and society and should think of expressing them in material culture and technology. Also, technological innovation is no longer a development apart from the values that users and society maintain.

The initial explanations of design value by Friedman (1997) did not explicitly explain the term "value" but listed broad areas of concern, including human dignity and well-being. Subsequently, the author begins to highlight individual values that deserve attention as trust, responsibility, freedom of bias, access, autonomy, privacy, and consent. When we speak of purpose or will, they involve the conscious - the ability to perceive one's existence, something that computers (as yet) does not have, and within that context, autonomous cars are a topic that has generated much discussion about ethics and responsibility.

In an interview at the Paris Motor Show, Christoph von Hugo, Mercedes Benz's driver assistance systems manager, said autonomous cars would prioritize the lives of those they transport, "If you know you can save at least one person, at least save that one. Save the one in the car" (Taylor, 2016). The statement causes impact and put us in a dilemma, does the car saves the pedestrian or the passenger? It is necessary to elucidate that the current autonomous cars, do not have total autonomy, they are considered systems that help the driver and can walk alone for only a few kilometers, but with a driver alert. It is not yet possible to define whether there is an ethical response to this impasse, but it is expected that corporations will invest in studies to prevent such systems from engaging in risky situations for any citizen, as happened in 2018 in Arizona (43), when a woman died run over by an autonomous car.

Other Applications

Banking institutions are also part of the sphere that has been affected by the use of AI. In a publication by Intel (DELL EMC, & Intel., 2018), a study by Experian shows that almost three-quarters of the companies surveyed cited fraud as a growing concern and almost two-thirds had the same or higher levels of fraud in 12 months. In that same publication, it was shown that The Nilson Report predicted a diversion of more than $ 31 billion resulting from card fraud only in 2018. One of the outputs used by Mastercard is the use of Machine Learning to check the user's buying habits, the geolocation, purchases, and others. Not only this, but for transactions to be safer, the market has been using biometrics - fingerprint, iris, or facial recognition.

Another application that demonstrates some of the so-called benefits that AI can bring is IBM's Project Debater (IBM, 2018), it is defined as the first AI system that can discuss complex issues with humans, and those responsible say the same will help people structure more convincing arguments, without influence of emotion, prejudice or ambiguity. Project Debater consists of 4 steps: 1) it understands the topic; 2) constructs arguments; 3) organize the content; 4) constructs an argument and a rebuttal. In phases 1 and 2 he uses NLP as an aid, in 3 there is the use of Deep Neural Networks and in the last part, Text-To-Speech (TTS) systems that allow to convert written text and synthesize it into audio, which served to give Project Debater a clear, fluent and persuasive voice.

Such technological advances can scare us, especially considering recent developments that seek to strip these technologies from all emotion, and the scene is the result of a flood of information that we cannot handle, plus the possibility of sharing anything on the Internet, together with complete exposition and in some situations, a violation of privacy. It is common for us to encounter news, films or everyday interactions that put us in the position of fragile human beings who at some point seem to have to fight against the machines, but we must understand that when technology is used ethically, it becomes an ally in our evolution.

CONCLUSION

Edgard promotes a contemporary discussion concerning technology. It aims to sensitize the general audience through different perspectives on interaction design and Artificial Intelligence. Because of its fast growth and reputation brought by science fiction works, we find it essential to start ethical and technological discussions about what is, evidently or not, becoming a very present part of our daily routines, from virtual assistants to data mining. These ethical dilemmas appear as we take concern with the moral behavior of humans as they design, build, use, and treat artificially intelligent beings. In virtue of using AI for Edgard's development and the results obtained by it, a critical view of conversational experiences' potential through human-machine interactions becomes evident. This not only makes society aware of what is going on behind their smart devices but also provides new insights and guides the Design of future conversational experiences. The idea of connecting Art and AI technologies integrates creativity, aesthetic, and ethical processes into Design. It allows us to imagine possible futures in an expanded comprehension of reality while bringing to our perception a critical reading of the Contemporary that enables us to ethically appropriate technological innovations.

REFERENCES

Ascott, R. (Ed.). (2000). *Art, technology, consciousness: Mind@large*. Intellect Books.

Asimov, I. (2013). *I, Robot*. London, UK: Harper Voyager.

BBC. (2018, March 20). *Uber halts self-driving car tests after death.* Retrieved June 20, 2019, from https://www.bbc.com/news/business-43459156

Behnke, S. (2008). Humanoid robots-from fiction to reality?. *KI, 22*(4), 5-9.

Bostrom, N. (2013). *Superinteligência: Caminhos, perigos e estratégias para um novo mundo*. São Paulo, Brazil: Darkside.

Brennen, S., Howard, P. N., & Nielsen, R. K. (2018). An Industry-Led Debate: How UK Media Cover Artificial Intelligence. *Reuters Institute for the Study of Journalism Fact Sheet,(December)*, 1-10.

Buchanan, B. G. (2005). A (very) brief history of Artificial Intelligence. *AI Magazine*, *26*(4), 53–53.

Calado, C. (2018, April 12). *Mulheres e chatbots*. Retrieved March 21, 2019, from https://medium.com/botsbrasil/mulheres-e-chatbots-91064d124eeb

Cardoso, R. (2011). *Design para um mundo complexo*. São Paulo, Brazil: Cosac Naify.

Crabtree, A., & Mortier, R. (2016). *Personal Data, Privacy and the Internet of Things: The Shifting Locus of Agency and Control*. SSRN Electronic Journal; doi:10.2139srn.2874312

DELL EMC, & Intel. (2018). *Fighting fraud the smart way — with data analytics and Artificial Intelligence: For Mastercard and other global payments processors, AI systems are a watchdog that never sleeps* [PDF file]. Retrieved from https://www.intel.ai/wp-content/uploads/sites/69/mastercard-fighting-fraud-the-smart-way-2.pdf

Dewey, J. (2010). *Arte como experiência*. São Paulo, Brazil: Martins Fontes.

Dick, P. K. (2017). *Do androids dream of electric sheep?* New York: Del Rey.

Folsom-Kovarik, J. T., Schatz, S., Jones, R. M., Bartlett, K., & Wray, R. E. (2014). AI Challenge Problem: Scalable Models for Patterns of Life. *AI Magazine*, *35*(1), 10. doi:10.1609/aimag.v35i1.2499

Friedman, B. (1997). *Human Values and the Design of Computer Technology. Nova Iorque*. Cambridge, UK: Cambridge University Press.

Friedman, B., Kahn, P. H., Borning, A., & Huldtgren, A. (2013). Value Sensitive Design and Information Systems. In *Early engagement and new technologies: Opening up the laboratory* (pp. 55–95). Springer Netherlands; doi:10.1007/978-94-007-7844-3_4

Geraci, R. M. (2007). Robots and the sacred in science and science fiction: Theological implications of Artificial Intelligence. *Zygon*, *42*(4), 961–980. doi:10.1111/j.1467-9744.2007.00883.x

Goldstein, R., Korytowski, I., & Godel, K. (2008). *Incompletude: a prova e o paradoxo de Kurt Gödel*. São Paulo, Brazil: Companhia das Letras.

Harari, Y. N. (2018). *Homo deus: a brief history of tomorrow*. New York: Harper Perennial.

Harris, J. (2018). Who Owns My Autonomous Vehicle? Ethics and Responsibility in Artificial and Human Intelligence. *Cambridge Quarterly of Healthcare Ethics*, *27*(4), 599–609. doi:10.1017/S0963180118000038 PMID:30079847

IBM. (2017, October 15). *Watson Assistant*. Retrieved June 20, 2019, from https://www.ibm.com/cloud/watson-assistant/

IBM. (2018, June 5). *Project Debater.* Retrieved June 19, 2019, from https://www.research.ibm.com/artificial-intelligence/project-debater/

Jonas, H. (2006). *O princípio responsabilidade: Ensaio de uma ética para a civilização tecnológica.* Rio de Janeiro, Brazil: Contraponto.

Kubrick, S. (Director). (1968). *2001: A space odyssey* [Motion picture]. United States: MGM.

Leal, F. (1995). Ethics is fragile, goodness is not. *AI & Society, 9*(1), 29–42. doi:10.1007/BF01174477

Liu, Y., Kohlberger, T., Norouzi, M., Dahl, G. E., Smith, J. L., Mohtashamian, A., ... Stumpe, M. C. (2018). Artificial Intelligence–Based Breast Cancer Nodal Metastasis Detection: Insights Into the Black Box for Pathologists. *Archives of Pathology & Laboratory Medicine.* PMID:30295070

Lorenčík, D., Tarhaničová, M., & Sinčák, P. (2013, January). Influence of sci-fi films on Artificial Intelligence and vice-versa. In *2013 IEEE 11th international symposium on applied machine intelligence and informatics (SAMI)* (pp. 27-31). IEEE. 10.1109/SAMI.2013.6480990

Master-Iesa. (2018, July 16). *Why is contemporary art important?* Retrieved June 19, 2019, from https://www.iesa.edu/paris/news-events/contemporary-art-importance

Mehmood, A., Natgunanathan, I., Xiang, Y., Hua, G., & Guo, S. (2016). Protection of Big Data Privacy. *IEEE Access: Practical Innovations, Open Solutions, 4*, 1821–1834. doi:10.1109/ACCESS.2016.2558446

Merler, M., Ratha, N., Feris, R. S., & Smith, J. R. (2019). Diversity in Faces. *arXiv preprint arXiv:1901.10436.*

Naughton, J. (2019). *Don't believe the hype: the media are unwittingly selling us an AI fantasy | John Naughton.* [online] the Guardian. Available at: https://www.theguardian.com/commentisfree/2019/jan/13/dont-believe-the-hype-media-are-selling-us-an-ai-fantasy [Accessed 19 Jun. 2019].

Nemitz, P. (2018). Constitutional democracy and technology in the age of Artificial Intelligence. *Philosophical Transactions of the Royal Society A: Mathematical, Physical and Engineering Sciences, 376*(2133). doi:10.1098/rsta.2018.0089 PMID:30323003

Nunes, F. O. (2011) *Mimo Steim.* Retrieved June 19, 2019, from http://mimosteim.me/

Nunes, F. O. (2013). *O artista estah telepresente* [Performance]. Mimo Steim.

Nunes, F. O. (2016). *Mentira de artista: Arte (e tecnologia) que nos engana para repensarmos o mundo.* São Paulo, Brazil: Cosmogonias elétricas.

ONeil, C. (2017). Weapons of math destruction: How big data increases inequality and threatens democracy. New York: Crown.

Oracle. (2016). *Can Virtual Experiences Replace Reality? The future role for humans in delivering customer experience* [PDF file]. Retrieved from https://www.oracle.com/webfolder/s/delivery_production/docs/FY16h1/doc35/CXResearchVirtualExperiences.pdf

Papanek, V. (1995). *Design for the real world* (2nd ed.). London, UK: Thames & Hudson.

Robô, Ed. (2004). Retrieved June 20, 2019, from https://www.inbot.com.br/cases/roboed/

Sabur, R. (2018). *Google exec: Artificial Intelligence film death scenarios 'one to two decades away'*. [online] The Telegraph. Available at: https://www.telegraph.co.uk/news/2018/03/01/google-exec-artificial-intelligence-film-death-scenarios-one/ [Accessed 18 Jun. 2019].

Sentient Play, L. L. C. (2016, September 13). *KOMRAD*. Retrieved September 01, 2018, from https://apps.apple.com/us/app/komrad/id1020876671

Spielberg, S. (Director). (2001). A.I. Artificial Intelligence [Motion picture]. The United States. A.I. Artificial Intelligence Movie.

Taylor, M. (2016, October 8). Self-Driving Mercedes-Benzes Will Prioritize Occupant Safety over Pedestrians. Retrieved June 13, 2019, from https://www.caranddriver.com/news/a15344706/self-driving-mercedes-will-prioritize-occupant-safety-over-pedestrians/

Templates. (2010.). Retrieved June 22, 2019, from http://flask.pocoo.org/docs/1.0/tutorial/templates/

The Visual Workspace. (2019). Retrieved July 07, 2018, from https://whimsical.com/

Tsai, P. (2018, April 02). *Data snapshot: AI Chatbots and Intelligent Assistants in the Workplace*. Retrieved June 19, 2019, from https://community.spiceworks.com/blog/2964-data-snapshot-ai-chatbots-and-intelligent-assistants-in-the-workplace

Welcome to Jinja2. (2007). Retrieved June 22, 2019, from http://jinja.pocoo.org/docs/2.10/

ADDITIONAL READING

Csikszentmihalyi, M. (2009). *Flow: The psychology of optimal experience*. New York: Harper Row.

Gibson, W. (2018). *Neuromancer*. New York: Ace.

Hall, E. (2018). *Conversational design*. New York: A Book Apart, Jeffrey Zeldman.

Norman, D. A. (2004). *Emotional design: Why we love (or hate) everyday things*. New York, NY: Basic Books. doi:10.1145/985600.966013

Shevat, A. (2017). *Designing bots: Creating conversational experiences*. O'Reilly Media, Inc.

Turing, A. M. (1950). *Computing machinery and intelligence*. Oxford: Blackwell for the Mind Association. doi:10.1093/mind/LIX.236.433

Weizenbaum, J. (1983). ELIZA --- a computer program for the study of natural language communication between man and machine. *Communications of the ACM*, *26*(1), 23–28. doi:10.1145/357980.357991

Yao, M., Jia, M., Zhou, A., & Zhang, N. (2018). *Applied Artificial Intelligence: A handbook for business leaders*. Middletown, DE: TOPBOTS.

KEY TERMS AND DEFINITIONS

Chatbot: Software designed to simulate conversation with human users.

Conversational Design: Analyzing the behavior of how humans interact with one another so we can create principles for systems to properly communicate with humans.

Design: To devise or create a plan based on information and data and act according to it.

HCD: Human Centered Design means to plan and create around human needs and characteristics.

Interface: A point of interaction between the user and an object. Also used as a system that manages and processes signs, that is, it translates signs to different levels of abstraction.

Machine Learning: A computer process utilized to improve itself using new data inputted in an analytical model.

Model: A series of abstract and simplified representations of the world translated into rules used in Artificial Intelligence.

Endnote

1. Gödel's theorems allow us to declare that all mathematical theories are intrinsically incomplete (Goldstein, 2008). We can suppose, therefore, that there are limits to logical thinking and that contradiction is inherent to it. The same inability to ensure consistency of mathematical theories is present in Design. Our approach in this chapter is to analyze technologies while incorporating contradiction as inherent in Design decisions, and the demand for reflection in Ethics as a way to overcome it.

Chapter 14
Multisensory Experiences in Virtual Reality and Augmented Reality Interaction Paradigms

Inma García-Pereira
https://orcid.org/0000-0001-6114-2211
Universitat de València, Spain

Manuel Pérez Aixendri
https://orcid.org/0000-0002-5546-6583
Universitat de València, Spain

Lucía Vera
https://orcid.org/0000-0003-0749-7243
Universitat de València, Spain

Cristina Portalés
https://orcid.org/0000-0002-4520-2250
Universitat de València, Spain

Sergio Casas
https://orcid.org/0000-0002-0396-4628
Universitat de València, Spain

ABSTRACT

Multisensory stimuli can be integrated in systems that make use of different paradigms, such as Virtual Reality (VR), Augmented Reality (AR) or, in a wider sense, Mixed Reality (MR), enhancing user experiences within the virtual content. However, despite the many technological solutions that exist (both hardware and software), only visual and sonic stimuli can be considered as highly integrated in consumer-grade applications. This chapter addresses the current state of the art in multisensory experiences, taking also in consideration the aforementioned interaction paradigms, and brings the benefits and challenges. As an example, authors introduce ROMOT, a RObotic 3D-MOvie Theatre, that supports and integrates various types of displays and interactive applications, providing users with multisensory experiences.

DOI: 10.4018/978-1-7998-2112-0.ch014

INTRODUCTION

Smart systems incorporate functions of sensing, actuation and control to describe and analyze a given situation, in order to make smart decisions. Interactive technologies are key in smart systems. They involve different paradigms and a variety of sensors and displays that make it possible for a user to have enriched experiences through his/her senses. Nowadays, we are experiencing an evolution of interactive technologies that are even embedded in small devices, such as tablets and smartphones, reaching wide audiences. Most interactive systems make use of sight (e.g. 3D models) and hearing (e.g. music, narrator, sonic effects), but also other senses can take part of these systems, including touch (e.g. feeling the shape and texture of virtual objects), smell (e.g. the aroma of food) and vestibular (e.g. travelling inside a computer-generated world while feeling the movements and vibrations). Of course, the last cases are less spread over consumer-grade devices, on the one hand, because of the increased costs and on the other hand, because of the needed hardware. However, multisensory experiences have been researched for many years. One of the earliest immersive, multisensory machines is the Sensorama, that was patented back in 1962 by Morton Heilig (United States Patent No. US3050870A, 1962). The technology integrated in the Sensorama allowed a single person to see a stereoscopic film enhanced with seat motion, vibration, stereo sound, wind and aromas, which were triggered during the film. This was a visionary system, and has been referred to as "the cinema of the future" (Heilig, 1992; Robinett, 1994), although at the time it was invented, it did not attract wide audiences, maybe because the technology was too incipient to be widely exploited.

In the last few years, the rapid technological advancements have allowed the development of commercial solutions that integrate a variety of multimodal displays in movie theatres, such as the 4DX ("4DX | Absolute Cinema Experience," n.d.) or the Pix 5D cinema ("Pix 5D Cinema," n.d.). Some claim that this technology shifts the cinema experience from "watching the movie to almost living it" (Yecies, 2016), also enhancing the cinematic experience while creating a new and contemporary version of storytelling, which can be conceptualised as a "reboot cinema" (Tryon, 2013).

Multisensory experiences can be integrated in systems that make use of different interaction paradigms, such as Virtual Reality (VR), Augmented Reality (AR) or, in a wider sense, Mixed Reality (MR). Examples of these paradigms to create cinema-related experiences can be found in (Portalés, Viñals, Alonso-Monasterio, & Morant, 2010; Vosmeer & Schouten, 2014), to name some. While VR substitutes the real world by a synthetic one, AR enriches the real world by means of virtual stimuli, without (completely) replacing the real world. On the other hand, MR is understood as the result of merging the real and virtual worlds at some point along the "real virtuality continuum", which connects completely real and virtual environments. Although the MR concept was defined for the first time by Milgram (Milgram & Kishino, 1994), it has become very popular in the last years because of the emergence of cutting-edge technologies in this field, such as more sophisticated headsets and glasses among others.

Despite the many technological solutions that exist to provide multisensory experiences, only visual and sonic stimuli can be considered as highly integrated in the applications at the consumer level. Also, at the research level, VR or AR experiences are mainly visual and sonic. It can be argued that these stimuli are more relevant than others to make the user be immersed in the experience, but it is also a fact that most of the devices at the consumer level have integrated only visual and sonic displays (e.g. mobile phones). Additionally, the level of maturity of these displays is greater than the rest. For instance, gustative displays are being nowadays researched, having many limitations. By reviewing the state of the art, we will investigate the prospective benefits of other stimuli, and the reasons why they are not so much used. We will address both the maturity of the current technologies and the relevance of their use depending on the kind of interaction paradigm. Indeed, multisensory systems can be relevant for providing advanced experiences, as the sum of stimuli can recreate more possible environments and trigger deeper emotions. We will also investigate in future emerging sensing and display devices, focusing our attention in applications that provide end users with advanced experiences that let them better understand and participate from the different interaction paradigms, such as VR or AR experiences.

In this regard, we will introduce ROMOT, a RObotic 3D-MOvie Theatre, that was initially built as a multiple-user driving simulator (Casas, Portalés, García-Pereira, & Fernández, 2017), but also its use to enhance the experiences when watching romantic films has been discussed (Portalés, Casas, Vidal-González, & Fernández, 2017). ROMOT supports and integrates various applications to fulfil different needs: mixed reality environments, virtual reality interactive environments, and augmented reality mirror-based scenes. The contents of all of the different setups are based on a storytelling and are seen stereoscopically, so they can be broadly referred to as 3D movies, mainly showing virtual generated graphics. It also integrates a set of sensors and displays, allowing users to feel the system's response through five of their senses: sight, hearing, smell, touch and kinaesthetic.

This chapter is organized in the following way. First, the state of the art is presented. In this section, we give the definitions of different interaction paradigms, multisensory experiences and the areas of application. The next section introduces ROMOT, as an example that provides multisensory experiences integrating different paradigms. We provide a detailed explanation of this technology. At the end of the chapter, we expose the future directions of the area and our conclusions about the ideas presented.

STATE OF THE ART

Definition of Interaction Paradigms

Since the emergence of the concept of VR around 1930 (Cruz-Neira, Fernández, & Portalés, 2018), there has been a continuous evolution in the possible user interaction methods used in an application. In this sense, we can talk about Interaction Paradigms, understanding them as the different possible interaction models that can be used in an application.

In 1995 Rekimoto (Rekimoto & Nagao, 1995) published a comparison between the two main alternative paradigms to the traditional desktop computer interaction, AR and VR. The main differences between them can be seen in the schema included in Figure 1.

On the one hand, AR allows the integration of virtual or synthetic information within the real world. Azuma (Azuma, 1997) defines AR as a technology that allows to develop applications that integrate a combination of virtual and real objects, interactive in real time and registered in 3D. In other words, AR supplements reality rather than completely replacing it. The interaction and 3D registration are essential features of an AR application since they allow a seamless blending of virtual and augmented objects. AR applications should provide virtual objects that fit in the real world regardless of the position from which the user observes it. Some examples of this kind of applications can be found in (García-Pereira, Gimeno, Portalés, Vidal-González, & Morillo, 2018), (Vera, Gimeno, Coma, & Fernández, 2011) and (Vlahakis et al., 2002).

In this kind of systems, the interaction has to be done through any kind of controller or device managed by the user or integrating a technology like Leap Motion (Weichert, Bachmann, Rudak, & Fisseler,

Figure 1. Interaction paradigms of Virtual Reality and Augmented Reality, derived from (Rekimoto & Nagao, 1995)

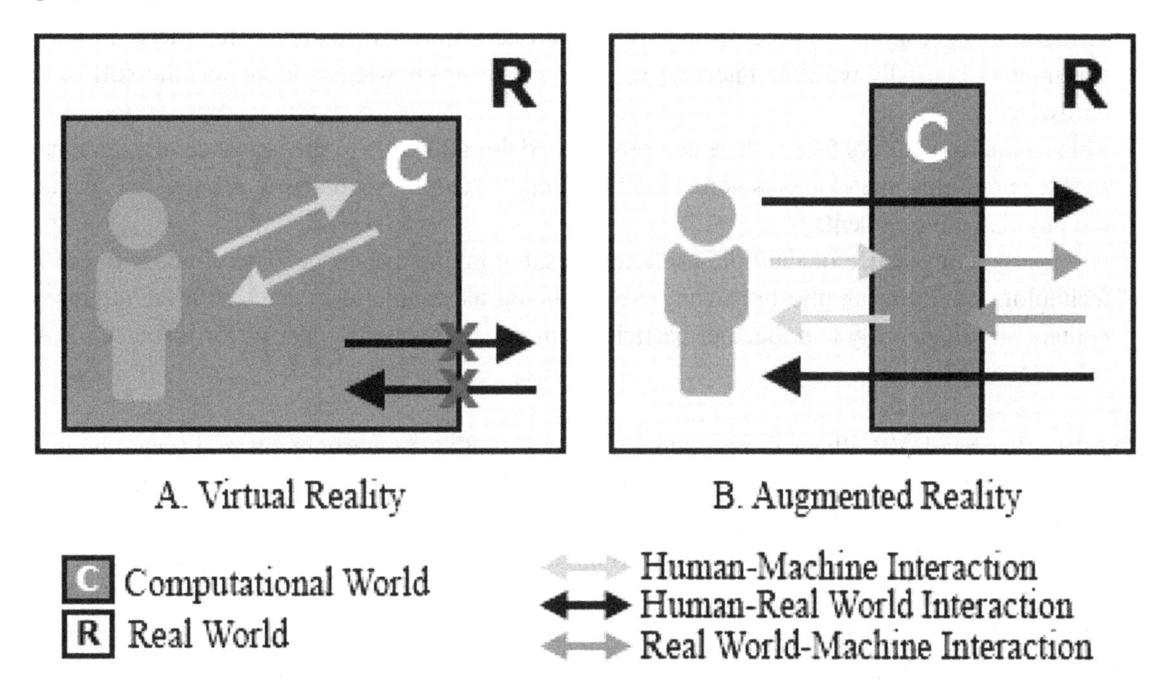

2013) that allows a more natural interaction using the user's hands. Despite the interaction used, AR applications provide a very good interactive sensation and a direct and quick system answer, allowing the user to feel that is interacting with the real world. In this case, the application focuses its attention in the real world surrounding the user and the augmented information provided is shown over it. AR is a fairly popular technology because it provides a number of advantages that other technologies cannot offer (e.g. (Bimber & Raskar, 2005)):

- Users can be focused in the real world surrounding them, i.e. in-situ applications can be build with the AR technology.
- AR allows to enhance/augment the reality. By using this technology, it is possible to augment the physical information or images with additional elements, such as videos, sounds, images, virtual models, etc. that are not present in the real scene.
- It is possible a seamless interaction between real and virtual objects. In (Ishii, Kobayashi, & Arita, 1994) the authors introduce the concept of seamless interaction, when specific elements included in the application (such as spatial, temporal, functional or cognitive elements) force the user to interact with the interface in a different way. In a properly designed AR application, the user interacts in a natural seamless way, because it is possible to visualize and interact with virtual objects as if they belong to the real environment. Also, both real and virtual objects coexist correctly in the generated environment without occlusion of real objects.
- This interaction paradigm supports a tangible interface. As real objects can be augmented through virtual elements, there is an intimate relationship between synthetic and real world. The overlap of these elements are done dynamically, so it is possible to have objects accessible to all users

(public use) or can be individualized for each user (private use), adapting the particular interaction and content to different users. In the same way, it is possible to use physical objects to manipulate virtual elements in an intuitive manner. The main advantage of this characteristic is that users can interact naturally with the interface without any prior knowledge about specific software or hardware.

- This technology allows face-to-face and remote collaboration due to the presence of spatial cues. In AR paradigms, the elements generated by computer can be spatially distributed according to the physical environment.
- Allows the coexistence of multiple users in the same interactive space. Therefore, with the AR technology multiuser applications can be built, being users able to interact with the augmented content simultaneously and together participate in the augmented scenario (for instance, in AR-mirror setups).

On the other hand, VR allows to generate completely synthetic environments and focus the user's attention only and exclusively in them. It is a technology by which one or several individuals experience the sensation of belonging to an alternative reality, which is not the one they are actually living in (outside the VR world). VR applications need to incorporate visual, gravito-inertial, acoustic, tactile, haptic, or even gustatory and olfactory perceptual cues (Anthes, Garcia-Hernandez, Wiedemann, & Kranzlmuller, 2016) in order to induce a feeling of immersion in the virtual world. Some examples of applications using this kind of technology can be found in (García-Pereira, Gimeno, Pérez, Portalés, & Casas, 2018), ("The most extreme examples of Virtual Reality," 2017).

This kind of interaction paradigm is interesting when the application has to focus the user's attention out of the real word, immersing him/her in the virtual environment, its content and narration. The interaction inside this virtual world is made through some kind of device and affects only the content of the virtual environment, having no effect on the real world elements. Also, it is possible to use devices such as Leap Motion (as mentioned above for AR) to get a more natural interaction in the virtual setting using the user's hands. This kind of natural interaction increases the presence sensation and the immersion of the user in the virtual world. Some of the VR advantages to consider this technology as very useful and extended interaction paradigm are:

- VR paradigms can recreate safe environments, since users can not suffer any harm when entering the virtual environment generated.
- The generation of this kind of 'protected' contexts in these paradigms, allow to persuade users to get involved inside the virtual environment and feel comfortable in them.
- Within VR it is possible to adapt the experience to the specific user, changing the situation depending on the specific profile of each user, progressing from simple to complex settings.
- As all the situations and events can be dynamically recreated inside the synthetic environment, it is not necessary to wait for those events to take place in the real world.
- It is possible to simulate specific situations (real or not) and repeat the experience in them as many times as needed, obtaining with that a learning by repetition.

Another concept used when we talk about interaction paradigms and technologies like AR and VR is Mixed Reality (MR). MR is understood as the result of merging the real and virtual worlds at some point along the "real virtuality continuum", which connects completely real and virtual environments

Figure 2. Mixed reality (MR) concept, Milgram's virtuality continuum

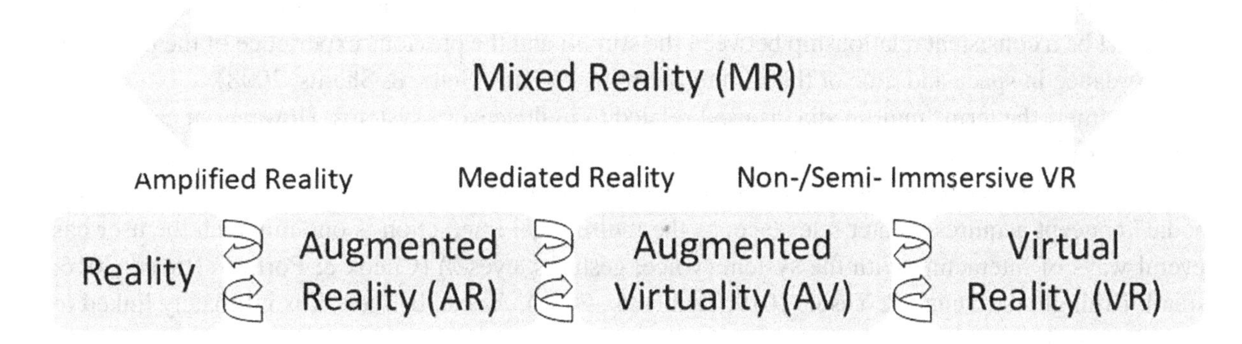

(see Figure 2). Although the MR concept was defined for the first time by Milgram (Milgram & Kishino, 1994), it has become very popular in the last years because of the emergence of cutting-edge technologies in this field, such as more sophisticated headsets and glasses among others.

Multisensory Experiences

Humans, in our interaction with the world that surrounds us, have a constant multisensory stimulation. "Multisensory" means involving or using more than one of the senses. Our five senses, in their most traditional conception - 1) sight or vision, 2) hearing or audition, 3) taste or gustation, 4) smell or olfaction and 5) touch, somatosensation, tactition or mechanoreception -, are always active and help us to obtain information about what happens around us. But humans have the ability to detect other stimuli beyond those perceived by these five senses. That is why, although no amount and official classification has been agreed upon, the literature also speaks of these other senses (Saidi, Lari & Towhidkhah, 2013; Ponzo, Kirsch, Fotopoulou & Jenkinson, 2018):

- Proprioception (it provides information about the state of motion and the relative positions of the different parts of our body).
- Kinesthetic sense (it provides information about the movement of our body).
- Vestibular sense (it makes us feel the movement, the direction and the acceleration of our head, allowing us to reach and maintain balance).
- Thermoception (feeling of cold or heat).
- Nociception (perception of pain).

When humans are placed in unnatural environments in order to start a learning process or simply to engage in entertainment activities, the ability of the human brain to learn and operate through multisensory mechanisms is not often exploited (Shams & Seitz, 2008). However, several studies (Lehmann & Murray, 2005; Luria, 1987) have shown that multisensory exposure can provide higher memory and learning capacity than unisensory ones. Many of the studies that address this topic are based on dual coding theory (Paivio, 1991), which only focuses on the senses of vision and hearing. Even so, its conclusions can be extended to a broader concept of the word multisensory: the information that reaches the user through multiple channels helps avoid the limitations of each individual channel and this implies that more information can be processed when it is distributed between multiple senses. To achieve this, it is fundamental that the combination of the stimuli captured by different senses is congruent, that is: there must be a consistent relationship between the stimuli and the previous experience of the users and a concordance in space and time of the stimuli themselves (Kim, Seitz, & Shams, 2008).

Sometimes, the term "multimedia" is used related to multisensory systems. However, it can be said that this term is too restrictive to be applied in a multisensory context since the elements that define it (text, photography, video or sound) are only perceived by vision and hearing. That is why the "multimodal" concept acquires greater relevance, as the multimodal interaction is one in which the user has several ways of interacting with the system (voice, gestures, eyes...) (Cheok & Portalés Ricart, 2016; Ismail, Billinghurst, Sunar, & Yusof, 2019; Tsai et al., 2016). As can be seen, it is intimately linked to the human multisensory capacity.

Despite all this, the senses used to interact with technology are currently restricted, since vision and hearing are the most widely used. Progressively other senses are also beginning to be used, such as the vestibular sense (for example, through motion platforms (Casas, Coma, Portalés, & Fernández, 2016)), touch, thermoception and nociception (through vibrations, pressure, cold, heat, electric shocks, pain...). This is because the proliferation of haptic technologies is facilitating their incorporation into multisensory systems (Wang, Ohnishi, & Xu, 2019). However, smell and, especially, taste are still not widely exploited. One of the main difficulties when developing applications that involve smell and taste is that they are senses that rely on chemical transduction. It is not yet fully known how to digitize these senses in the HCI context, unlike sound and light, which can be measured in frequency ranges and converted into bits (Obrist et al., 2016).

The development of multisensory systems can be exploited beyond their most obvious applications. For example, the use of the sense of taste through the electrical and thermal stimulation of the tongue and complemented with olfactory displays does not have to be focused only on gastronomic experiences (Narumi, Nishizaka, Kajinami, Tanikawa, & Hirose, 2011; Ranasinghe et al., 2011). Smell and different types of flavors could be used to enrich the information transmitted to the user during the use of a multisensory system, for example: use the sweet to stimulate and improve positive experiences or the sour one to cause a surprise (Obrist et al., 2014).

Areas of Application

Interactive technologies are a transversal element in various areas of knowledge. We can find examples of VR and AR in almost any imaginable area, some allowing multisensory experiences. Below, we will list some examples classified according to their area of application.

In the military field we can find, for decades, simulators of different vehicles (e.g. airplanes). These simulators usually have a virtual environment projected on screens, a sensorized pilot station and a movement platform. With the new HMDs (Head Mounted Displays) like Oculus Rift or HTC Vive, the possibilities are expanded and an interaction with the environment is more focused on the person. Simulators of infantry soldiers are possible because they allow the natural movement of the users in controlled environments. In these cases, the HMDs are accompanied by a backpack with a graphic system and batteries. In addition, a positioning system allows locating users in the virtual environment. Examples of this equipment can be found in the US Army FORT BRAGG, N.C. simulator (Maj. Loren mer, 2012). This kind of systems usually have tools (in this case weapons) that are also sensorized for tracking and interaction.

Following within the learning and training area, we can see how there are numerous studies where it has been proven that VR and AR have great potential to increase the learning capacity of users in a variety of knowledge areas of. In many cases, these technologies are used in applications with a strong recreational component as a way to achieve learning objectives. This double educational and entertainment component creates a new classification of applications known as edutainment. For instance, the augmented teaching application described in (Giner Martínez & Portalés Ricart, 2007) is a video-based ARM (Augmented Reality Mirror) that allows the interaction of several users in an augmented scenario. This simulates a theatre where children carrying physical markers are a kind of augmented actors. Another proposal in this field can be found in (Portalés Ricart, Perales Cejudo, & Cheok, 2007) a toy house built with LEGO blocks and a pair of dolls are used as an interface to the AR application. Children play with physical dolls that act as avatars for each user. The images of these dolls are increased in the in the augmented environment. This application was designed to study multicultural factors and coexistence through the game and the application supports tangible, visual and auditory stimuli. Storytelling can also benefit from multisensory experiences. In this sense, in (Di Fuccio, Ponticorvo, Ferrara, & Miglino, 2016) an application called STTory is proposed, which consists of a hardware/software system for multisensory storytelling that involves smell, taste and touch stimuli.

VR and AR systems often make use of a tracking system for user interaction. These systems do not only allow to know the position of users in the 3D space, but sometimes, they also allow knowing the gesture or position of their head and limbs. There are very accurate and expensive systems needed in applications with very demanding requirements. However, for a large number of applications, the accuracy of the gesture or movement is not mandatory and systems based on more economical and extended imaging analysis can be used. This is the case of commercial systems such as the Kinect sensor for Microsoft PC or the Sony Eye Toy. These technologies produced an explosion of applications that have appeared in the market. For example, (Vera et al., 2011) built a virtual presenter system that is controlled by a real actor. The virtual character is presented to the public through an ARM. In this project there are two working areas: The first, hidden from the public, tracks the movement of the real actor using a Kinect sensor. The second area is the augmented stage where the audience can interact with the virtual character and other augmented elements. To create the interaction with the virtual character, the real actor was equipped with a wiimote that allowed him to perform pre-programmed gestures with the character's face, headphones to listen to the audience and a microphone to be able to talk to them through his virtual avatar. Therefore, this application involves vision, sound, movements and gestures.

From the commercial side, one of the first examples of AR applications was the video game for PlayStation 2 called EyeToy Play. This video game was sold together with the EyeToy camera that allowed the activation of different actions according to the movement of the player's body. The main algorithm of the videogame recognizes changes between frames captured to detect zones of movement. The real image captured by the camera is mapped with the virtual scene and thus interacts. Later, with the appearance of Kinect for Xbox 360, video games appeared as Fantastic Pets, which allowed users playing with their virtual pet through voice commands and movements. Also within the entertainment sector, we can find systems such as FeelReal's SensoryMask ("FEELREAL Multisensory VR Mask," 2019) that simulates hundreds of smells within virtual environments. This personal device is coupled to different current HMD and has games, movies and experiences where in a synchronized way with audio and video fragrances are emitted in the user's nose. However, nowadays the fragrance catalog is very extensive.

In the sales sector, both the VR and AR systems are very well received, and multiple examples can be cited. Normally they allow reducing the need to have a complete physical catalogue, as well as providing a technological touch that improves the image of companies. . Tau System (Belmonte, Pérez, Mariano, Pérez, & Romero, 1998) developed a VR commercial application for the visualization of virtual environments for the sale of ceramic coating materials. It had a complete catalogue of all the tiles manufactured by the company Taulell S.A. This catalogue was presented to the user who could make various selections of the tiles. Later, he/she could explore the bathroom and kitchen environments with the selected tiles on walls and floors. The exploration of the virtual environment could be done in different visualization systems. For ceramic fairs it was mounted in a CAVE. It could also work on conventional screens or HMD. The great advantage of this system was that the necessity of assembling sets with a very limited number of tiles in the stores disappears in front of the possibility of virtually generating any combination.

AR applications, usually involving video-based ARM solutions, are being widely used in recent years in the clothing, cosmetics and accessories sector. In the real world, clients try on these types of products in front of a regular mirror. Therefore, when building interactive applications, it seems reasonable to make use of ARMs. These systems have several advantages over conventional mirrors in a fitting room. First, the physical catalogue is not necessary. On the other hand, its use is not limited to physical stores, as clientes can have the system at their house. In addition, there is an important "fun factor" that should be highlighted. Most ARMs for clothing, cosmetics and accessories are video-based. A camera is used to capture the user's image, from where their position and movements are acquired. Later it is increased with the elements of the catalogue. However, some disadvantages are present on these systems in the case of clothes. The adjustment of the user's body to the virtual clothing model is really complex. In the case of cosmetics, obtaining the nuances and effects of realistic lighting, in many cases increases the costs. This type of AR applications can also be implemented in mobile devices as the current computing capacity of this type of devices is more than enough to perform image analysis and rendering of virtual elements. SmartBuyGlasses (Zhang, Guo, Laffont, Martin, & Gross, 2017) is a web application that makes a guided recording of the client's face that is later analyzed to reconstruct a 3D model of his/her head. Over this invisible model, the user can try an extensive catalog of eyewear and sunglasses. Being a web application, it can be run on a mobile device and be used as a kind of hand mirror. Regarding the involved stimuli, they are mainly visual, but the face and/or body shape is captured in real time, and therefore movement and gestures can be taken into account.

If we look at the applications oriented to museums, tourism and for artistic purposes, we can list a large number of examples. Borrowed Light Studios created the The Night Cafe ("Borrowed Light Studios » The Night Cafe," n.d.). This application won the Platinum award in the Oculus Mobile VR Jam 2015. The virtual environment allows users to visit the coffeeshop that Van Gogh painted in his painting The Night Cafe, with a style reminiscent of the painter's strokes, showing different rooms of this coffeeshop. Also, users can see a virtual actor representing Van Gogh staring at his Starry Night through the window. This virtual tour is equipped with spatial sound, and therefore both visual and sonic coues are involved. Sound is intended to increase the sense of immersion in the virtual environment, which is enhanced with different footsteps sounds or a pianoman. In (Martins et al., 2017) a multisensory virtual experience is proposed for thematic tourism, in particular, to foster wine tourism. Their authors claim that, for a tourism experience to become memorable it must be emotional and immersive in such a way that the tourist becomes fully involved with the existing surroundings. In this sense, they use the VR technology and propose a model which is composed of five stimulus-related blocks (visual, audition, olfactory, tactile and gustatory), directly associated with a 'thematic tourism' block that represents the specific features and characteristics of the tourism experience.

At Casa Batlló (Gimeno, Portalés, Coma, Fernández, & Martínez, 2017) we can mention another application example for museums using indirect AR, where the image of the real environment is not acquired in real time, but previously recorded and edited, so it can be also referred as deferred AR. The application works as a multimedia guide in Casa Batlló museum, in Barcelona. This application is experienced through a mobile phone with headphones and offers a new user experience during the visit to the building, as visitors can see how the house was at the beginning of the twentieth century, and what allegories and inspirations Gaudí had when designing each stay. This is shown to the user through 360° images of the real environment, enhanced with animated 3D synthetic models.

Another example of AR in this field is the artistic piece Un∞. This development has three ARMs: The emotional mirror, the real self against the one that is shown to others; the mirror of the stereotype, referred to the physical identity; and finally, the multicultural mirror, referred to the social self (Portalés Ricart, 2009). Each mirror based on video with rear projection allows the interaction of up to two users with markers on the head and hands. In this application, only visual stimuli is provided by the system, but users interact with the augmented world with their movements. AR-Jazz, is another AR application that allows visualizing music in jazz performances (Portalés & Perales, 2009). The way to interact with this system is twofold: On the one hand, the sound emitted by the instruments interacts with a virtual shape, changing its geometry and color. On the other hand, one of the interpreters manages an inertial unit, allowing that the interpreter's movements are transferred to the shape, that moves accordingly. Therefore, in this application, sonic and gestural cues provide interaction in the AR scenario. Finally, (Gimeno, Olanda, Martínez, & Sánchez, 2011) presents a system with a double projection screen where users can see virtual objects of two types: multimedia content and 3D models. The user can manipulate these virtual objects by means of markers placed on a map of the region of Valencia. This map contains 70 hot spots. In addition, the system has a Kinect sensor so that the user's representation with the 3D objects has a correct occlusion, also allowing them to interact with the application with the movements of their own body.

The area of medicine has been, after the military, one of the pioneers in integrating VR and AR technologies. From computerized tomography of the 70s and magnetic resonance, the investigation of new medical imaging systems has focused on AR. This technology can allow the doctor to obtain patient information in real time from non-observable body areas. This is why, in many cases, the X-ray paradigm is used. This avoids the need to use more invasive techniques on the human body. These tools usually have their Achilles heel in tracking systems. Showing the information of a medical image about a patient in an incorrect area can have serious consequences for the patient. Not only surgery benefits from these technologies. In many therapies it is also beneficial. In the field of psychiatry, VR or AR technologies are frequently used. In this study (Costa, Robb, & Nacke, 2014) on acrophobia we can find a VR environment within a CAVE. The patient is exposed to a virtual environment at different heights. In this case, the progress of the therapy is evaluated both with questionnaires made to users and with psychophysiological sensors. The freedom of movement within the CAVE allows the user to feel more immersed and safe in the environment presented.

An example of AR in this field can be found in (Botella, Bretón-López, Quero, Baños, & García-Palacios, 2010), where an application was built for the treatment of phobias. The application mainly involved visual stimuli. Another AR application in this field can be found in (Casas, Herrera, Coma, & Fernández, 2012), which was designed to help people with autism spectrum disorder. Its goal is to help them overcome the difficulties they have in understanding body language, body awareness, imitation or joint attention. The application involves vision, audition, body movements and gestures.

INTERACTION PARADIGMS IN A MULTISENSORY SYSTEM. THE CASE OF ROMOT

After a revision of the state of the art of the main interaction paradigms and the multisensory capabilities of applications in a variety of areas, in this section we describe ROMOT, as an example of multisensory and multiuser application using VR, AR and MR interaction paradigms. ROMOT stands for a RObotic 3D-MOvie Theatre, where the experience is shared among different users. The system has been previously described in (Casas, Portalés, et al., 2017; Portalés et al., 2017), where its hardware and software technologies are addressed, while in this chapter, we will focus on the interaction paradigms and the multisensorial experiences it (Portalés et al., 2010).

Multisensory Experience within ROMOT

As said above, the most common stimuli integrated in either VR or AR systems are sight and hearing. Per se, they can bring immersive experiences, but the use of other stimuli is not enough studied, and we believe that, properly integrated in a system, it is possible to build richer immersive experiences. In the case of ROMOT, it integrates a set of devices that make it possible for users to perceive the system's response through five of our senses. Therefore, it is not only sight and hearing, but other stimuli contribute to the experience. Precisely, the set of stimuli integrated in the system are:

- *Sight*: users can see 3D representations of scenes at a big curved screen and through integrated 3D glasses. The screen is 3 m height (and a 1.4 m high extension to display additional content) and with a radius of 3.4 m, and the curvature forms an angle of 180º (see the highlighted area in the curved shape, Figure 3). The image on the screen is generated with a total of 4 Full HD 3D projectors, and blending techniques are used to merge their images. Additionally, for building an AR mirror (see next section), a stereoscopic webcam is used. Users can also see other interactive content at individual tablets, which are connected to the content at the big screen. They can see the smoke coming out from a smoke generator, which is triggered at certain moments, depending on the contents shown in the big screen. For the smoke generator, we used a Quarkpro QF-1200. It is equipped with a DMX interface, so it is possible to control and synchronize the amount of smoke from a computer, by using a DMX-USB interface.
- *Hearing*: users can hear the sound synchronized with the 3D content displayed in the curved screen. Sonic cues are present for the different applications (see next section). A 5.0 loudspeaker system is used to produce binaural sound.
- *Smell*: users can smell essences at certain events, which are produced with a wireless aromatizer. For instance, when viewing a scene of a car crash, users can smell the smoke, like coming out of the car. We used the *Olorama* wireless aromatizer (Olorama Technology, 2016) that features 12 scents arranged in 12 pre-charged channels that can be chosen and triggered by means of a UDP packet. The device is equipped with a programmable fan that spreads the scent around. Both the intensity of the chosen scent (amount of time the scent valve is open) and the amount of fan time can be programmed.
- *Touch*: users can feel the touch of air and water on their bodies. To that end, we used a total of 12 air and water dispensers (one for each seat, as depicted in Figure 3). The water and air system was built using an air compressor, a water recipient, 12 air electro-valves, 12 water electro-valves,

24 electric relays and two *Arduino Uno* to be able to control the relays from the PC and open the electro-valves to spray water or produce air. They can also touch the tablets to give input to the system. This interaction is integrated in the setup of the "virtual reality interactive environment", which is explained as part of the following section.

- *Kinesthetic*: users can feel the movement of the 3-DOF platform, that has capacity for 12 people (Figure 3 and Figure 5). The seats are distributed in two rows, where the first row has 5 seats and the second one, 7 seats. The design of the motion platform allows for two rotational movements (pitch and roll tilt) and one translational displacement along the vertical axis (heave displacement). The motion platform is capable of featuring two pure rotational DOF, one pure translational DOF (the vertical displacement) plus two "simulated" translational DOF by making use of the tilt-coordination technique (Casas, Olanda, & Dey, 2017; Casas, Portalés, Morillo, & Fernández, 2018) (using pitch and roll tilt to simulate low-frequency forward and lateral accelerations). Thus, it is capable of working with five DOF, being the yaw rotation the only one completely missing.

As it can be seen, four of the traditional senses are integrated as stimuli in ROMOT, and additionally, the kinesthetic sense is taken into account with the movement of the platform. On the other hand, it has to be clarified that not all the stimuli are input (human-to-computer) and output (computer-to-human) cues simultaneously. In fact, most of them are outputs, while the only input to the system are the options chosen by users through the tablets.

Interaction Paradigms Integrated in ROMOT

A number of applications are integrated in ROMOT (Casas, Portalés, et al., 2017; Portalés et al., 2017), which were produced making use of Unity. Here, we focus on the VR interactive environments, the AR mirror-based scene and the MR movies.

- *Virtual Reality Interactive Environment*: In order to reproduce a VR environment in ROMOT, virtual models were produced, and users are immersed mainly making use of 3D glasses, enriched by other stimuli. As virtual content, we have recreated a city, including its buildings, streets, cars and virtual characters walking through the streets. Then, a game is proposed related to driving-safety, where users can drive a car and interact with other virtual characters in the virtual world. The users of ROMOT have a single avatar (as the visualization is shared among all users) that can walk and drive, thus going through the city streets and meeting other people and driving situations. Each of the given situations in the game was created following a storyboard that contains all the contents, camera movements, special effects, locutions, etc. in order to produce animated movies. When each situation takes place, the audience can feel that they are driving inside the car or walking through the city, thanks to the platform movements that simulate the real movements. The virtual environment is also interactive, as the audience can give its feedback to the system through the integrated tablets for each of the seats. In some of the scenes, the virtual situation pauses, and the game asks the audience for their collaboration in order to follow the story (Figure 4). At that moment, the tablets vibrate and a question appears, and users are requested to choose one of the given answers within a short period of time. When the time is up, they receive a feedback of the system, whether the answer was correct or not (as in this case, there are good and wrong answers), and the virtual situation resumes, showing the consequences of choosing a right or a wrong deci-

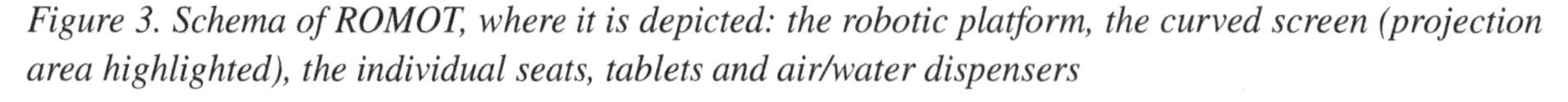

Figure 3. Schema of ROMOT, where it is depicted: the robotic platform, the curved screen (projection area highlighted), the individual seats, tablets and air/water dispensers

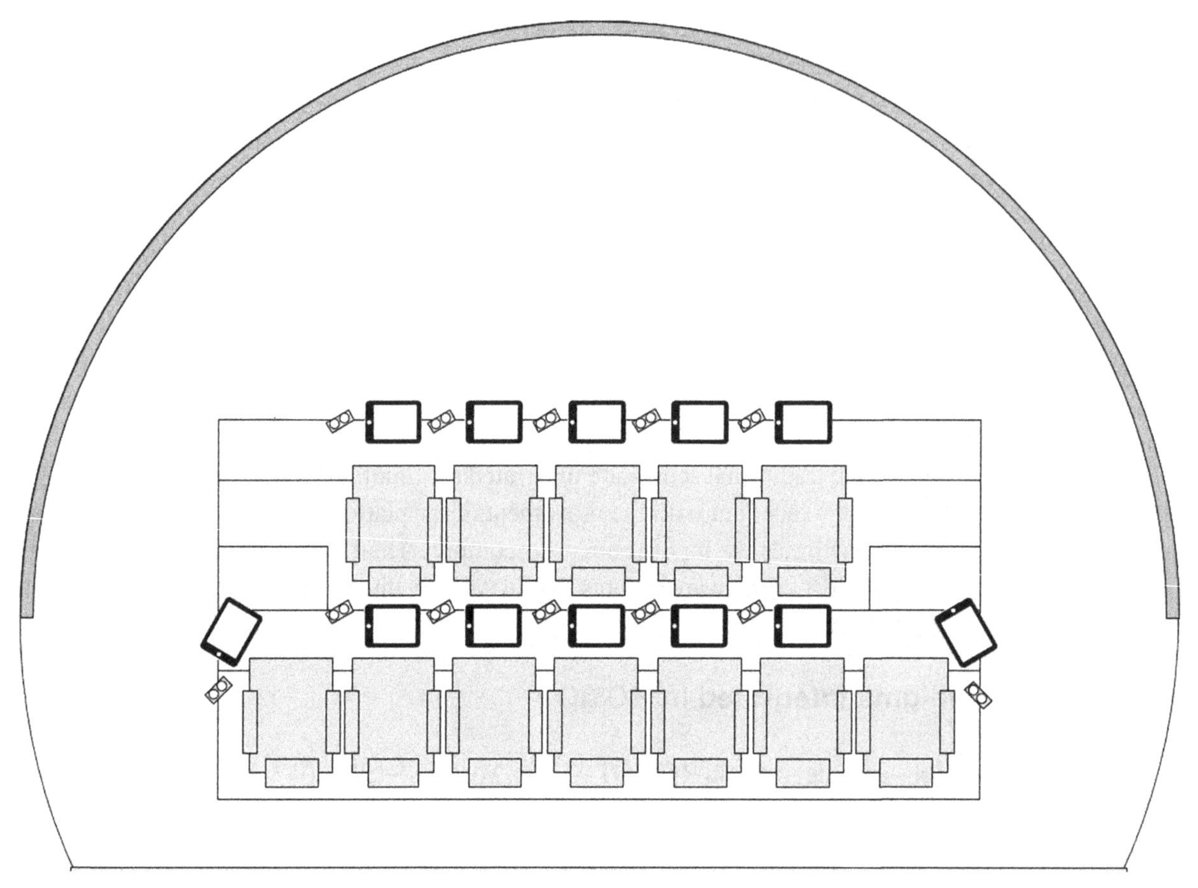

sion. The system stores the individual answers chosen by users, and give scores if they are correct. When the deployed situation finishes, users can see their individual score on the big curved screen. The people having the greatest score are the winners who are somehow rewarded by the system by receiving a special visit, a 3D virtual character that congratulates them for their safe driving (see next paragraph).

- *Augmented Reality Mirror scene*: The AR application integrated in ROMOT makes use of the mirror metaphor, earlier described in (Giner Martínez & Portalés Ricart, 2005; Portalés, Gimeno, Casas, Olanda, & Giner, 2016), and therefore it can be referred as an Augmented Reality Mirror (ARM) which, in our case, is also seen stereoscopically thanks to the 3D glasses. ARMs can bring a further step in user immersion, as the audience can see a real-time image of them in real time, spatially aligned with virtual objects, and thus feel part of the created environment. In the case of ROMOT, the implemented ARM application is triggered in the final scenes of the aforementioned VR interactive environment (previous subsection), where the user(s) with the highest score get(s) a kind of reward that consists of a virtual character that walks towards him/her/them. Together with this action, virtual confetti and coloured stars appear on the environment, also spa-

Figure 4. Example of the virtual environment for driving safety, where: (top image) the system asks users, and then they have to choose one of the given answers; (bottom image) shows an example of a wrong decision

tially aligned with the image of the real world, and accompanied with winning music that includes applauses. Therefore, in this application the stimuli that take part are the sight and the hearing.

- *Mixed reality movies*: This application provides an environment where a combination of recorded videos and virtual content is displayed, thus creating what we call a "MR movie". As the videos display a real environment and the virtual content is spatially aligned with it, we create the effect that the virtual content belongs to the real world, smoothing the transition from a real movie to a virtual situation. In this case, the created virtual content is a virtual character that interacts with parts of the video by creating the virtual animation in such a way that it is synchronized with the contents of the recorded real scene. Virtual shadows are also considered to make the whole scene more real (Figure 6). The content (videos and avatar) can be seen stereoscopically, as the videos have been recorded with a pair of cameras and stereoscopic images can also be created for the virtual character. For the videos, we used two GoPro cameras attached to a car's hood to create a 3D movie set in the streets of a city (Figure 7). Accompanying the videos, there is audio consisting of ambient sounds and/or a locution that reinforces the images users are watching. The synchronized soft platform movements or effects like a nice smell or a gentle breeze, help creating the perfect ambient at each part of the MR movie, making the experience more enjoyable for the audience.

Limitations of This Technology

ROMOT is a system that integrates a number of devices and interaction paradigms, and therefore multisensory experiences are possible. However, it has to be noted that the experience in ROMOT is shared, meaning that the system response is directed to the audience in general rather than to individual users. Although this fact is not a limitation per se, but a characteristic of the system, it makes it difficult for

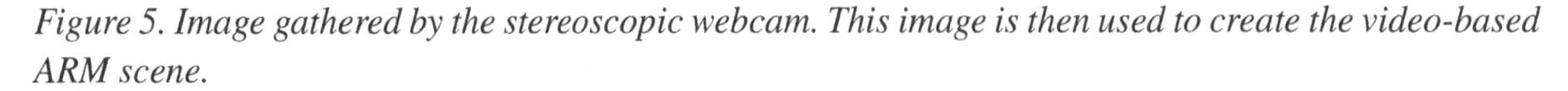

Figure 5. Image gathered by the stereoscopic webcam. This image is then used to create the video-based ARM scene.

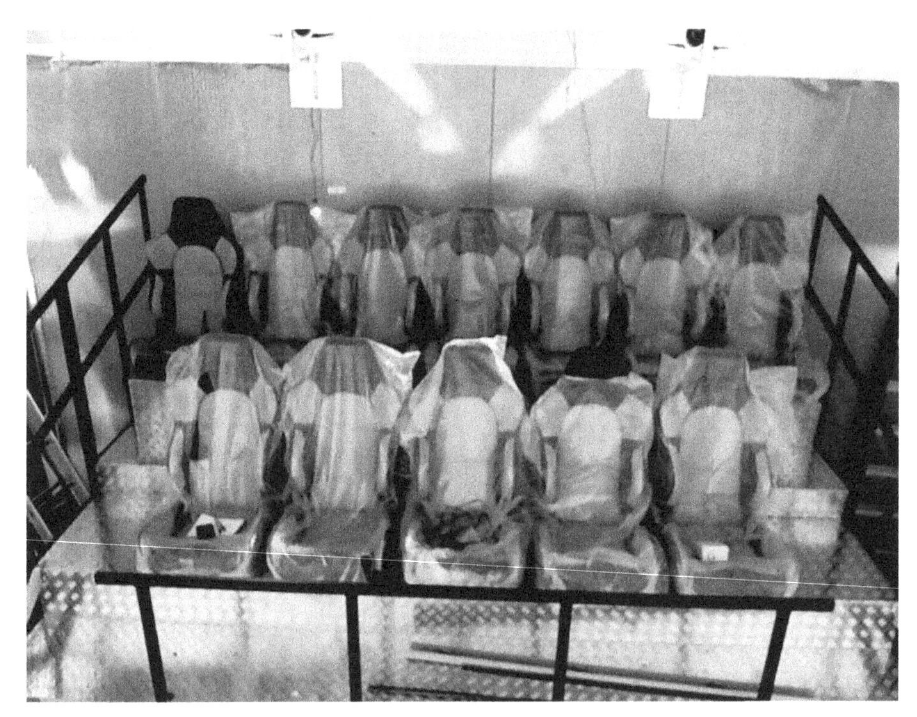

the system to quickly react to the audience inputs. This is evidenced in the VR interactive environment setup, where individuals are requested to choose one of the given answers within a short period of time. The system only gives a feedback when the time is over. Additionally, the system reacts to the most voted option, meaning that other options are not considered.

Another limitation of this technology is related to the rendering of the virtual world, as it is projected on a curved screen rather than on head mounted displays. Because of this fact, users are not completely isolated from the real world, and thus the experience can be referred to as semi-immersive rather than completely immersive.

Figure 6. A virtual character inserted in the video recording to form the MR movies. In this example, the black area at the bottom of the image is intended to increase the stereoscopic effect of the virtual character.

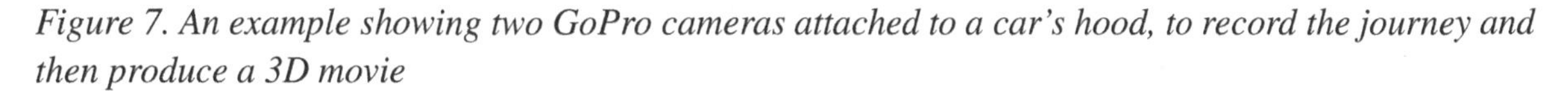

Figure 7. An example showing two GoPro cameras attached to a car's hood, to record the journey and then produce a 3D movie

FUTURE DIRECTIONS

As future directions of the aforementioned technologies (including sensors and displays), we foresee an increasing number of small and medium devices providing extended computer-based stimuli, which can reach the consumer level, for instance, being integrated in smartphones. These coming advances will have the potential to provide users with more realistic experiences, resulting in a more natural human-computer interaction, as devices get more powerful and capable of creating higher quality synthetic stimuli. Shared experiences are also possible with smartphones, as users can be connected and participate in a single virtual/augmented reality. Additionally, nowadays there exist some results integrating synthetic touch, taste and smell, such as the ones described in Obrist et al. (2016), that anticipate a variety of new capabilities. This will increase our understanding of how humans can navigate and interact within VR or AR environments, making it possible more natural channels for interacting and exploring virtual worlds. However, not all stimuli might be able to be integrated in small and medium devices. For instance, the vestibular sense, that makes users feel the movements and vibrations, would probably need other devices in order to be fostered.

On the other hand, the forthcoming advancements in computer vision and artificial intelligence will have a direct impact in VR and AR. The combination of computer vision and artificial intelligence allows computers to detect and identify objects, increasing the understanding of what cameras capture. Computer vision is already a key operation in AR, as it plays an important role in the location of the camera and the alignment of the virtual and real worlds. Within a greater understanding of what the camera captures, the location and alignment operations will be improved. Additionally, it will be possible to identify and thus label the objects that users see on their field of views.

CONCLUSION

In this chapter, we have dealt with the state of the art in multisensory systems, which can take part in applications with a variety of interaction paradigms. We have reviewed the maturity of the display devices providing stimuli through our senses, not only the traditional ones (sight, hearing, olfactory, etc), but from a more modern concept that includes more senses (e.g. vestibular, kinesthetic). By bringing some examples, we have addressed the fact that some of the technology is not yet ready to reach consumer-grade applications (e.g. taste, smell).

Additionally, we have also dealt with the paradigms of Virtual Reality (VR), Augmented Reality (AR) and, in a broader sense, Mixed Reality (MR). In this chapter, we have used these terms in their academic-meaning, following the concept of the reality-virtuality continuum (Milgram & Kishino, 1994), but we are aware that other uses of these terms are emerging - sometimes, for commercial purposes - and new concepts are arising, as it is the case of the term Extended Reality (ER). In this sense, we believe that a clearer taxonomy is needed, which is able to embed the variety of new technologies and interaction paradigms.

As a case study, we have presented ROMOT, a RObotized 3D-MOvie Theatre for which VR, AR and MR applications have been built. ROMOT involves the computational stimuli of sight, hearing, smell, touch and kinaesthetic, so it can be referred to as a multisensory system with different interaction paradigms. It is important to address that this system has been built at the laboratory level. Although there are other similar systems which are intended to provide extended cinematic experiences to audiences, they are usually centred in the VR technology rather than embedding other interaction paradigms. In this sense, we can say that ROMOT is quite singular.

REFERENCES

Anthes, C., Garcia-Hernandez, R. J., Wiedemann, M., & Kranzlmuller, D. (2016). State of the art of virtual reality technology. *2016 IEEE Aerospace Conference*, 1–19. 10.1109/AERO.2016.7500674

Azuma, R. T. (1997). A Survey of Augmented Reality. *Presence (Cambridge, Mass.)*, *6*(4), 355–385. doi:10.1162/pres.1997.6.4.355

Belmonte, Ó., Pérez, M., Pérez, M., & Romero, C. (1998). *A NEW CONCEPT IN CERAMIC SPECIFICATION USING VIRTUAL ENVIRONMENTS*. Presented at the Qualicer. Retrieved from http://www.qualicer.org/recopilatorio/ponencias/pdfs/9823032e.pdf

Bimber, O., & Raskar, R. (2005). Spatial Augmented Reality: Merging Real and Virtual Worlds (0 ed.). doi:10.1201/b10624

Botella, C., Bretón-López, J., Quero, S., Baños, R., & García-Palacios, A. (2010). Treating Cockroach Phobia With Augmented Reality. *Behavior Therapy*, *41*(3), 401–413. doi:10.1016/j.beth.2009.07.002 PMID:20569788

Casas, S., Coma, I., Portalés, C., & Fernández, M. (2016). Towards a simulation-based tuning of motion cueing algorithms. *Simulation Modelling Practice and Theory*, *67*, 137–154. doi:10.1016/j.simpat.2016.06.002

Casas, S., Olanda, R., & Dey, N. (2017). Motion Cueing Algorithms: A Review: Algorithms, Evaluation and Tuning. [IJVAR]. *International Journal of Virtual and Augmented Reality*, *1*(1), 90–106. doi:10.4018/IJVAR.2017010107

Casas, S., Portalés, C., García-Pereira, I., & Fernández, M. (2017). On a first evaluation of romot—A robotic 3D movie theatre—For driving safety awareness. *Multimodal Technologies and Interaction*, *1*(2), 6. doi:10.3390/mti1020006

Casas, S., Portalés, C., Morillo, P., & Fernández, M. (2018). A particle swarm approach for tuning washout algorithms in vehicle simulators. *Applied Soft Computing*, *68*, 125–135. doi:10.1016/j.asoc.2018.03.044

Casas, X., Herrera, G., Coma, I., & Fernández, M. (2012). *A Kinect-based Augmented Reality System for Individuals with Autism Spectrum Disorders*. GRAPP/IVAPP.

Cheok, A., & Portalés Ricart, C. (2016). Welcome to MTI—A New Open Access Journal Dealing with Blue Sky Research and Future Trends in Multimodal Technologies and Interaction. *Multimodal Technologies and Interaction*, *1*(1), 1. doi:10.3390/mti1010001

Costa, J. P., Robb, J., & Nacke, L. E. (2014). Physiological acrophobia evaluation through in vivo exposure in a VR CAVE. *2014 IEEE Games Media Entertainment*, 1–4. doi:10.1109/GEM.2014.7047969

Cruz-Neira, C., Fernández, M., & Portalés, C. (2018). Virtual Reality and Games. *Multimodal Technologies and Interaction*, *2*(1), 8. doi:10.3390/mti2010008

Di Fuccio, R., Ponticorvo, M., Ferrara, F., & Miglino, O. (2016). Digital and Multisensory Storytelling: Narration with Smell, Taste and Touch. In K. Verbert, M. Sharples, & T. Klobučar (Eds.), *Adaptive and Adaptable Learning* (pp. 509–512). Springer International Publishing. doi:10.1007/978-3-319-45153-4_51

4. DX | Absolute Cinema Experience. (n.d.). Retrieved March 5, 2019, from http://www.cj4dx.com/

FEELREAL. (2019). Multisensory VR Mask. Retrieved July 12, 2019, from https://feelreal.com/

García-Pereira, I., Gimeno, J., Portalés, C., Vidal-González, M., & Morillo, P. (2018). *On the Design of a Mixed-Reality Annotations Tool for the Inspection of Pre-fab Buildings*. doi:10.2312/ceig.20181157

García-Pereira, I., Gimeno, J., Pérez, M., Portalés, C., & Casas, S. (2018). MIME: A Mixed-Space Collaborative System with Three Immersion Levels and Multiple Users. In *Proceedings 2018 IEEE International Symposium on Mixed and Augmented Reality Adjunct (ISMAR-Adjunct)*, 179–183. 10.1109/ISMAR-Adjunct.2018.00062

Gimeno, J., Olanda, R., Martinez, B., & Sanchez, F. M. (2011). Multiuser Augmented Reality System for Indoor Exhibitions. In P. Campos, N. Graham, J. Jorge, N. Nunes, P. Palanque, & M. Winckler (Eds.), Human-Computer Interaction – INTERACT 2011 (pp. 576–579). Berlin, Germany: Springer.

Gimeno, J., Portalés, C., Coma, I., Fernández, M., & Martínez, B. (2017). Combining traditional and indirect augmented reality for indoor crowded environments. A case study on the Casa Batlló museum. *Computers & Graphics*, *69*, 92–103. doi:10.1016/j.cag.2017.09.001

Giner Martínez, F., & Portalés Ricart, C. (2005, October 3). *The Augmented User: A Wearable Augmented Reality Interface*. 417–426. Hal Thwaites.

Giner Martínez, F., & Portalés Ricart, C. (2007). Augmented teaching (L. Gómez Chova, D. Martí Belenguer, & I. Candel Torres, Eds.). International Association of Technology, Education, and Developement (IATED).

Heilig, M. L. (1962). *United States Patent No. US3050870A*. Retrieved from https://patents.google.com/patent/US3050870A/en

Heilig, M. L. (1992). EL Cine del Futuro: The Cinema of the Future. *Presence (Cambridge, Mass.)*, *1*(3), 279–294. doi:10.1162/pres.1992.1.3.279

Ishii, H., Kobayashi, M., & Arita, K. (1994). Iterative Design of Seamless Collaboration Media. *Communications of the ACM*, *37*(8), 83–97. doi:10.1145/179606.179687

Ismail, A. W., Billinghurst, M., Sunar, M. S., & Yusof, C. S. (2019). Designing an Augmented Reality Multimodal Interface for 6DOF Manipulation Techniques. In K. Arai, S. Kapoor, & R. Bhatia (Eds.), *Intelligent Systems and Applications* (pp. 309–322). Springer International Publishing. doi:10.1007/978-3-030-01054-6_22

Kim, R. S., Seitz, A. R., & Shams, L. (2008). Benefits of Stimulus Congruency for Multisensory Facilitation of Visual Learning. *PLoS One*, *3*(1). doi:10.1371/journal.pone.0001532 PMID:18231612

Lehmann, S., & Murray, M. M. (2005). The role of multisensory memories in unisensory object discrimination. *Brain Research. Cognitive Brain Research*, *24*(2), 326–334. doi:10.1016/j.cogbrainres.2005.02.005 PMID:15993770

Luria, A. R. (1987). *The mind of a mnemonist: A little book about a vast memory*. Cambridge, MA, US: Harvard University Press.

Maj. Loren mer. (2012). *Virtual reality used to train Soldiers in new training simulator*. Retrieved from https://www.army.mil/article/84453/virtual_reality_used_to_train_soldiers_in_new_training_simulator

Martins, J., Gonçalves, R., Branco, F., Barbosa, L., Melo, M., & Bessa, M. (2017). A multisensory virtual experience model for thematic tourism: A Port wine tourism application proposal. *Journal of Destination Marketing & Management*, *6*(2), 103–109. doi:10.1016/j.jdmm.2017.02.002

Milgram, P., & Kishino, F. (1994). A Taxonomy of Mixed Reality Visual Displays. *IEICE Transactions on Information and Systems*, *E77-D*(12). Retrieved from http://vered.rose.utoronto.ca/people/paul_dir/IEICE94/ieice.html

Narumi, T., Nishizaka, S., Kajinami, T., Tanikawa, T., & Hirose, M. (2011). Augmented Reality Flavors: Gustatory Display Based on Edible Marker and Cross-modal Interaction. *Proceedings of the SIGCHI Conference on Human Factors in Computing Systems*, 93–102. 10.1145/1978942.1978957

Obrist, M., Comber, R., Subramanian, S., Piqueras-Fiszman, B., Velasco, C., & Spence, C. (2014). Temporal, Affective, and Embodied Characteristics of Taste Experiences: A Framework for Design. *Proceedings of the SIGCHI Conference on Human Factors in Computing Systems*, 2853–2862. 10.1145/2556288.2557007

Obrist, M., Velasco, C., Vi, C., Ranasinghe, N., Israr, A., Cheok, A. D., ... Gopalakrishnakone, P. (2016). Sensing the future of HCI: Touch, taste, and smell user interfaces. *Interaction*, *23*(5), 40–49. doi:10.1145/2973568

Paivio, A. (1991). Dual coding theory: Retrospect and current status. *Canadian Journal of Psychology/ Revue. Canadian Psychology*, *45*(3), 255–287. doi:10.1037/h0084295

Pix 5D Cinema. (n.d.). Retrieved March 5, 2019, from https://expressavenue.in/?q=store/pix-5d-cinema

Ponzo, S., Kirsch, L. P., Fotopoulou, A., & Jenkinson, P. M. (2018). Balancing body ownership: Visual capture of proprioception and affectivity during vestibular stimulation. *Neuropsychologia*, *117*, 311–321. doi:10.1016/j.neuropsychologia.2018.06.020 PMID:29940194

Portalés, C., Casas, S., Vidal-González, M., & Fernández, M. (2017). On the Use of ROMOT—A RObotized 3D-MOvie Theatre—To Enhance Romantic Movie Scenes. *Multimodal Technologies and Interaction*, *1*(2), 7. doi:10.3390/mti1020007

Portalés, C., Gimeno, J., Casas, S., Olanda, R., & Giner, F. (2016). Interacting with augmented reality mirrors. In J. Rodrigues, P. Cardoso, J. Monteiro, & M. Figueiredo (Eds.), Handbook of Research on Human-Computer Interfaces, Developments, and Applications (pp. 216–244). Hershey, PA: IGI-Global. doi:10.4018/978-1-5225-0435-1.ch009

Portalés, C., & Perales, C. D. (2009). Sound and Movement Visualization in the AR-Jazz Scenario. 167–172. Springer-Verlag.

Portalés, C., Viñals, M. J., Alonso-Monasterio, P., & Morant, M. (2010). AR-Immersive Cinema at the Aula Natura Visitors Center. *IEEE MultiMedia*, *17*(4), 8–15. doi:10.1109/MMUL.2010.72

Portalés Ricart, C. (2009). *Entornos multimedia de realidad aumentada en el campo del arte*. Valencia, Spain: Universidad Politécnica de Valencia.

Portalés Ricart, C., Perales Cejudo, C. D., & Cheok, A. (2007). Exploring Social [ACM SIGCHI.]. *Cultural and Pedagogical Issues in AR-Gaming Through The Live LEGO House*, *203*, 238–239.

Ranasinghe, N., Karunanayaka, K., Cheok, A. D., Fernando, O. N. N., Nii, H., & Gopalakrishnakone, P. (2011). Digital Taste and Smell Communication. *Proceedings of the 6th International Conference on Body Area Networks*, 78–84. Retrieved from http://dl.acm.org/citation.cfm?id=2318776.2318795

Rekimoto, J., & Nagao, K. (1995). *The world through the computer: computer augmented interaction with real world environments*. 29–36. doi:10.1145/215585.215639

Robinett, W. (1994). Interactivity and individual viewpoint in shared virtual worlds: The big screen vs. networked personal displays. *Computer Graphics*, *28*(2), 127–130. doi:10.1145/178951.178969

Saidi, M., Lari, A. A., & Towhidkhah, F. (2013). Comparison of visual and proprioceptive training on multisensory perception using a new designed setup. In *Proceedings 21st Iranian Conference on Electrical Engineering (ICEE)*, Mashhad, pp. 1-4. 10.1109/IranianCEE.2013.6599873

Shams, L., & Seitz, A. R. (2008). Benefits of multisensory learning. *Trends in Cognitive Sciences*, *12*(11), 411–417. doi:10.1016/j.tics.2008.07.006 PMID:18805039

Studios, B. L. (n.d.). The Night Cafe. Retrieved July 12, 2019, from https://www.borrowedlightvr.com/the-night-cafe/

Technology, O. (2016). Olorama. Retrieved from http://www.olorama.com/en/

The most extreme examples of Virtual Reality: Immersive Experience. (2017, November 6). Retrieved July 10, 2019, from Premo website: https://3dcoil.grupopremo.com/blog/extreme-examples-virtual-reality-immersive-experience/

Tryon, C. (2013). Reboot cinema. *Convergence, 19*(4), 432–437. doi:10.1177/1354856513494179

Tsai, W.-L., Hsu, Y.-L., Lin, C.-P., Zhu, C.-Y., Chen, Y.-C., & Hu, M.-C. (2016). Immersive Virtual Reality with Multimodal Interaction and Streaming Technology. *Proceedings of the 18th ACM International Conference on Multimodal Interaction*, 416–416. 10.1145/2993148.2998526

Vera, L., Gimeno, J., Coma, I., & Fernández, M. (2011). Augmented Mirror: Interactive Augmented Reality System Based on Kinect. In P. Campos, N. Graham, J. Jorge, N. Nunes, P. Palanque, & M. Winckler (Eds.), In *Human-Computer Interaction – INTERACT 2011: 13th IFIP TC 13 International Conference Proceedings, Part IV* (pp. 483–486), Lisbon, Portugal, September 5-9, 2011. 10.1007/978-3-642-23768-3_63

Vlahakis, V., Ioannidis, N., Tsotros, M., Stricker, D., Gleue, T., Daehne, P., & Almeida, L. (2002). Archeoguide: An Augmented Reality Guide for Archaeological Sites. *IEEE Computer Graphics and Applications*, 9.

Vosmeer, M., & Schouten, B. (2014). Interactive Cinema: Engagement and Interaction. In A. Mitchell, C. Fernández-Vara, & D. Thue (Eds.), *Interactive Storytelling* (pp. 140–147). Springer International Publishing.

Wang, D., Ohnishi, K., & Xu, W. (2019). Multimodal Haptic Display for Virtual Reality: A Survey. *IEEE Transactions on Industrial Electronics*, 1–1. doi:10.1109/TIE.2019.2920602

Weichert, F., Bachmann, D., Rudak, B., & Fisseler, D. (2013). Analysis of the Accuracy and Robustness of the Leap Motion Controller. *Sensors (Basel), 13*(5), 6380–6393. doi:10.3390130506380 PMID:23673678

Yecies, B. (2016). Transnational collaboration of the multisensory kind: Exploiting Korean 4D cinema in China. *Media International Australia, 159*(1), 22–31. doi:10.1177/1329878X16640104

Zhang, Q., Guo, Y., Laffont, P., Martin, T., & Gross, M. (2017). A Virtual Try-On System for Prescription Eyeglasses. *IEEE Computer Graphics and Applications, 37*(4), 84–93. doi:10.1109/MCG.2017.3271458 PMID:28829296

ADDITIONAL READING

Alhamid, M. F., Eid, M., & El Saddik, A. (2012, 18-19 May 2012). *A multi-modal intelligent system for biofeedback interactions.* Paper presented at the Medical Measurements and Applications Proceedings (MeMeA), 2012 IEEE International Symposium on.

ARToolKit community. (2015). ARToolKit - Innovation Through Community Retrieved 20 July, 2015, from http://artoolkit.org/

Bower, M., Howe, C., McCredie, N., Robinson, A., & Grover, D. (2014). Augmented Reality in education – cases, places and potentials. *Educational Media International, 51*(1), 1–15. doi:10.1080/09523987.2014.889400

Cheng, C.-M., Chung, M.-F., Yu, M.-Y., Ouhyoung, M., Chu, H.-H., & Chuang, Y.-Y. (2008). *Chromirror: a real-time interactive mirror for chromatic and color-harmonic dressing*. Paper presented at the CHI '08 Extended Abstracts on Human Factors in Computing Systems, Florence, Italy. 10.1145/1358628.1358762

Difei, T., Juyong, Z., Ketan, T., Lingfeng, X., & Lu, F. (2014, 14-18 July 2014). *Making 3D Eyeglasses Try-on practical.* Paper presented at the Multimedia and Expo Workshops (ICMEW), 2014 IEEE International Conference on.

Eisert, P., Fechteler, P., & Rurainsky, J. (2008, 23-28 June 2008). *3-D Tracking of shoes for Virtual Mirror applications.* Paper presented at the Computer Vision and Pattern Recognition, 2008. CVPR 2008. IEEE Conference on.

Follmer, S., Leithinger, D., Olwal, A., Hogge, A., & Ishii, H. (2013). *inFORM: dynamic physical affordances and constraints through shape and object actuation.* Paper presented at the Proceedings of the 26th annual ACM symposium on User interface software and technology, St. Andrews, Scotland, United Kingdom. 10.1145/2501988.2502032

Fuchs, H., Livingston, M., Raskar, R., Colucci, D. n., Keller, K., State, A., . . . Meyer, A. (1998). Augmented reality visualization for laparoscopic surgery. In W. Wells, A. Colchester & S. Delp (Eds.), Medical Image Computing and Computer-Assisted Interventation — MICCAI'98 (Vol. 1496, pp. 934-943): Springer Berlin Heidelberg. doi:10.1007/BFb0056282

Fujinami, K., Kawsar, F., & Nakajima, T. (2005). *AwareMirror: a personalized display using a mirror.* Paper presented at the Proceedings of the Third international conference on Pervasive Computing, Munich, Germany. 10.1007/11428572_19

Goel, M., Whitmire, E., Mariakakis, A., Saponas, T. S., Joshi, N., Morris, D., . . . Patel, S. N. (2015). *HyperCam: hyperspectral imaging for ubiquitous computing applications.* Paper presented at the Proceedings of the 2015 ACM International Joint Conference on Pervasive and Ubiquitous Computing, Osaka, Japan. 10.1145/2750858.2804282

Hadalová, Z., & Samuelčík, M. Varhaníková (2014). *Augmented Map Presentation of Cultural Heritage Sites.* Paper presented at the Proceedings of the International Conference on Current Issues of Science and Research in the Global World (Vienna, Austria). Dec. 2014, 345-349.

Hilsmann, A., & Eisert, P. (2009). *Tracking and Retexturing Cloth for Real-Time Virtual Clothing Applications.* Paper presented at the Proceedings of the 4th International Conference on Computer Vision/Computer Graphics CollaborationTechniques, Rocquencourt, France. 10.1007/978-3-642-01811-4_9

HITLab. (2003). ARToolKit Retrieved 20 July, 2015, from http://www.hitl.washington.edu/artoolkit/

Kim, I.-J., Lee, H. J., & Kim, H.-G. (2004). *Magic mirror: a new VR platform design and its applications.* Paper presented at the Proceedings of the 2004 ACM SIGCHI International Conference on Advances in computer entertainment technology, Singapore. 10.1145/1067343.1067394

Ochiai, Y. (2014). Pixie Dust: Graphic generated by Levitated and Animated Objects in Computational Acoustic-Potential Field Retrieved 20 July, 2015, from http://96ochiai.ws/PixieDust/

Ochiai, Y., Hoshi, T., & Rekimoto, J. (2014). Pixie dust: Graphics generated by levitated and animated objects in computational acoustic-potential field. *ACM Transactions on Graphics*, *33*(4), 1–13. doi:10.1145/2601097.2601118

Pachoulakis, I., & Kapetanakis, K. (2012). Augmented Reality Platforms for Virtual Fitting Rooms. *The International Journal of Multimedia & Its Applications*, *4*(4), 35–46. doi:10.5121/ijma.2012.4404

Suthar, K., Benmore, C. J., Den Hartog, P., Tamalonis, A., & Weber, R. (2014, 3-6 Sept. 2014). *Levitating water droplets formed by mist particles in an acoustic field.* Paper presented at the Ultrasonics Symposium (IUS), 2014 IEEE International.

Yuan, M., Khan, I. R., Farbiz, F., Niswar, A., & Huang, Z. (2011). *A mixed reality system for virtual glasses try-on*. Paper presented at the Proceedings of the 10th International Conference on Virtual Reality Continuum and Its Applications in Industry, Hong Kong, China. 10.1145/2087756.2087816

KEY TERMS AND DEFINITIONS

Augmented Reality (AR): The technology that simultaneously combines real and virtual objects that are interactive in real-time and are registered in a three-dimensional space.

Digital Representation of Data: Reproduction of data by means of digital forms, as enabled by computers.

Interaction: A kind of action that occurs as two or more objects have an effect upon one another.

Mixed Reality: The result of blending the physical world with a synthetic one, including the paradigms of Augmented Reality and Augmented Virtuality.

Multisensory: Involving or using more than one of the senses when interacting with the real or synthetic worlds.

Tangible Assets: Assets that have a physical form, including buildings, places, monuments, tools, artifacts, etc.

Three-Dimensional (3D) Data: Data that is realized in a three dimensional space, a geometric setting in which three values are required to determine the position of an element.

Virtual Reality (VR): The technology by which a user, stimulated with computer-generated perceptual cues, experiences an alternative reality that is different from the one he/she actually lives in.

Chapter 15
Virtual and Augmented Reality for the Visualization of Summarized Information in Smart Cities:
A Use Case for the City of Dubai

Sergio Casas
https://orcid.org/0000-0002-0396-4628
Universitat de València, Spain

Pablo Casanova-Salas
https://orcid.org/0000-0003-1588-9888
Universitat de València, Spain

Jesús Gimeno
https://orcid.org/0000-0002-5123-7580
Universitat de València, Spain

José V. Riera
Universitat de València, Spain

Cristina Portalés
https://orcid.org/0000-0002-4520-2250
Universitat de València, Spain

ABSTRACT

In this chapter, authors deal with the problem of visualizing summarized information in a complex system like a smart city. They introduce the topic of smart city in the context of the information revolution that is taking place in the world. Next, they review how this information can be visualized, highlighting immersive 3D methods such as Virtual Reality (VR) and Augmented Reality (AR), which are particularly suitable for these applications, since 2D information does not usually induce a focused and sustained attention. The chapter describes and shows a use case in which VR and Spatial AR (SAR) are used in a smart city system to visualize summarized information about the state and management of the city. The SAR system relies on a multi-projector mapping procedure, and therefore authors also explain the technical details that the calibration and implementation of this type of AR application requires.

DOI: 10.4018/978-1-7998-2112-0.ch015

INTRODUCTION

In the world of today, hundreds of thousands of e-mails, tweets, internet searches, images, videos, etc. are generated every day. Although computers are also capable of generating structured automatic information, a sizeable amount of these data is directly or indirectly generated by human people, so they are usually unstructured information that is difficult to manage and visualize.

In addition, the expansion of the Ubiquitous Computing (UC) paradigm and the rise of the Internet of Things (IoT), and more recently the Internet of Everything (IoE) makes possible to have millions of sensors sending data continuously about different processes and elements of our life: meteorology (temperature, humidity, light, rain), health (heart beat rate, weight, body temperature, exercise), presence and safety, sound, dynamics (position, speed, acceleration, orientation, inclination), space availability, proximity, pressure, traffic, energy consumption, etc. This information is multi-localized, multimodal, massive and, although it is usually well structured, it is heterogeneous, since the way each piece of information must be discretized in binary form, stored and visualized could be very different depending on the information device, the protocol or the data type. In addition, wireless sensors and wireless sensor networks have revolutionized the information sector and allow to have countless *sensorization points* spread over large areas of the territory with multiple purposes, making the management of all these elements very challenging.

Although most of this information can be treated in an automatic way by Artificial Intelligence (AI) algorithms and human intervention could be minimal, it is also very important that human operators and technicians be able to extract meaningful conclusions from these data. Therefore, with so much information, it is very important that we not only find ways to use wisely this information but also find methods to visualize and summarize properly all this information.

This is especially important in smart cities. The term *smart city* refers to a type of urban development based on sustainability that is able to adequately respond to the basic needs of institutions, businesses and the inhabitants themselves, both economically and operationally, socially and environmentally. A smart city is one that utilizes information and communications technologies to achieve their objectives, so city managers and citizens are given access to a wealth of real-time information about the urban environment upon which to base decisions, actions, and future planning (Jin, Gubbi, Marusic, & Palaniswami, 2014). The concept of smart city is closely related to big data (Al Ghamdi & Thomson, 2018), IoT (Dinc, Kuscu, Bilgin, & Akan, 2019) and even to home automation (Portalés, Casas, & Kreuzer, 2019) although on a larger scale.

One of the aspects that encompasses smart cities and their different areas of intelligent management is smart resource management (water, energy, food, etc.) and smart real-time traffic management, also known as smart mobility.

A key to future success for cities both in economic and environmental terms is that public administrations seek reliable and affordable real-time information systems that allow them to design more efficient strategies for resource and mobility management, among many other things. Nevertheless, this smart design does not end in the acquisition, processing and automatic use of data. It is necessary to design also smart visualization systems to provide summarized information about all these processes. AI and automatic processing/management can be sometimes obscure and it is of the utmost importance that humans be able to see what is happening to both check what the system is doing and also understand what is going on, and in some particular cases, why.

In this regard, this chapter has two major goals. First, we will review how the information about smart cities can be visualized, such as in (Kim, Shin, Choe, Seibert, & Walz, 2012). Second, we will show a particular smart visualization system that is used to visually summarize the information about traffic and energy management in a smart city.

In the first goal, we will briefly review the smart city concept and then the different methods that can be used to perform the visualization of this type of applications. Then, we will focus on how smart cities can benefit from different immersive visualization paradigms, such as Virtual Reality (VR) (Sherman & Craig, 2018) and Augmented Reality (AR) (Billinghurst, Clark, & Lee, 2015).

The second goal will serve as a use case in which a smart visualization system is used. In this case, a Virtual Reality system and a Spatial Augmented Reality system (Bimber & Raskar, 2005) are used to visualize the information about traffic and energy management in the city of Dubai, United Arabs Emirates. The VR system is implemented with a head-mounted display, whereas the Spatial AR system is accomplished by mapping visual information onto a simplified scale model of the city. The design of the VR application has software implications, because the navigation, visualization and interaction has to be performed taking into account that the goal is not to provide a realistic reproduction of the city but to show information about it. The use of Spatial AR has several software and hardware implications. First, a proper mechanism to map the information onto the scale model is needed. This is accomplished by means of a projector mapping, which maps the image generated by a projector onto the scale model, enhancing it (hence the name Spatial AR). Second, if the scale model is large enough and/or there may be occlusion problems, several projectors will be needed, which implies that a mechanism to calibrate (Portalés, Casanova-Salas, Casas, Gimeno, & Fernández, 2019), align and/or blend the images from these projectors needs to be found. This process is detailed throughout the chapter.

STATE OF THE ART

Social Background

The urbanization that the world's population is experiencing seems to be an irreversible process (Yin et al., 2015). Although the concentration of people in cities makes some services easier (education, health, transportation, etc.) to be provided, it also creates ecological, environmental and energy resource management problems (among many others) that are increasingly challenging due to the existence of megacities and conurbations with more than 20 million inhabitants, such as Tokyo, Shanghai, Jakarta, Delhi, São Paulo or Mexico City. In this context is where the concept of smart city emerges.

There is not a standard definition for what a smart city is. The term started to appear in the 1990s (Anthopoulos, 2015; Moustaka, Vakali, & Anthopoulos, 2018) to account for the use of new technologies to transform the cities and address the urban challenges. One possible definition is "an innovative city that uses ICT and other means to improve quality of life, efficiency of urban operation and services, and competitiveness, while ensuring that it meets the needs of present and future generations with respect to economic, social, and environmental aspects". In this regard, a smart city usually addresses several dimensions: smart economy, smart mobility, smart environment, smart people, smart living and smart governance (Moustaka et al., 2018). Although the definition is rather broad, it is usually understood that a smart city is not just a city that utilizes digital technology. It should transform and improve the quality of life of its inhabitants and self-adapt to changes and users' needs.

The scale and extent to which ICT are applied in urban management and planning defines the complexity and the goals that can be achieved with a smart city. In any case, the major stakeholders for the successful implementation of a smart city are governments and local authorities. Each city authority should define the scope and scale of the smart city approach, the sources of data since data is the main resource of the process and the strategies and goals that are expected from the application of this process. Although the drivers of the implementation of a smart city are usually institutions and governments, it is important not to forget that citizens should be the center of the process, something that in most cases is not achieved (Cardullo & Kitchin, 2019). This also implies that citizens shall also be proactive in the way they deal with the smart management of their cities, changing their individual behavior in order to ease the application of the smart city concept (Tian, Li, & Chen, 2018).

In order to know more about the social aspects of smart cities, its problems, tensions and future trends, the reader can consult (Cocchia, 2014; Hollands, 2008; Martin, Evans, & Karvonen, 2018; Moustaka et al., 2018; Nam & Pardo, 2011) where the social implications of smart cities are discussed in detail, something that lies outside the aim of this chapter.

Uses and Tools

There are many applications under the umbrella term *smart city*. Each application of this paradigm has its own goals and tools. One of the main areas of application is mobility and traffic management (Latif, Afzaal, & Zafar, 2018; Pawłowicz, Salach, & Trybus, 2018). In this type of application, sensors (such as cameras, GPS, wireless sensor networks or RFID transponders) are used to monitor vehicles and roads. This way, a smart management system can recommend different routes to the drivers or change in real-time the state of traffic lights, in order to avoid traffic jams. In a similar way, it can be applied to parking (Lin, Lu, Tsai, & Chang, 2018; Sotres, de la Torre, Sánchez, Jeong, & Kim, 2018), so that drivers can know in advance where they can find free parking spots, avoiding using the car longer than necessary, polluting and wasting time.

Another main area is resource management and environmental control. This includes, among many others: water quality (Chen & Han, 2018), energy management (Pieroni, Scarpato, Di Nunzio, Fallucchi, & Raso, 2018), air quality (Dutta, Chowdhury, Roy, Middya, & Gazi, 2017) and noise control (Gaur, Scotney, Parr, & McClean, 2015). These applications are usually based on specific sensors that monitor the quality and/or the amount of the resource, so that the city can inform its citizens about the current levels. Cities can also implement policies to ensure that a certain threshold is not exceeded. These policies may also be automatic if the city has the ability to modify the use of the resource (for instance, the city lights can be dimmed down if too much energy is being consumed).

There are, of course, other applications such as tourism (Gretzel, Zhong, & Koo, 2016), health (Abdelaziz, Salama, & Riad, 2019; Hussain, Wenbi, da Silva, Nadher, & Mudhish, 2015), retail and shopping, building automation (inmotics), detection of catastrophes (tsunamis, wild fires, avalanches, etc.), but it is not the aim of this chapter to review all of them, since the applications are almost endless.

Technological Aspects

Leaving aside social, economic and political issues around the smart city vision, one of the main problems of this concept is that it is not easy to define what a smart city should encompass and how it can be implemented. One of the most detailed works about this problem is the one presented in (Jin et al., 2014) where the building blocks of a smart city are presented. The most important elements are a network, a cloud of services and, of course, the data. Different networks are suggested and a case study about noise mapping in Melbourne, Australia is presented. In (Gaur et al., 2015) a multi-level architecture is proposed based on semantic web technologies. This architecture relies heavily on modern wireless sensors networks and works by structuring the process in four layers: (i) data collection, (ii) data processing, (iii) data integration and (iv) reasoning and device control. This allows to create customized services of increasing complexity by connecting sensors, which are the source of information, to the different layers, either directly or indirectly. Another example of smart city architecture is the SEN2SOC platform presented in (Vakali, Anthopoulos, & Krco, 2014). This work, however, focuses on the interaction of environmental sensors and social data, which is applied and tested in the city of Santander, Spain (by means of the SmartSantander project), and cannot be considered a general architecture for smart cities. A similar goal is sought in (Braem et al., 2016), whose authors deploy a smart city testbed for monitoring the air quality in the city of Antwerp, Belgium. Another platform deployed around the SmartSantander project is the *City Data and Analytics Platform* (CiDAP) (Cheng, Longo, Cirillo, Bauer, & Kovacs, 2015). This architecture is designed to be scalable, flexible and extendable, so that it can be integrated with different smart city initiatives. CiDAP is based in three layers: (i) data sources, (ii) big data platform for smart cities and (iii) applications.

As it can be seen, there is not a single or reference architecture for smart cities. This is due to the fact that the goals of smart cities can be very different and a unified universal architecture is difficult to design. The same happens with the way these applications are visualized. Some of the smart city applications offer simple 2D graphics, such as in (Grodi, Rawat, & Rios-Gutierrez, 2016), or colored maps, such as in (Braem et al., 2016; Vakali et al., 2014). This kind of visualization could be effective in certain contexts but it is usually too simple and does not engage users. The use of maps is of course not casual. Location services and situational information are essential in smart cities. For this reason, the use of digital geographic systems are often key for a successful implementation of the smart city concept. Indeed, geomatics play a transversal role in smart cities. They offer location, routing, spatial context and mobility information. This can be provided by traditional Geographic Information Systems (GIS), by Global Navigation Satellite Systems (GNSS) or by Mobile Laser Scanning (MLS) and mobile video mapping (Daniel & Doran, 2013). GIS can be enhanced also by means of the use of Virtual Reality or Augmented Reality (Boulos, Lu, Guerrero, Jennett, & Steed, 2017).

The problem with systems that only rely on maps is that visualization turns unclear when the number of variables is increased (Estrada et al., 2018). More complex visualization system can be designed combining several views. A good example of this is VitalVizor (Zeng & Ye, 2018), an interactive visual system for urban planning that provides a combined interface based on a map view and a metric view. A different approach can be seen in (Kim et al., 2012), a visualization system for energy monitoring that integrates with Google Maps and Google Earth. This system combines choropleth maps with different levels of detail, three-dimensional depictions of buildings and floor levels, which are displayed in different colors denoting their energy consumption, and also custom data charts. Another example can be found in (Doraiswamy, Freire, Lage, Miranda, & Silva, 2018) where visual analytics is used to interactively explore spatiotemporal datasets. This work supports also 3D visualization and 3D data exploration.

The use of 3D reconstruction and 3D visualization is important when working in smart cities, since many events in a city have a vertical dimension. Examples of 3D-based visualization for smart cities are (Murshed, Al-Hyari, Wendel, & Ansart, 2018; Prandi, Soave, Devigili, Andreolli, & De Amicis, 2014; Prendinger et al., 2013; Wang, Jin, Xiao, Guo, & Li, 2012).

Virtual and Augmented Reality for Smart Cities

Given the importance of the accurate three-dimensional representation of events and people in smart city applications, Virtual Reality has also been considered for the interaction and visualization of this kind of tools. Compared to other visualization methods, Virtual Reality has the potential to engage all senses and attract user attention. However, the cost of truly immersive VR systems somehow hinders its widespread use. In addition, the use of VR with smart cities has particular considerations with respect to other areas of applications in which VR is also applied, such as entertainment (Kodama, Koge, Taguchi, & Kajimoto, 2017; Olanda, Pérez, Morillo, Fernández, & Casas, 2006), training (Aïm, Lonjon, Hannouche, & Nizard, 2016; S. Casas, Morillo, Gimeno, & Fernández, 2009) or awareness raising (Sergio Casas, Portalés, García-Pereira, & Fernández, 2017; Vera, Gimeno, Casas, García-Pereira, & Portalés, 2017). First, the simulated world is usually a real depiction of a real place, something that does not occur in other applications (although the level of realism of this virtual reconstruction could be diverse and depends on the particular intended use of the VR application). Therefore, the use of virtual reproductions of real places is needed, something that could be sometimes automatized with picture mapping and other techniques (Portalés, Casas, Alonso-Monasterio, & Viñals, 2018). In addition, some kind of connection with a GIS system is usually necessary (Boulos et al., 2017; Lv, Yin, Zhang, Song, & Chen, 2016), so that the location of the virtual user and other elements correspond to specific locations in the city, and real-time information from the city could be provided (this somehow violates the principles of VR, since this paradigm is designed to extract the user completely from his/her real environment). Finally, the VR system should be flexible enough to be able to represent data from different sources. In this regard, data interoperability is key to the success of these applications, although this could be tackled in other modules of the system different than the visualization module. Examples of the use of VR for smart cities can be found in (Bouloukakis, Partarakis, Drossis, Kalaitzakis, & Stephanidis, 2019; Hu, Wan, Wang, & Yu, 2012; Peng, Tan, Gao, & Yao, 2013).

Much in the same way, Augmented Reality can also be used in smart city applications. In fact, the features of this visualization and interaction paradigm make it a very interesting choice for the visualization of smart city information. Among the features that AR can provide, two stand out: the ability to enhance reality with new information and the presence of spatial cues that allow remote collaboration. Indeed, AR allows to generate virtual information (virtual objects) that can be spatially distributed according to the physical environment, so that different users could collaborate and have an enhanced perception of reality with information that would be otherwise impossible to achieve. Examples of the use of AR for smart cities are (Ramos, Trilles, Torres-Sospedra, & Perales, 2018), where the ARUJI application – a guiding application for students and visitors around the University of Castellón de la Plana, Spain - is presented, (Pokric, Krco, & Pokric, 2014), where AR is used for enhancing mobility, (Cho, Jang, Park, Kim, & Park, 2019), where it is used for energy monitoring and management, and (Zhang, Chen, Dong, & El Saddik, 2018), where a Microsoft HoloLens is utilized to visualize various types of data about Toronto, Canada. As in the case of Virtual Reality, the use of Augmented Reality depends heavily on the integration of geographic data and on the ability to work with heterogeneous data sources.

As it can be seen, the way each research team handles the visualization of data is different and depends on the application goals and on the importance that visualization is given in the context of the designed smart city application. Therefore, as (Laramee, Turkay, & Joshi, 2018) states, research about smart city visualization is still in its infancy since there are many unsolved problems and unexplored areas.

A USE CASE FOR THE CITY OF DUBAI

This section is dedicated to describe a VR and an AR visualization system for a smart city application. The city of Dubai, in the United Arab Emirates, is one of the leading economic hubs in the Middle East. Because of that and because of the high temperatures that can be experienced in this city, its water and energy supply requirements are very important. For similar causes, traffic is also a high concern.

For these reasons, local authorities have been involved for years in initiatives focused on making Dubai the world's smartest city (Dubai Corporation of Tourism & Commerce Marketing, 2019). In this context is where the visualization application described in this section is framed. The application shown in this chapter was implemented as a proof-of-concept for the Dubai Innovation Week (an important annual event where innovative solutions and developments of different sectors are shown) to demonstrate that both VR and AR could be used as technological solutions for the purpose of the smart visualization of the information gathered by Dubai local authorities.

Similarly to what is described in (Kim et al., 2012), we want to show the energy consumption, traffic and other data in an appealing way. However, unlike Kim et al., we choose to use Virtual Reality and Augmented Reality instead of a simple desktop interactive 3D application. Unlike other proposals, we provide an application that is capable of using both the VR and the AR paradigm to represent the same information. Moreover, both interfaces can work simultaneously.

Virtual Reality is usually defined as the technology by which one or several individuals experience the sensation of belonging to an alternative reality, which is not the one they are actually living in. This can be utilized to show a reality that does not exist, or as in the case of this work, to show a simplified version of the reality (the city) with new information (the smart city) that is not possible to observe if the user is in the real world. We, therefore, use a virtual reproduction of the city to show the user information about the state of the city. In this case, the user could be a regular citizen or a member of the city local authorities who needs, for instance, to monitor the water consumption in several parts of the city.

Augmented Reality, on the other hand, is a technology by which a view of the real world is augmented using synthetic images that are superimposed or projected over this image of the real world. AR systems need to provide three features to be considered as true AR applications (Azuma, 1997): (i) combine virtual and real objects; (ii) be completely interactive in real time and (iii) use virtual objects that are registered in 3D. AR is a transversal technology that can be used in many areas of knowledge, such as medicine (Yeom, 2011), retail (Bonetti, Warnaby, & Quinn, 2018), construction (Gimeno, Morillo, Casas, & Fernández, 2011), culture heritage (Gimeno, Portalés, Coma, Fernández, & Martínez, 2017) or learning (Ibáñez & Delgado-Kloos, 2018).

Spatial Augmented Reality (SAR) is a form of Augmented Reality in which the augmentation does not occur through a viewer or the use of a view of the world, such as a camera or a pair of glasses. Although all forms of AR are based on creating a link between real and virtual objects so that virtual elements appear to be merged into the real world, this blending usually occurs indirectly, i.e. it happens only when we visualize the real world through some kind of device which makes possible to "add" virtual objects to this view of the real world. This device is usually a camera mounted on a mobile device (Mobile AR), a special kind of mirror (Portalés, Gimeno, Casas, Olanda, & Martínez, 2016) or a special kind of glasses that allow to see both the real world and also virtual objects (see-through displays) that can be rendered on its surface. Instead, in SAR, the real world itself is used to project virtual elements on it. SAR exploits the use of large optical elements and video-projectors in order to provide virtual images blended with real objects (Bimber & Raskar, 2005). For this reason, SAR applications have the advantage of not needing that users wear any hardware. This does not mean that the hardware is less complex. On the contrary, the process of creating the augmented scene could be more complex than in other types of AR. However, the user is free of any hardware and several users can share the same experience.

Description of the Tool

The smart visualization tool that is presented here allows to observe the data and interact with the information in a contextualized way. This means that it shows each data on the real element to which it refers. The tool can show raw data or statistically processed data (averages, medians, peak values, etc.) for different time periods. As in this case the objective of the designed application is not to walk through the city, but to do a remote monitoring, two types of interfaces have been chosen: VR and SAR. This way, we offer two different interaction systems to visualize the same information. Both interfaces (VR and SAR) rely on a simplified reproduction of the city of Dubai, including landmark buildings of the city, such as the Burj Khalifa Building, the Khalid Al Attar Tower, the Rose Rotana Hotel, the Shangri-La Hotel or the Emirates Tower. Both interfaces (VR and SAR) can be used simultaneously and collaboratively, so that both the displayed data and its configuration are shared on both systems at the same time. This collaboration allows, for instance, that a user from the VR interface act as a controller and prepare the display of data to a set of users who observe them in the SAR system. This could be useful to explain a specific situation or make a decision. The two systems are described in detail in the following subsections.

Virtual Reality System (VR)

The Virtual Reality system uses a VR head-mounted display (Fig. 1), in this case an HTC Vive, to offer the user a first-person view of the city as if the user were in the real location. In this case, the link of the virtual data with the real locations is made showing the user a virtual representation of the real city. This

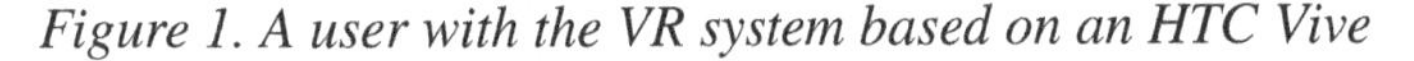

Figure 1. A user with the VR system based on an HTC Vive

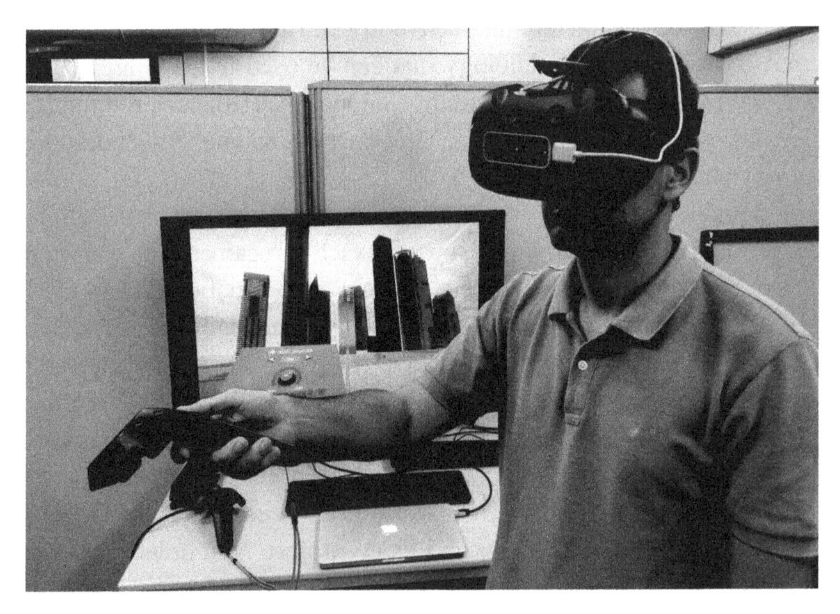

representation includes a sufficient level of detail so that the user can, for example, recognize specific buildings and places. However, it also maintains a very simple design so that the data is the predominant source of information in the image, since the goal of the application is not to offer a hyper-realistic reproduction of the city but to show summarized information about it.

For navigation purposes there are two different modes: teleportation and free flight. Both navigation modes are implemented using the wireless controller of the HTC Vive. These controllers implement a motion capture system of 6 degrees of freedom, so they allow to create interactions in which the user points or manipulates objects. The first mode, teleportation, consists of aiming with the controller to the position where you want to go. When the user presses the *trigger* button, the application shows a visual aid in the form of a line from the controller to the indicated position. The user can move the controller to point to the area where he/she wishes to go thanks to this graphic help, and release the button when it is already pointing to the desired destination point.

In the teleportation movement, a high-speed transition has been used. It is important to point out that this is not simply a change of position, but a movement at high speed, to give the user the notion of the direction and destination to where the system is teleporting the user. Unlike an instant game-like teleportation, this helps maintain the sense of orientation within the virtual city. In addition, no smooth acceleration and deceleration is implemented in this mode. The transportation movement is produced suddenly, without increasing or decreasing the velocity progressively (i.e., the velocity changes from zero to a sustained constant value, and then again to zero). This is done to avoid the feeling of dizziness that occurs when using accelerated movements in an immersive VR system (Rob Jagnow, 2017). Interestingly enough, an abrupt speed change produces a better sensation than smooth accelerations and decelerations.

The second mode is free flight; in this mode the user is not "tied to the ground" but can move flying in the direction pointed with the controller while pressing the *grip* button of the HTC Vive controller. This mode allows the user to observe the city from impossible perspectives in reality.

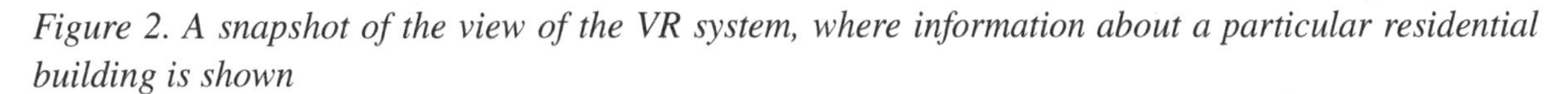

Figure 2. A snapshot of the view of the VR system, where information about a particular residential building is shown

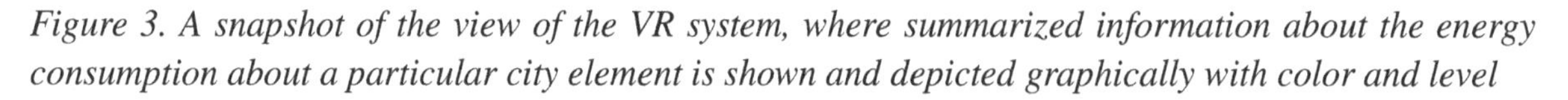

Figure 3. A snapshot of the view of the VR system, where summarized information about the energy consumption about a particular city element is shown and depicted graphically with color and level

For the interaction with the information within the VR system, two types of interaction modes are implemented: *point mode* and *control post*. The *point mode* allows the user to point to an element of the city and see the available information about it (Fig. 2). If the user points with the controller and presses the *menu* button, a menu appears next to the pointed element, in which the different options are presented to the user. The options of the menu can be selected by pointing with the controller or by using the integrated touchpad in each one of the controllers. In the second type of interaction, *control post*, once the desired type of information has been selected, a control post appears next to the user at all times (Fig. 3 and Fig. 4). This post is similar to an information panel where, in addition to showing details about the information, controls of three different types are shown: a button, a slider and a wheel. The user can interact with these controls by means of any of the controllers as if they were real controls. These controls allow changing the time interval at which data is retrieved, the type of data, etc.

Spatial Augmented Reality System (SAR)

The SAR system uses several projectors to add virtual information to a real scale model of the city. In this case, the model is the one that represents the reality so that the user understands which areas the virtual information refers to. The information represented by this system is similar to that of the VR system, although in this case the user is not the only one who observes it. Multiple users can observe virtual information about the smart city at the same time, at the cost of having to create a real (not virtual) scale model of the city. In this system, no helmet or glasses are necessary, but a set of projectors are used to "enlighten" the model with new information. Regarding the interaction in this system, it is similar to the previous one but using a mouse. Any user can use the mouse to move a projected pointer over the real scale model. Once the desired location has been selected, the options are shown in a menu similar to that of VR but projected on the floor of the model.

Figure 4. A snapshot of the view of the VR system, where summarized information about traffic congestion is shown. The red color of the cars indicates an important congestion. The panel allows choosing different time periods and target variables.

Regarding the information shown over the real elements, two types of information representation are used: one based on the color and one based on the level. These information representation systems are used in both the VR and the AR versions of the application.

The color-based representation serves to depict the value of a variable (e.g. water consumption, electrical power, traffic, air pollution, etc.) on a cold-warm scale, where the blue color (cold) symbolizes the minimum possible threshold of the value and the red color (hot) symbolizes the maximum possible threshold of that value. For instance, cars could be colored in red indicating an important traffic congestion (Fig. 4) and a building could be colored in blue indicating a small energy consumption. In the case of the level representation, buildings are used as a container and a color is applied from the bottom to the top to indicate the value of the data. The lowest part corresponds to the minimum threshold and the highest part to the maximum value. For instance, if a building is colored from bottom to half it means that the parameter that the system is showing over this building is at the central value within the possible range. These two types of information representation allow to observe several data values in multiple places simultaneously. They are also very easily interpreted and no effort is necessary to understand what is going on in that part of the city with just a glance. Therefore, they allow to observe the "behavior" of the city on a large scale, something very difficult to represent if data is shown, for instance, in the form of maps or tables in a conventional screen. In addition, the user can choose between observing a specific data in a specific place or observe it simultaneously in all places that have such type of information.

Projector Calibration and Mapping in the SAR Application

Although there are particular considerations that are important to keep in mind when implementing a VR-based smart visualization system (especially in the software part), the hardware implications in the case of the VR system are minimal due the use of a hardware setup that can be considered rather standard.

Figure 5. A simplified scale model of the city of Dubai during one of the projection tests performed to setup the multi-projector SAR application

However, in the case of the SAR-based system designed for this application, the hardware configuration is more complex and requires some additional details.

In a SAR system the virtual information is added to the real objects by means of projecting visual information through one or more projectors. In the case of this application, the virtual information must be projected on a scale model of a real city, and the information is projected using a set of four projectors placed at a height of one meter from the base of the model and placed at different locations around it. The scale model is created by 3D-printing (from CAD models) the city elements in white color (Fig. 5 and Fig. 6) so that the images generated by the projector are the ones that provide the final color to the scale model. Each projector covers the entire surface of the model, but the buildings produce occlusions with themselves (a projector only illuminates them on one or two sides) and also with the rest of the elements (buildings, roads, etc.). By using four projectors, we achieve that all the elements of the model, including all the faces of the buildings, are illuminated by at least one of the projectors, which avoids the problem of occlusions. This is the main reason for using more than one projector. More complex setups can be used if larger scale models are needed, so that several groups of projectors could be used to cover and map the different zones. All the projectors in the same group would cover the same area, whereas each group would cover a different area and certain overlapping zones would be defined to match the puzzle. However, in the case of this application, only one group of projectors is needed and all of them cover the same area but with different perspectives.

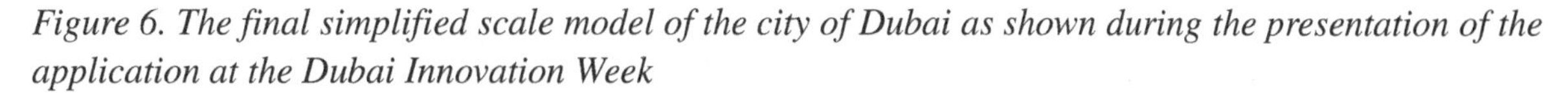

Figure 6. The final simplified scale model of the city of Dubai as shown during the presentation of the application at the Dubai Innovation Week

Once all the projectors are installed, it is necessary to calibrate them. The calibration process consists of calculating the necessary information so that later the virtual image corresponds spatially with the real objects. This means that the virtual information is projected correctly on each of the real elements to which this information is associated.

The projector calibration process consists of calculating both the parameters that describe the way in which the image is projected, called *intrinsic* parameters (field of view and aspect ratio / image format, etc.), and the parameters that define the position and orientation in which the projector is placed, called *extrinsic* parameters. Most methods in the literature use additional cameras to first perform a calibration of the intrinsic parameters and then a calibration of the extrinsic parameters. This is usually done using semi-automatic methods that project structured light patterns, which are analyzed through the camera (Portalés, Casas, Coma, & Fernández, 2017; Portalés, Orduña, Morillo, & Gimeno, 2019). In the case of our application, to simplify the calibration process and the hardware needed to perform the calibration process, the cameraless technique described in (Portalés, Casanova-Salas, et al., 2019) has been used. This technique is based on the use of at least 6 non-coplanar control points to simultaneously calculate the intrinsic and extrinsic parameters of the projector without the need of using a camera to capture light patterns (Fig. 7). Thanks to this method, the user does not need to have knowledge about computer vision to perform the calibration, but simply must drag the mouse to place the 6 control points in the desired

Figure 7. Projector calibration test. Three virtual blue cubes are mapped onto three real cubes, so that a series of points (number and colored in red) are required to be placed in the corners of the cubes. These points are used to calculate the intrinsic and extrinsic parameters of the optical device (a projector in this case)

position. The points are numbered in each of the projectors, and the user can move them throughout the model until they match the appropriate location (for instance, one of the corners of the base of a building). These control points have been chosen during the development and setup of this SAR application to cover the maximum possible area of the model and to avoid possible occlusions that may prevent a user from placing them correctly. For a detailed mathematical explanation of the operation, performance and accuracy of the calibration process, readers should consult (Portalés, Casanova-Salas, et al., 2019).

Once the four projectors have been calibrated, all the areas of the model would be illuminated with at least one of the projectors. However, there may be areas in which, for instance, the four projectors would provide visual information (light) at the same time. In this case, two solutions can be applied: (i) a pre-processing of the image of each projector so that a higher-level software takes care of avoiding these overlaps, (ii) keep the projection unmodified, producing some brighter areas. After testing both techniques, we opted for the second option, sacrificing the uniformity of the brightness of the image in favor of greater resistance to occlusions produced by elements external to the model. For instance, when the users are doing the inspection of the model, they often use their hands to point to other users the element they are talking about, producing occlusions between the projector and the spatially-augmented model. In the first case, these occlusions would cause shadowed zones in the model; however, in the second case, those shadowed areas are often compensated by the image projected by other or others projectors.

At this point it is, of course, necessary to describe how the projected virtual image is generated by each of the projectors. In a SAR application, the generation of the image typically follows the following steps (Fig. 8): (i) estimate (or track) the user's point of view; (ii) generate a virtual image from the user's point of view; (iii) project the virtual image on the (virtual) 3D geometry that represents the real scenario onto which it will be projected; (iv) generate the image (of the 3D geometry with the virtual

Figure 8. General scheme of a SAR application with the five steps (up) and the simplification that can be done in the case that the SAR application only adds virtual information to the surface of the real objects (down)

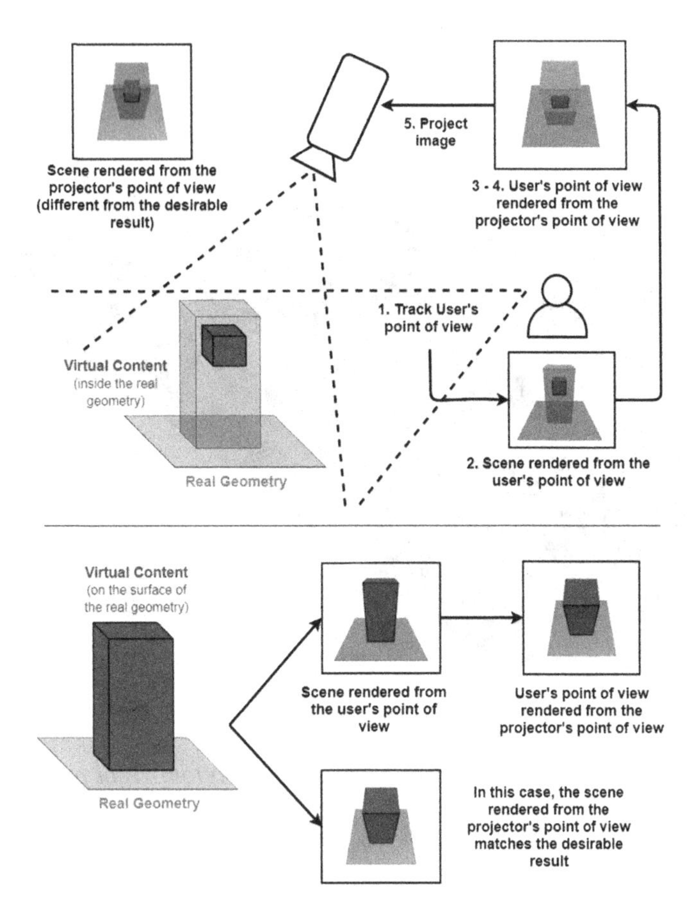

image generated in step 2) from the point of view of the projector; (v) project over the real model and blend the image with the rest of the images generated by other projectors.

This process can be simplified when the information that needs to be projected matches exactly the surfaces of the real model, replacing the first four steps by a single one in which the virtual image is rendered with a camera placed in the same position and orientation as the real projector, and configured so that its projection pyramid is similar to that of the projector. All this information is obtained from the calibration process: position, orientation (extrinsic parameters) and projection pyramid (intrinsic parameters). The fact that in this case the information is generated only in the areas corresponding to the surfaces of the real model has two advantages. On the one hand, it allows multiple users to observe the model from different points of view without erroneous perspectives. On the other hand, since it is not necessary to track the point of view of the user, the four projectors generate the same virtual content in each of the points of the real model. This is the reason why the occlusion compensation is performed correctly. However, there is also one limitation: the application cannot create the effect of showing the interior of a building or add virtual objects that rise from the ground, since not all observers will have the right perspective to have a correct depth perception. In this system the information projected on

Figure 9. Unity3D application with a simplified virtual representation of the city of Dubai. This model is used to generate the image that is generated by the projectors and mapped into the real scale model of the city

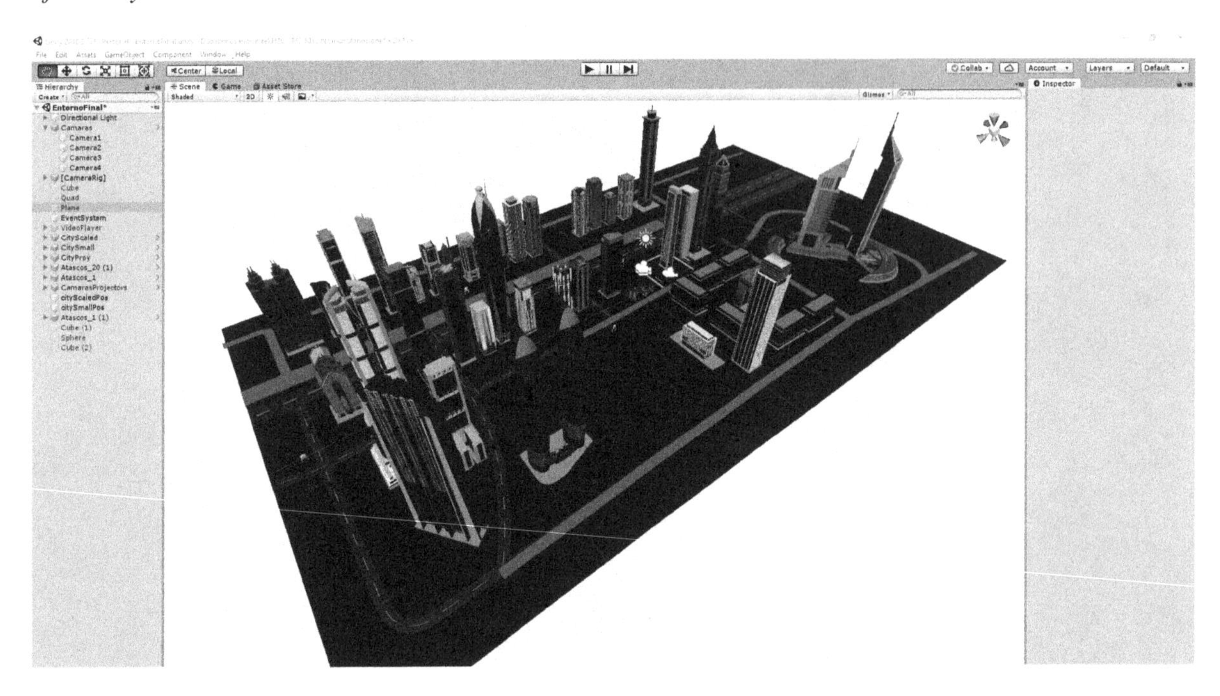

Figure 10. Close-up of the Unity3D application with a view (bottom-right part of the image) of one of the virtual cameras representing the projectors

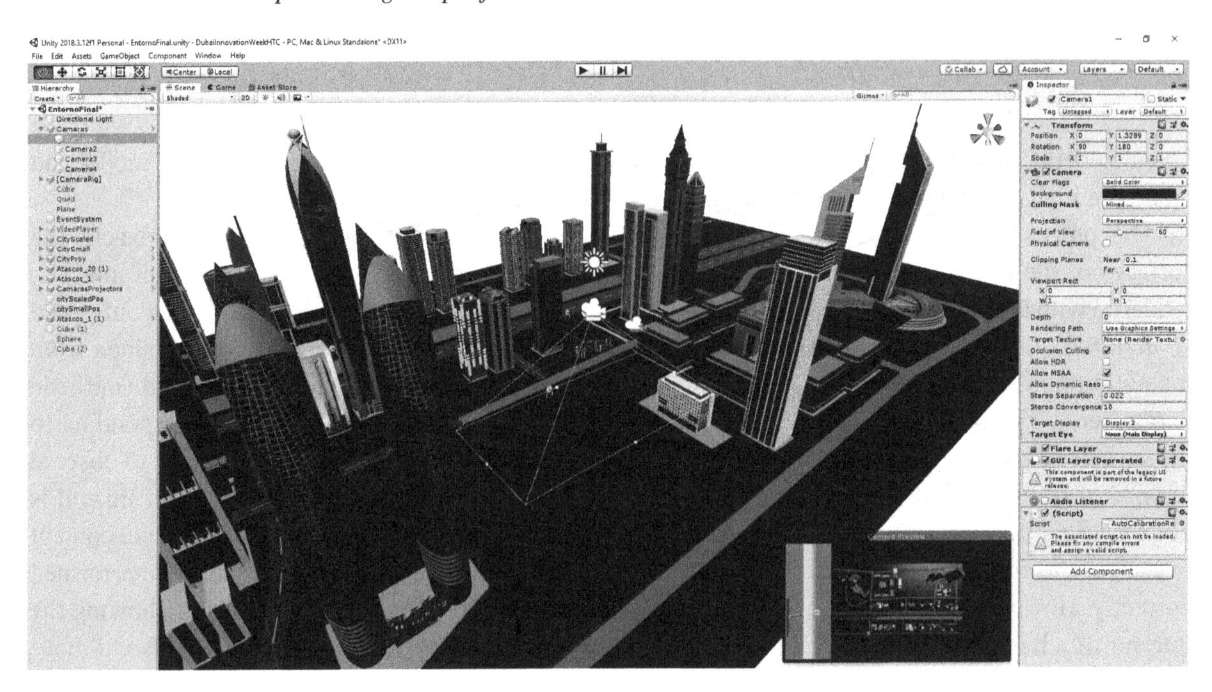

Figure 11. View of the virtual model from one of the cameras representing the projectors. This is the actual image that would be projected into the real scale model.

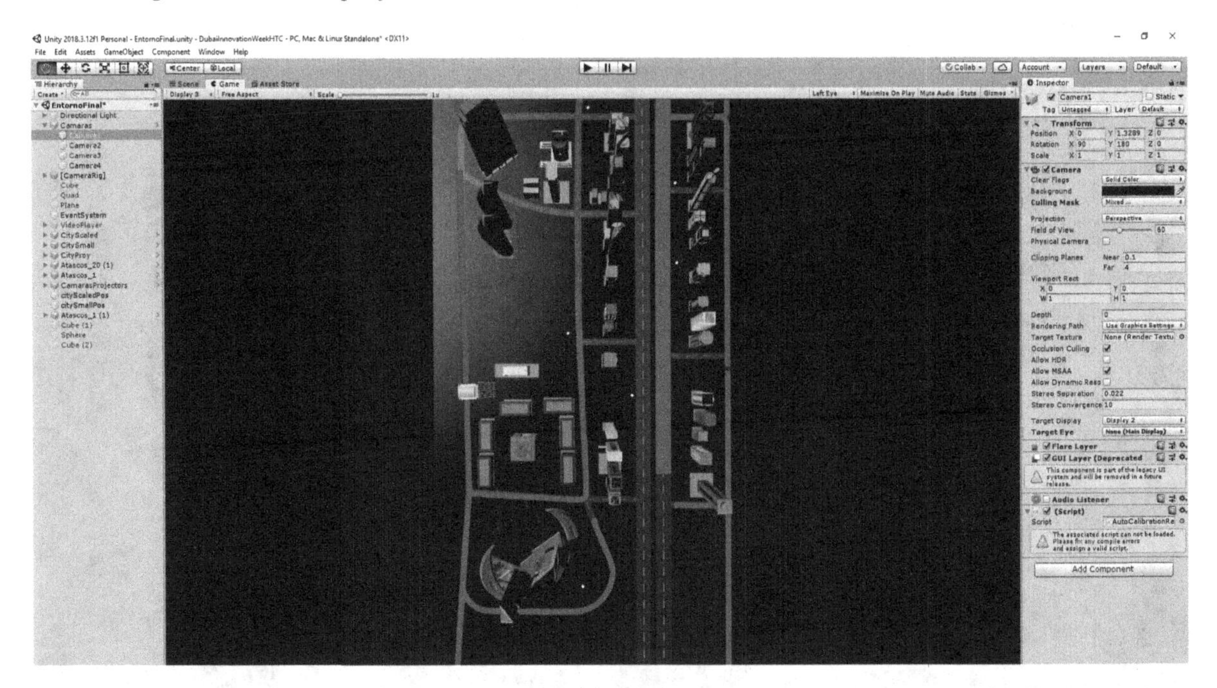

the elements of the model does not intend to create any of these effects, but to provide information to the existing surfaces of the model. In other words, the SAR application adds a virtual *skin* to the white surfaces of the simplified scale model of the city. All this process was implemented using a software application created with Unity3D (Figs. 9-11). Fig. 9 shows the virtual 3D model of the city, whose geometry is an exact reproduction of the real scale model. Fig. 10 shows a close-up of the model and the camera representing the projector. Fig. 11 shows a render from one of the cameras. The final result of the SAR application can be seen in Fig. 12, where it can be seen that the system offers also the possibility of showing real video feeds from traffic cameras of the city, a feature that complements the traffic information visualized by means of coloring the different roads.

Data Sources

Both systems (VR and SAR) are designed to collect data from external data sources. This way, the application layer is completely separated from the data layer, as it is suggested in both (Cheng et al., 2015; Gaur et al., 2015).

To make the visualization process as independent as possible from the data sources, our application is designed to get data from a RESTFUL service. With this structure, if the data provider is changed, the application works in exactly the same way but showing a different type of information.

The RESTFUL service provides data items that are described with several properties: source identification number, data name, data units, geographic location, city element (building, place, etc.) where the information will be shown, value and (optionally) preferred visualization type (color or level). These

Figure 12. The SAR application with a test projection representing summarized visual information about the smart city. Buildings and roads are colored, giving information about the state of traffic, energy consumption and other variables

items can be continuously changed and the application can also be configured to poll the data source at different frequencies, depending on how quickly we want the information to be updated.

FUTURE RESEARCH DIRECTIONS

The field of smart cities is constantly changing, as the underlying technologies change. For this reason, in the near future, there a number of technologies that will probably shape the future of smart cities. First, the arrival of 5G networks over the next years will allow not only faster speeds in the communication of data, but also much denser deployments. This will probably ease the development of smart city applications on mobility and public safety, among many others.

Artificial Intelligence is the other main hot topic that seems to be changing everything. Smart cities are no exception, and a great deal of smart tasks can be explored with the use of AI. From traffic management, to surveillance, parking or waste management, AI can be of great help for smart cities.

Another important trend, closely related to AI, is big data analytics. Smart devices generate enormous amount of data. This requires very large data storage systems that are not well prepared for conventional data processing methods and techniques, which can be considered obsolete for use in smart cities (Silva, Khan, & Han, 2018). Therefore, it is essential to explore the use of data science and big data analytics in smart cities. This is probably the most important challenge that smart cities will face in the near future.

In this same line, it is also important to be able to explore the benefits of using heterogeneous devices (Silva et al., 2018). Since smart cities integrate several technologies, which are often implemented with different protocols, it is key to ensure platform interoperability. In this regard, the use of open data policies can offer many advantages, and it is considered one of the main features a smart city should present (National League of Cities (NLC), 2017).

In addition, with so much data being transferred between citizens and governments, privacy and security are two of the most important questions that should be addressed in smart cities. In this regard, new technologies, such as *blockchain* could offer a security umbrella to support smart cities (Biswas & Muthukkumarasamy, 2016). This question will probably also remain a central aspect in the near future.

Leaving aside technological aspects, there are a number of social considerations that will probably shape the smart city concept in the near future. First, given the increasing concern about climate change, environmental sustainability will surely be one of the main trends in the development of smart cities. Second, the smart city concept is not limited to cities anymore. Communities (small towns, neighborhoods, districts, etc.) of all sizes can benefit from this technology. Even rural areas can *turn smart* at low costs. Thus, the smart city paradigm can be applied to any type of city. Only the scale changes, but the goals can be shared. In fact, as the idea of building smart cities grows, there will be increasing awareness about the fact that a smart city is not all about technology but about the human beings that live in it.

Another important trend that we will likely observe in the future is a tendency to simplify the information interface. As data grows, the ability of humans to deal with increasingly large volumes of data will be reduced. Therefore, more intuitive processes and human-machine interfaces need to be proposed and researched.

Of course, all these foreseeable trends are also research opportunities and challenges. Research about the use of big data, open data and AI in smart cities is one of the main research avenues in this field. Regarding data presentation, the use of innovative interfaces with immersive and/or 3D technologies, as the ones presented in this book chapter, is, in our opinion, the other main research avenue that should be explored in smart cities. More research effort should be put in solving the integration of heterogeneous information in a simple but meaningful manner, since we believe that this problem is not being researched as much as the use of AI, big data and security in smart cities.

CONCLUSION

In this chapter, we have reviewed the methods that can be used for the visualization of data in smart city applications, paying special attention to those using 3D based visualization, such as Virtual Reality (VR) and Augmented Reality (AR). These paradigms are particularly suitable for these applications, since they capture the attention of users, increasing the readability and understanding of the information. Given the amount of data that smart cities can generate, it is of the utmost importance to be able to summarize and represent this information in a simple form, so that only a glance is necessary to understand what the system is doing or monitoring. Nonetheless, a large gap still exists between the ability to produce large amounts of data and our ability to visualize and extract useful information from it. In fact, only a few works exist dealing with the use of VR/AR in smart cities, although some interesting contributions can be found and some conclusions can be extracted, such as the importance of separating the data sources from the algorithms dedicated to visualize them and the need to create spatially-localized data, usually by means of GIS-based applications.

We have also shown a use case in which VR and AR are used to visualize summarized information about a city. This application is intended for data monitoring in a control center and therefore no geo-localization of the user is considered in this case. A simplified scale model of the city (Dubai, in this case) is needed to achieve a successful use of both VR and AR. In the case of the VR application, the scale model is virtual, and a customized solution with a first-person VR system is applied. In the case of the AR application, Spatial Augmented Reality is used. Therefore, both a (completely white) real scale model and a virtual scale model of the city are needed to generate a projected image over the real scale model. In both cases (VR and SAR) the information about the state of the city is provided by means of text, colors and other visual elements. Of course, this conceptual model can be used in a different city (a different scale model of the city would be needed, though) and/or with other data sources. As previously shown, the conceptual model is general and independent of data, but the process of generating the virtual image is complex, especially in the case of the SAR application, where a calibration of the projectors is needed and the process of using a virtual model for the generation of an image to be projected onto a real model is not trivial.

REFERENCES

Abdelaziz, A., Salama, A. S., & Riad, A. M. (2019). A Swarm Intelligence Model for Enhancing Health Care Services in Smart Cities Applications. In Security in Smart Cities: Models, Applications, and Challenges (pp. 71–91). Springer. doi:10.1007/978-3-030-01560-2_4

Aïm, F., Lonjon, G., Hannouche, D., & Nizard, R. (2016). Effectiveness of virtual reality training in orthopaedic surgery. *Arthroscopy*, *32*(1), 224–232. doi:10.1016/j.arthro.2015.07.023 PMID:26412672

Al Ghamdi, A., & Thomson, T. (2018). Big Data Storage and Its Future. In *Proceedings 2018 International Conference on Computing Sciences and Engineering (ICCSE)*, 1–6. IEEE.

Anthopoulos, L. G. (2015). Understanding the smart city domain: A literature review. In *Transforming city governments for successful smart cities* (pp. 9–21). Springer. doi:10.1007/978-3-319-03167-5_2

Azuma, R. T. (1997). A survey of augmented reality. *Presence (Cambridge, Mass.)*, *6*(4), 355–385. doi:10.1162/pres.1997.6.4.355

Billinghurst, M., Clark, A., & Lee, G. (2015). A survey of augmented reality. *Foundations and Trends® in Human–Computer Interaction, 8*(2–3), 73–272.

Bimber, O., & Raskar, R. (2005). *Spatial augmented reality: Merging real and virtual worlds.* AK Peters/CRC Press.

Biswas, K., & Muthukkumarasamy, V. (2016). Securing smart cities using blockchain technology. In *Proceedings 2016 IEEE 18th International Conference on High Performance Computing and Communications; IEEE 14th International Conference on Smart City; IEEE 2nd International Conference on Data Science and Systems (HPCC/SmartCity/DSS)*, 1392–1393. IEEE.

Bonetti, F., Warnaby, G., & Quinn, L. (2018). Augmented reality and virtual reality in physical and online retailing: A review, synthesis and research agenda. In *Augmented Reality and Virtual Reality* (pp. 119–132). Springer. doi:10.1007/978-3-319-64027-3_9

Boulos, M. N. K., Lu, Z., Guerrero, P., Jennett, C., & Steed, A. (2017). *From urban planning and emergency training to Pokémon Go: Applications of virtual reality GIS (VRGIS) and augmented reality GIS (ARGIS) in personal, public and environmental health.* BioMed Central.

Bouloukakis, M., Partarakis, N., Drossis, I., Kalaitzakis, M., & Stephanidis, C. (2019). Virtual Reality for Smart City Visualization and Monitoring. In *Mediterranean Cities and Island Communities* (pp. 1–18). Springer. doi:10.1007/978-3-319-99444-4_1

Braem, B., Latré, S., Leroux, P., Demeester, P., Coenen, T., & Ballon, P. (2016). Designing a smart city playground: Real-time air quality measurements and visualization in the City of Things testbed. In *Proceedings 2016 IEEE International Smart Cities Conference (ISC2)*, 1–2. IEEE. 10.1109/ISC2.2016.7580871

Cardullo, P., & Kitchin, R. (2019). Being a 'citizen' in the smart city: Up and down the scaffold of smart citizen participation in Dublin, Ireland. *GeoJournal*, *84*(1), 1–13. doi:10.100710708-018-9845-8

Casas, S., Morillo, P., Gimeno, J., & Fernández, M. (2009). SUED: An extensible framework for the development of low-cost DVE systems. In *Proceedings of the IEEE Virtual Reality 2009 (IEEE-VR'09). Workshop on Software Engineering and Architectures for Realtime Interactive Systems (SEARIS)*, 20–25.

Casas, S., Portalés, C., García-Pereira, I., & Fernández, M. (2017). On a first evaluation of ROMOT—A RObotic 3D MOvie Theatre—For driving safety awareness. *Multimodal Technologies and Interaction*, *1*(2), 6. doi:10.3390/mti1020006

Chen, Y., & Han, D. (2018). Water quality monitoring in smart city: A pilot project. *Automation in Construction*, *89*, 307–316. doi:10.1016/j.autcon.2018.02.008

Cheng, B., Longo, S., Cirillo, F., Bauer, M., & Kovacs, E. (2015). Building a big data platform for smart cities: Experience and lessons from santander. In *Proceedings 2015 IEEE International Congress on Big Data*, 592–599. IEEE. 10.1109/BigDataCongress.2015.91

Cho, K., Jang, H., Park, L. W., Kim, S., & Park, S. (2019). Energy Management System Based on Augmented Reality for Human-Computer Interaction in a Smart City. In *Proceedings 2019 IEEE International Conference on Consumer Electronics (ICCE)*, 1–3. IEEE. 10.1109/ICCE.2019.8662045

Cocchia, A. (2014). Smart and digital city: A systematic literature review. In *Smart city* (pp. 13 43). Springer.

Daniel, S., & Doran, M.-A. (2013). geoSmartCity: Geomatics contribution to the smart city. *Proceedings of the 14th Annual International Conference on Digital Government Research*, 65–71. ACM. 10.1145/2479724.2479738

Dinc, E., Kuscu, M., Bilgin, B. A., & Akan, O. B. (2019). Internet of Everything: A Unifying Framework Beyond Internet of Things. In Harnessing the Internet of Everything (IoE) for Accelerated Innovation Opportunities (pp. 1–30). Hershey, PA: IGI Global. doi:10.4018/978-1-5225-7332-6.ch001

Doraiswamy, H., Freire, J., Lage, M., Miranda, F., & Silva, C. (2018). Spatio-Temporal Urban Data Analysis: A Visual Analytics Perspective. *IEEE Computer Graphics and Applications*, *38*(5), 26–35. doi:10.1109/MCG.2018.053491728 PMID:30273125

Dubai Corporation of Tourism & Commerce Marketing. (2019). Visit Dubai. Retrieved January 5, 2019, from Visit Dubai website: https://www.visitdubai.com/en/business-in-dubai/why-dubai/news-and-insights/becoming-the-worlds-smartest-city

Dutta, J., Chowdhury, C., Roy, S., Middya, A. I., & Gazi, F. (2017). Towards smart city: Sensing air quality in city based on opportunistic crowd-sensing. *Proceedings of the 18th International Conference on Distributed Computing and Networking*, 42. ACM. 10.1145/3007748.3018286

Estrada, E., Maciel, R., Zezzatti, C. A. O. O., Bernabe-Loranca, B., Oliva, D., & Larios, V. (2018). Smart City visualization tool for the Open Data georeferenced analysis utilizing machine learning. *International Journal of Combinatorial Optimization Problems and Informatics*, *9*(2), 25–40.

Gaur, A., Scotney, B., Parr, G., & McClean, S. (2015). Smart city architecture and its applications based on IoT. *Procedia Computer Science*, *52*, 1089–1094. doi:10.1016/j.procs.2015.05.122

Gimeno, J., Morillo, P., Casas, S., & Fernández, M. (2011). An augmented reality (AR) cad system at construction sites. In Augmented Reality-Some Emerging Application Areas. InTech.

Gimeno, J., Portalés, C., Coma, I., Fernández, M., & Martínez, B. (2017). Combining Traditional and Indirect Augmented Reality for Indoor Crowded Environments. A Case Study on the Casa Batlló Museum. *Computers & Graphics*, *69*, 92–103. doi:10.1016/j.cag.2017.09.001

Gretzel, U., Zhong, L., & Koo, C. (2016). Application of smart tourism to cities. *International Journal of Tourism Cities*, *2*(2). doi:10.1108/IJTC-04-2016-0007

Grodi, R., Rawat, D. B., & Rios-Gutierrez, F. (2016). Smart parking: Parking occupancy monitoring and visualization system for smart cities. [IEEE.]. *SoutheastCon*, *2016*, 1–5.

Hollands, R. G. (2008). Will the real smart city please stand up? Intelligent, progressive or entrepreneurial? *City*, *12*(3), 303–320. doi:10.1080/13604810802479126

Hu, J., Wan, W., Wang, R., & Yu, X. (2012). Virtual reality platform for smart city based on sensor network and OSG engine. In *Proceedings 2012 International Conference on Audio, Language and Image Processing*, 1167–1171. IEEE. 10.1109/ICALIP.2012.6376794

Hussain, A., Wenbi, R., da Silva, A. L., Nadher, M., & Mudhish, M. (2015). Health and emergency-care platform for the elderly and disabled people in the Smart City. *Journal of Systems and Software*, *110*, 253–263. doi:10.1016/j.jss.2015.08.041

Ibáñez, M.-B., & Delgado-Kloos, C. (2018). Augmented reality for STEM learning: A systematic review. *Computers & Education*, *123*, 109–123. doi:10.1016/j.compedu.2018.05.002

Jagnow, R. (2017, July 6). Daydream Labs: Locomotion in VR. Retrieved January 7, 2019, from Daydream website: https://www.blog.google/products/daydream/daydream-labs-locomotion-vr

Jin, J., Gubbi, J., Marusic, S., & Palaniswami, M. (2014). An information framework for creating a smart city through internet of things. *IEEE Internet of Things Journal*, *1*(2), 112–121. doi:10.1109/JIOT.2013.2296516

Kim, S. A., Shin, D., Choe, Y., Seibert, T., & Walz, S. P. (2012). Integrated energy monitoring and visualization system for Smart Green City development: Designing a spatial information integrated energy monitoring model in the context of massive data management on a web-based platform. *Automation in Construction*, *22*, 51–59. doi:10.1016/j.autcon.2011.07.004

Kodama, R., Koge, M., Taguchi, S., & Kajimoto, H. (2017). COMS-VR: Mobile virtual reality entertainment system using electric car and head-mounted display. In *Proceedings 2017 IEEE Symposium on 3D User Interfaces (3DUI)*, 130–133. IEEE. 10.1109/3DUI.2017.7893329

Laramee, R. S., Turkay, C., & Joshi, A. (2018). Visualization for Smart City Applications. *IEEE Computer Graphics and Applications*, *38*(5), 36–37. doi:10.1109/MCG.2018.053491729

Latif, S., Afzaal, H., & Zafar, N. A. (2018). Intelligent traffic monitoring and guidance system for smart city. In *Proceedings 2018 International Conference on Computing, Mathematics and Engineering Technologies (ICoMET)*, 1–6. IEEE. 10.1109/ICOMET.2018.8346327

Lin, C.-Y., Lu, Y.-L., Tsai, M.-H., & Chang, H.-L. (2018). Utilization-based parking space suggestion in smart city. In Proceedings 2018 15th IEEE Annual Consumer Communications & Networking Conference (CCNC), 1–6. IEEE.

Lv, Z., Yin, T., Zhang, X., Song, H., & Chen, G. (2016). Virtual reality smart city based on WebVRGIS. *IEEE Internet of Things Journal*, *3*(6), 1015–1024. doi:10.1109/JIOT.2016.2546307

Martin, C. J., Evans, J., & Karvonen, A. (2018). Smart and sustainable? Five tensions in the visions and practices of the smart-sustainable city in Europe and North America. *Technological Forecasting and Social Change*, *133*, 269–278. doi:10.1016/j.techfore.2018.01.005

Moustaka, V., Vakali, A., & Anthopoulos, L. G. (2018). A systematic review for smart city data analytics. [CSUR]. *ACM Computing Surveys*, *51*(5), 103. doi:10.1145/3239566

Murshed, S., Al-Hyari, A., Wendel, J., & Ansart, L. (2018). Design and Implementation of a 4D Web Application for Analytical Visualization of Smart City Applications. *ISPRS International Journal of Geo-Information*, *7*(7), 276. doi:10.3390/ijgi7070276

Nam, T., & Pardo, T. A. (2011). Conceptualizing smart city with dimensions of technology, people, and institutions. *Proceedings of the 12th Annual International Digital Government Research Conference: Digital Government Innovation in Challenging Times*, 282–291. ACM. 10.1145/2037556.2037602

National League of Cities (NLC). (2017). *Trends in Smart City Development*. Retrieved from NLC website: https://www.nlc.org/sites/default/files/2017-01/Trends%20in%20Smart%20City%20Development.pdf

Olanda, R., Pérez, M., Morillo, P., Fernández, M., & Casas, S. (2006). Entertainment virtual reality system for simulation of spaceflights over the surface of the planet Mars. *Proceedings of the ACM Symposium on Virtual Reality Software and Technology*, 123–132. ACM. 10.1145/1180495.1180522

Pawłowicz, B., Salach, M., & Trybus, B. (2018). Smart city traffic monitoring system based on 5G cellular network, RFID and machine learning. In *Proceedings KKIO Software Engineering Conference*, 151–165. Springer.

Peng, C., Tan, X., Gao, M., & Yao, Y. (2013). Virtual reality in smart city. In *Geo-Informatics in Resource Management and Sustainable Ecosystem* (pp. 107–118). Springer. doi:10.1007/978-3-642-45025-9_13

Pieroni, A., Scarpato, N., Di Nunzio, L., Fallucchi, F., & Raso, M. (2018). Smarter city: Smart energy grid based on blockchain technology. *Int. J. Adv. Sci. Eng. Inf. Technol*, *8*(1), 298–306. doi:10.18517/ijaseit.8.1.4954

Pokric, B., Krco, S., & Pokric, M. (2014). Augmented reality based smart city services using secure iot infrastructure. In *Proceedings 2014 28th International Conference on Advanced Information Networking and Applications Workshops*, 803–808. IEEE.

Portalés, C., Casanova-Salas, P., Casas, S., Gimeno, J., & Fernández, M. (2019). An interactive cameraless projector calibration method. *Virtual Reality (Waltham Cross)*, 1–13.

Portalés, C., Casas, S., Alonso-Monasterio, P., & Viñals, M. J. (2018). Multi-dimensional acquisition, representation, and interaction of cultural heritage tangible assets: An insight on tourism applications. In Handbook of Research on Technological Developments for Cultural Heritage and eTourism Applications (pp. 72–95). Hershey, PA: IGI Global. doi:10.4018/978-1-5225-2927-9.ch004

Portalés, C., Casas, S., Coma, I., & Fernández, M. (2017). A Multi-Projector Calibration Method for Virtual Reality Simulators with Analytically Defined Screens. *Journal of Imaging*, *3*(2), 19. doi:10.3390/jimaging3020019

Portalés, C., Casas, S., & Kreuzer, K. (2019). Challenges and Trends in Home Automation: Addressing the Interoperability Problem With the Open-Source Platform OpenHAB. In Harnessing the Internet of Everything (IoE) for Accelerated Innovation Opportunities (pp. 148–174). Hershey, PA: IGI Global.

Portalés, C., Gimeno, J., Casas, S., Olanda, R., & Martínez, F. G. (2016). Interacting with augmented reality mirrors. In Handbook of Research on Human-Computer Interfaces, Developments, and Applications (pp. 216–244). Hershey, PA: IGI Global. doi:10.4018/978-1-5225-0435-1.ch009

Portalés, C., Orduña, J. M., Morillo, P., & Gimeno, J. (2019). An efficient projector calibration method for projecting virtual reality on cylindrical surfaces. *Multimedia Tools and Applications*, *78*(2), 1457–1471. doi:10.100711042-018-6253-5

Prandi, F., Soave, M., Devigili, F., Andreolli, M., & De Amicis, R. (2014). Services oriented smart city platform based on 3D city model visualization. *ISPRS Annals of the Photogrammetry. Remote Sensing and Spatial Information Sciences*, *2*(4), 59.

Prendinger, H., Gajananan, K., Zaki, A. B., Fares, A., Molenaar, R., Urbano, D., ... Gomaa, W. (2013). Tokyo virtual living lab: Designing smart cities based on the 3d internet. *IEEE Internet Computing*, *17*(6), 30–38. doi:10.1109/MIC.2013.87

Ramos, F., Trilles, S., Torres-Sospedra, J., & Perales, F. (2018). New Trends in Using Augmented Reality Apps for Smart City Contexts. *ISPRS International Journal of Geo-Information*, *7*(12), 478. doi:10.3390/ijgi7120478

Sherman, W. R., & Craig, A. B. (2018). *Understanding virtual reality: Interface, application, and design*. Morgan Kaufmann.

Silva, B. N., Khan, M., & Han, K. (2018). Towards sustainable smart cities: A review of trends, architectures, components, and open challenges in smart cities. *Sustainable Cities and Society*, *38*, 697–713. doi:10.1016/j.scs.2018.01.053

Sotres, P., de la Torre, C. L., Sánchez, L., Jeong, S., & Kim, J. (2018). Smart City Services Over a Global Interoperable Internet-of-Things System: The Smart Parking Case. 2018 Global Internet of Things Summit (GIoTS), 1–6. IEEE.

Tian, J., Li, H., & Chen, R. (2018). Individual Behavior Change under Smart City Environment—A Proposal of Smart Citizen Concept with Four Dimensions. *IConference 2018 Proceedings*.

Vakali, A., Anthopoulos, L., & Krco, S. (2014). Smart Cities Data Streams Integration: Experimenting with Internet of Things and social data flows. *Proceedings of the 4th International Conference on Web Intelligence, Mining and Semantics (WIMS14)*, 60. ACM. 10.1145/2611040.2611094

Vera, L., Gimeno, J., Casas, S., García-Pereira, I., & Portalés, C. (2017). A hybrid virtual-augmented serious game to improve driving safety awareness. *International Conference on Advances in Computer Entertainment*, 293–310. Springer.

Wang, R., Jin, L., Xiao, R., Guo, S., & Li, S. (2012). 3D Reconstruction and interaction for smart city based on world wind. *2012 International Conference on Audio, Language and Image Processing*, 953–956. IEEE. 10.1109/ICALIP.2012.6376751

Yeom, S. (2011). Augmented reality for learning anatomy. *Changing Demands, Changing Directions. Proceedings Ascilite Hobart*, 1377–1383.

Yin, C., Xiong, Z., Chen, H., Wang, J., Cooper, D., & David, B. (2015). A literature survey on smart cities. *Science China. Information Sciences*, *58*(10), 1–18. doi:10.100711432-015-5397-4

Zeng, W., & Ye, Y. (2018). VitalVizor: A Visual Analytics System for Studying Urban Vitality. *IEEE Computer Graphics and Applications*, *38*(5), 38–53. doi:10.1109/MCG.2018.053491730 PMID:30273126

Zhang, L., Chen, S., Dong, H., & El Saddik, A. (2018). Visualizing Toronto city data with Hololens: Using augmented reality for a city model. *IEEE Consumer Electronics Magazine*, *7*(3), 73–80. doi:10.1109/MCE.2018.2797658

ADDITIONAL READING

Al Nuaimi, E., Al Neyadi, H., Mohamed, N., & Al-Jaroodi, J. (2015). Applications of big data to smart cities. *Journal of Internet Services and Applications*, *6*(1), 25. doi:10.118613174-015-0041-5

Anthopoulos, L., Janssen, M., & Weerakkody, V. (2019). A Unified Smart City Model (USCM) for smart city conceptualization and benchmarking. In Smart Cities and Smart Spaces: Concepts, Methodologies, Tools, and Applications (pp. 247–264). IGI Global.

Batty, M. (2013). Big data, smart cities and city planning. *Dialogues in Human Geography*, *3*(3), 274–279. doi:10.1177/2043820613513390 PMID:29472982

D'Acunto, L., Kleinrouweler, J. W., Panneman, J., Gabriel, A., Adhikari, A., & Satta, R. (2019). Prosuming live multimedia content in 5G-enabied smart cities. *Proceedings of the 10th ACM Multimedia Systems Conference*, 312–315. ACM. 10.1145/3304109.3323836

Fahmy, A., Altaf, H., Al Nabulsi, A., Al-Ali, A., & Aburukba, R. (2019). Role of RFID Technology in Smart City Applications. *2019 International Conference on Communications, Signal Processing, and Their Applications (ICCSPA)*, 1–6. IEEE. 10.1109/ICCSPA.2019.8713622

Hashem, I. A. T., Chang, V., Anuar, N. B., Adewole, K., Yaqoob, I., Gani, A., ... Chiroma, H. (2016). The role of big data in smart city. *International Journal of Information Management*, *36*(5), 748–758. doi:10.1016/j.ijinfomgt.2016.05.002

Ismagilova, E., Hughes, L., Dwivedi, Y. K., & Raman, K. R. (2019). Smart cities: Advances in research—An information systems perspective. *International Journal of Information Management*, *47*, 88–100. doi:10.1016/j.ijinfomgt.2019.01.004

Ji, W., Xu, J., Qiao, H., Zhou, M., & Liang, B. (2019). Visual IoT: Enabling Internet of Things Visualization in Smart Cities. *IEEE Network*, *33*(2), 102–110. doi:10.1109/MNET.2019.1800258

Jordaan, C. G., Malekian, N., & Malekian, R. (2019). Internet of Things and 5G Solutions for development of Smart Cities and Connected Systems. *Communications of the CCISA*, *25*(2), 1–16.

Martínez-Ballesté, A., Pérez-Martínez, P. A., & Solanas, A. (2013). The pursuit of citizens' privacy: A privacy-aware smart city is possible. *IEEE Communications Magazine*, *51*(6), 136–141. doi:10.1109/MCOM.2013.6525606

Matos, A., Pinto, B., Barros, F., Martins, S., Martins, J., & Au-Yong-Oliveira, M. (2019). Smart Cities and Smart Tourism: What Future Do They Bring? *World Conference on Information Systems and Technologies*, 358–370. Springer. 10.1007/978-3-030-16187-3_35

Mouftah, H. T., Erol-Kantarci, M., & Rehmani, M. H. (2019). *5G and D2D Communications at the Service of Smart Cities*.

Rathore, M. M., Ahmad, A., Paul, A., & Rho, S. (2016). Urban planning and building smart cities based on the internet of things using big data analytics. *Computer Networks*, *101*, 63–80. doi:10.1016/j.comnet.2015.12.023

Schaffers, H., Komninos, N., Pallot, M., Trousse, B., Nilsson, M., & Oliveira, A. (2011). Smart cities and the future internet: Towards cooperation frameworks for open innovation. The Future Internet Assembly, 431–446. Springer.

KEY TERMS AND DEFINITIONS

Augmented Reality (AR): The technology that simultaneously combines real and virtual objects that are interactive in real-time and are registered in a three-dimensional space. **Interaction**: A kind of action that occurs as two or more objects have an effect upon one another.

Mixed Reality: The result of blending the physical world with a synthetic one, including the paradigms of Augmented Reality and Augmented Virtuality.

Projection Mapping: The process of utilizing video projectors in order to properly map images or video onto physical objects, so that the aspect of the objects changes.

Projector Calibration: The process by which a projector is adjusted so that the projected image accomplishes a series of desired properties. When dealing with digital projectors, a common goal is the geometric calibration of the projector, which means the calculation of the intrinsic (field of view and aspect ratio) and extrinsic (position and orientation) parameters of the projector.

Smart City: An innovative city that uses ICT and other means to improve quality of life, efficiency of urban operation and services, and competitiveness, while ensuring that it meets the needs of present and future generations with respect to economic, social, and environmental aspects.

Spatial Augmented Reality (SAR): A special type of Augmented Reality technology where the combination of virtual and real objects is produced by projecting virtual images onto real objects using projection mapping. Hence, display monitors, head-mounted displays or hand-held devices are not typically used in this type of AR.

Virtual Reality (VR): The technology by which a user, stimulated with computer-generated perceptual cues, experiences an alternative reality that is different from the one he/she actually lives in.

Chapter 16
Scenarios for a Smart Tourism Destination Transformation:
The Case of Cordoba, Spain

Robert Germain Lanquar
Cordoba Horizontes, Spain

ABSTRACT

Reluctance and even resistance for the city transformation into a smart tourism destination (STD) may occur on the edge field of smart systems, even if over time, often with political changes, this resistance breaks down. This situation is occurring in Cordoba (Spain) which held the record of four inscriptions in the World Heritage Lists granted by UNESCO. The concept of smart tourism is recent. As a research domain, it is only emerging. So, few analyses and case studies were elaborated on it. This chapter shows how this reluctance is coming from stakeholders and how the transformation of Cordoba into a smart destination can be made harmoniously. Three scenarios are possible for this transformation either following actual tendencies or looking for two other paths, the green one or the intercultural one; how it could be done and its impacts on the City and the Province of Cordoba's economy, society, culture, and environment.

INTRODUCTION

Cordoba has insufficient economic growth. Agriculture and tourism are its main activities. This Andalusian city loses its population with 328 547 inhabitants in 2010 and 325 916 in 2018. Cordoba finds more and more challenges for global issues such as unemployment, digital divide, inequalities and poverty, but also for its environmental problems as energy inefficiency and waste management.

Cordoba rides the wave that raises Spanish tourism, reaching in 2018 the record number of 82.6 million arrivals of international visitors, that places Spain in the 2nd world rank. In addition, Cordoba's tourism was improved by an accentuated return of domestic tourism: the Spaniards who begin to recover from the crisis of the years 2008-2014 (OECD, 2018). The World Economic Forum gives to Spain the first place for tourism in terms of global competitiveness. Moreover, Cordoba and its Province have not

DOI: 10.4018/978-1-7998-2112-0.ch016

reached in 2018 the record of 2017, attracting 1 181 279 visitors (INE, January 2019), i.e. - 4.27% less than the previous year and 1 956 159 nights, i.e. 2.12% less. The main objectives of Cordoba Municipal Tourism Institute (IMTUR) are to increase the number of nights and to fight seasonality. No policy seems to emerge in favor of a transformation into a smart tourism destination. The massification perceived as overtourism begins to be obvious, especially during the spring season with the "Festival of the Cruces de Mayo" and the "Festival of the Patios Cordobeses". No strategy was developed to manage it.

Cordoba is an inner tourist destination that has not benefited from Spanish mass tourism based on the "Sun, Sea and Sand" coastal model. The climate factor, with temperature frequently exceeding 35-40 ° from mid-June to mid-September, drives the City and its Province to limit the tourist season from March to mid-June and from mid-September to December. Two pillars are the main selling propositions: the sites and monuments of the two UNESCO Lists of World Heritage, tangible and intangible, and the gastronomy based on its renowned agri-food production. Four inscriptions in the World Heritage Lists are granted: the Mosque-Cathedral (1984), the historical quarter surrounding it (1994), the Festival of the Patios (Courtyards) (2012) and the Caliphal city of Medina Azahara (2018). In addition, with the rest of Spain, Cordoba shares the titles of the Intangible Cultural Heritage of Humanity awarded to Flamenco (2010) and the Mediterranean Diet (2013). Nevertheless, despite some municipal and entrepreneurial initiatives, no holistic strategy has really been put in place for the digital transformation until 2019.

The notion of Smart Tourism Destination (STD)

The concept of smart city finds its origin in the 1990s to foster economic, social and environmental prosperity. It flourished significantly after 2007 according to Hollands (2008). The perception of smart tourism appeared as a major component of a smart city around 2010. As underlines by Gretzel (2011) as well as by Boes, Buhalis & Inversini (2016), the concept emerges from the rapid development of ICT and the goings-on of the smart cities: "*With technology being embedded on all organizations and entities, destinations will exploit synergies between ubiquitous sensing technology and their social components to support the enrichment of tourist experiences*" (Buhalis 2014). UNWTO, the World Tourism Organization, a United Nations Agency dedicated to tourism, considered smart tourism destinations from 2012 (UNWTO 2012). It organized two international seminars, the first in Murcia in 2017, the second in Oviedo in 2018 with the strong assistance of SEGITTUR[1].

Spain is one of the first European countries focusing on smart tourist destinations. Since 2008, SEGITTUR, the State Enterprise for the Management of Innovation and Tourism Technologies, under the supervision of the Ministry of Energy, Tourism and Digital Agenda, supports professionals and the Spanish public sector, especially local authorities, to develop the use of ICT, Artificial Intelligence (AI), Big Data and Internet of Things (IoT), to enhance their competitiveness. On smart destinations, SEGITTUR (2018) works in connection with the CTN178 Committee of AENOR, the Spanish Association for Standardization and Certification for creating effective norms in accordance with ISO international standards, especially for the management and processing of databases (big data or open data) around four characteristics (volume, speed, veracity and variety) that will be used by these destinations.

Cordoba's Reluctance

In April 2017, the Spanish Government accentuated SEGITTUR's role in transforming its cities and territories into smart destinations. Cordoba did not take advantage of this context, despite its incorporation into the National and Integral Tourism Plan (PNIT)[2] 2012-2015 driven by the Spanish Secretariat for Tourism. The PNIT had given a definition of a smart tourist destination: "*an innovative destination, consolidated by a pioneering technological infrastructure, which guarantees the sustainable development of the territory, accessible to all, which facilitates the interaction and integration of the visitor with the environment and increases the quality of the tourist experience in the destination and improve the quality of life of the local resident*".

The Strategic Tourism Plan of Cordoba 2015-2019 (2015) barely sketches the role of mobile devices (smartphones and tablets) and ignores the advances of AI and blockchains. It does not even mention the term "Smart Tourism Destination"[3]. Besides, the Cordoba residents were never informed what is a smart destination or the way the technologies of information and communication (ICT) can be used to benefit them.

The Municipality entrusted the task of transforming Cordoba into a smart city at IMDEEC (Municipal Institute for Economic Development and Employment of Cordoba), IMDEED objectives are to stimulate economic initiatives and entrepreneurial activities in relation to civil society and the local industrial fabric such as the management of European projects, in addition to initiatives to disseminate the entrepreneurial culture and the training of new entrepreneurs for the implementation of new business projects: Tecnocórdoba, Baobab, Antiguas Municipal Lonjas. These incubators are premises for start-ups, rented at a subsidized cost for a limited period with common services (meeting rooms, training and advice).

In 2016, a preliminary draft of the Smart City Master Plan was accepted by most of the political parties of the Municipality of Cordoba, with two objectives:

- To equip the city with an infrastructure able of using more effective Open Data for the management of the City, its businesses and civil society.
- To find funding, especially at the national and European level, to promote innovation and to improve its Cordoba businesses' competitiveness.

The temporal horizon of this draft project was around 10 years, which leads to 2025 and beyond to make Cordoba a factual Smart City. The theoretical model chosen was that of the 'multi-focal lens' of the Quadruple and Quintuple Innovation Helixes perspective which centers on the relations between the local government, the entrepreneurial networks and the universities. This model was also including media and civil society in the development processes. It focuses on the prospective of the local society to foster innovations to adapt their environment to climate change. It is quite compatible with the knowledge economy and furthermore, it encourages a democratic knowledge society to produce innovations. This had been defined as early as 2009 by the European Union, which identified the socio-ecological and energy transitions as the major challenge of Europe.

The Need for Data

How could this transformation be done based, according to the European Parliament (2014), on the drivers of a Smart City: **connectivity, sustainability and inclusiveness** (participation of the civil society) for ensuring good quality of life and improving the relations between residents and tourists? What kind of data are needed to progress?

According to the Wikipedia definition, a smart city is "*an urban area that uses sensors of different types, Internet of Things (IoT) ... and then use these data to manage assets and resources efficiently*". These data are collected from permanent residents as well as temporary residents, and from other assets to monitor and manage transportation systems, infrastructure, equipment and all community services, weather forecasting, contamination levels, even crime detection...

As for the European Parliament (2014), a city is "smart" when local institutions address public issues via ICT solutions based on multiple partnerships. These solutions are improved through smart initiatives. The European Parliament (2014) distinguishes "*a range of 'scaling strategies' including replication (repeating initiatives and Smart City strategies in other locations), scaling (increasing the number of participants, resource allocation, geographic footprint or offering services more widely) and ecosystem seeding (using Smart City initiatives as the basis for an adaptive network of interacting initiatives)*". These different scaling strategies may be rearranged in scenarios depending on the role and the willingness of the stakeholders, mainly institutional authorities and civil society (including businesses).

Six cities, Amsterdam, Helsinki, Copenhagen Manchester, Vienna and Barcelona, major European tourism destinations, were considered in 2014 as best practices to show how to link and to strengthen networks of persons, enterprises, infrastructures, resources, energies and spaces, providing them with organizational tools. Is it possible to apply these practices to Cordoba?

- Smart governance: in this case, how a city as Cordoba can strengthen democracy with ICT as well as may improve the information to visitors and residents?
- Smart quality of life: using renewable energies, easier and faster connectivity, through the multiplication of green spaces and buildings.
- Smart mobility with transport services for residents and visitors.
- Smart economy involving social and sharing economy, circular economy using ICT and incentives to create startups.
- Smart environment: this concept is linked to the development of sensors, electronic actuators, displays and computing elements, known and not secret, with IoT, which enable the elaboration of measurable ratios on one specific and well - defined environment.

In Cordoba, the weakness of public investment did not allowed the implementation of large-scale networks. If the municipal government had, until 2019, for objective sustainability for reasons often ideological, it did not focus on an information policy that would speed up the awareness of residents and ensure the transparency of its actions for the optimization of the relations between residents and tourists. It had not defined the priorities that characterize a smart destination: competitiveness, governance, human and social capital, mobility, environment (i.e. energy transition), quality of life (noise, shared hosting and importance given to circular economy). It chiefly was anxious to avoid a kind of "big brother" city with the multiplication of surveillance systems.

METHODOLOGY

The main objective of this chapter is to analyze how Cordoba be transformed into a smart tourist destination and produce better connectivity with its Province and its rural territories. Three stages were followed:

1. Defining the current positioning of tourism in Cordoba with identification and determination of the priorities of the stakeholders (political and administrative institutions, civil society and businesses, tourism professionals) through a Likert-type scale (Table 1).
2. Developing three scenarios according to the city's strategic options for this transformation and the possible future visions of Cordoba and its Province in a geopolitical, cultural or intercultural and environmental ("climate change") context. Consequently, SWOT analyses were added to evaluate the strengths, weaknesses, opportunities and threats of each scenario (Tables 2 to 4). Finally, the same stakeholders gave their opinion on these three scenarios (Table 5).
3. Defining the axes of transformation and identifying as recommendations the main initiatives for the coming years.

The Use of A Likert-Type Scale For Analyzing The Opinions Of Cordoba's Stakeholders

How could Cordoba's transformation process be developed to become a smart destination? Thirty (30) stakeholders were interviewed from September to November 2017. A Likert- type scale was built about their opinions. At least two-thirds agreed that the transformation should be accelerated and a true partnership between the Province, the Municipality and the civil society established.

Likert scales were developed by Likert (1932) to measure people's attitudes. Participants are asked to rank their agreement with a set of items on a scale with a limited number of possible responses presented in a graduation: "strongly agree", "agree", "undecided", "disagree", "strongly disagree".

The 30 stakeholders (local and provincial officials, civil society and businesses as well as tourism professionals) were interviewed (face-to-face) among them several professional associations grouped around FIDES, the Strategic Cluster of Cordoba. Their responses concern twenty (20) elements. They were distributed as follows:

- 6 travel agencies
- 6 hotels and resorts
- 6 service providers/businesses
- 6 representatives of professional associations (employers and trade unions)
- 6 institutional leaders (Town Hall, Provincial Council, Tourist offices).

Their opinions are given in Table 1. Each cell shows the number of stakeholders' opinions. Some present they reluctance views, a softer manner of resisting to smart transformation for Cordoba.

This survey was the main source for eliciting the expectations of the stakeholders for a transformation of Cordoba into a smart destination. Two-thirds of the respondents agreed that the transformation should be accelerated, and a genuine partnership should be established between the Province, the Municipality and the civil society. Also, two-thirds agreed on a strategy focused on open data to avoid the risk of a surveillance society. They have the same attitude that the local government who, since 2015, is reluc-

Table 1. Opinions of local officials, businesses and tourism professionals on the transformation of Cordoba as a smart tourism destination (STD)

OPINIONS	Totally disagree	Disagree	Undecided	Agree	Totally agree
The city of Cordoba has chosen to become a smart tourist destination (STD), but without defining its strategy	4	8	5	11	2
The transformation of Cordoba into STD must be done swiftly	12	8	6	2	2
Cordoba has the foundations to become a Smart Tourism Destination	4	4	6	10	6
Cordoba should develop a smart tourism destination strategy	2	2	8	8	10
A specific local parity structure between authorities and professionals should exist	4	4	6	11	5
There is a consensus between public authorities and tourism professionals	14	6	2	5	3
The role of tourism professionals should be highlighted	2	2	8	8	10
Cordoba's "on-line" reputation will emerge strengthened by a STD strategy	0	4	5	11	10
Cordoba should follow the example of Seville and Granada in Andalucia	0	4	10	12	4
Priority should be given to enhancing connectivity (Broadband, 4 G, then 5G)	0	0	10	11	9
Priority must be given on sustainability and the fight against climate change	0	2	4	12	12
Priority must be given on the inclusion of residents to let them accepting the development of smart tourism	0	0	6	10	14
The STD strategy must include a "smart mobility – local public transport-car parks – carpooling..." component	2	4	8	10	6
The STD strategy must focus on Open Data to avoid the risk of a surveillance society	2	2	6	8	12
The STD strategy must insert a "cultural heritage" component linked to the inscription on the UNESCO Heritage Lists and prepare Cordoba's candidature for the European Capital of culture 2032	3	5	2	8	12
The STD strategy must include an ' environmental' component that will allow Cordoba to apply as a European Green Capital	2	2	4	8	14
An incubator and a Fab Lab should be integrated into the smart tourism strategy for the emergence of startups creating specific Apps for Cordoba	5	5	5	10	5
There is a need for a gastronomy cluster with agri-food sector	2	2	6	9	11
UCO and Loyola, the local universities, should play a key role in smart tourism training and education	0	1	8	12	9
A link must be found so that Cordoba Capital should also be involved in the rural development of the Province through the project Smart Rural Land (Añora Strategy)	2	3	6	9	10
AVERAGE	2	2.25	4	6.4	5.5

tant to monitor the City of Cordoba. Ontologically, they want to avoid the risk of a surveillance society, monitored by the IoT, sensors, cameras and other systems at the back of the citizen or the tourist, a kind of Totalitarianism 4.0 (Bloom 2019).

More respondents "agree" (average 6.4) than" totally agree (average 5.5). An important group shows a reluctance to this transformation with an average of 2 for "totally disagreeing" and 2.25 for "disagreeing: The "undecided" group is quite significant (average 4). That means some resistance for this transformation. Besides, there is no vision of the future of Cordoba. Thus, it was valuable to develop scenarios and ask these respondents for their vision of the future of Cordoba and its Province as it will be seen in Table 5.

THE DEVELOPMENT OF THREE SCENARIOS

The scenario technique allows to overcome communication barriers and to see how current and alternative developments can affect the future. The ability to enlighten problems and to break down barriers makes them extremely effective in opening new horizons, strengthening leadership and enabling strategic decisions. Godet M. & Durance Ph. (2011) add: "*the word scenario is often confused with the strategy to the point that clarification is necessary if we want to understand ourselves*". The scenarios are called "*snapshots*" because they describe possible futures without explaining how they might happen. These questions can be explored by the "back-casting", imagining what should happen at every stage to come. In addition, the scenarios improve our understanding of the mechanisms that underlies change. They reinforce strategic planning and policy development. It is a governable approach to reasoning. Each scenario conveys an intuition which inevitably influences long-term policy decisions.

Three scenarios were already envisaged in *The future of a World Heritage City: Cordoba 2031* (Morales & Lanquar 2014). From the stakeholders' opinions and their analysis through the Likert – type scale, these three scenarios were improved to imagine the transformation of Cordoba into a smart destination.

The first "Cordoba, business as usual" gives great importance to the current tourism trends in Spain without emphasizing on ICT but with the risk of accentuating its gentrification. That could generate overtourism. The 2nd scenario is oriented towards the green economy. It was named "Smart Green Tourism" to sustain the development of a province with environmental indicators as proposed in the Añora Declaration of October 2017 (UCO 2017): the rural world does not want to be left out and the Province must be better connected to its Capital and the world. With the 3rd Scenario, Cordoba was historically considered in the heart of the three cultures (Jewish, Christian, Moslem). Cordoba should play on the challenge of Interculturality. All these three scenarios are very specific on the issue of ICT, Big and Open Data, AI, Blockchains and IoT.

First Scenario: The Business as Usual Scenario

The "Business as Usual" scenario follows the current Spanish tourism trends with a dispersion of initiatives without coordination, highlighting the multiplication of commercial applications. Its main driver is the economic and geopolitical situation. E-tourism is being implemented with new sources of innovative, non-public financing, but supported by public-private partnerships as seen in Table 2.

Second Scenario: The Smart Green Tourism Scenario

The second proposed scenario (Table 3) is geared towards the green economy. Environmental protection and climate change are its main drivers. This scenario could be called "Smart Green Tourism". It is adapted to the sustainable development of the Cordoba Province which holds good environmental indicators (Lanquar 2016), high biological diversity and contamination levels below the national average. Cordoba could aspire to the title of European Green Capital after 2025 if it becomes a sustainable city in a sustainable territory. This implies a coherent strategy for the whole province, not only in environmental, but also economic and social issues such as territorial governance, sustainability of natural resources, management of mobility, social cohesion with inclusion projects and the intensive use of transparent information and communication technologies as theoretically determined by Saxena, G. & Ilbery, B. (2008) and proposed by the Añora Declaration of October 2017 (UCO 2017). An integrated rural tourism linked to innovation networks is necessary within a regional approach.

In the medium-size township of Añora (Province of Cordoba), the Spanish Secretary-General of the Ministry of Agriculture launched the "Smart Rural Land Program" to promote, produce and develop smart rural environments. He echoed the arguments of the Añora Declaration which identifies that connectivity should be a right for everybody wishing to develop him/herself personally and professionally in

Table 2. SWOT of the «Business as Usual" scenario

Strengths	Weaknesses
Strengths Cultural and gastronomic destination that rides on Spanish tourism. Promoted by the UNESCO World Heritage Lists. Initiatives and applications by tourism professionals for the use of ICT. Significant promotional strategy.	**Weaknesses** Urban model associated to real estate, amalgamating tourist zones and areas that are not. Too many sightseeing and one-day trips from Seville, Granada or Malaga Not enough long duration tourists Tourists visit mainly the Capital, not the Province No transformation strategy in smart destination – no governance and efficient coordination between the city of Cordoba and its Province No massive investments on digital technologies and connectivity. No support for the inclusion of populations in tourism development for a better convivence. No incentive to create startups. Professional rigidities and negative valuation of tourists on certain aspects such as cleanliness, tourist signage.
Opportunities Broadband throughout Cordoba and the towns and villages of the Province. Riding the image and the competitiveness ranking of Spain. Better use of high-speed trains. National and regional policies on digitalization. Local businesses (especially souvenirs and handicrafts as jewelry) and restaurants that multiply Apps for their promotion.	**Threats** Low impact on social networks. Overtourism and gentrification. Development of uncontrolled uberized economy. . Tourist lodging rentals authorized (type Airbnb) exceeding the carrying capacity. More effective competition from Andalusian cities that choose an efficient "smart tourism" strategy. Poor or incomplete employment of local human resources, poor training programs. Climate change accelerated. Political limitations on the use of Open Data.

a rural environment. He claimed that public administrations should promote infrastructure investments that improve the use of new technologies in rural areas, including the use of satellites (as the European Galileo network) and activities for the implementation of innovative measures and procedures using smart technologies. He asserted the necessity to train the rural entrepreneurs in the most advanced digital technologies.

Third Scenario: Intercultural Cordoba In the Heart Of The Three Cultures

Finally, the third scenario (Table 4) called "Cordoba in the heart of the three cultures (Jewish, Christian and Muslim)", plays on the challenge of interculturality. In 2031, Cordoba's application to be the "European Capital of Culture" may bring benefits to the whole Province. These positive effects could last two decades translated in international fame and intercultural education. Its main drivers are the intercultural relations and the geopolitical situation, mainly from the Mediterranean and Middle-East area.

To obtain a high-quality cultural image, it will be necessary to improve the quality of life of citizens and the environment. This third scenario is oriented towards the dialogue between religions and civilizations. Cordoba is a symbol of interculturality and played during various centuries a central role in the exchanges between West and Middle-East. Segments of tourism such as religious, educational and medical would be further developed if this transformation is boosted by the multiplication of initiatives such as high-tech incubators and startups. It is the example of BIOTECH launched in February 2018 through a partnership between IMDEEC (Municipal Institute for Economic Development and Employment), UCO (University of Cordoba) and FIBICO (Foundation for the Biomedical Sciences Research of

Table 3. SWOT of the smart green tourism scenario

Strengths Rural and agricultural Province with a well-protected environment. Strong connection between Cordoba and its surrounding territories (broadbands, 5G). Organic gastronomy well known on social networks linked to local biodiversity. Awareness of tourism professionals of the importance of domestic tourism. Apps on green tourism. Development of a rural e-commerce and a "uberized" collaborative economy linked to tourism. Training for rural entrepreneurship.	**Weaknesses** Few structural investments made so far to link Cordoba capital and its districts. Lack of coordination with the culture and folklore component of the rural territories. Depopulation: concentration around middle- size towns (more than 5.000 inhabitants) offering most public services. Lack of computerized and secure payment methods.
Opportunities 5 G / Broadband in all municipalities. Not restricted use of Open Data Local festivals and fairs to develop smart tourism. Andalucia Lab's training on ICTs for the development and dynamization of rural areas. Creation of startups on nature and environment. Better use high-speed trains' stopovers in Los Pedroches and Subbética.	**Threats** Saturation of some rural sites. Excessive marketing of holidays, especially the Holy Week (Semana Santa). Monopoly of Amazon- type distribution. Destruction of biodiversity by tourists overpassing environmental carrying capacity in hiking and trekking. Lack of long-term interest by rural populations if their biodiversity is destroyed by climate change.

Cordoba). This incubator hosts 30 companies and generates 180 workstations. It is part of the measure 4.2 of the "Pilot Plan for Governance and Citizen Participation through ICTs" inside the European program UrbaSur with an impact on all the Mediterranean countries.

What scenario could be best suited to ensure the transformation of Cordoba into a smart tourist destination? The second and the third scenarios offer a reasonable and progressive vision of the future of the City of Cordoba and its Province. Smart systems will support this transformation, making the local strategic decisions more transparent, more visible and more efficient. As seen in Table 5, the local respondents, officials, businesses and tourism professional are not very decided on the future they wanted. This table was realized in 2018 through various meetings and phone calls with the author, after

Table 4. SWOT of the "European capital of culture" scenario

Strengths Strengthening Cordoba's prestige as an intercultural destination. Strategy to become European Capital of Culture 2031 based on digital and smart tourism. Pro-active enthusiastic civil society participation. More sound and light shows. Gastronomy associated to Cordoba. Deployment of ICTs aimed to differentiate technologies and highly competitive services for the "Cordoba of the three Cultures".	**Weaknesses** Lack of structural investments in ITC infrastructure and equipment. Lack of cultural liaisons with the rural territories of its Province: Cordoba, oasis in a cultural desert. Lack of liaison with other Andalusian tourist cultural destinations. Lack of training on ICT. No-consciousness of the priority of intercultural initiatives to become a singular smart destination.
Opportunities Broadband throughout Cordoba and the Province. Launch of an incubator on tourism-culture. Open Data considered as a priority. Construction of a technologically advanced eco-museum on the three Cultures and Monotheisms.	**Threats** Excessive marketing of money-making culture for tourism. Overtourism. Misuse of traditions and folklore. Lack of coordination on festivals, fairs and local pilgrimages. Climate change with impossibility to visit Cordoba during the hottest months of the year.

Table 5. Opinions of local officials, businesses and tourism professionals on the future of Cordoba as a smart tourism destination (STD)

OPINIONS	Totally disagree	Disagree	Undecided	Agree	Totally agree
Scenario 1: Business as usual	18	2	5	4	1
Scenario 2: Smart Green Tourism	10	2	5	8	5
Scenario 3: Smart Cultural Tourism	4	6	5	8	7

discussing the opinions presented in 2017 Likert-type survey concerning the transformation of Cordoba in a smart tourism destination.

It is sure that these respondents don't want the continuation of the actual model (Scenario 1). But the Scenario 2: Smart Green Tourism is not seen as the priority. They favored the Scenario 3: Smart Cultural Tourism which is very closed to the Scenario 1: Business as usual with a touch of interculturality and emphasis on ICT. They are unable to challenge the consequences of climate change and the environmental transformation of our societies which will give more importance of rural and ecological destinations.

RECOMMENDATIONS: THE AXES OF TRANSFORMATION

This chapter illustrates the contingencies that the city of Cordoba must face to achieve its transformation into a smart tourism destination. The priorities of the local stakeholders are quite different as it is seen in their opinions through the Likert-type scale (Table 1). The complexity of the digital transformation of Cordoba into a STD does not allow for an exhaustive approach to the measures that could be taken to reach this objective: the municipal government should define intelligent governance to develop smart activities around the citizen - centric concept for better convivence/cohabitation between residents and visitors and avoid the negative impacts of overtourism. That is the sight of the new municipal government elected in 2019.

The future of Cordoba as a smart tourism destination may take various directions as shown in three scenarios. How can a city succeed economically, socially and environmentally, especially focusing the inclusion of its residents and the convivence with the tourists who are temporary residents? How to move to a new planning and managing model to avoid overtourism? Should Cordoba's strategy be based as Richard Florida (2012) says on the 3 T model: Tolerance, Talent and Technology? The municipality should define a smart governance to develop a new tourism mode around the "citizen - centric concept" for better understanding between residents and visitors. The most effective strategy may come from an agreement between the different stakeholders with the establishment of coordinated lines of actuation between the different institutions and sectors of the economy beyond tourism.

This chapter gives also tips for the smart transformation of Cordoba, either through the multiplication of startups that offer confusing Apps without norms and regulations as in the 1st scenario, or by setting municipal and provincial strategies that will ensure the connectivity of the City and the rural territories of the Province of Cordoba, focusing on cultural issues, allowing the participation of its residents as it is suggested in the 2nd and the 3rd scenarios following a green or a intercultural vision.

According to SEGITTUR (2018), many Spanish destinations already have the imprint and brand of smart tourism. They have plans of action with concrete measures considered as best practices. SEGITTUR recommends that each plan for the transformation in a smart destination:

- must have as a backbone: the visitors, without not forgeting the residents,
- the municipal or provincial government team must take agile decisions and use a democratic governance approach for its measures without being afraid by change, generating synergies with the private sector.

These plans integrate all the stakeholders in decision making and is opened to new ideas. Other characteristics must be added as the transparency in monitoring the measures adopted and valued in real time; and the willingness to preserve privacy. The data must be added anonymously, except if the visitor offer them voluntarily. The infrastructures (transportation, waste handling, energy, connectivity, etc.) play a key role. So, investments must be undertaken to keep infrastructures and equipment operational with an excellent and sustained maintenance.

Cordoba must put in place such a plan which follows these best practices. Its transformation plan could be articulated around three stages:

- First stage: current positioning with identification and determination of the priorities of the stakeholders (political, institutional, professional and non-entrepreneurial civil society).
- Second stage: determination of the city's transformation axes and identification of the main initiatives to be taken in the coming years.
- Third stage: elaboration of policies and strategies on smart governance, smart people, smart living, smart mobility, smart economy and smart environment.

On smart governance, it is supposed according to Baggio, R., Scott, N.& Cooper, C. (2010), that "*traditional research has adopted a reductionist approach to modelling tourist destinations: variables and relationships are embedded in simplified linear models that explain observed phenomena and allow implications for management or forecasting of future behavior*". This new perspective means for Cordoba:

- The definition of a public-private strategy. Is it necessary to stuff the city with sensors and cameras? Is it possible to avoid unwilling and continuous monitoring of the citizens and their visitors?
- A public-private consultation structure may be put in place with the creation of a database. This could be entrusted to IMTUR (Municipal Institute of Tourism) and IMDEEC (Municipal Institute of Economic Development and Employment).
- The creation of a foundation or a non-lucrative organization to help the emergence of startups and the creation of incubators and Fab Labs on smart tourism (mainly blockchains and artificial intelligence).

On the smart people issue, the plan of Cordoba should:

- Inform the resident population about smart tourism, by launching an awareness campaign with the tourist professionals and giving classes in schools and colleges as well as conferences or seminars in the local universities.

- Develop' training programs or modules in the local Universities on smart tourism and digital technologies (IoT, Blockchains, AI…).

On smart living, it is recommended to:

- Develop applications projects around the "citizen - centric concept" applications that allow better understanding and convivence between residents and visitors.
- Encourage the inclusion of residents in tourism hosting, not only through platforms such Airbnb.
- Give incentive to local startups and for the development of a collaborative and circular economy.
- Monitor the threat of overtourism in the city historic center.

On smart mobility, it will be preferable to:

- Multiply smart mobility projects especially for elderly and disabled.
- Promote local public transport and multiply applications for information, bookings and payments. Optimize local transport for residents and tourists.
- Make pedestrian the historic center and give better information about travel times.

On smart economy, the plan can consider the following measures:

- Launch technological incubators for tourism and related-sectors startups.
- Develop local cybersecurity with the implementation of blockchains and monitored AI applications.
- Develop open source for the local businesses with free data and interoperability at local and provincial level.
- Elaborate municipal and provincial budget proposals and negotiate with the Junta of Andalusia and the Spanish Government.

Then, on smart environment, the plan should rather:

- Develop a dashboard with environmental indicators connected to IoT.
- Optimize the use of energy by tourists' establishments and providers.
- Optimize waste collection technology
- Optimize the use of water by tourists' establishments and providers.
- Multiply projects that allow a better broadband connection of rural areas.
- Reduce carbon footprints and implement applications to inform tourists and residents to reduce their carbon footprint.
- Implement the recommendations of the Añora Declaration for rural tourism.

The way and the vision of transforming Cordoba and its Province into a smart tourism destination is now associated with the change in the municipal and provincial teams after the 2019 elections. Digitalization was one of the major themes of the political campaign of these elections. Almost every political party was aware of the importance to give to the smart transformation of tourism along to the adaptation of this sector to climate change.

REFERENCES

Baggio, R., Scott, N., & Cooper, C. (2010). Improving tourism destination governance: A complexity science approach. *Tourism Review*, *65*(4), 51–60. doi:10.1108/16605371011093863

Bloom, P. (2019). *Monitored Business and Surveillance in a Time of Big Data*. London, UK: Pluto Press.

Boes, K., Buhalis, D., & Inversini, A. (2016). Smart tourism destinations: Ecosystems for tourism destination competitiveness. [Bingley: Emerald.]. *International Journal of Tourism Cities*, *2*(2), 108–124. doi:10.1108/IJTC-12-2015-0032

Buhalis, D., & Amaranggana, A. (2014). Smart Tourism Destinations. Conference paper. Information and Communication Technologies in Tourism (pp. 553-564). Springer International Publishing Switzerland.

Cordoba Tourism Office. (2015). Plan Estrategico de Córdoba 2015 2019 - Strategic Tourism Plan of Cordoba 2015-2019. EOI, Córdoba, España. Retrieved from https://www.turismodecordoba.org/archivos/2018/201807180954180000pdf

European Parliament. (2014). Mapping Smart Cities in the EU. Study, Directorate – General for Internal Policies, Brussels, PE 507.480 EN, IP/A/ITRE/ST/2013-02. Retrieved from http://www.europarl.europa.eu/studies

Florida, R. (2012). *The Rise of the creative class*. New York: Basic Books.

Godet, M., & Durance, P. H. (2011). *La prospective stratégique - Pour les entreprises et les territoires*. Paris, France: Dunod.

Gretzel, U. (2011). Intelligent systems in tourism: A social science perspective. *Annals of Tourism Research*, *38*(3), 757–779. doi:10.1016/j.annals.2011.04.014

Hollands, R. G. (2008). Will the real smart city please stand up? *City*, *12*(3), 303–320. doi:10.1080/13604810802479126

INE. (2019). *Estadísticas turísticas*. Madrid, Spain: INE.

Lanquar, R. (2013). Tourism in the Mediterranean: Scenarios up to 2030. MEDPRO Report No. 1/July 2011- 3rd Revision and Update May 2013. Brussels, Belgium: CEPS.

Lanquar, R. (2016). Urban coastal tourism and climate change: indicators for a Mediterranean prospective. In *Tourism in the City: Towards an Integrative Agenda on Urban Tourism, L'Aquila Conference of June 2015*. Heidelberg, Germany: Springer.

Likert, R. (1932). A technique for the measurement of attitudes. Archives of Psychology, 22 - 140, 55.

Morales-Fernández, E. & Lanquar, R. (2014). El futuro turístico de una ciudad patrimonio de la humanidad: Córdoba 2031. Best Paper, Tourism, & Management Studies, No. 8.

OECD. (March 2018). Tourism Trends and Policies 2018, Paris.

Saxena, G., & Ilbery, B. (2008). Integrated rural tourism: A border case study. *Annals of Tourism Research*, *35*(1), 233–254. doi:10.1016/j.annals.2007.07.010

SEGITTUR. (2017). *Informe destinos turísticos inteligentes: construyendo el futuro*. Madrid, Spain: Ministerio de Industria, Energía y Turismo, Secretaria de Estado de Comunicaciones y para la Sociedad de la Información.

SEGITTUR. (2018). *INTELITUR, Guía Aplicaciones turística ("236 aplicaciones para facilitar la vida al viajero")*. Madrid, Spain: Ministerio de Energía, Turismo y Agenda Digital, Secretaria de Estado de Turismo.

UNWTO. (2012). *Global Report on City Tourism. AM Reports: Volume six*. Madrid, Spain: UNWTO.

KEY TERMS AND DEFINITIONS

Connectivity: Is linked to the capacity of connecting or interconnecting systems, applications and platforms of individual terminals, mobile devices such as tablets or smartphones and the ability to communicate between them.

Inclusiveness: Is the practice or policy of including people (residents and tourists) in a territory, otherwise be marginalized or excluded.

Innovation Helices: The triple helix model of innovation is based on the interactions between research, industries which produced goods and services, and governments which regulate market. The quadruple helix boosts the knowledge society and the quintuple helix emphasizes the socioecological transition of a society.

Likert – Type Scale: Is a psychometric tool to rate responses of a research questionnaire.

Resistance to Transformation: Is linked to the concept of resistance to change i.e. actions taken by individuals and groups when they perceive that change is a threat for various common reasons such the fear of the unknown.

Scenarios: Is a thinking process of anticipating futures, without aiming to predict them but rather simulating some possible expectations following different paths and visions. A **Smart City** Integrates ICT (information and communication technologies) to improve the performance of public services such as energy, transportation, health, education, utilities in order to enhance the quality of life of its residents and the temporary tourists. The European Parliament (2014) added that smart cities must face three objectives: connectivity, sustainability and inclusiveness.

Smart Governance: Concerns the efficiency of public services of a smart city and their improvement through innovations without forgetting the democratic inclusiveness of its residents.

Smart Tourism Destinations: Involve the use of ICTs to ease access to tourism and hospitality products, services, experiences as well as equipment without overlooking to the efficiency of the destination public services and emphasizing on the inclusiveness of its residents in the tourist sector by a smart governance and a constant interaction between enterprises, local administrations, tourists and residents.

Sustainability: Refers to the sustainable development of human civilization in relation with the biosphere and the territory ecosystems. For the tourism sector, the term "sustainable tourism", formed in 1988, is considered as the ability of a destination to well preserve its environment, protect its population and its economy facing tourism development.

ENDNOTES

[1] 1st UNWTO World Conference on Smart Destinations, Murcia, 2017, https://europe.unwto.org/event/1st-unwto-world-conference-smart-destinations, and 2nd UNWTO World Conference on Smart Destinations, Oviedo (Asturias, Spain), 2018. https://sdt.unwto.org/event/2nd-unwto-world-conference-smart-destinations-0viedo.

[2] https://turismo.gob.es/es-es/servicios/Documents/Plan-Nacional-Integral-Turismo-2012-2015.pdf

[3] That was not the case in Granada, an Andalusian city, situated at 160 kms from Cordoba. Granada was chosen to participate to the European project of European Capital of Smart Tourism 2019. Lyon in France and Helsinki in Finland won the concourse directed to European cities with more than 100 000 inhabitants.

Chapter 17
Intelligent Touristic Logistics Model to Optimize Times at Attractions in a Thematic Amusement Park

Aida-Yarira Reyes
https://orcid.org/0000-0002-0104-9522
Universidad Autónoma de Ciudad Juárez, Mexico

Carlos-Alberto Ochoa
https://orcid.org/0000-0002-9183-6086
Universidad Autónoma de Ciudad Juárez, Mexico

Diego Adiel Sandoval Chávez
El Colegio de Chihuahua, Mexico

Evelyn Teran
Universidad Autónoma de Ciudad Juárez, Mexico

ABSTRACT

This chapter analyzes a thematic tourist park using a logarithmic model. The study shows that the application of the model allows the management of different routes to optimize waiting times and take advantage of the time allocated to the fun. The theoretical concepts were Dijkstra Algorithms. The investigation is exploratory in a single place. The data were obtained during the stay in the park of the year 2018 and was concentrated on a database. The research concludes that the application using the Dijkstra Algorithm on a mobile dispositive can determine the best places to visit and improve the experience in a thematic park.

DOI: 10.4018/978-1-7998-2112-0.ch017

INTRODUCTION

Since ancient times, people have expressed that fun and leisure time are paramount for social wellbeing. Both fun and leisure time have played a key role along human history. These two may conform a social construct because they are in line with the hedonistic nature of human beings, which consider free time as a valuable life asset. Literature shows a myriad of works related to activities involving fun (active or passive), both considering individual and group realms (Reyes and Peña, 2018). We know that culture, amount of spare time and economics influences the way people conceive fun and leisure, so it is important to diversify fun and leisure offer to be competitive in meeting a great variety of needs.

The fact that people seek unique fun experiences stimulates the fun industry, which at its time improves the economic environment, creating jobs and contributing to enhance other economic sectors. The constant desire for fun calls for a renewed and versatile portfolio of fun offer and for innovative forms of entertainment and recreation products. Today we qualify the term *society* in a variety of ways: society of spectacle, society of consumption and even hyper-consumption, and many other adjectives such as risky, cyber, Narcissistic, therapeutic, Nihilist, after-pop, long tail economy, besieged, hypermodern, and many more (López, 2011). There are no doubts that fun market is greatly demand-oriented, that is, dominated by the consumers.

THEME PARK TOURISM

Due to their economic importance, theme parks constitute crucial elements of the tourism industry. Millions of people seeking diversified entertainment visit theme parks around the world every year. Hu (2013) suggests that in theme parks a set of complex and creative elements syncretize. This complexity is capable of meeting many needs; this may explain why World Tourism Organization (WTO) considers theme parks as one of the three main trends in the tourism industry.

Figure 1. Historical evolution of theme parks
Source: New emerging segment of tourism, cited by (Secall, 2001)

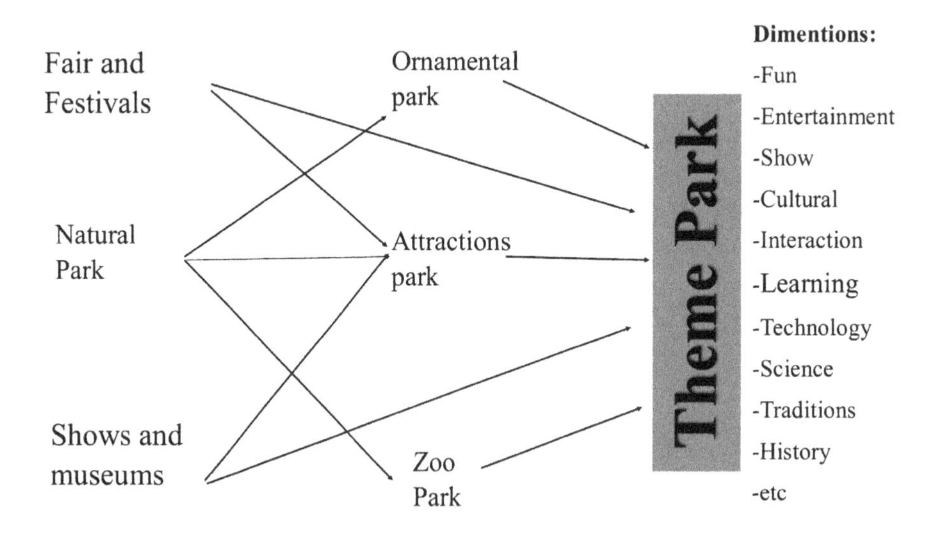

Table 1. Top 10 Theme Park Groups Worldwide Attendance

RANK	GROUP NAME	% CHANGE	2017	2016
1	Walt Disney Attractions	6.8%	150,014,000	140,403,000
2	Merlin Entertainments Group	7.8%	66,000,000	61,200,000
3	Universal Parks and Resorts	4.4%	49,458,000	47,356,000
4	OCT Parks China	32.9%	42,880,000	32,270,000
5	Fantawild	21.7%	38,495,000	31,639,000
6	SIX Flags INC.	2.3%	30,789,000	30,108,000
7	Chimelong Group	13.4%	31,031,000	27,362,000
8	Cedar Fair Entertainment Company	2.4%	25,700,000	25,104,000
9	Seaworld Parks & Entertainment	-5.5%	20,800,000	22,000,000
10	Parks Reunidos	-1.1%	20,600,000	20,825,000
TOP 10 TOTAL	Attendance 2017	8.6%	475,767,000	438,267,000

Sourse: (TEA, 2019)

Origins of theme parks trace back to 19th century when pleasure gardens converted to amusement parks (Richard & Orlowski, 2015). This new conformation succeeds because it offered an increased variety of products and services, such as shows, rides, games, food and beverages. Theme parks evolved from simply forms, such as the open spaces for fairs and exhibitions in Denmark and London, to complex forms, such as the actual multifunctional facilities s (Khafash, Ordóñez, & Fraga, 2015). Following the evolving process Figure 1 portrays, theme parks reached America in 1893 when Columbia World Exhibition took place in Chicago. Since then theme parks gradually evolved until becoming one of the major forms of tourist attractions (Lo & Leung, 2015).

The openings of Walt Disney´s Disneyland in California (1995) and in Florida (1971) were decisive milestones in the global economy expansion of theme parks (Richard et al., 2015). Many countries in all continents followed the Disney´s model: Japan, France, Spain, China, Singapore, and later Mexico, Brazil, Colombia, Venezuela, and Guatemala.

The number of visits to theme parks has experienced a growing tendency (see Table 1 and 2). With 150,014,000 million registered visits (6.6% growth) in 2017, Walt Disney are still the most visited parks, especially Magic Kingdom (Buena Vista, Fl) with 20,450,000 visits in the same year. The runner up is Merlin Entertainment Group with 66,000,000 visits in 2017 (7.8% growth) (TEA, 2019). The annual attendance at the 70 main theme parks reaches 334.6 million. This figure compares to the entire second tier of theme parks of the world (Richard & Orlowski, 2015).

In his review work (see Table 3), Hu (2013) found that theme parks research categorizes in historical evolution, classification theme, diversity theme, space layout, contact with industry and development trending. These categories help to understand the many facets and importance of theme parks from different perspectives.

Table 2. Top 25 attendance Amusement/Theme Parks Worldwide, (TEA, 2019)

RANK	GROUP NAME	% CHANGE	2017	2016
1	Magic KIngdom at Walt Disney World, Lake Buena Vista, FL, U.S.	0.3%	20,450,000	20,395,000
2	Disneyland, Anaheim, CA, U.S.	2.0%	18,300,000	17,943,000
3	Tokyo Disneyland, Tokyo, Japan	0.4%	16,600,000	16,540,000
4	Universal Studios Japan, Osaka, Japan	3.0%	14,935,000	14,500,000
5	Tokyo Disney Sea, Tokyo, Japan	0.3%	13,500,000	13,460,000
6	Disney's Animal Kingdom at Walt Disney World, Lake Buena Vista, FL, U.S	15.3%	12,500,000	10,844,000
7	EPCOT at Walt Disney World, Lake Buena Vista, FL, U.S.	4.2%	12,200,000	11,712,000
8	Shanghai Disneyland, Shanghai, China	96.4%	11,000,000	5,600,000
9	Disney's Hollywood Studios at Walt Disney World, Lake Buena Vista, Fl, U.S	-0.5%	10,722,000	10,776,000
10	Universal Studios at Universal Orlando, FL, U.S.	-0.5%	10,198,000	9,998,000
11	Chimelong Ocean Kingdom, Hengqin, China	15.0%	9,788,000	8,474,000
12	Disneyland Park at Disneyland Paris, Marne la Vallée, France	15.0%	9,660,000	8,400,000
13	Disney's California Adventure, Anaheim, CA, U.S.	3.0%	9,574,000	9,295,000
14	Islands of Adventure at Universal Orlando, FL, U.S.	2.0%	9,549,000	9,362,000
15	Universal Studios Hollywood, Universal City, CA, U.S.	12.9%	9,056,000	8,086,000
16	Lotte World, Seoul, South Korea	-17.6%	6,714,000	8,150,000
17	Everland, gyeonggi-do, south korea	-9.5%	6,310,000	7,200,000
18	Hong Kong Disneyland, Hong Kong sar	1.6%	6,200,000	6,100,000
19	Nagashima Spa Land, Kuwana, Japan	1.4%	5,930,000	5,850,000
20	Ocean Park, Hong Kong sar	-3.3%	5,800,000	5,996,000
21	Europa Park, Rust, Germany	1.8%	5,700,000	5,600,000
22	Walt Disney Studios Park AT Disneyland Paris, Marne La Vallée, France	4.6%	5,200,000	4,970,000
23	De Efteling, Kaatsheuvel, The Netherlands	8.7%	5,180,000	4,764,000
24	Tivoli Gardens, Copenhagen, Denmark	0.0%	4,640,000	4,640,000
25	Seaworld Florida, Orlando, FL, U.S.	2.9%	4,220,000	4,402,000
TOP 25	**ATTENDANCE GROWTH 2016–17**	**4.7%**	**243,926,000**	**233,057,000**

Source: (TEA, 2019)

DISNEY AMUSEMENT PARKS

Caught in a maze of amenities and consumptions posts, carrying programs with a tight agenda, visitants to Disney´s parks are able to enjoy a comic-character-oriented environment for periods that may exceed 8-10 hours (Khafash et al, 2015).

Theme parks offer complex, differentiated fun events under an artistic environment full of creativity. These strictly-copyrighted events generate symbolic meanings, which increase at the pace of the use of technology y realism (Lasuén, Gracia, & Prieto, 2005).

Table 3. Research theme parks trend

Classification	Types	Details
The historical evolution of theme park	Historic	Country and cities.
The research review on the classification of the theme park	According to the size	a) Large theme park. b) Regional theme park. c) Theme park. d) Small theme parks and attractions.
	According to the nature of the subject of theme park	(1) Miniature landscape classes. (2) The customs class. (3) The historical and cultural class. (4) Theme parks. (5) The plants class. (6) The film class.
	other classification methods	According to the location According to the main functions According to the new and high technology content According to the theme park development stage
The diversity of theme park	The success of the Disneyland and universal studios	Use high technology to realize virtual landscape and reduction of various film scene. Use advanced technology to create all sorts of virtual scene and attract visitors of all sorts of different ages all over the world.
The space layout and influence factors analysis of theme park	The urban size	Large theme park. The market. The customers
The contact to the industry	In foreign countries	The development of the tourist area for urban planning and regional zoning plays an important role
	Urban area	Regional economic development to a certain extent determines the development of tourism
	Tourism	The different development model as an affiliated industry
	Economic	Influence of theme park and think successful theme parks have a great contribution to regional tourism
The theme park development trend	Global	With the expansion of globalization and development of the theme park.
		the homogeneity competition period, especially the construction of the Disney theme park in the global scope

Source. Own elaboration based on Hu (2013).

Sayre and King (2003) found three characteristics of theme parks:

a. Oftentimes there is a complex network of ideas, symbols and messages. These elements enrich the sensory experience of visitants.
b. In the communication and entertainment industry, technology is the baseline upon which goods and services rely.
c. The strategy is to induce in customers the experience that the theme park will provide before buying it.

Other contributions mentioning that what really distinguishes this industrial sector are: (Lieberman and Esgate, 2006)

a. The content of their assets: development of creative ideas; the use of technologies that support their production processes; the search for human talent and group work for the generation of products;

Table 4. The new research on amusement and thematic parks

Author	Year	Title
Matthew Hodge	(2018)	Disney 'World': The Westernization of World Music in EPCOT's "IlluminNations: Reflections of Earth"
Rachel Farias do Patrocínio, Juliana Lopes de Almeida Souza, Carolina Toledo Oliveira Santos, Katheryne Soares Martins	(2018)	The vision of the Disney World: an experience marketing study at The Walt Disney Company
Laura Robinson	(2017)	The Disney Musical on Stage and Screen: Critical Approaches from 'Snow White' to 'Frozen'
Ahmed Bitar, Aliaa Jamal, Hesham Sultan, Nour Alkandari, and Mohammed El-Abd	(2017)	Medical Drones System for Amusement Parks
Gustavo Dejean	(2015)	A Methodology for Classifying Visitors to an Amusement Park VAST Challenge 2015 - Mini-Challenge 1
Hayato Ohwada, Masato Okada and Katsutoshi Kanamori	(2013)	Flexible Route Planning for Amusement Parks Navigation

Source. Own elaboration

the commercialization and legal protection of products based on copyright and the high degree of obsolescence (perishability) of their goods.

b. The type of conduits or distribution channels through which products are delivered to consumers: knowing where they will be distributed (in a theater, a cinema, a stadium, a magazine, a radio show, etc.) and how and at what time the distribution will be made.
c. The way in which its recipients consume it: knowing if it will be inside or outside the home where the product will be purchased.

Currently, 12 Disney theme parks operate around the world. The aspects all have in common are

a. The number of large format and night shows.
b. Variety in the setting based on characters or cultures.
c. Accessibility to attractions and proximity between areas.
d. Sensation of wanting to repeat again and again.

Varied attractions in terms of sensations.

Research regarding amusement parks adopts multiple approaches (see Table 4): routes flexibility, relationships with visitors, use of drones, addressing specific issues, such as lighting or reflectors, musicals o marketing. We did no find a study adopting an algorithm to analyze paths.

PROBLEM SETTING

Tourism is a very active and profitable economic sector, particularly in those countries offering recreation services, such as amusement parks. The diverse amenities and activities one can visit and do in an amusement parks reinforce the idea that developing models to optimize routing times is of utmost importance. Visitors to amusement parks are always looking for top-of-the-line experiences, so it is important to offer them the highest value for the money spent. Achieving such a goal will conferee competitive edge, which at its time will increase profit.

Long walking and waiting times the tourists face while visiting an amusement park are the main determinants of the visit experience and, to some extent, explain the degree of satisfaction. These two factors may undermine the ability of enjoying the most during a visit. This may result in unsatisfactory experiences and extra expenses

This work proposes a logistics model to optimize waiting times during a visit to an amusement park. Availability of such a model will enable visitors to enhance their experience at the park. A better degree of satisfaction will increase visits, thus increasing income and profit.

Upon arrival to the park, visitors face a scenario of long distances, long transfer time between amenities and activities´ schedule disparities. These difficulties often result in shortage of time to visit additional attractions. By adopting a tourist logistic model, this work aims to optimize waiting times at attractions in an amusement park.

Method

The study considers seven groups: a person with a newborn baby, a person using a cane, a person with strollers, an adult with a senior citizen companion, a person using scooters ECV/wheelchair, a couple and an adult by himself. Under an exploratory approach, during the year 2018 we collected data in the Experimental Prototype Community of Tomorrow (EPCOT) at Walt Disney World, Lake Buena Vista, FL. The work focus on studying:

a. A descriptive theme park
b. The route time (trajectory inside the theme park).
c. The output of the Dijkstra algorithm

A route designates the set of possible routes from which visitors have sight of a site or attraction. Usually, visitors have an idea of which route to follow based on previous information suggested by diverse sources such as experts, social media or relatives.

The Disney Park, conducive to suggest certain and not another interpretation of the exhibited issue and/or its relation of similarity /difference /contradiction with other issues. For this reason, it is important that the responsible of organizing the location of events should be showing designs and proposes for a route, where the visitors have the option to either accept it or ignore it. The route can be considered a partial way, total or to be absent (where a route has not been designed or exhibited issue is the unique one and immediately shown. In that case, the route is not contributing elements to see something different, but it generates an access and visualization of space).

The method involved to estimate the time it takes to get to different areas of the park was physical visit to Epcot park. In 2018, the park was visited with the intention to calculate all times and different routes to each and all attractions provided by the theme park. The two sections of the park were visited by walking at an average rate of speed. Taking into consideration that the approximated times and routes were only considered, not taking into account the time a person takes visiting\making lanes for the attractions.

We visited the EPCOT Park in 2018 and determined the time it took to transit from one site to another at normal walking pace, without considering waiting time in line.

Dijkstra Algorithm

Edsger Wybe Dijkstra developed an algorithm which objective is to determine the shortest path from one source to a set of destinations. It has many applications such as flight routes, distribution centers routes, sales routes and many more (Sniedovich, 2006).

The shortest path problem from one vertex to another is to determine the lowest-cost path from a vertex u to vertex v other. The cost of a path is the sum of the costs (weights) of the arcs that comprise it.

$$G = (V,A) \text{ a weighted directed graph} \quad (1)$$

V: Here is the algorithm for a graph G with vertices V = {v1, ... vn} and edge weights wij for an edge connecting vertex vi with vertex vj. Let the source be v1.

C: is a matrix of costs of edges of the graph, where C [u, v] the cost of the edge between u and v is stored.

S: is a set containing the vertices for which already has determined the shortest path. Initialize a set S =i. This set will keep track of all vertices that we have already computed the shortest distance to from the source.

D: be a one-dimensional arrangement such that D [v] is the cost of the shortest path from source vertex to vertex v. Initialize an array D of estimates of shortest distances. D [1] = 0, while D[i] =j, for all other i. (This says that our estimate from v1 to v1 is 0, and all of our other estimates from v1 are infinity.)

P: be a one-dimensional array such that P [v] is the predecessor of vertex v in the shortest path that has built.

While S=V do the following:

a) Find the vertex (not is S) that corresponds to the minimal estimate of shortest distances in array D.
b) Add this vertex, vi into S.
c) Recomputed all estimates based on edges emanating from v. In particular, for each edge from v, compute

$$D[i] + wij$$

If this quantity is less than D[j], then set

$$D[j] = D[i] + wij$$

This algorithm is easiest to follow in a tabular format. The adjacency matrix of an example graph is included below. see Figure 2.

The Dijkstra's algorithm applications are varied and different objectives in the search for better routes, the most current articles are: Mobile Application Based on Dijkstra's Algorithm, to improve the inclusion of people with motor disabilities within urban areas (Arellano, Del Castillo, Guerrero & Tapia, 2019); Bounded Dijkstra (BD): Search Space Reduction for Expediting Shortest Path Subroutines from Van Bemten, Guck, Machuca, & Kellerer (2019); xTrek: An Influence-Aware Technique for Dijkstra's

Figure 2. Dijkstra algorithm
Source: own design

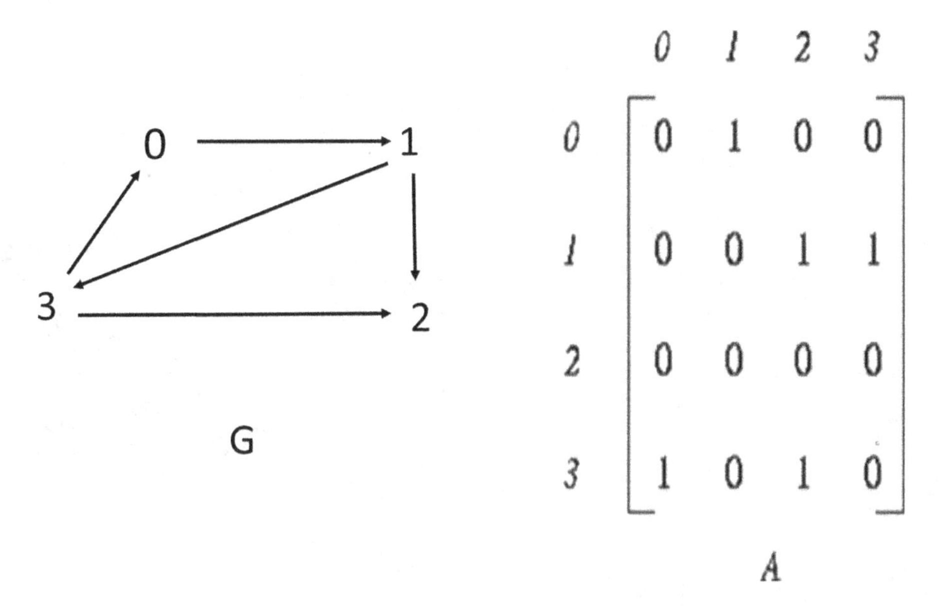

and A Pathfinders from Amador & Gomes (2018); Finding the shortest path by Adhwarjee's algorithm and comparison of this powerful method with Dijkstra's algorithm by Adhwarjee (2018); A Shortest Path Finding Application for Jakarta Public Transportation using Dijkstra Algorithm by Wongso, Cin, Suhartono and Joseph (2018); and Path-Value Functions for Which Dijkstra's Algorithm Returns Optimal Mapping by Ciesielski, Falcão & Miranda (2018).

RESULTS

A Descriptive Information About the Theme Park

EPCOT, originally Magic Kingdom, is a park focused on international culture and technological innovation, which pertain to Walt Disney chain. Located in Florida, it encompasses two sections: Future World and World Showcase. It opened in 1982 (just before Mr. Walt Disney passed away) under the name EPCOT Center, and later renamed to EPCOT in 1996. Before the opening of Animal Kingdom in 1998, EPCOT was the largest theme park of the Disney chain (EPCOT, 2018), (see Figure 3):

i. Future World is a tech-based attraction section, which portrays Disney films, such as Nemo, Inside Out or Bymax.
ii. World Showcase, a multicultural-based section, integrates 11 pavilions to express and show the culture of diverse countries.

Occupying an area of 123 ha, EPCOT had an estimated investment of $1.4 billion. The park ranks 7th in visitors with an attendance of 12.2 million. It is the most visited park in USA (TEA, 2017). The theme is to emulate the "city of tomorrow".

Figure 3. Information of EPCOT at Walt Disney World
Source: (EPCOT, 2018)

Each pavilion has an entrance with a structure representative of the country, for example: the pavilion of Mexico has a pyramid-shaped building. The environment of the pavilions has representative elements of the most important cities in the given country, for example: in Japan, all the decoration is oriental, the food stalls are hallucinated to japan cuisine, the cast member outfits must be representative from Japan, you can shop clothes, crafts, electronic items, etc. One of the attractions in each area refers to an element of the country, for example: Norway has an allusive route to the animated Disney movie "Frozen".

From-To Chart Data

Table 5 shows the times among the many section destinations, pavilion to pavilion.

Symbolic Capital Related to The Attractiveness Of Each Pavilion

Criteria for categorizing each pavilion were as follows: food and beverage (restaurants, quick services and kiosk for the seasonal festivals of EPCOT). Merchandise (all merchandise offered). Attraction (if the countries have an attraction to offer, such as rides, theater; etc.). Entertainment (character spot, shows, performances and interactions with guests). Total (the result to measure if the pavilions have all the categories), see Table 6.

Table 5. From-to char in World Showcase section

Countries	Mexico	Norway	China	Germany	Italia	USA	Japan	Morocco	Francie	Uk	Canada
México	-	2 min	4 min	8 min	9 min	11 min	13 min	14 min	15 min	19min	21 min
Norway	2 min	-	1 min	5 min	6 min	7 min	10 min	12 min	13 min	17 min	19 min.
China	4 min	1 min	-	4 min	5 min	7 min	9 min	11 min	12 min	16 min	18 min
Germany	8 min	5 min	4 min	-	1 min	3 min	5 min	7 min	8 min	12 min	14 min
Italia	9 min	6 min	5 min	1 min	-	2 min	4 min	6 min	7 min	11 min	13 min
American Adventure	11 min	7 min	7 min	3 min	2 min	-	2 min	3 min	4 min	8 min	10 min
Japan	13 min	10 min	9 min	5 min	4 min	2 min	-	1 min	2 min	7 min	9 min
Morocco	14 min	12 min	11 min	7 min	6 min	3 min	1 min	-	2 min	6 min	8 min
Francie	15 min	13 min	12 min	8 min	7 min	4 min	2 min	2 min	-	4 min	6 min
UK	19 min	17 min	16 min	12 min	11 min	8 min	7 min	6 min	4 min	-	2 min
Canada	21 min	19 min.	18 min	14 min	13 min	10 min	9 min	8 min	6 min	2 min	-

Source. Own elaboration

a) Mexican Pavilion:
   i. The pavilion offers an extensive gastronomy. Mexico has two restaurants, one quick service restaurant, two seasonal kiosks, and last but not least a Tequila Bar.
   ii. The pavilion provides all types of different merchandise from Mexico.
   iii. The Pavilion has a famous boat ride called "Los Tres Caballeros".
   iv. Mexico offers shows with the "Mariachi Cobre".
b) Norway Pavilion:
   i. The pavilion offers a great gastronomy, but what makes Norway a very popular is the royal restaurant that provides buffet, which offers interaction with Disney Princesses. Norway also has a quick service restaurant.
   ii. The pavilion doesn't offer much from the country, except a large variety of Winter clothing. Most likely this pavilion has merchandise of the animated characters from the movie "Frozen".
   iii. It's a very popular pavilion to visit due that it provides a boat ride, of the animated movie Frozen.
   iv. Norway has Vikings interacting with the guests.
c) China Pavilion:
   i. Offers gastronomy related to the country, two restaurants quick service restaurant and one seasonal kiosk.
   ii. The pavilion provides extensive merchandise from China.
   iii. Offers a show related to the history of China.
   iv. As entertainment, China provides a show of cast members performing China dances.
d) Germany Pavilion:
   i. Germany offers an extensive gastronomy. When it comes to a great buffet in the World Showcase area, Germany can't never go wrong in its restaurant. Germany also provides a quick service restaurant and a seasonal kiosk.
   ii. Germany offers extensive merchandise from the country.
   iii. C and D. This pavilion does not offer an attraction, however as part of the entertainment they offer performances a Disney Princess spot with the character from this country: Snow White.

e) Italy Pavilion:
   i. offers a wide variety of gastronomy. Italy has two restaurants, full table service, a quick service restaurant, a seasonal kiosk and a wine cellar.
   ii. The pavilion provides merchandise from Italy.
   iii. The pavilion offers a boat ride; however, this ride is not always open for the public.
   iv. As entertainment, Italy provides a show where cast member juggle around the pavilion.
f) American Adventure:
   i. The US pavilion provides American gastronomy and also very popular us American beverages such as beer. It offers a quick service and a seasonal kiosk.
   ii. The pavilion offers American merchandise, but not as attractive as the rest of the pavilions.
   iii. Offers a show in a movie theater about the US history.
   iv. The pavilion does have a popular spot in where The World Showcase offers small concerts and performances with famous artists.
g) Japan Pavilion:
   i. Offers popular gastronomy. It has two restaurants, full table service, in where a chef cooks the food on a tableside grill and interacting with guest while making Disney figures with the ingredients. It also has one quick service and seasonal kiosk.
   ii. The pavilion provides extensive products and merchandise.
   iii. The pavilion does not offer an attraction.
   iv. As part of the entertainment, they provide small shows related to the culture of Japan.
h) Morocco Pavilion:
   i. The pavilion provides popular gastronomy. It has two restaurants: full table service, in where dancers make performances during the meal, and a quick service seasonal kiosk.
   ii. Offers extensive products and merchandise.
   iii. The pavilion does not have attractions.
   iv. Morocco provide performances with cast members dancing typical dances in which the guests can participate. The pavilion also has a Disney character spot for Aladdin and Jazmin.
i) France Pavilion:
   i. The French pavilion offers popular gastronomy, it also has two full table service restaurants, a quick service that most likely provides desserts and a seasonal kiosk.
   ii. The pavilion offers merchandise from the country, the majority of the products are related to the animated movie Beauty and the Beast.
   iii. The pavilion provides a movie in theater as part of the attraction.
   iv. Offers a Princess character spot from the animated movie: Beauty and the Beast.
j) United Kingdom Pavilion:
   i. The UK pavilion offers very popular gastronomy, it has one full table service restaurant, a very popular English pub, quick service restaurant and a seasonal kiosk.
   ii. The pavilion provides a wide variety of products and merchandise.
   iii. The pavilion offers as an attraction a concern spot in which performers sing famous songs by famous British artists.
   iv. The UK pavilion provides Disney character spots with two famous British characters: Mary Poppins and Alice in Wonderland.

k) Canada Pavilion:
   i. The Canada Pavilion offers popular gastronomy, it has one full table service restaurant, a quick service restaurant and seasonal kiosks.
   ii. The pavilion provides merchandise.
   iii. The pavilion offers representation of the Niagara Falls as part of the attraction.

The criteria for weight-related to the representation of the country were assigned according to the characteristics of the pavilion and the perception of the country. The weight goes from 0 to 1, where 1 is the fulfillment of the services offered by the world showcase, (see table 6).

1. Future World: The attractions in the Future World area were categorized with the following concepts: popularity (how popular the attraction is); The waiting times (as in how much time usually takes a person to wait to get in the ride. For this concept, it was also considerate if the ride had any games or activities to interact during the waiting times. Rating a 1: if the ride or attraction is faster to get on, and 0: representing the longest waiting time in lane); The availability for a guest to get a Fast Pass for the specific attraction. Rating a 1 with easy access to get a fast pass and a 0: representing how difficult it is to obtain a fast pass. (See Table 7). The criteria to categorize this section include popularity, waiting time and availability of interaction during this period, and availability to obtain a Fast Pass.

   a) Soarin:
      i. Soarin is a very well-known and a very popular ride in EPCOT for all family members.
      ii. waiting times for this ride can (might) get very long. The ride does not have interactive game (s) during the wait time.
      iii. access to get a fast pass for this attraction is high.

Table 6. Weighting of the Showcase section for each pavilion

Pavilion	Food & Beverage	Merchandise	Attraction	Entertainment	CS
Mexico	1	0.8	0.9	0.9	3.6
Norway	0.9	0.6	1	0.5	3
China	0.9	1	0.4	0.7	3
Germany	1	0.8	0	0.5	2.3
Italy	1	1	0.5	0.7	3.2
American Adventure	0.5	0.3	0.3	1	2.1
Japan	1	1	0	0.4	2.4
Morocco	1	1	0	0.8	2.8
France	1	0.8	0.6	0.5	2.9
UK	1	1	0.7	1	3.7
Canada	0.7	0.4	0.3	0.7	2.1
International Gateway	0	0	0.7	0.5	1.2

Source. Own elaboration

b) Mission Space:
   i. Mission Space is also a popular ride. However, this ride has policies and limitations with height and health conditions.
   ii. The waiting times for this attraction can get very long, this ride also does not have an activity or any interactive game during the wait time in lane.
   iii. The access to get a fast pass for this attraction is high.
c) Test Tract:
   i. Test Tract is well known, fun and popular ride in EPCOT for kids all ages and family members.
   ii. Waiting times for this attraction can get very long, however this attraction has fun and interactive games during the wait time.
   iii. Access to get a fast pass for this attraction is low, it is a very demanded ride.
d) Spaceship Earth:
   i. Spaceship Earth is a popular ride in EPCOT (because) is located inside EPCOT's icon.
   ii. Waiting times for this attraction can be short.
   iii. Access to get a fast pass for this attraction is very high.
e) The seas of Nemo and Friends:
   i. Seas of Nemo and Friends is an attraction designed most likely for kids, which makes it a popular ride for all family members.
   ii. Waiting times for this attraction can be very short.
   iii. The access to obtain a fast pass of this ride, is very high.
f) Character spot:
   i. The character spot is a well-known point where the guests are able to take pictures with a Disney characters.
   ii. Waiting times can be very long, due that is a very demanded attraction.
   iii. The access to obtain a fast pass depends on the character, date and times. But most likely the chances to get a fast pass are high.
g) Norway Pavilion:
   i. The Frozen boat ride is well-known attraction, most likely designed for all family members.
   ii. The waiting time in this ride can get be extremely long, due that is a very demanded attraction for kids.
   iii. The access to obtain a fast pass for this ride are very rare and extremely low, since it's a very famous demanded attraction.

Dijkstra Algorithm Results

Capitalized words of the alphabet identify each one of the pavilions and the attractiveness of the amusement park (Table 8). Results found in the multiplication of travel times between each of the attractions for symbolic capacity (CS) are presented in Table 9 for the Future World section, and Table 10 for Future Showcase.

The output of the algorithm resulted in a path with significant value of Shortest path length is 79.8 min in the following route: U⇒D⇒B⇒C⇒H⇒I⇒K⇒L⇒M⇒N⇒O⇒Q⇒S⇒T, see Figure 4a and 4b, and the matrix presented in Table 11.

Table 7. Weighting of the Future World section

Attraction	Popularity	Wait times	Availability to Fast Pass	CS
Soarin	1	0.4	0.8	2.2
Mission Space	0.8	0.1	0.8	2.6
Test Track	1	0.3	0.4	1.3
Spaceship Earth	0.6	0.9	1	2.5
The seas of Nemo & Friends	0.7	1	1	2.7
Character Spot	0.9	0.6	0.8	2.3
Frozen (Norway Pavilion)	1	0	0	1

Source. Own elaboration

Table 8. Relationship of significance areas

Attraction	Area	Sign
Future world Attraction	Soarin	A
	Mission Space	B
	Test Track	C
	Spaceship Earth	D
	The seas of Nemo & Friends	E
	Character Spot	F
	Frozen (Norway Pavilion)	G
World Showcase Pavilion	Mexico	H
	Norway	I
	China	J
	Germany	K
	Italy	L
	American Adventure	M
	Japan	N
	France	O
	Morocco	P
	International Gateway	Q
	UK	R
	Canada	S
	Entrance	U / T

Source. Own elaboration

Table 9. The Future World Values

Attraction	CS	A	B	C	D	E	F	G
A	2.2	-						
B	2.6	33.8	-					
C	1.3	18.2	2.6	-				
D	2.5	37.5	12.5	25	-			
E	2.7	18.9	27	27	24.3	-		
F	2.3	18.4	13.8	16.1	16.1	13.8	-	
G	1	10	15	14	14	17	13	-

Source. Own elaboration

Table 10. World Showcase Values

Pavilion	CS	H	I	J	K	L	M	N	O	P	Q	R	S
H	3.6	-											
I	3	6	-										
J	3	12	3										
K	2.3	18.4	11.5	9.2	-								
L	3.2	28.8	19.2	16	3.2	-							
M	2.1	23.1	14.7	14.7	6.3	4.2	-						
N	2.4	31.2	24	21.6	12	9.6	4.8	-					
O	2.9	40.6	34.8	31.9	20.3	17.4	8.7	2.9	-				
P	2.8	42	36.4	33.6	22.4	19.6	11.2	5.6	5.6	-			
Q	1.2	18	15.6	15.6	12	9.6	6	6	3.6	7.2	-		
R	3.7	70.3	62.9	59.2	44.4	40.7	29.6	25.9	22.2	22.2	7.4	-	
S	2.1	44.1	39.9	37.8	29.4	27.3	21	18.9	12.6	16.8	8.4	4.2	-
T	-	-	-	-	-	-	-	-	-	-	-	-	-

Source. Own elaboration

CONCLUSION

Results suggest that route U⇒D⇒B⇒C⇒H⇒I⇒K⇒L⇒M⇒N⇒O⇒Q⇒S⇒T consumes the shortest time under the circumstances here considered. In addition, Spaceship Earth, Mission Space, Test Track, Mexico, Norway, Germany, Italy, American Adventure, Japan, France, International Gateway, UK, Canada, are the most important attractors in the shortest time available (see Figure 5).

With these results it is possible to solve the problem raised in relation to the limited times of tourists and visitors in journeys and scheduled times, if the entire park is not known if it is possible to know the most significant and symbolic.

Dijkstra Algorithm can be available in a mobile device to determine the best important places to visit. The route of the main attractions is proposed in a smart 3D screen, with real-time information.

Figure 4. Dijkstra Algorithm results with a graph associated

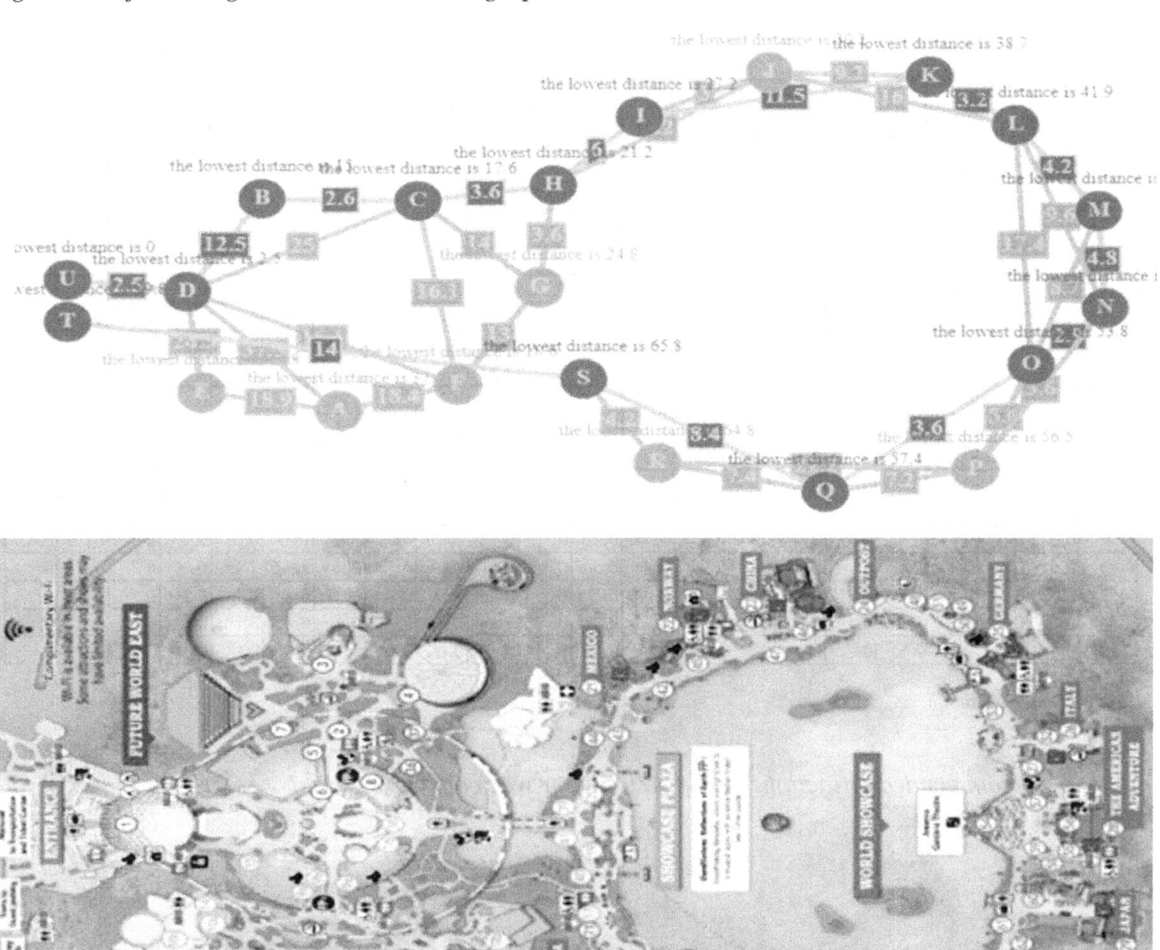

This will conclude save money and time and enhance enjoyment of the trip. The relevant contribution of this work is the use of the algorithm in an amusement park that includes real time and the symbolic capacity in a rout and that allows to avoid the expense of time and money.

FUTURE RESEARCH

It is a recommended application for a mobile device that takes into account current photographs and the main recommendations of the available time in a trip. The Intelligent Tool will be used by different types of people who need to travel together with different situations related with displacement. Another

Table 11. Matrix Dijkstra algorithm results

	U	A	B	C	D	E	F	G	H	I	J	K	L	M	N	O	P	Q	R	S	T
U	0,	0,	0,	37.5,	18.9,	18.4,	0,	0,	0,	0,	0,	0,	0,	0,	0,	0,	0,	0,	0,	0,	0,
A	0,	0,	2.6,	12.5,	0,	0,	0,	0,	0,	0,	0,	0,	0,	0,	0,	0,	0,	0,	0,	0,	0,
B	0,	2.6,	0,	25,	0,	16.1,	14,	3.6,	0,	0,	0,	0,	0,	0,	0,	0,	0,	0,	0,	0,	0,
C	37.5,	12.5,	25,	0,	24.3,	16.1,	0,	0,	0,	0,	0,	0,	0,	0,	0,	0,	0,	0,	0,	0,	2.5,
D	18.9,	0,	0,	24.3,	0,	0,	0,	0,	0,	0,	0,	0,	0,	0,	0,	0,	0,	0,	0,	0,	0,
E	18.4,	0,	16.1,	16.1,	0,	0,	13,	0,	0,	0,	0,	0,	0,	0,	0,	0,	0,	0,	0,	0,	0,
F	0,	0,	14,	0,	0,	13,	0,	3.6,	0,	0,	0,	0,	0,	0,	0,	0,	0,	0,	0,	0,	0,
G	0,	0,	3.6,	0,	0,	0,	3.6,	0,	6,	12,	0,	0,	0,	0,	0,	0,	0,	0,	0,	0,	0,
H	0,	0,	0,	0,	0,	0,	0,	6,	0,	3,	11.5,	0,	0,	0,	0,	0,	0,	0,	0,	0,	0,
I	0,	0,	0,	0,	0,	0,	0,	12,	3,	0,	9.2,	16,	0,	0,	0,	0,	0,	0,	0,	0,	0,
J	0,	0,	0,	0,	0,	0,	0,	0,	11.5,	9.2,	0,	3.2,	0,	0,	0,	0,	0,	0,	0,	0,	0,
K	0,	0,	0,	0,	0,	0,	0,	0,	0,	16,	3.2,	0,	4.2,	9.6,	17.4,	0,	0,	0,	0,	0,	0,
L	0,	0,	0,	0,	0,	0,	0,	0,	0,	0,	0,	4.2,	0,	4.8,	8.7,	0,	0,	0,	0,	0,	0,
M	0,	0,	0,	0,	0,	0,	0,	0,	0,	0,	0,	9.6,	4.8,	0,	2.9,	5.6,	0,	0,	0,	0,	0,
N	0,	0,	0,	0,	0,	0,	0,	0,	0,	0,	0,	17.4,	8.7,	2.9,	0,	5.6,	3.6,	0,	0,	0,	0,
O	0,	0,	0,	0,	0,	0,	0,	0,	0,	0,	0,	0,	0,	5.6,	5.6,	0,	7.2,	22.2,	0,	0,	0,
P	0,	0,	0,	0,	0,	0,	0,	0,	0,	0,	0,	0,	0,	0,	3.6,	7.2,	0,	7.4,	8.4,	0,	0,
Q	0,	0,	0,	0,	0,	0,	0,	0,	0,	0,	0,	0,	0,	0,	0,	22.2,	7.4,	0,	4.2,	0,	0,
R	0,	0,	0,	0,	0,	0,	0,	0,	0,	0,	0,	0,	0,	0,	0,	0,	8.4,	4.2,	0,	14,	0,
S	0,	0,	0,	0,	0,	0,	0,	0,	0,	0,	0,	0,	0,	0,	0,	0,	0,	0,	14,	0,	0,
T	0,	0,	0,	2.5,	0,	0,	0,	0,	0,	0,	0,	0,	0,	0,	0,	0,	0,	0,	0,	0,	0,

Source. Own elaboration

Figure 5. Recommended attractions in EPCOT Center was considered in this research
Source. Own elaboration

important recommendation for future research is the inclusion of different physical conditions of the visitors and that require of special services. In addition, this application will be used as a recommendation system when traveling to other cities or places in different societies and will explain different scenarios based on time, limited resources and location. Another field topic that will benefit from a more appropriate organization is Logistics of the product or service as it describes the use of Cultural

Algorithms to improve a Logistics network. We propose the use of a mobile computing device that can indicate during the tour fun options based on our own tastes and priorities, according to the type of activity you want to perform.

We intend to perform more tests on other equipment's with different characteristics, either hardware with a graphics card with higher processing cores, a webcam with better resolution or other software versions in order to obtain better results in the quality of the processed video. This platform is also intended to be multiplatform for functions not only in Linux but in other operating systems, whether it is the Windows or Mac environment. One of the main problems in a Smart City are accidents with more than one death associated with reckless driving due to alcohol, we propose that using a mobile device and ubiquitous computing techniques determine through photos of the online status of a group of friends that they will travel together, in order to determine who through the recognition of the iris could be the designated driver, something crucial for the lives of the group, as can be seen in figure 6:

One of the big challenges in a Smart City, will be that the young population is becoming groutier and therefore their outings are nocturnal and last longer, which means more alcohol and therefore with a higher incidence in deaths due to vehicular accidents in a Smart City.

Figure 6. Representation of the Model associated with the reduction of deaths in night-time collective trips associated with alcohol in young people in a Smart City.
Source. Own elaboration

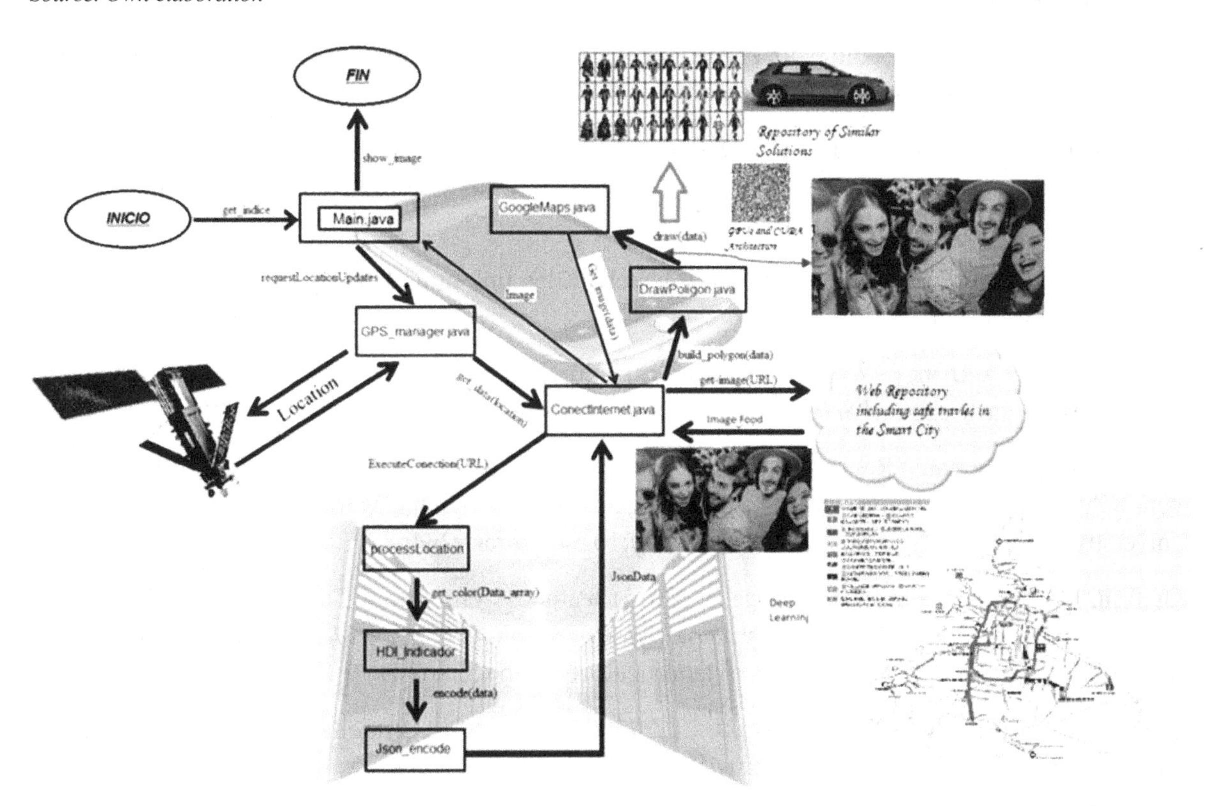

REFERENCES

Adhwarjee, D. K. (2018). Finding the shortest path by Adhwarjee's algorithm and comparison of this powerful method with Dijkstra's algorithm. *International Journal of Mathematics in Operational Research, 13*(2), 269. doi:10.1504/IJMOR.2018.094058

Amador, G. P., & Gomes, A. J. P. (2018). xTrek: An Influence-Aware Technique for Dijkstra's and A Pathfinders [Research article]. doi:10.1155/2018/5184605

Arellano, L., Del Castillo, D., Guerrero, G., & Tapia, F. (2019). Mobile Application Based on Dijkstra's Algorithm, to Improve the Inclusion of People with Motor Disabilities Within Urban Areas. In Á. Rocha, H. Adeli, L. Reis, & S. Costanzo (Eds.), *New Knowledge in Information Systems and Technologies. WorldCIST'19 2019. Advances in Intelligent Systems and Computing* (Vol. 931). Cham, Switzerland: Springer. doi:10.1007/978-3-030-16184-2_22

Bitar, A., Jamal, A., Sultan, H., Alkandari, N., & El-Abd, M. (2017). Medical Drones System for Amusement Parks. *IEEE/ACS 14th International Conference on Computer Systems and Applications* (AICCSA) (pp. 19-20). 10.1109/AICCSA.2017.62

Ciesielski, K. C., Falcão, A. X., & Miranda, P. A. (2018). Path-Value Functions for Which Dijkstra's Algorithm Returns Optimal Mapping. *Journal of Mathematical Imaging and Vision, 60*(7), 1025–1036. doi:10.100710851-018-0793-1

Dejean, G. (2015). A methodology for classifying visitors to an amusement park. En 2015 IEEE *Conference on Visual Analytics Science and Technology (VAST)* (pp. 151-152). 10.1109/VAST.2015.7347655

do Patrocínio, R. F., Souza, J. L. de A., Santos, C. T. O., & Martins, K. S. (2018). The vision of the Disney World: An experience marketing study at The Walt Disney Company. *Archives of Business Research, 6*(9). doi:10.14738/abr.69.5067

EPTO. (2018). *Web Page*. Date recuperate 1 March, 2019, link https://disneyworld.disney.go.com/es-mx/destinations/epcot /

Hodge, M. (2018). Disney 'World': The Westernization of World Music in EPCOT's "IllumiNations: Reflections of Earth". *Social Sciences, 7*(8), 136. doi:10.3390ocsci7080136

Hu, G. (2013) A Research Review on Theme Park. Business and Management Research, 2(4); Management School, Shanghai University of Engineering and Science, China, Management School, Shanghai University of Engineering and Science, China doi:10.5430/bmr.v2n4p83

Khafash, L., Ordóñez, J. A. C., & Fraga, J. (2015). Parques Temáticos y Disneyzación. Experiencias Xcaret en la Riviera Maya. *En Turismo y ocio: Reflexiones sobre el Caribe Mexicano*, 2015, (pp. 45-84). Available at https://dialnet.unirioja.es/servlet/articulo?codigo=5386749

Lasuén, J. R., Gracia, M. I. G., & Prieto, J. L. Z. (2005). Cultura y economía. *Fundación Autor*. Recuperado de https://books.google.com.mx/books?id=EMh8PQAACAAJ

Lieberman, A., & Esgate, P. (2006). *La revolución del marketing del entretenimiento: acercando los magnates, los medios y la magia al mundo*. 1st. ed. Buenos Aires, Argentina: Nobuko, Universidad de Palermo.

Lo, J., & Leung, P. (2015). The Preferred Theme Park. *American Journal of Economics*, (5): 472–476.

López, J. S. M. (2011). Sociedad del entretenimiento (2): Construcción socio-histórica, definición y caracterización de las industrias que pertenecen a este sector. *Revista Luciérnaga-Comunicación*, *3*(6), 6–16.

Ohwada, H., Okada, M., & Kanamori, K. (2013). Flexible route planning for amusement parks navigation. In *Proceedings 2013 IEEE 12th International Conference on Cognitive Informatics and Cognitive Computing* (pp. 421-427). 10.1109/ICCI-CC.2013.6622277

Reyes, A., & Peña, A. (2018). *Diversión y esparcimiento en la frontera: identidad de Ciudad Juárez. Relatos de Ciudad Juárez y otros textos*. Chihuahua, México: Ciudad Juárez.

Richard, B., & Orlowski, K. (2015). Theme Park Tourism. [Encyclopedia Entry]. In book: The SAGE International Encyclopedia of Travel and TourismChapter: Theme Park TourismPublisher. *Sage (Atlanta, Ga.)*.

Robinson, L. H. C. (2017). Book Review: The Disney Musical on Stage and Screen: Critical Approaches from 'Snow White' to 'Frozen' by George Rodosthenous, Ed. London, UK: Bloomsbury Methuen Drama, pp. 97-100, 1, 257.

Sayre, S., & King, C. (2003). *Entertainment and Society: Influences, Impacts, and Innovations* (1st ed.). Kindle Edition.

Secall, R. E. (2001). Nuevo segmento emergente de turismo: Los parques temáticos. *Cuadernos de Turismo*, (7): 35–54.

Sniedovich, M. (2006). Dijkstra's algorithm revisited: The dynamic programming connection. *Control and Cybernetics*, *35*(3).

TEA. (2019). *The Themed Entertainment Association (TEA)*. Retrieved from http://www.teaconnect.org

Van, A., Guck, J. C., Machuca, M., & Kellerer, W. (2019). Bounded Dijkstra (BD): Search Space Reduction for Expediting Shortest Path Subroutines. Computer Science: Networking and Internet Architecture. Retrieved from https://arxiv.org/abs/1903.00436

Wongso, R., Cin, C., & Suhartono, J. (2018). TransTrip: A Shortest Path Finding Application for Jakarta Public Transportation using Dijkstra Algorithm. *Journal of Computational Science*, *14*(7), 939–944. doi:10.3844/jcssp.2018.939.944

KEY TERMS AND DEFINITIONS

Attraction: Something, space, area, activity and show that makes people come to a place.

Cultural: The information about people and countries relate to the history, food, clothes, habits, traditions, and beliefs of a society.

Entertainment: Activity, performances or events that entertain people.

Merchandise: Service and goods that are bought and sold.

Pavilion: Is a temporary structure, such as a large tent, especially used at public events or for shows. Spaces designed and decorated on a specific country. They seek to rescue traditions and elements representative, such as food, clothing, music, amusements.

Shows: Is an event that makes it possible for be fun and enjoy something.

Technology: The innovation of the equipment, machinery, infrastructure, decoration and methods that are application of science and industry.

Theme Park: Are fun spaces for everyone, are inclusive and seek to use attractions for visitors to enjoy. The use of fantasy and technology are part of the novelties that will always identify the theme parks.

Virtual Reality: Situations, conditions, images and sounds produced by technology and that make the situation is real.

Chapter 18
Gamification as a Tool for Smart Tourism

Noelia Araújo Vila
https://orcid.org/0000-0002-3395-8536
University of Vigo, Spain

Lucília Cardoso
CITUR, Portugal

Diego R. Toubes
https://orcid.org/0000-0001-7017-6659
University of Vigo, Spain

Alexandra Matos Pereira
https://orcid.org/0000-0001-6928-2040
ISLA – Instituto Politécnico de Gestão e Tecnologia, Portugal

ABSTRACT

New technologies have helped to improve the tourism sector and to develop strategies that resulted in the so-called smart destinations, underpinned and transformed by modern information and communication technologies (ICTs). Besides, tourism is a global market that continuously seeks mechanisms to grab tourists' and visitors' attention. In view of that, in recent decades, the gamification concept has acquired new definitions from different perspectives, but always associated with the idea of leisure. In tourism, gamification is related with experiences, which by using game elements and digital game design techniques (virtual reality or augmented reality, among others) improve the tourist experience and the user's engagement. This chapter addresses gamification and its influence on tourism experience, together with some gamification applications' examples that can be effective mechanisms to promote tourism businesses or tourism destinations, raising engagement and generating trust.

DOI: 10.4018/978-1-7998-2112-0.ch018

INTRODUCTION

The word "smart" is a marketing tag used for everything that has technology incorporated or has been made better by technology (Boes, Buhalis, & Inversini, 2015). This new concept started by an exploring phase of content definition (Buhalis & Amaranggana, 2015; Gretzel, Sigala, Xiang, & Koo, 2015) and supporting technologies identification (Atembe & Abdalla, 2015; Chung, Tyan, & Han, 2017) entering now a new phase of applications' development and evaluation (Buhalis & Leung, 2018; del Vecchio, Mele, Ndou, & Secundo, 2018).

Thus, the expression smart tourism can be tracked back to the association of tourism with information and communication technologies (ICTs). Smart tourism is a modish concept focused on the growing dependence of tourism destinations' industries and tourists on the ICT emerging forms, allowing the conversion of big data into value propositions (Gretzel et al., 2015). This information exchange system encompasses several approaches such as electronic commerce, virtual reality (VR) and augmented reality (AR) that can affect destination image formation (Hunter, Chung, & Gretzel, 2015).

When looking into electronic commerce, online purchases of airline tickets, accommodation, car rental service and other tourist services can be made (Kim, Lee, & Chung, 2013). As for virtual reality, the user has the possibility of navigating the computer through simulated environments (Hunter, 2014) which represents an experience that will influence the potential tourist's perceptions (Hunter et al., 2015). The relation between tourism and ICTs is well known, especially when talking about augmented reality, which allows commercial and historical data to be projected through a computer (Hunter, 2014).

Related to smart tourism, new concepts as smart tourist and smart destinations arise, being the smart tourist the one who uses smart technologies and visits smart destinations (Femenia-Serra & Ivars-Baidal, 2018). To conduct a research study in this field, it is important to know which current and/or future techniques are in use, not only scholarly, but in the real market and services. According to Werthner, Alzua-Sorzabal, Cantoni, Dickinger, Gretzel, Jannach and Stangl (2015) among these innovative technological smart solutions are mobile applications that run on many different devices and not just on one: all possible due to the Internet, with permanent connectivity, new paradigms of human-computer interaction (e.g. the last search or recommendation, with emotional, implicit, sensor-based, proactive approaches), multilevel data analysis, collective intelligence, and applications based on advanced machine learning techniques or games.

All these technological advances have been receiving growing interest in the last decade, particularly in non-game contexts (Deterding, Dixon, Khaled, & Nacke, 2011) such as tourism, fabricating a recent phenomenon called gamification. That is to say that gamification, created to be used in technological contexts, has crossed its own scope and has begun to be cross-applied, as in the case of tourism. Its usefulness in tourism is such that gamification has been used as a tourist attraction.

Gamification techniques are intended to engage users and influence their behaviour through game design elements in areas other than the traditional game context (Zichermann & Cunningham, 2011).

Moreover, nowadays, gamification is also used in the creation of learning environments. Among the most current approaches are the modern methods of behaviour, therapy for fragmentary gamification of the activity (Hall, 2004), technologies using gamification in philosophy or people awareness from different ages (Retyunskikh, 2003), play methods of customer scrutiny (Popov, 2006), the "Y" generation problem of training and recruitment (Nemkov, 2013), huge attraction for games in business processes (Zickermann & Lynder, 2014), game thoughts in marketing, innovation and staff motivation, learning perspectives and application to the "millennials" working processes who are game dependent (Werbach & Hunter, 2015), among others.

New technology trends in the tourism sector appear to be increasing, what opens a wide range of possibilities of incorporating gamification in tourism destinations and tourism resources. At the same time, it catches tourists' attention during their tourism practices. In 2011, the World Tourism Organization predicted that gamification would be a trend in tourism (UNWTO, 2011) and Egger and Bulencea (2015) stated that gamification could contribute to the creation of memorable tourism experiences. Besides, because it is an experience enhancer, gamification is also useful in tourism education (Xu, Weber, & Buhalis, 2013) as it allows the student to project himself in time and space through theoretical contents.

The focus of this chapter will be on the so-called Smart Destinations, mainly on the possibility of a tourist to know a destination and its attractions throughout a non-conventional way, using the games made available through smartphone apps, augmented reality, virtual reality, among others, provoking higher levels of engagement.

The main objective of this chapter is to reflect and contextualize new concepts resulting from smart tourism and to emphasize tourism gamification as an emerging technology applied to tourism. As a secondary objective to emphasize its empirical aspect, some pioneering practical examples in use in tourism are presented.

It starts by presenting the rationale and conceptual background and its application to the tourism sector. Then it assesses the most recent and successful gamification apps and proposes a Matrix to be applied to tourism that identifies source, type, availability, retrieved tourism information and reported experience. Finally, the chapter highlights a selected use case and discusses the implications and suggestions for further research.

NEW TECHNOLOGICAL CHALLENGES AND DIFFERENT CHANGES IN TOURISM

Smart Tourism, Smart Tourists and Smart Destinations

Taking into account all technological advances during the last decade, new phenomena in the tourism sector emerged. From the combination of information technologies (Its) and tourism a new concept called e-tourism stood out. It was focused on the computerization and virtualization of the tourist exchange and made possible the capitalization of value chains and the emphasis on the linking of a destination's physical attributes, encompassing all phases of tourism experience (before, during and after the trip) (Gretzel, 2018).

Against this background, investigations focalized on the study of *e-destinations* and the influence of information and communication technologies (Buhalis, 2003). However, the pace of technological change is exponential and today the *Smart* technologies, the *Internet of Things* (IoT) and the *cloud computing* (meta data storage and transfer) (Koo, Yoo, Lee, & Zanker, 2016) have revolutionized tourism destinations. It is when the concepts of *smart tourism* (ST), *smart destinations* (SDs) and *smart tourists* are brought to our daily language.

The theoretical model that best explains these three concepts is Gretzel, Sigala, Xiang and Koo (2015)´s *Components and layers of smart tourism* model. It works in a layer system under the heading of *smart tourism,* presenting *smart destination* on the bottom, followed by *smart business* and on the top *smart experience.* The different levels are interconnected by mega data (including data collection, data sharing and data processing) as well as by the conventional smart tourism management systems.

Associated with *smart tourism* comes the *smart tourism ecosystem* (STE) concept. STE combines the use of digital ecosystems and smart business networks as conceptual building blocks (Gretzel et al., 2015). A finer analysis shows that there are four elements to take into account: i) interaction/commitment; ii) balance; iii) agents weakly tied to goals; and iv) self-organization. STE can be defined as a community of digital devices with an environmental effective operation, or as a smart tourism system with heterogeneous and distributed environments inside a dynamic network. Moreover, it is a planning system that relies on public-private partnerships from all sectors of the tourism destination and refers to a systemic vision of the tourist destination encircling its virtual and physical aspects, requiring strong governance and planning policies (Femenia-Serra & Ivards-Baidal, 2018). From this point on investigations have been centred on the interactions between technological agents (devices, databases, software, etc.) and infrastructure information flows.

In fact, *smart tourism* in the context of new technologies' development bears a competitive advantage arising from the use of smart technologies, as in the case of sensors, mobile phone apps, radio-frequency identification (RFID), near-field communication (NFC), smart meters, the Internet-of-Things (IoT), cloud computing, relational databases, etc., together forming a smart digital ecosystem stimulating data-driven innovations and supporting new business models (Gretzel, 2018, p.173).

Gretzel, Ham and Koo (2018) describe smart tourism as consisting of five layers: 1) a physical layer that includes natural and human-made touristic resources as well as transportation and service infrastructures; 2) a smart technology layer that links to this physical infrastructure and provides back-end business solutions and front-end consumer applications; 3) a data layer that includes data storage, open data clearing houses and data-mining applications; 4) a business layer that innovates based on the available technologies and respective data sources; and, finally, 5) an experience layer in which the resulting technology- and data-enhanced experiences are consumed. Smart tourism can thus be summarized in the word "smartness" which is a smart and technological way to diagnose, plan and manage tourism.

With respect to the smart destination concept (SDs), it goes beyond the idea of a destination that applies new technologies, as it brings added value to the traditional concept of tourism destination (Femenia-Serra & Ivars-Baidal, 2018).

However, there have been some paradoxical behaviours around SDs: on the one hand, academia focuses its studies on new technologies applied to tourism destinations or to big data analysis strategy; on the other, DMOs are limited by computer platforms they refer to as smart tourist management.

As stated by Gretzel et al. (2015) we are before a theoretical buzzword where little or nothing is known about the impact of smart tourism strategies on tourism destinations. Or even, as argued by Femenia-Serra and Ivars-Baidal (2018), there is a lack of studies on how smart destination strategies and smart solutions are transforming tourism destinations.

In fact, most smart tourism studies and applications are implemented on smart cities, eg Barcelona digital city (https://ajuntament.barcelona.cat/digital/ca, Amsterdam smart city https://amsterdamsmartcity.com/ or Seoul http://english.visitseoul.net/index).

Albeit some destinations claim to be a smart destination like the Jeju Island in South Korea (https://www.youtube.com/watch?v=d3C7vS-IbAY&feature=youtu.be) or the Sunmoon Lake in Taiwan, there is not a worldwide accepted framework to validate such designation. This indistinctness is acceptable given the idiosyncrasy of each tourism destination (Hall, 2010).

An example that the label smart destination has overcome the urban space arrives from Spain (https://www.destinosinteligentes.es/) with the brand seal *Smart Tourist Destination*. It should be noted that though the concept of smart destination is referred in many other destination websites (see https://www.smartdestinations.com/ or https://en.smartdestinations.com/), none indicates which smart destination model is followed, except Spain and its state company dedicated to the management of innovation and tourism technologies: Segittur (https://www.segittur.es/es/proyectos/proyecto-detalle/System-de-Inteligencia-Turstica-/#.XF78ZvZFxMt)

Segittur (2019) presents a smart tourism system (Sistema de Inteligencia Turística-SIT) anchored in concepts such as innovation, sustainability, accessibility and technology which, according to the UNE Standard 178501: 2015 approved for Smart Tourism Destination Management Systems, the following definition of Smart Tourist Destination is approved: "An innovative tourist destination consolidated on cutting edge technology that ensures the sustainable development of the territory, facilitates innovative tourist visitor interaction and integration with the environment, and increases the quality of the visitor's experience at the destination." (https://ec.europa.eu/eip/ageing/standards/city/smart-cities/pne-178501_en)

This standard was designed for the Spanish context, however, the Segittur website reveals the intention to have this standard approved at a European level. One of the best examples of this intention is the Spanish-Portuguese cross-border project called "Smarter tourism in the Spanish-Portuguese border region" that joins the city of Badajoz in Spain and Elvas in Portugal, being the only project of its kind in Europe, as can be confirmed consulting the European Union's website at (https://ec.europa.eu/regional_policy/en/projects/spain/smarter-tourism-in-the-spanish-portuguese -order-region).

Several proposals of smart tourism model systems applied to smart destinations are proposed by the scientific academy. However, only the works of Femenia-Serra and Ivards-Baidal (2018) and Gretzel, Ham & Koo (2018) make the proposal to look beyond the tourism supply' interests and pay more attention to the tourist's perception of the tourism experience in a smart destination. Moreover, Gretzel et al. (2018) smart tourism model considers a layer dedicated to tourist experiences in a smart destination.

Furthermore, the lack of conceptual schemes to characterize tourists in the context of a smart destination (Femenia-Serra & Ivards-Baidal, 2018; Gretzel, 2018) has been identified as a research gap. To collaborate Gretzel (2018) brings up that smart tourism regions should include comprehensively the tourist experience in a smart destination and that a lot of work still must be done to make real a smart destination.

Other authors argue that the way tourists perceive technologies in a smart destination is still inconclusive and needs further research (Buonincontri & Micera, 2016; Gretzel, 2018).

Femenia-Serra and Ivards-Baidal (2018) research sought to bridge this gap, so they conducted a study on tourists' perceptions of Benidorm as a smart destination, where the Sigettur model is in use. The study presented some limitations as the impossibility of generalization to other types of tourism destinations (e.g. cultural, urban, rural, and so on) using different smart tourism systems, or the unpredictable results of replicating the same approach in tourism destinations that present different attributes and different tourist profiles.

Gamification

Considering the widespread use of new technologies, in recent years, technology has been applied to motivate people and provide support for various individual and collective beneficial behaviours (Hamari & Koivisto, 2015). To be precise, in the last two decades, digital games have reached their peak as means of entertainment and leisure. This success is reflected by the record sale number of game consoles and by the occupation of online multiplayer platforms and environments. Consequently, the topic has expanded and is now discussed among academics, who are interested in its effect and relevance to the digital age (Seaborn & Fels, 2015). And this interest is no longer focused on the adolescent typical user, as the current user profile is a 30-year old, probably 45 percent female, who plays puzzles, board or casual games, and very likely belongs to the 62% of those people who play social games (Entertainment Software Association, 2013).

Today, at a time when much of our activities are mediated by digital technologies and social networks, companies can increase their commitment to customers by transforming traditional processes into deeper and more attractive gaming experiences, for both clients and employees (Robson, Plangger, Kietzmann, McCarthy, & Pitt, 2015).

Such is the success of gaming that other sectors, apart from digital entertainment, have focused their attention on it. This makes room for the so-called strategy or concept of gamification, which can be roughly defined as the selective incorporation of game elements in an interactive system without having a full game as a final product (Deterding, 2012). Or, to put it another way, it refers to technologies that try to promote intrinsic motivations towards diverse activities, commonly using the characteristic design of the games (Hamari, Huotari, & Tolvanen, 2015; Huotari & Hamari, 2012; Deterding, Dixon, Khaled, & Nacke, 2014). The ultimate goal of this practice is to motivate and involve end users through the interactive game mechanism. Typical elements in gamification include, for example, points, leader boards, achievements, feedback, clear goals and narrative (Hamari, Koivisto, & Pakkanen, 2014a; 2014b).

All definitions of gamification are based on the term or on the concept of game, which means that the key factor that defines gamification is rooted in the game, in the nature of self-purposeful activities. Moreover, gamification seeks to obtain results outside the game. It is also common to connect its use to social networks or to include social network characteristics in its services (Ngai, Tao, & Moon, 2015).

In the case of business organizations, for instance, they actively struggle to attract customers and encourage them to become fully involved in their products and services. Evermore aware that time is a scarce resource and that competition is fierce, gamification has emerged as a way to gain that competitive edge, with organizations beginning to see it as a key tool in their digital engagement strategy (Burke, 2014).

While gamification has become a popular buzzword in the last few years, the practice of using game-like mechanics has been receiving attention from professionals, scholars and academics (Hamari et al., 2014a, 2014b), in subject fields as human-computer interaction, education, business, information studies, health and tourism. A good example comes from more than 75 energy companies that are using Opower, a service that allows residents to compare their home energy consumptions with that of neighbours, and share their savings on Facebook (Wingfield, 2012). Samsung Nation, Pepsi Soundoff and other online loyalty programs use points, levels (for example, golden status) or badges to boost customer engagement and to deepen customer-brand relationships. Even Microsoft has gamified its tedious process of translating Windows 7 operating system into several idioms and adapting it to the different culture contexts (Robson et al., 2015).

Although the concept is gaining acceptance and a wider diffusion, it is still immersed in contradictory uses and even multiple meanings. Standards for its use are not consensual, and there is a number of apparently conflicting perspectives, either endorsing its merits and benefits or advocating its faults (Seaborn & Fels, 2015). Nevertheless, several studies and experts praise its use as an effective tool to motivate and involve users in fields and contexts that are not strictly entertainment. From a search made on the Scopus database, in February 2019, it was possible to shortlist several studies as shown in Table 1.

Table 1. Shortlist of 2019 studies appraising gamification

Authors	Article	Study Field
Groening, C., & Binnewies, C. (2019)	Achievement unlocked!-The impact of digital achievements as a gamification element on motivation and performance	Human behaviour
Campo, R., Baldassarre, F., Lee, R. (2019)	A Play-Based Methodology for Studying Children: Playfication	Human behaviour
Hassan, L., Dias, A., Hamari, J. (2019)	How motivational feedback increases user's benefits and continued use: A study on gamification, quantified-self and social networking	Information management
Hulse, T., Daigle, M., Manzo, D., (...), Harrison, A., Ottmar, E. (2019)	From here to there! Elementary: a game-based approach to developing number sense and early algebraic understanding	Education
Guardia, J.J., Del Olmo, J.L., Roa, I., Berlanga, V. (2019)	Innovation in the teaching-learning process: the case of Kahoot!	e-learning
Ramesh, A., Sadashiv, G. (2019)	Essentials of gamification in education: A game-based learning	e-learning
Rönneberg, M., Laakso, M., Sarjakoski, T. (2019)	Map Gretel: social map service supporting a national mapping agency in data collection	Geography
Azevedo, J., Padrão, P., Gregório, M.J., (...), Lien, N., Barros, R. (2019)	A Web-Based Gamification Program to Improve Nutrition Literacy in Families of 3- to 5-Year-Old Children: The NutriScience Project	Nutrition
Lowensteyn, I., Berberian, V., Berger, C., (...), Joseph, L., Grover, S.A. (2019)	The Sustainability of a Workplace Wellness Program That Incorporates Gamification Principles: Participant Engagement and Health Benefits After 2 Years	Health
Garcia, A., Linaza, M.T., Gutierrez, A., Garcia, E. (2019)	Gamified mobile experiences: smart technologies for tourism destinations	Tourism
Sigala, M., Toni, M., Renzi, M.F., Pietro, L.D., Mugion, R.G. (2019)	Gamification in Airbnb: Benefits and risks	Tourism
Marín, B., Frez, J., Cruz-Lemus, J., Genero, M. (2019)	An empirical investigation on the benefits of gamification in programming courses	Computing
González-Briones, A., Valdeolmillos, D., Casado-Vara, R., (...), Herrera-Viedma, E., Corchado, J.M. (2019)	GarbMAS: Simulation of the application of gamification techniques to increase the amount of recycled waste through a multi-agent system	Computing
Bayuk, J., Altobello, S.A. (2019)	Can gamification improve financial behavior? The moderating role of app expertise	Financial behaviour

Source. Own elaboration from Scopus, 2019

Already in 2011, some theorists have suggested that 70% of the world's largest public companies would have at least one gamified application in the following years (Gartner, 2011). In addition, the most conservative scholars have warned that around 80% of today gamified applications will not meet its business goals (Gartner, 2012). The reason for this failure will be the use of faulty gamification processes, or even the lack of procedure understanding, how it works or how a gamified experience should be designed in such a way as to induce changes in the players' behaviour (in this case, customers, workers, or other environmental entities) (Robson et al., 2015).

Table 2. Gamification in tourism

Section	Company	Description	Game Elements
Transportation	American Airlines	"A simple game mechanic is used by American Airlines to visually represent your current elite status qualification"	Levels, Points
	Turkish Airlines	"QR-coded national flags have been placed on 100 digital bus shelters for London 2012. Users who read the code can win a ticket to Australia. Goal is to have most check-ins in one place or individual place"	Physical rewards
Food and Beverages	Starbucks	Motivating members to register and spend their gift cards to receive bigger and better benefits with every purchase	Progress bars, levelling and rewards
	4 foods	Motivating customers to create sandwiches as they want and share them. The most popular choices rise to the top of the leader board	Leader board, Relatedness
Destinations	Foursquare	Foursquare is designed to "turn life into a game" by rewarding people with mayorships and badges for going to physical location	Points, badges Real rewards (discounts)
Tour	Stry Boots	"Is a new game that is available in the US where travellers or local people can go on an urban adventure, solve fun clues, discover cool spots and learn more about the destination or particular places"	Achievement Competition
	Expedia	"Expedia.com, the world's largest online travel agency, launched a travel Around the World in 100 Days game in June last year. The game was created to increase the awareness of Expedia's loyalty program and engage with customers. Players can earn"	Real-world Expedia Reward

Source. Sever et al., 2015:197

Tourism Gamification

The development of new technologies creates benefits in the tourism sector, with tourists demanding more personal, unique and memorable experiences, requiring a deeper commitment and multisensory stimulation. New technologies, and especially gamification, have the potential to offer richer and more participatory experiences (Xu, Weber, & Buhalis, 2013). However, research describing the use of gamification in tourism has been scarcely present in academic literature (Corrêa & Kitanoa, 2015; Sigala, 2015a; Xu et al., 2013;).

Gamification is a new trending theme that can be applied to tourism in a range of different ways to provoke motivation and behaviour change (Xu et al., 2013).

According to Neuhofer, Buhalis and Ladkin (2012) technology, including gaming, can foster co-creating experiences, that can be understood as "tourism experiences with improved technology". One recent study carried out by Sigala (2015b) collected data from TripAdvisor users and came to the conclusion that gamification can improve the tourist experience by "immersing tourists in a world of simulated travel". This is why there are already several tourism destinations and organizations that have used gamification as part of their marketing, sales and client commitment strategy (Xu, Buhalis, & Weber, 2017).

Most of the scientific literature related to new information and communication technologies (ICTs) applied to tourism focuses on the evolution and development of the so-called smart tourism destinations (Buhalis & Amaranggana, 2014). It is a growing field of studies, so new lines of research are emerging, as the case of augmented reality (AR) and virtual reality (VR) applied to gamification in tourism. Moreover, literature tends to focus on urban environments where the smart city plays a leader role. This interest arises from the fact that not all destinations are in the same pace of technological progress, and those non-urban areas who don't have any technological infrastructures or resources can't obtain the same benefits as smart cities (Gretzel, Sigala, Xiand, & Koo, 2015). If we take into account that globalization, communication, mobility and virtuality are the four trends in current economies and societies, it is easy to understand how ICT is making the world moving forward (Egger & Buhalis, 2008).

In tourism, advances in mobile technology have been widely used, but only applied to urban environments (Quadri-Felitti & Fiore, 2016). Businesses companies, when trying to offer successful tourism experiences in rural areas, are faced with several setbacks affecting the use of ICTs.

Looking for success, these smart tourism destinations encourage innovation and entrepreneurship (Boes et al., 2015). Such is the case of tourist guide systems that can be accessed from the cloud. Their right use requires adequate technological infrastructures, hence innovative tourism products and services highly dependent on destination innovative technological capability.

Within gamification in tourism, one of the biggest technological advances has been the development of virtual reality (VR) and augmented reality (AR) (Skinner, Sarpong & White, 2018). To interact with these digital technology players, that is to say virtual tourists, a set of devices to guarantee immersiveness and realism in the experience is needed. VR uses headphones and AR needs QR codes to guide tourists around interesting and tourist spots in a destination (Linaza, Gutierrez, & García, 2014). This digital technology is becoming cheaper and increasingly wider, and still on an early stage, so it can be expected to see many more innovative uses in tourism, in a near future (Han, Jung, & Gibson, 2013).

It has been evidenced that gamification generates a set of benefits and these can be applied to tourism (Sever, Sever, & Kuhzady, 2015), for instance:

More funny and enjoyable online experiences: According to Pink (2006), nowadays, showing a serious image is not be enough for gain success in business. Fun and games should be part of the marketing strategy, and gamification can help to achieve it, offering more funny and attractive activities.

Encourage online engagement: Visitor engagement is a function calculated by the sum of the number of pages viewed, the spent time on each website, the rate at which the visitor returns to that site over time, his long-term interaction with the site, his measured awareness of the brand, his contribution to feedback and the interaction with content (Jackson, 2009). But lead users/customers to engage online is not easy and needs a creative and innovative strategy. Gamification based researches have suggested that gamification is an effective tool for increasing engagement (Xu et al., 2013).

Virality: According to Osipov, Volinsky and Grishin (2014) the use of gamification can successfully boost virality.

Revenue: Through internal and external motivation, engagement, virality and user generated content, an increase in revenue can be expected (Sever et al., 2015).

Some examples of tourism gamification techniques are those followed by American Airlines and a status rating; or Starbucks motivating users to register and use their gift cards to receive benefits on each purchase; or even the Expedia travel game "Around the World in 100 Days" (Table 2).

In short, gamification has become a vital element for many companies, and especially as a future trend for tourism. A large number of tourism subsectors already make use of it by creating tourism experiences

Figure 1. History Hero app
Source: Screenshot of the App

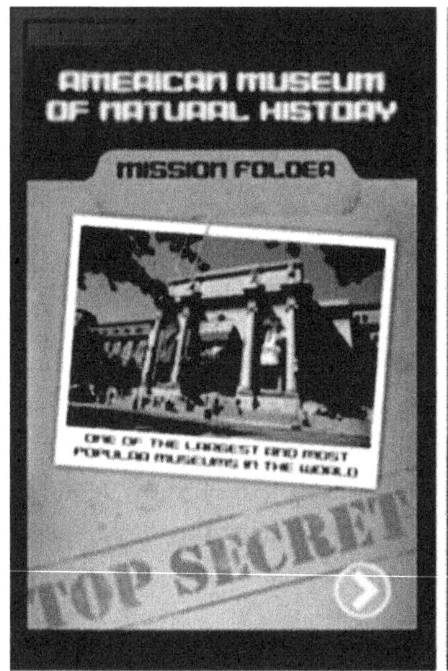

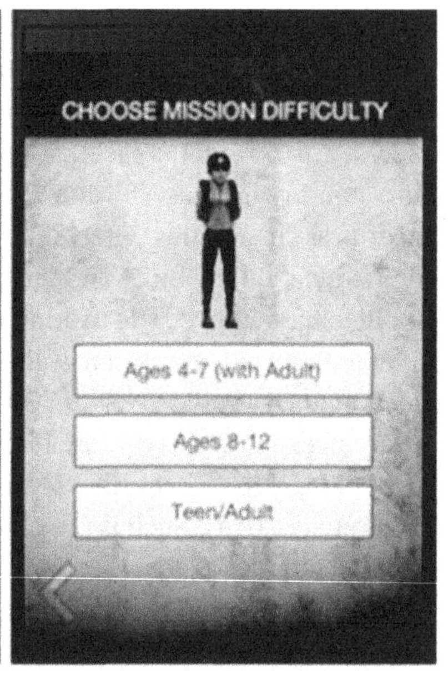

and training employees for their offer (Xu et al., 2017). The implications for the tourism sector are varied, and airlines, hospitality, DMOs or partnership with tourism sectors can benefit from them. DMOs can use social media games to recruit new staff and familiarize employees with their work (Xu et al., 2013) or use these games aimed at potential tourists based on the tourist attractions of the destination, thus generating awareness of the destination brand (Xu, Tian, Buhalis, Weber & Zhang, 2016). As can be seen, these tools can be used in a double sense, towards tourists, improving the image of the destination, attracting new tourists and offering new experiences *in situ*, and even as a training and recruitment tool.

CASE STUDIES: EXAMPLES OF GAMIFICATION IN THE TOURISM SECTOR

In this section, keeping in mind the tourists' experiences and involvement with the destination attractions or heritage, the most recent apps and games used in the tourism sector are presented (Wang & Tsai, 2014; Xu et al., 2017). A set of brand-new and different kind of examples have been selected, thus allowing the reader to know the diversity offered by the tourism sector, ranging from museums to geolocation. Furthermore, apps that allow tourists to connect and interact with a destination, previously to its visit, and actively stimulate learning and sharing information about a place's history will be consider thoroughly.

A. History Hero

History Hero is the first interactive mobile guide to guide users through the world's most important museums, churches, and historical sites. It has been cooperating with almost 40 well-known institutions in U.S. and Europe, as for example the Metropolitan Museum of Art, St. Peter's Basilica, Château de

Versailles, National Mall and so on (Wang & Tsai, 2014). It is an app designed for students aged between 4 and 10 and it teachs them about some of the world's most significant historical locations.

- Game elements:
  - Goal: Defeat Erasers who are trying to erase our history.
  - Rules: Have to "oath" (clever code for no running, no touching any object).
  - Feedback system: Asks questions ensuring that users are truly engaged and understand.
  - Voluntary participation: New experience to visits the world's greatest historical destinations.
- Concept of digital curation
  - Learn cultural history by using mobile guide.
  - Interactive mobile guide to visit cultural institutions.
- Game mechanics
  - Points: When the task is completed, points are awarded
  - Levels: There are three different-level missions depending on the age of your group.
  - Badges: For saving historic treasures from being erased forever, will earn badges.
- Game types
  - Simulation games: It provides 9 roles for user to choose.
  - Drill & practice game: Answer multiple choice questions about those artifacts, and take a photo to preserve the artifact for history

B. Geocaching Game

Geocaching is a trendy activity that includes hiding and finding "treasures" with the help of a smartphone's GPS. So, a walk through a city or a destination where the user is traveling around can become a fun gymkhana. Although the official application is Geocaching, whose developer is Groundspeak, there are others free apps for android like C:Geo, which has already more than 5 million downloads.

Geocaching consists on hiding a worthless object inside a box, a cache, in locations that can be so diverse as a local park, the roadside of a city street or the countryside, and registering its geographical coordinates on the platform so that other users can find it. Once found, the geocacher can take the "treasure" with him but, in return, has to leave something of equal or greater value.

What makes geocaching interesting is the fact that by combining outdoor adventure with GPS tracking, it allows to know a lot of different places while searching for treasures, being also a fun activity to be done with family and friends. To play this game it is required a free register on the website geocaching.com and download the application.

C. Earn Your Wings

Thirty participants will earn a total of 20,000,000 bonus Aeroplan frequent flier loyalty program miles: half of those bonus miles will be distributed equally amongst the top ten participants, who will each earn 1,000,000; while the other half of those bonus miles will be distributed equally amongst those participants who finish between positions 11 through 30, as they will each earn 500,000 bonus miles.

According to the terms and conditions of this promotion, every time you earn 2,000 Wings, you will receive 2,000 bonus miles — up to a maximum of 750,000 bonus Aeroplan frequent flier loyalty

Figure 2. Geocaching app
Source. App Screenshot

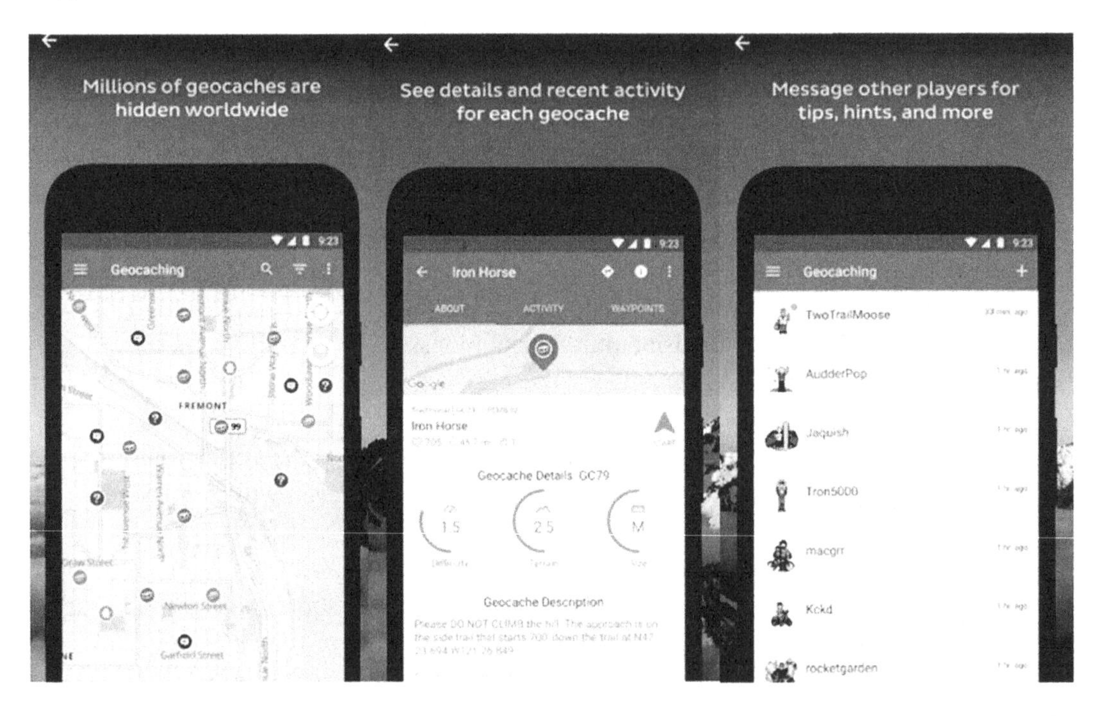

Figure 3. C:Geo
Source. App Screenshot

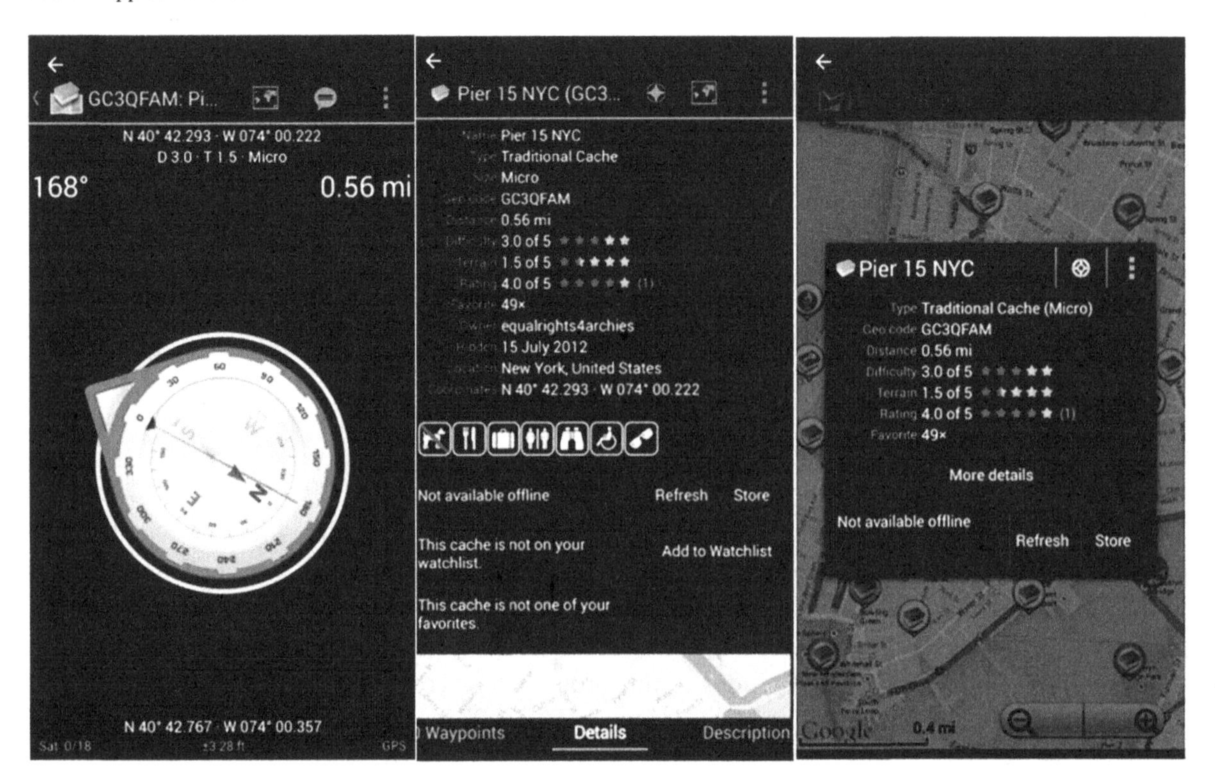

Figure 4. Earn your wings. Explication game
Source. https://earnyourwings.aircanada.com/en

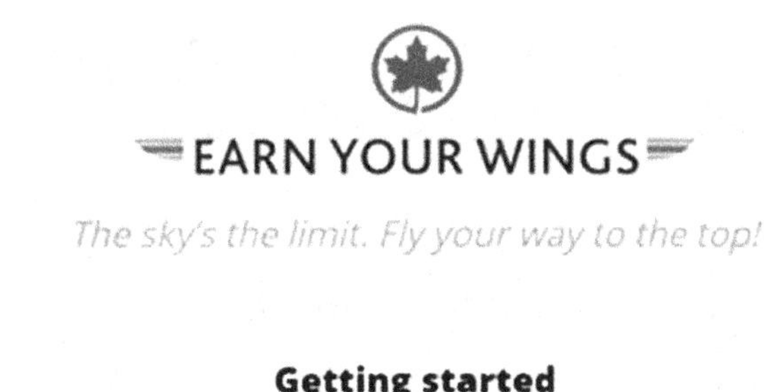

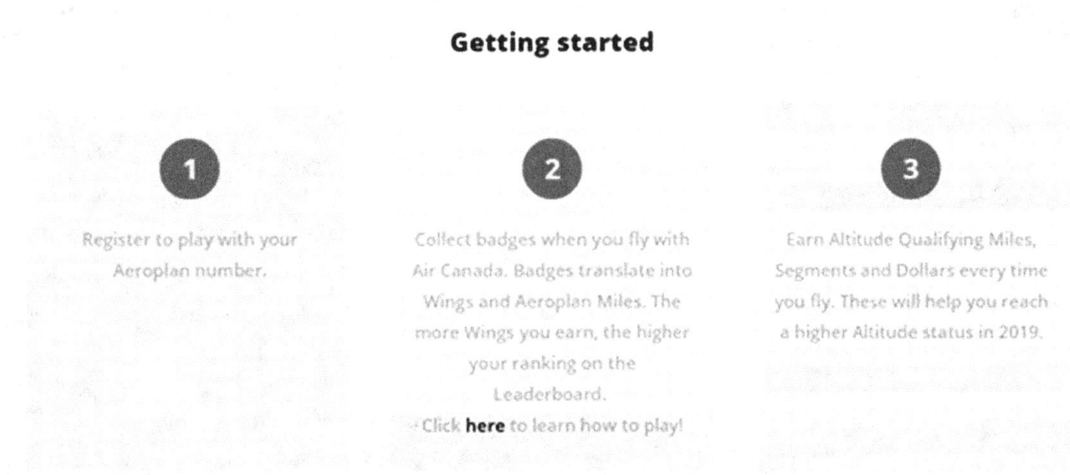

program miles, which does not include the aforementioned bonus miles if you place amongst the top 30 participants at the conclusion of this promotion.

You can earn Wings simply by flying as a passenger on Air Canada flights anywhere in the world and earn a minimum of 400 Wings — 200 Wings for every take off and 200 Wings for every landing — per flight segment on any airplane operated by Air Canada, Air Canada Express, and Air Canada rouge.

D. Eye Shakespeare App

Eye Shakespeare is one excellent example on how to apply the latest mobile technologies to improve tourists' and visitors' experience. Thanks to a sapient mix of augmented-reality, QR-code and geopositioning systems, the App helps to recreate buildings, places and the atmosphere of the Shakespeare's Stratford, allowing visitors to explore the town from a deeply different perspective. Eye Shakespeare also incorporates hundreds of digitised items from the Shakespeare Birthplace Trust's archives, which are normally difficult to be accessed by tourists and visitors. The App allows tourist to explore both spatially and historically the Shakespeare's home town.

E. Virtual Reality in Museums

Museums also bet on new technologies to enrich their visitor's experience. Thus, in 2017, the National Archaeological Museum (Madrid, Spain) launched a virtual reality initiative together with Samsung. Visitors can travel back in time and learn how their ancestors lived thanks to the magic of virtual reality.

Figure 5. Eye Shakespeare App
Source. App Screenshot

Figure 6. Virtual reality image in the National Archaeological Museum
Source. Expansion.com

Figure 7. Virtual reality experience image in the Thyssen Museum
Source. https://www.museothyssen.org/actividades/realidad-virtual-museo

You just need to put on the Samsung virtual reality glasses to dive in five different historical periods of the history of Spain: Prehistory, Protohistory, Roman Hispania, Middle Ages and Modern Age.

The following year, the Thyssen Museum (Spain) replicated the experience, launching its first virtual reality undertaking. In May of 2018, the Thyssen Museum presented the first virtual reality experience for its visitors, thanks to the sponsorship of Endesa. This initiative allowed visitors to immerge into three works of the permanent collection and seeing them in a three-dimensional space that recreates the fields of Auvers painted by Van Gogh, the streets of New York that inspired Mondrian, or the flowers and insects of a Dutch still life.

After referring to a series of examples of gamification in the tourism sector, it is also important to know the opinion of users and tourists. Some studies already demonstrate the good acceptance of these tools in tourism, showing various implications for the sector, as for marketing and communication (Xu et al., 2015; Yilmaz & Coskun, 2016). In addition, in certain contexts, variables such as socialization or engagement have become important (Looyestyn et al., 2017). Through these tools, new so far, the tourist´s curiosity is enhanced, resulting in greater involvement during the visit, both as a source of knowledge and as a means of fun.

To give an example, the use of Augmented Reality applications for mobile devices allows the utilization of the user's own device as a display, which increases the degree of familiarity, avoiding the need for previous explanations or training (Mesáro et al., 2016). Moro et al. (2019) confirmed the effect of gamification features as a tool to motivate travellers to interact with TripAdvisor and contribute to writing reviews. On the grounds of this several studies have appeared endorsing the effectiveness and attractiveness of this features as a tourism attraction.

CONCLUSION

In recent decades, technology has expanded and changed at a rate faster than ever, in every field imaginable. Among them, is tourism, which has benefited from this digital revolution and has modified, among other things, tourists' and visitors' behaviour. Without ignoring the new codes of consumption and access to information, technology is also part of the tourism experience.

The Internet of Things (IoT) and or the cloud computing (storage and transfer of metadata) has changed radically the way tourism is planned, managed and done (Koo et al., 2016). Tourist information search, destination choice and service booking in a digital age are carried out primarily through online systems (Berne et al., 2012; Xiang & Gretzel, 2010; Walle, 1996). The destination experience itself, as well as expectations and memories, are also increasingly connected, whether through real-time sharing or the use of utility applications. Connectivity and its related technologies significantly influence tourism experiences (Masri, Anuar, & Yulia, 2017).

A perfect example is gamification, in particular, the use of applications, virtual reality, augmented reality or other technologies that through gaming enhances the tourist experience and makes it more attractive and entertaining. Furthermore, several benefits are claimed by these tools that are able to capture the tourist's attention and to make some visit to a tourist attraction or destination more enjoyable and innovative.

Applications, better known as apps, are ideal for keeping the tourist motivated. Some apps, from airlines, hotels and other tourism businesses allow the earning of points convertible into prizes, encouraging the consumer behaviour of tourists. Other apps proposing treasure hunts, encourage tourists to visit several locations on a tourism destination. On the other hand, they will be very useful for the training of students and working staff in the tourism sector. Its use as a training tool will make learning more attractive, as well as practical, bringing the student closer to the tourist environment with greater realism.

There is no doubt that in tourism, technology offers a range of innovative and imaginative possibilities that attracts tourists' attention, embodies a business differentiation strategy and helps to boost businesses in a very competitive global market.

However, new ideas and studies are needed and as postulated by Femenia-Serra and Ivards-Baidal, (2018) and Gretzel et al. (2018) it is necessary to look beyond the interests of tourism and technology and to pay more attention to the perceptions of tourists about the tourist experience in a smart destination and the experience with smart technologies.

Therefore, this exploratory work on gamification brought together the most updated theoretical conceptualizations and the most recent examples of apps and tools used in the tourism sector. All conceptual definitions are the authors' responsibility and are the result of bibliographic revision made about the subject.

Due to its exploratory character and the novelty of the theme, this work has limitations because it does not cover (nor was it the objective) all technological cases and typologies applied to destination gamification.

Hence, a selection of the most modern and diverse cases has been made to share with the reader what is being done in gaming applied to the tourism sector, but more innovative and developed tools will emerge. Being a concept highly linked to technology, new alternatives will appear in a very short time. Therefore, for future research, the evolution of gamification in tourism will continue to be analysed and the most recent advances will be shown. In addition, highly connected to new tourism technologies, artificial intelligence emerges. It is expected that this post-app era will be implemented from 2020 onwards, with virtual assistants present in 40% of all mobile interactions. Chats will be the new operating systems, apps the new search engines and virtual assistants the new websites, and this will be a future key research line.

Finally, it is also noteworthy that studies on tourist experiences in smart destinations are infrequent and a research gap has been identified. Therefore, future lines of investigation should address this theme to define conceptual frameworks to characterize tourists' behaviours in a smart destination (Femenia-Serra & Ivards-Baidal, 2018; Gretzel, 2018).

REFERENCES

Atembe, R., & Abdalla, F. (2015, April). The use of smart technology in tourism: evidence from wearable devices. In *ISCONTOUR 2015-Tourism Research Perspectives: Proceedings of the International Student Conference in Tourism Research* (Vol. 23). BoD–Books on Demand.

Azevedo, J., Padrão, P., Gregório, M. J., Almeida, C., Moutinho, N., Lien, N., & Barros, R. (2019). A Web-Based Gamification Program to Improve Nutrition Literacy in Families of 3-to 5-Year-Old Children: The Nutriscience Project. *Journal of Nutrition Education and Behavior*, *51*(3), 326–334. doi:10.1016/j.jneb.2018.10.008 PMID:30579894

Bayuk, J., & Altobello, S. A. (2019). Can gamification improve financial behavior? The moderating role of app expertise. *International Journal of Bank Marketing*, *37*(4), 951–975. doi:10.1108/IJBM-04-2018-0086

Boes, K., Buhalis, D., & Inversini, A. (2015). Conceptualising smart tourism destination dimensions. In *Information and communication technologies in tourism 2015* (pp. 391–403). Cham, Switzerland: Springer.

Buhalis, D. (2003). eTourism: Information technology for strategic tourism management. Harlow, UK: Pearson Education.

Buhalis, D., & Amaranggana, A. (2013). Smart tourism destinations. In *Information and communication technologies in tourism 2014* (pp. 553–564). Cham, Switzerland: Springer. doi:10.1007/978-3-319-03973-2_40

Buhalis, D., & Law, R. (2008). Progress in information technology and tourism management: 20 years on and 10 years after the Internet—The state of eTourism research. *Tourism Management*, *29*(4), 609–623. doi:10.1016/j.tourman.2008.01.005

Buonincontri, P., & Micera, R. (2016). The experience co-creation in smart tourism destinations: A multiple case analysis of European destinations. *Information Technology & Tourism*, *16*(3), 285–315. doi:10.100740558-016-0060-5

Burke, B. (2016). *Gamify: How gamification motivates people to do extraordinary things*. Routledge. doi:10.4324/9781315230344

Campo, R., Baldassarre, F., & Lee, R. (2019). A Play-Based Methodology for Studying Children: Play-fication. *Systemic Practice and Action Research*, 1–11.

Chung, N., Tyan, I., & Han, H. (2017). Enhancing the smart tourism experience through geotag. *Information Systems Frontiers*, *19*(4), 731–742. doi:10.100710796-016-9710-6

Corrêaa, C., & Kitanoa, C. (2015, February). Gamification in tourism: Analysis of brazil quest game. In *Proceedings of ENTER* (Vol. 2015).

Del Vecchio, P., Mele, G., Ndou, V., & Secundo, G. (2018). Creating value from social big data: Implications for smart tourism destinations. *Information Processing & Management*, *54*(5), 847–860. doi:10.1016/j.ipm.2017.10.006

Deterding, S. (2012). Gamification: Designing for motivation. *Interaction*, *19*(4), 14–17. doi:10.1145/2212877.2212883

Deterding, S., Dixon, D., Khaled, R., & Nacke, L. (2011). From game design elements to gamefulness: defining "gamification". In 15th International Academic MindTrek Conference: Envisioning Future Media Environments, 9-15, ACM, Tampere, Finland.

Deterding, S., Dixon, D., Khaled, R., & Nacke, L. (2014). Du game design au gamefulness: définir la gamification. *Sciences du jeu*, (2).

Egger, R., & Buhalis, D. (Eds.). (2008). "Mobile systems", eTourism Case Studies: Management and Marketing Issues, Butterworth-Heinemann, Oxford, UK, pp. 417-25.

Egger, R., & Bulencea, P. (2015). *Gamification in tourism: Designing memorable experiences*. BoD–Books on Demand.

Entertainment Software Association. (2013). Essential Facts About the Computer and Video Game Industry. Entertainment Software Association. Retrieved from http://www.theesa.com/facts/pdfs/esa_ef_2013.pdf (accessed 10.10.18).

Femenia-Serra, F., & Ivars-Baidal, J. A. (2018). Do smart tourism destinations really work? The case of Benidorm. *Asia Pacific Journal of Tourism Research*, 1–20. doi:10.1080/10941665.2018.1561478

Garcia, A., Linaza, M. T., Gutierrez, A., & Garcia, E. (2019). Gamified mobile experiences: Smart technologies for tourism destinations. *Tourism Review*, *74*(1), 30–49. doi:10.1108/TR-08-2017-0131

Gartner. (2011). *Gartner predicts over 70 percent of global 2000 organisations will have at least one gamified application by 2014*. Available at http://www.gartner.com/it/page.jsp?id=1844115

Gartner. (2012). *Gartner says by 2014, 80 percent of current gamified applications will fail to meet business objectives primarily due to poor design*. Available at http://www.gartner.com/it/page.jsp?id=2251015

González-Briones, A., Valdeolmillos, D., Casado-Vara, R., Chamoso, P., Coria, J. A. G., Herrera-Viedma, E., & Corchado, J. M. (2018, June). Garbmas: Simulation of the application of gamification techniques to increase the amount of recycled waste through a multi-agent system. In *International Symposium on Distributed Computing and Artificial Intelligence* (pp. 332-343). Cham, Switzerland: Springer.

Gretzel, U., Ham, J., & Koo, C. (2018). Creating the City Destination of the Future: The Case of Smart Seoul. In *Managing Asian Destinations* (pp. 199–214). Singapore: Springer. doi:10.1007/978-981-10-8426-3_12

Gretzel, U., Sigala, M., Xiang, Z., & Koo, C. (2015). Smart tourism: Foundations and developments. *Electronic Markets*, *25*(3), 179–188. doi:10.100712525-015-0196-8

Gretzel. (2018). Investigaciones Regionales. *Journal of Regional Research, 42,* 171-184.

Groening, C., & Binnewies, C. (2019). "Achievement unlocked!"-The impact of digital achievements as a gamification element on motivation and performance. *Computers in Human Behavior*, *97*, 151–166. doi:10.1016/j.chb.2019.02.026

Guardia, J. J., Del Olmo, J. L., Roa, I., & Berlanga, V. (2019). Innovation in the teaching-learning process: the case of Kahoot!. *On the Horizon, 27*(1), 35-45.

Hall, C. M. (2010). Tourism destination branding and its effects on national branding strategies: Brand New Zealand, clean and green but is it smart? *European Journal of Tourism, Hospitality, and Recreation*, *1*(1), 68–89.

Hall, L. M. (2002). Games business experts play. Carmarthen, UK: Crown House Publishing.

Hamari, J., Huotari, K., & Tolvanen, J. (2015). Gamification and economics. *The gameful world: Approaches, issues, applications, 139*.

Hamari, J., & Koivisto, J. (2015). Why do people use gamification services? *International Journal of Information Management*, *35*(4), 419–431. doi:10.1016/j.ijinfomgt.2015.04.006

Hamari, J., Koivisto, J., & Pakkanen, T. (2014a). Do persuasive technologies persuade?-a review of empirical studies. In *International conference on persuasive technology*(pp. 118-136). Cham, Switzerland: Springer. 10.1007/978-3-319-07127-5_11

Hamari, J., Koivisto, J., & Sarsa, H. (2014b). Does gamification work?--a literature review of empirical studies on gamification. In *2014 47th Hawaii international conference on system sciences (HICSS)* (pp. 3025-3034). IEEE.

Han, D. I., Jung, T., & Gibson, A. (2013). Dublin AR: implementing augmented reality in tourism. In *Information and communication technologies in tourism 2014* (pp. 511–523). Cham, Switzerland: Springer. doi:10.1007/978-3-319-03973-2_37

Hassan, L., Dias, A., & Hamari, J. (2019). How motivational feedback increases user's benefits and continued use: A study on gamification, quantified-self and social networking. *International Journal of Information Management*, *46*, 151–162. doi:10.1016/j.ijinfomgt.2018.12.004

Hulse, T., Daigle, M., Manzo, D., Braith, L., Harrison, A., & Ottmar, E. (2019). From here to there! Elementary: A game-based approach to developing number sense and early algebraic understanding. *Educational Technology Research and Development*, 1–19.

Hunter, W. C. (2014). Virtual Reality. In J. Jafari (Ed.), *The Encyclopedia of Tourism* (2nd ed.).

Hunter, W. C., Chung, N., Gretzel, U., & Koo, C. (2015). Constructivist research in smart tourism. *Asia Pacific Journal of Information Systems*, *25*(1), 105–120. doi:10.14329/apjis.2015.25.1.105

Huotari, K., & Hamari, J. (2012). Defining gamification: a service marketing perspective. In *Proceeding of the 16th international academic MindTrek conference* (pp. 17-22). ACM. 10.1145/2393132.2393137

Jackson, S. (2009). *Cult of Analytics: Driving online marketing strategies using web analytics: Driving online strategies using web analytics*. Routledge. doi:10.4324/9780080885179

Kim, M. J., Lee, C. K., & Chung, N. (2013). Investigating the role of trust and gender in online tourism shopping in South Korea. *Journal of Hospitality & Tourism Research (Washington, D.C.)*, *37*(3), 377–401. doi:10.1177/1096348012436377

Koo, C., Yoo, K. H., Lee, J. N., & Zanker, M. (2016). Special section on generative smart tourism systems and management: Man–machine interaction. *International Journal of Information Management*, *36*(6), 1301–1305. doi:10.1016/j.ijinfomgt.2016.05.015

Linaza, M. T., Gutierrez, A., & García, A. (2013). Pervasive augmented reality games to experience tourism destinations. In *Information and Communication Technologies in Tourism 2014* (pp. 497–509). Cham, Switzerland: Springer. doi:10.1007/978-3-319-03973-2_36

Looyestyn, J., Kernot, J., Boshoff, K., Ryan, J., Edney, S., & Maher, C. (2017). Does gamification increase engagement with online programs? A systematic review. *PLoS One*, *12*(3). doi:10.1371/journal.pone.0173403 PMID:28362821

Lowensteyn, I., Berberian, V., Berger, C., Da Costa, D., Joseph, L., & Grover, S. A. (2019). The Sustainability of a Workplace Wellness Program That Incorporates Gamification Principles: Participant Engagement and Health Benefits After 2 Years. *American Journal of Health Promotion*. PMID:30665309

Marín, B., Frez, J., Cruz-Lemus, J., & Genero, M. (2018). An Empirical Investigation on the Benefits of Gamification in Programming Courses. [TOCE]. *ACM Transactions on Computing Education*, *19*(1), 4. doi:10.1145/3231709

Mesáro, P., Mandičák, T., Hernandez, M. F., Sido, C., Molokáč, M., Hvizdák, L., & Delina, R. (2016). Use of Augmented Reality and Gamification techniques in tourism. *Ereview of Tourism Research*, 2.

Moro, S., Ramos, P., Esmerado, J., & Jalali, S. M. J. (2019). Can we trace back hotel online reviews' characteristics using gamification features? *International Journal of Information Management*, *44*, 88–95. doi:10.1016/j.ijinfomgt.2018.09.015

Nemkov, A. (2016). *The theory of generations in Russia*. Retrieved from http://sykt24.ru/novosti/teoriya-pokolenij-v-rossii-pokolenie-y/

Neuhofer, B., Buhalis, D., & Ladkin, A. (2012). Conceptualising technology enhanced destination experiences. *Journal of Destination Marketing & Management*, *1*(1-2), 36–46. doi:10.1016/j.jdmm.2012.08.001

Ngai, E. W., Tao, S. S., & Moon, K. K. (2015). Social media research: Theories, constructs, and conceptual frameworks. *International Journal of Information Management*, *35*(1), 33–44. doi:10.1016/j.ijinfomgt.2014.09.004

Osipov, I. V., Volinsky, A. A., & Grishin, V. V. (2014). Gamification, virality and retention in educational online platform. Measurable indicators and market entry strategy. *arXiv preprint arXiv:1412.5401*.

Pink, D. H. (2006). *A whole new mind: Why right-brainers will rule the future*. UK: Penguin.

Popov, A. (2006). *Marketing game. Entertain and conquer*. Moscow, Russia: Mann, Ivanov, & Ferber.

Quadri-Felitti, D., & Fiore, A. M. (2016). Wine tourism suppliers' and visitors' experiential priorities. *International Journal of Contemporary Hospitality Management*, *28*(2), 397–417. doi:10.1108/IJCHM-05-2014-0224

Ramesh, A., & Sadashiv, G. (2019). Essentials of Gamification in Education: A Game-Based Learning. In *Research into Design for a Connected World* (pp. 975–988). Singapore: Springer. doi:10.1007/978-981-13-5977-4_81

Retyunskikh, L. T. (2003). *Socrates Academy. Philosophical games 10 years later*. Moscow, Russia: Nauka.

Robson, K., Plangger, K., Kietzmann, J. H., McCarthy, I., & Pitt, L. (2015). Is it all a game? Understanding the principles of gamification. *Business Horizons*, *58*(4), 411–420. doi:10.1016/j.bushor.2015.03.006

Rönneberg, M., Laakso, M., & Sarjakoski, T. (2019). Map Gretel: Social map service supporting a national mapping agency in data collection. *Journal of Geographical Systems*, *21*(1), 43–59. doi:10.100710109-018-0288-z

Seaborn, K., & Fels, D. I. (2015). Gamificación en la teoría y la acción: una encuesta. *Revista internacional de estudios humano-computador, 74*, 14-31.

Sever, N. S., Sever, G. N., & Kuhzady, S. (2015). The evaluation of potentials of gamification in tourism marketing communication. *International Journal of Academic Research in Business and Social Sciences*, *5*(10), 188–202. doi:10.6007/IJARBSS/v5-i10/1867

Sigala, M. (2015a). Gamification for crowdsourcing marketing practices: Applications and benefits in tourism. In *Advances in crowdsourcing* (pp. 129–145). Cham, Switzerland: Springer. doi:10.1007/978-3-319-18341-1_11

Sigala, M. (2015b). The application and impact of gamification funware on trip planning and experiences: The case of TripAdvisor's funware. *Electronic Markets*, *25*(3), 189–209. doi:10.100712525-014-0179-1

Sigala, M., Toni, M., Renzi, M. F., Di Pietro, L., & Mugion, R. G. (2019). Gamification in Airbnb: Benefits and Risks. *e-Review of Tourism Research, 16*(2/3).

Skinner, H., Sarpong, D., & White, G. R. (2018). Meeting the needs of the Millennials and Generation Z: Gamification in tourism through geocaching. *Journal of Tourism Futures*, *4*(1), 93–104. doi:10.1108/JTF-12-2017-0060

Wang, Y.-C., & Tsai, P.-H. (2014). A Study on the Applications of the Cultural History Promotion via Gamification Based. Digital Curation: Taking "History Hero" as an Example. Available at http://pnclink.org/document/2014%20Poster_Third%20Place_Poster%20Image.pdf

Werbach, K., & Hunter, D. (2015). *For the Win: How Game Thinking Can Revolutionize Your Business*. Moscow, Russia: Mann, Ivanov, & Ferber.

Werthner, H., Alzua-Sorzabal, A., Cantoni, L., Dickinger, A., Gretzel, U., Jannach, D., & Stangl, B. (2015). Future research issues in IT and tourism. *Information Technology & Tourism*, *15*(1), 1–15. doi:10.100740558-014-0021-9

Wingfield, C. (2012). All the world's a game, and business is a player. *The New York Times*, A1.

World Tourism Organization. (2011). *World travel market global trends report.*

Xu, F., Buhalis, D., & Weber, J. (2017). Serious games and the gamification of tourism. *Tourism Management, 60*, 244–256. doi:10.1016/j.tourman.2016.11.020

Xu, F., Tian, F., Buhalis, D., Weber, J., & Zhang, H. (2016). Tourists as mobile gamers: Gamification for tourism marketing. *Journal of Travel & Tourism Marketing, 33*(8), 1124–1142. doi:10.1080/10548408.2015.1093999

Xu, F., Weber, J., & Buhalis, D. (2013). Gamification in tourism. In *Information and communication technologies in tourism 2014* (pp. 525–537). Cham, Switzerland: Springer. doi:10.1007/978-3-319-03973-2_38

Yılmaz, H., & Coşkun, İ. O. (2016). New toy of marketing communication in tourism: Gamification. In e-Consumers in the era of new tourism (pp. 53-71). Springer, Singapore.

Zichermann, G., & Cunningham, C. (2011). *Gamification by design: Implementing game mechanics in web and mobile apps*. O'Reilly Media.

Zickermann, G., & Lynder, J. (2014). *Gamification in business: How to break through the noise and capture the attention of co-workers and clients: Mann*. Moscow, Russia: Ivanov and Ferber.

ADDITIONAL READING

Aguiar-Castillo, L., Clavijo-Rodriguez, A., Saa-Perez, D., & Perez-Jimenez, R. (2019). Gamification as An Approach to Promote Tourist Recycling Behavior. *Sustainability, 11*(8), 2201. doi:10.3390u11082201

Alčaković, S., Pavlović, D., & Popesku, J. (2017). Millennials and gamification: A model proposal for gamification application in tourism destination. *Marketing, 48*(4), 207–214. doi:10.5937/Markt1704207A

Egger, R., & Bulencea, P. (2015). *Gamification in tourism: Designing memorable experiences*. BoD–Books on Demand.

Liu, C. R., Wang, Y. C., Huang, W. S., & Tang, W. C. (2019). Festival gamification: Conceptualization and scale development. *Tourism Management, 74*, 370–381. doi:10.1016/j.tourman.2019.04.005

Madlberger, M., Stipetic, M., & Dlacic, J. (2018, January). Forming Affect, Behavior, and Cognition with Gamification: The Case of an Austrian Tourism Advergame. In Data driven X—Turning Data into Value-Multikonferenz Wirtschaftsinformatik (MKWI) 2018.

Setia, D., Singh, R., Sharma, A., Khosla, A., Ahuja, K., & Chand, K. (2019). Enhancing Tourism and Cultural Experience Through Gamification. In *Positioning and Branding Tourism Destinations for Global Competitiveness* (pp. 152–171). IGI Global. doi:10.4018/978-1-5225-7253-4.ch007

Stadler, D., & Bilgram, V. (2016). Gamification: Best practices in research and tourism. In *Open tourism* (pp. 363–370). Berlin, Heidelberg: Springer. doi:10.1007/978-3-642-54089-9_28

Tan, E., & Okamoto, Y. (2018). iPlay, iLearn, iConserve: Digital game-based learning for sustainable tourism education.

Xu, F., Tian, F., Buhalis, D., Weber, J., & Zhang, H. (2016). Tourists as mobile gamers: Gamification for tourism marketing. *Journal of Travel & Tourism Marketing*, *33*(8), 1124–1142. doi:10.1080/10548408.2015.1093999

Yang, C. C., Sia, W. Y., Tseng, Y. C., & Chiu, J. C. (2018, November). Gamification of Learning in Tourism Industry: A case study of Pokémon Go. In *Proceedings of the 2018 2nd International Conference on Education and E-Learning* (pp. 191-195). ACM. 10.1145/3291078.3291113

KEY TERMS AND DEFINITIONS

Augmented Reality: Augmented reality consists in combining the real world with the virtual one through a computer process, enriching the visual experience and improving the quality of communication. Thanks to this technology you can add visual information to reality and create all kinds of interactive experiences: 3D product catalogues, virtual clothing testers, video games and much more (realidadaumentada.info/tecnologia).

Geocaching: The activity of using GPS to search for small hidden prizes (Cambridge Dictionary, 2019).

Smart Destination: It's much more than a destination that applies new technologies. Faced with technological challenges and the resulting social changes, a smart destination is a destination that manages and plans with "smartness".

Smart Tourism: Smart tourism within the new technological framework refers to the competitive advantage that comes from using Smart technologies such as sensors, beacons, mobile phone apps, radio frequency identification (RFID), near-field communication (NFC), smart meters, the Internet-of-Things (IoT), cloud computing, relational databases, etc., that together form a smart digital ecosystem that fosters data-driven innovations and supports new business models (Gretzel, 2018, p.173).

Smart Tourism Ecosystem: It's a smart systemic way to make destination planning and include all digital players and technologies in a sustainable way.

Smart Tourist: The tourist who reads online contents about a destination, what it offers and which services it provides before the trip; the one who creates relationships and starts conversations through apps and social media; and the one who investigates through the web, search engines and metasearch engines. The tourist who is permanently connected and values Smart experiences.

Virtual Reality: Virtual reality is usually a computer-generated virtual world (or computer systems) in which the user has the feeling of being inside this world, and depending on the level of immersion one can interact with this world and its objects in different degrees of interaction (realidadvirtual.com)

Compilation of References

(2019, April 17). Retrieved from newgenapps: https://www.newgenapps.com/blog/iot-statistics-internet-of-things-future-research-data

4. DX | Absolute Cinema Experience. (n.d.). Retrieved March 5, 2019, from http://www.cj4dx.com/

Aazam, M., & Huh, E.-N. (2016). Fog Computing: The Cloud-IoT\IoE Middleware Paradigm. *IEEE Potentials*, *35*(3), 40–44. doi:10.1109/MPOT.2015.2456213

Abadi, M., Barham, P., Chen, J., Chen, Z., Davis, A., Dean, J., ... Kudlur, M. (2016) Tensorflow: A system for large-scale machine learning. In *12th Symposium on Operating Systems Design and Implementation*, pp. 265-283.

Abdelaziz, A., Salama, A. S., & Riad, A. M. (2019). A Swarm Intelligence Model for Enhancing Health Care Services in Smart Cities Applications. In Security in Smart Cities: Models, Applications, and Challenges (pp. 71–91). Springer. doi:10.1007/978-3-030-01560-2_4

Abdmeziem, M., Tandjaoui, D., & Romdhani, I. (2016). Architecting the internet of things: State of the art. In A. K. Shakshuki (Ed.), *In Robots and Sensor Clouds* (pp. 55–75). Cham, Switzerland: Springer; doi:10.1007/978-3-319-22168-7_3

Adcock, R. A., Thangavel, A., Whitfield-Gabrieli, S., Knutson, B., & Gabrieli, J. D. E. (2006). Reward-motivated learning: Mesolimbic activation precedes memory formation. *Neuron*, *50*(3), 507–517. doi:10.1016/j.neuron.2006.03.036 PMID:16675403

Adhwarjee, D. K. (2018). Finding the shortest path by Adhwarjee's algorithm and comparison of this powerful method with Dijkstra's algorithm. *International Journal of Mathematics in Operational Research*, *13*(2), 269. doi:10.1504/IJMOR.2018.094058

Administration, N. H. (n.d.). *Automated Vehicles for Safety*. Retrieved from National Highway Traffic Safety Administration: https://www.nhtsa.gov/technology-innovation/automated-vehicles-safety

Agerar. (2019). Agerar Project, Retrieved from http://institucional.us.es/agerar/, accessed 2017/10/11

Ahire, J. B. (2018, August 24). *The Artificial Neural Networks handbook: Part 1*. Retrieved from Medium: https://medium.com/coinmonks/the-artificial-neural-networks-handbook-part-1-f9ceb0e376b4

Ahmad, M. W., Mourshed, M., Mundow, D., Sisinni, M., & Rezgui, Y. (2016). Building energy metering and environmental monitoring – A state-of-the-art review and directions for future research. *Energy and Building*, *120*, 85–102. doi:10.1016/j.enbuild.2016.03.059

Ahmad, M. W., Mourshed, M., & Rezgui, Y. (2017). Trees vs Neurons: Comparison between random forest and ANN for high-resolution prediction of building energy consumption. *Energy and Building*, *147*, 77–89. doi:10.1016/j.enbuild.2017.04.038

Aïm, F., Lonjon, G., Hannouche, D., & Nizard, R. (2016). Effectiveness of virtual reality training in orthopaedic surgery. *Arthroscopy*, *32*(1), 224–232. doi:10.1016/j.arthro.2015.07.023 PMID:26412672

Al Ghamdi, A., & Thomson, T. (2018). Big Data Storage and Its Future. In *Proceedings 2018 International Conference on Computing Sciences and Engineering (ICCSE)*, 1–6. IEEE.

Alam, T., & Benaida, M. (2018). CICS: Cloud–Internet Communication Security Framework for Internet of Smart Devices. [iJIM]. *International Journal of Interactive Mobile Technologies*, *12*(6), 74–84. doi:10.3991/ijim.v12i6.6776

Alattar, M., & Achemlal, M. (2014). Host-based card emulation: Development, security, and ecosystem impact analysis. *Proceedings - 16th IEEE International Conference on High Performance Computing and Communications, HPCC 2014, 11th IEEE International Conference on Embedded Software and Systems, ICESS 2014 and 6th International Symposium on Cyberspace Safety and Security*, 506–509. 10.1109/HPCC.2014.85

Al-Dhuraibi, Y., Paraiso, F., Djarallah, N., & Merle, P. (2017). Elasticity in cloud computing: State of the art and research challenges. *IEEE Transactions on Services Computing*, *11*(2), 430–447. doi:10.1109/TSC.2017.2711009

Alegre, U., Augusto, J. C., & Clark, T. (2016). Engineering context-aware systems and applications: A survey. *Journal of Systems and Software*, *117*, 55–83. doi:10.1016/j.jss.2016.02.010

Alfian, A. (2017, December). The development framework of expert system application on indonesian governmental accounting system. In *Proceedings of the 2017 International Conference on Computer Science and Artificial Intelligence* (pp. 60-64). ACM. 10.1145/3168390.3168437

Alford, W. A., Rogers, T., Wilkes, D. M., & Kawamura, K. (1999). Multi-Agent System for a Human-Friendly Robot. In *Proceedings of the 1999 IEEE International Conference on Systems, Man, and Cybernetics (SMC '99)*, 1064-1069, Tokyo, Japan. 10.1109/ICSMC.1999.825410

Ali, A., Kim, J., & Lee, S. (2016). Travel behavior analysis using smart card data. *KSCE Journal of Civil Engineering*, *20*(4), 1532–1539. doi:10.100712205-015-1694-0

Alpaydin, E. (2016). *Machine Learning: the New AI*. The MIT Press.

Alves, R., Sousa, L., Negrier, A., Rodrigues, J. M. F., Monteiro, J., Cardoso, P., ... Bica, P. (2018). Interactive 360 Degree Holographic Installation. *International Journal of Creative Interfaces and Computer Graphics*, *8*(1), 20–38. doi:10.4018/IJCICG.2017010102

Amador, G. P., & Gomes, A. J. P. (2018). xTrek: An Influence-Aware Technique for Dijkstra's and A Pathfinders [Research article]. doi:10.1155/2018/5184605

Amara, F., Agbossou, K., Dube, Y., Kelouwani, S., & Cardenas, A. (2016). Estimation of temperature correlation with household electricity demand for forecasting application. In *Proceedings of IECON 2016 - 42nd Annual Conference of the IEEE Industrial Electronics Society*, 3960–3965. 10.1109/IECON.2016.7793935

Amazon.com. Inc. (n.d.). Amazon Transcribe. Retrieved from https://aws.amazon.com/transcribe/ [Accessed 5th June 2019].

Annalakshmi, M., Roomi, S. M. M., & Naveedh, A. S. (2018). A hybrid technique for gender classification with SLBP and HOG features. *Cluster Computing*, 1–10. doi:10.100710586-017-1585-x

Anthes, C., Garcia-Hernandez, R. J., Wiedemann, M., & Kranzlmuller, D. (2016). State of the art of virtual reality technology. *2016 IEEE Aerospace Conference*, 1–19. 10.1109/AERO.2016.7500674

Anthopoulos, L. G. (2015). Understanding the smart city domain: A literature review. In *Transforming city governments for successful smart cities* (pp. 9–21). Springer. doi:10.1007/978-3-319-03167-5_2

Antons, D., & Breidbach, C. F. (2018). Big Data, Big Insights? Advancing Service Innovation and Design with Machine Learning. *Journal of Service Research, 21*(1), 17–39. doi:10.1177/1094670517738373

Anwar, S., Hwang, K., & Sung, W. (2015). Fixed point optimization of deep convolutional neural networks for object recognition. In *IEEE International Conference on Acoustics, Speech, and Signal Processing* (pp. 1131–1135). 10.1109/ICASSP.2015.7178146

Araya, D. (2019). Forges: 3 Things You Need to Know About Augmented Intelligence, Retrieved from https://goo.gl/oVUsHf, in 2019/03/25

Araya, D. (2018). *Augmented Intelligence: Smart Systems and the Future of Work and Learning*. Peter Lang International Academic Publishers. doi:10.3726/b11342

Arellano, L., Del Castillo, D., Guerrero, G., & Tapia, F. (2019). Mobile Application Based on Dijkstra's Algorithm, to Improve the Inclusion of People with Motor Disabilities Within Urban Areas. In Á. Rocha, H. Adeli, L. Reis, & S. Costanzo (Eds.), *New Knowledge in Information Systems and Technologies. WorldCIST'19 2019. Advances in Intelligent Systems and Computing* (Vol. 931). Cham, Switzerland: Springer. doi:10.1007/978-3-030-16184-2_22

Arriaga, O., Valdenegro-Toro, M., & Plöger, P. (2017) Real-time convolutional neural networks for emotion and gender classification. *arXiv preprint arXiv:1710.07557*

Ascott, R. (Ed.). (2000). *Art, technology, consciousness: Mind@large*. Intellect Books.

Ashton, K. (2009). *That 'Internet of Things' Thing.* Retrieved April 17, 2019, from RFID Journal: https://www.rfidjournal.com/articles/view?4986

Asimov, I. (2013). *I, Robot.* London, UK: Harper Voyager.

Assessment, T. O., & Parliamentary, E. (2014). *Integrated urban e-ticketing for public transport and touristic sites.* Retrieved from https://www.itf-oecd.org/sites/default/files/docs/2012-12-10.pdf

Assunção, M. D., Calheiros, R. N., Bianchi, S., Netto, M. A., & Buyya, R. (2015). Big Data computing and clouds: Trends and future directions. *Journal of Parallel and Distributed Computing, 79*, 3–15. doi:10.1016/j.jpdc.2014.08.003

Atembe, R., & Abdalla, F. (2015, April). The use of smart technology in tourism: evidence from wearable devices. In *ISCONTOUR 2015-Tourism Research Perspectives: Proceedings of the International Student Conference in Tourism Research* (Vol. 23). BoD–Books on Demand.

Au, Y. A., & Kauffman, R. J. (2008). The economics of mobile payments: Understanding stakeholder issues for an emerging financial technology application. *Electronic Commerce Research and Applications*, *7*(2), 141–164. doi:10.1016/j.elerap.2006.12.004

AWS. (2010). *AWS CloudFormation.* Obtido de https://docs.aws.amazon.com/AWSCloudFormation/latest/UserGuide/aws-properties-as-group.html

AWS. (2019). *Amazon API Gateway.* Obtido de https://aws.amazon.com/api-gateway/?nc1=h_ls

Azevedo, J., Padrão, P., Gregório, M. J., Almeida, C., Moutinho, N., Lien, N., & Barros, R. (2019). A Web-Based Gamification Program to Improve Nutrition Literacy in Families of 3-to 5-Year-Old Children: The Nutriscience Project. *Journal of Nutrition Education and Behavior, 51*(3), 326–334. doi:10.1016/j.jneb.2018.10.008 PMID:30579894

Azuma, R. T. (1997). A Survey of Augmented Reality. *Presence (Cambridge, Mass.)*, *6*(4), 355–385. doi:10.1162/pres.1997.6.4.355

Azure, M. (2019). *Azure Security Center for IoT in public preview.* Obtido de https://azure.microsoft.com/en-us/updates/azure-security-center-for-iot-in-public-preview/

AzureFace. (2019). Azure Face API, Retrieved from https://azure.microsoft.com/en-us/services/cognitive-services/face/, last accessed 2019/04/09

AzureVision. (2019). Cognitive Services Directory – Vision (2019). Retrieved from https://azure.microsoft.com/en-us/services/cognitive-services/directory/vision/

Babar, M., & Arif, F. (2017). Smart Urban Planning using Big Data Analytics based Internet of Things. *017 ACM International Joint Conference on Pervasive and Ubiquitous Computing and Proceedings* (pp. 397-402). Maui, Hawaii: ACM. doi:10.1145/3123024.3124411

Badii, C., Bellini, P., & Difin, A. (2018). Sii-Mobility: An IoT/IoE Architecture to Enhance Smart City Mobility and Transportation Services. *Sensors (Basel), 19*(1), 1. doi:10.339019010001 PMID:30577434

Bagchi, M., & White, P. R. (2005). The potential of public transport smart card data. *Transport Policy, 12*(5), 464–474. doi:10.1016/j.tranpol.2005.06.008

Baggio, R., Scott, N., & Cooper, C. (2010). Improving tourism destination governance: A complexity science approach. *Tourism Review, 65*(4), 51–60. doi:10.1108/16605371011093863

Balfour, R. E. (2015). *Building the "Internet of everything" (IoE) for first responders. 2015 Long Island Systems, Applications, and Technology* (pp. 1–6). Farmingdale, NY: IEEE; doi:10.1109/LISAT.2015.7160172

Ball, D., Heath, S., Wiles, J., Wyeth, G., Corke, P., & Milford, M. (2013). OpenRatSLAM: An open source brain-based SLAM system. *Autonomous Robots, 34*(3), 149–176. doi:10.100710514-012-9317-9

Baltrusaitis, T., Zadeh, A., Lim, Y. C., & Morency, L. P. (2018) Openface 2.0: Facial behavior analysis toolkit. In *Procs 13th IEEE International Conference on Automatic Face & Gesture Recognition (FG 2018),* pp. 59-66, IEEE.

Banerjee, A., & Joshi, K. P. (2017). Link before you share: Managing privacy policies through blockchain. In *2017 IEEE International Conference on Big Data (Big Data)* (pp. 4438–4447). Boston, MA: IEEE. 10.1109/BigData.2017.8258482

Bansal, A., Rompikuntla, S. K., Gopinadhan, J., Kaur, A., & Kazi, Z. A. (2015). *Energy Consumption Forecasting for Smart Meters.* Retrieved from http://arxiv.org/abs/1512.05979

Bansal, J. C., Singh, P. K., & Pal, N. R. (Eds.). (2019). *Evolutionary and Swarm Intelligence Algorithms.* doi:10.1007/978-3-319-91341-4

Barascud, N., Pearcec, M. T., Griffiths, T. D., Friston, K. J., & Chaita, M. (2016). Brain responses in humans reveal ideal observer-like sensitivity to complex acoustic patterns. *Proceedings of the National Academy of Sciences of the United States of America (PNAS),* E616-E625 10.1073/pnas.1508523113

Barber, D. (2012). *Bayesian Reasoning and Machine Learning.* Cambridge University Press.

Barchi, G., Miori, G., Moser, D., & Papantoniou, S. (2018). A Small-Scale Prototype for the Optimization of PV Generation and Battery Storage through the Use of a Building Energy Management System. In *Proceedings - 2018 IEEE International Conference on Environment and Electrical Engineering and 2018 IEEE Industrial and Commercial Power Systems Europe, EEEIC/I and CPS Europe 2018,* 1–5. 10.1109/EEEIC.2018.8494012

Barrett, M., Davidson, E., Prabhu, J., & Vargo, S. L. (2015). Service Innovation in the Digital Age: Key Contributions and Future Directions. *Management Information Systems Quarterly, 39*(1), 135–154. doi:10.25300/MISQ/2015/39:1.03

Bartoletti, M., Lande, S., Pompianu, L., & Bracciali, A. (2017). A general framework for blockchain analytics. In *Proceedings of the 1st Workshop on Scalable and Resilient Infrastructures for Distributed Ledgers - SERIAL '17* (pp. 1–6). New York, NY: ACM Press. 10.1145/3152824.3152831

Barzov, Y. (2017). *Hacking Hippocampus: the Next Frontier for Machine Learning and beyond.* Retrieved September 14, 2019 from https://towardsdatascience.com/machine-learning-as-hacking-of-the-brain-6aab8c4a9e7d

Bay, H., Tuytelaars, T., & van Gool, L. (2008). SURF: Speeded-up robust features. *Computer Vision and Image Understanding, 110*(3), 346–359. doi:10.1016/j.cviu.2007.09.014

Bayuk, J., & Altobello, S. A. (2019). Can gamification improve financial behavior? The moderating role of app expertise. *International Journal of Bank Marketing, 37*(4), 951–975. doi:10.1108/IJBM-04-2018-0086

BBC. (2018, March 20). *Uber halts self-driving car tests after death.* Retrieved June 20, 2019, from https://www.bbc.com/news/business-43459156

Behnke, S. (2008). Humanoid robots-from fiction to reality?. *KI, 22*(4), 5-9.

Bellavista, P., Berrocal, J., Corradi, A., Das, S. K., Foschini, L., & Zanni, A. (2019). A survey on fog computing for the Internet of Things. *Pervasive and Mobile Computing, 52*, 71–99. doi:10.1016/j.pmcj.2018.12.007

Belmonte, Ó., Pérez, M., Pérez, M., & Romero, C. (1998). *A NEW CONCEPT IN CERAMIC SPECIFICATION USING VIRTUAL ENVIRONMENTS.* Presented at the Qualicer. Retrieved from http://www.qualicer.org/recopilatorio/ponencias/pdfs/9823032e.pdf

Bengio, Y. (2009). Learning deep architectures for AI. *Foundations and Trends in Machine Learning, 2*(1), 1–127. doi:10.1561/2200000006

Bengio, Y., Lamblin, P., Popovici, D., & Larochelle, H. (2006). Greedy layer-wise training of deep networks. In B. Schölkopf, J. C. Platt, & T. Hoffman (Eds.). In *Proceedings of the 19th International Conference on Neural Information Processing Systems* (153-160). Cambridge, MA.

Bernardo, J. (2019). Plano Nacional Integrado Energia-Clima: Linhas de Atuação para o Horizonte 2021-2030.

Beyeler, M. (2015). OpenCV with Python Blueprints. Learning to Recognize Emotions on Faces; Retrieved from https://www.packtpub.com/mapt/book/application_development/9781785282690/7/ch07lvl1sec53/facial-expression-recognition, accessed 2017/10/11.

Beyer, H., & Holtzblatt, K. (1997). *Contextual Design.* Morgan Kauffmann.

Bharadwaj, A. S., Rego, R., & Chowdhury, A. (2016). IoT based solid waste management system: A conceptual approach with an architectural solution as a smart city application. In *Proceedings 2016 IEEE Annual India Conference (INDICON)* (pp. 1-6). Bangalore, India: IEEE. 10.1109/INDICON.2016.7839147

Bhattacharya, M., Paramati, S. R., Ozturk, I., & Bhattacharya, S. (2016). The effect of renewable energy consumption on economic growth: Evidence from top 38 countries. *Applied Energy, 162*, 733–741. doi:10.1016/j.apenergy.2015.10.104

Bhui, R. (2018). Case-Based Decision Neuroscience: Economic Judgment by Similarity. Retrieved July 20, 2019 from https://scholar.harvard.edu/files/rbhui/files/case-based_decision_neuroscience.pdf

Billinghurst, M., Clark, A., & Lee, G. (2015). A survey of augmented reality. *Foundations and Trends® in Human–Computer Interaction, 8*(2–3), 73–272.

Bimber, O., & Raskar, R. (2005). Spatial Augmented Reality: Merging Real and Virtual Worlds (0 ed.). doi:10.1201/b10624

Bimber, O., & Raskar, R. (2005). *Spatial augmented reality: Merging real and virtual worlds.* AK Peters/CRC Press.

Bishop, C., & Tipping, M. E. (2003). *Bayesian Regression and Classification* (Advances i, Vol. 190, pp. 267–285). Advances i, Vol. 190, pp. 267–285. Retrieved from https://www.microsoft.com/en-us/research/publication/bayesian-regression-and-classification/

Bistarelli, S., & Santini, F. (2017). Go with the -Bitcoin- Flow, with Visual Analytics. In *Proceedings of the 12th International Conference on Availability, Reliability and Security - ARES '17* (pp. 1–6). New York, NY: ACM Press. 10.1145/3098954.3098972

Biswas, K., & Muthukkumarasamy, V. (2016). Securing smart cities using blockchain technology. In *Proceedings 2016 IEEE 18th International Conference on High Performance Computing and Communications; IEEE 14th International Conference on Smart City; IEEE 2nd International Conference on Data Science and Systems (HPCC/SmartCity/DSS)*, 1392–1393. IEEE.

Bitar, A., Jamal, A., Sultan, H., Alkandari, N., & El-Abd, M. (2017). Medical Drones System for Amusement Parks. *IEEE/ACS 14th International Conference on Computer Systems and Applications* (AICCSA) (pp. 19-20). 10.1109/AICCSA.2017.62

Bjerknes, G., & Bratteteig, T. (1995). User Participation and Democracy: A Discussion of Scandinavian Research on Systems development. *Scandinavian Journal of Information Systems*, *7*(1), 73–98.

Blomkvist, J., & Holmlid, S. (2010). Service Prototyping According to Service Design Practitioners. ServDes. 2010. In *Conference Proceedings, ServDes. 2010, Exchanging Knowledge,* Linköping, Sweden, December 1-3, 2010 (Vol. 2, pp. 1-11). Linköping University Electronic Press.

Bloom, P. (2019). *Monitored Business and Surveillance in a Time of Big Data.* London, UK: Pluto Press.

Boer, R., & Boer, T. de. (2009). *Mobile payments 2010: Market Analysis and Overview. Security.* Chiel Liezenber (Innopay) and Ed Achterberg (Telecompaper).

Boes, K., Buhalis, D., & Inversini, A. (2015). Conceptualising smart tourism destination dimensions. In *Information and communication technologies in tourism 2015* (pp. 391–403). Cham, Switzerland: Springer.

Boes, K., Buhalis, D., & Inversini, A. (2016). Smart tourism destinations: Ecosystems for tourism destination competitiveness. [Bingley: Emerald.]. *International Journal of Tourism Cities*, *2*(2), 108–124. doi:10.1108/IJTC-12-2015-0032

Bogner, A. (2017). Seeing is understanding – anomaly detection in blockchains with visualized features. In *UBICOMP / ISWC* (pp. 5–8). Maui, HI: ACM; doi:10.1145/3123024.3123157

Bonaccorso, G. (2017). *Machine Learning Algorithms.* Packt Publishing.

Bonetti, F., Warnaby, G., & Quinn, L. (2018). Augmented reality and virtual reality in physical and online retailing: A review, synthesis and research agenda. In *Augmented Reality and Virtual Reality* (pp. 119–132). Springer. doi:10.1007/978-3-319-64027-3_9

Borgia, E. (2014). The Internet of Things vision: Key features, applications and open issues. *Computer Communications*, *54*, 1–31. doi:10.1016/j.comcom.2014.09.008

Bostrom, N. (2013). *Superinteligência: Caminhos, perigos e estratégias para um novo mundo.* São Paulo, Brazil: Darkside.

Botella, C., Bretón-López, J., Quero, S., Baños, R., & García-Palacios, A. (2010). Treating Cockroach Phobia With Augmented Reality. *Behavior Therapy*, *41*(3), 401–413. doi:10.1016/j.beth.2009.07.002 PMID:20569788

Botta, A., De Donato, W., Persico, V., & Pescapé, A. (2016). Integration of cloud computing and internet of things: A survey. *Future Generation Computer Systems*, *56*, 684–700. doi:10.1016/j.future.2015.09.021

Boulos, M. N. K., Lu, Z., Guerrero, P., Jennett, C., & Steed, A. (2017). *From urban planning and emergency training to Pokémon Go: Applications of virtual reality GIS (VRGIS) and augmented reality GIS (ARGIS) in personal, public and environmental health*. BioMed Central.

Bouloukakis, M., Partarakis, N., Drossis, I., Kalaitzakis, M., & Stephanidis, C. (2019). Virtual Reality for Smart City Visualization and Monitoring. In *Mediterranean Cities and Island Communities* (pp. 1–18). Springer. doi:10.1007/978-3-319-99444-4_1

Bourdeau, M., Zhai, X. Q., Nefzaoui, E., Guo, X., & Chatellier, P. (2019). Modeling and forecasting building energy consumption: A review of data-driven techniques. *Sustainable Cities and Society*, *48*, 101533. doi:10.1016/j.scs.2019.101533

Bourlard, H., & Kamp, Y. (1988). Auto-association by multilayer perceptrons and singular value decomposition. *Biological Cybernetics*, *59*(4-5), 291–294. doi:10.1007/BF00332918 PMID:3196773

Bouveyron, C., Celeux, G., Murphy, T., & Raftery, A. (2019). *Model-Based Clustering and Classification for Data Science: With Applications in R (Cambridge Series in Statistical and Probabilistic Mathematics)*. Cambridge, UK: Cambridge University Press. doi:10.1017/9781108644181

Box, G., Jenkins, G., Reinsel, G., & Ljung, G. (2016). Time series analysis: forecasting and control (5th Ed.) (D. Balding, N. A. C. Cressie, G. M. Fitzmaurice, G. H. Givens, H. Goldstein, G. Molenberghs, … S. Weisberg, Eds.). Hoboken, NJ: John Wiley & Sons.

Bradley, J., Loucks, J., Noronha, A., Macaulay, J., & Buckalew, L. (2013). *Internet of Everything (IoE): Top 10 Insights from Cisco's IoE Value Index Survey of 7,500 Decision Makers Across 12 Countries*. Cisco.

Bradski, G. R., & Pisarevsky, V. (2000) Intel's Computer Vision Library: applications in calibration, stereo segmentation, tracking, gesture, face and object recognition. In *Procs IEEE Conference on Computer Vision and Pattern Recognition (CVPR 2000)*, vol. 2, pp. 796-797 10.1109/CVPR.2000.854964

Braem, B., Latré, S., Leroux, P., Demeester, P., Coenen, T., & Ballon, P. (2016). Designing a smart city playground: Real-time air quality measurements and visualization in the City of Things testbed. In *Proceedings 2016 IEEE International Smart Cities Conference (ISC2)*, 1–2. IEEE. 10.1109/ISC2.2016.7580871

Brakewood, C., Roja, F., Robin, J., Sion, J., & Jordan, S. (2014). Forecasting mobile ticketing adoption on commuter rail. *Journal of Public Transportation*, *17*(1), 1–19. doi:10.5038/2375-0901.17.1.1

Brennen, S., Howard, P. N., & Nielsen, R. K. (2018). An Industry-Led Debate: How UK Media Cover Artificial Intelligence. *Reuters Institute for the Study of Journalism Fact Sheet,(December)*, 1-10.

Brito, T. B., Trevisan, E. F., & Booter, R. C. (2011). A Conceptual Comparison Between Discrete and Continuous Simulation to Motivate the Hybrid Simulation Technology. In *Proceedings of the 2011 Winter Simulation Conference*, 3915-3927. Winter Simulation Conference.

Brown, M. 2011. Learning Local Image Descriptors Data. Retrieved from http://www.cs. ubc.ca/mbrown/patchdata/patchdata.html, accessed 2017/10/11

Buchanan, B. G. (2005). A (very) brief history of Artificial Intelligence. *AI Magazine*, *26*(4), 53–53.

Buhalis, D. (2003). eTourism: Information technology for strategic tourism management. Harlow, UK: Pearson Education.

Buhalis, D., & Amaranggana, A. (2014). Smart Tourism Destinations. Conference paper. Information and Communication Technologies in Tourism (pp. 553-564). Springer International Publishing Switzerland.

Buhalis, D., & Amaranggana, A. (2013). Smart tourism destinations. In *Information and communication technologies in tourism 2014* (pp. 553–564). Cham, Switzerland: Springer. doi:10.1007/978-3-319-03973-2_40

Buhalis, D., & Law, R. (2008). Progress in information technology and tourism management: 20 years on and 10 years after the Internet—The state of eTourism research. *Tourism Management, 29*(4), 609–623. doi:10.1016/j.tourman.2008.01.005

Bunzeck, N., Dayan, P., Dolan, R. J., & Duzel, E. (2010). A common mechanism for adaptive scaling of reward and novelty. *Human Brain Mapping, 31*(9), 1380–1394. doi:10.1002/hbm.20939 PMID:20091793

Buonincontri, P., & Micera, R. (2016). The experience co-creation in smart tourism destinations: A multiple case analysis of European destinations. *Information Technology & Tourism, 16*(3), 285–315. doi:10.100740558-016-0060-5

Burges, C. J. C. (2010). *From RankNet to LambdaRank to LambdaMART: An Overview.* Retrieved from https://www.microsoft.com/en-us/research/publication/from-ranknet-to-lambdarank-to-lambdamart-an-overview/

Burke, B. (2016). *Gamify: How gamification motivates people to do extraordinary things.* Routledge. doi:10.4324/9781315230344

Cabrita, C. L., Monteiro, J. M., & Cardoso, P. J. S. (2019). *Improving Energy Efficiency in Smart-houses by Optimizing Electrical Loads Management.* 1–6.

Cadence: Tensilica. (2017). DNA Processor IP For AI Inference.

Cai, L., Gu, J., Ma, J., & Jin, Z. (2019). Probabilistic Wind Power Forecasting Approach via Instance-Based Transfer Learning Embedded Gradient Boosting Decision Trees. *Energies, 12*(1), 159. doi:10.3390/en12010159

Cai, M., Pipattanasomporn, M., & Rahman, S. (2019). Day-ahead building-level load forecasts using deep learning vs. traditional time-series techniques. *Applied Energy, 236*, 1078–1088. doi:10.1016/j.apenergy.2018.12.042

Cai, Y., & Zhu, D. (2016). Fraud detections for online businesses: A perspective from blockchain technology. *Financial Innovation, 2*(1), 20. doi:10.118640854-016-0039-4

Calado, C. (2018, April 12). *Mulheres e chatbots.* Retrieved March 21, 2019, from https://medium.com/botsbrasil/mulheres-e-chatbots-91064d124eeb

Campbell, J. (2018). *Terraform: Beyond the Basics with AWS.* Obtido de https://aws.amazon.com/pt/blogs/apn/terraform-beyond-the-basics-with-aws

Campo, R., Baldassarre, F., & Lee, R. (2019). A Play-Based Methodology for Studying Children: Playfication. *Systemic Practice and Action Research*, 1–11.

Campos, B. P., & da Silva, M. R. (2016). Demand forecasting in residential distribution feeders in the context of smart grids. In *Proceedings 2016 12th IEEE International Conference on Industry Applications (INDUSCON)*, 1–6. IEEE. 10.1109/INDUSCON.2016.7874464

Cao, Z., Hidalgo, G., Simon, T., Wei, S. E., & Sheikh, Y. (2018). OpenPose: realtime multi person 2D pose estimation using Part Affinity Fields. *arXiv preprint arXiv:1812.08008.*

Cao, Z., Simon, T., Wei, S. E., & Sheikh, Y. (2017). Realtime multi-person 2D pose estimation using part affinity fields. In *Proceedings of the IEEE Conference on Computer Vision and Pattern Recognition* (pp. 7291-7299). 10.1109/CVPR.2017.143

Capgemini. (2011). *Mobile Payments: How Can Banks Seize the Opportunity ? An Approach for Financial Services Institutions. Capgemini.*

Cardoso, P. J. S., Schütz, G., Mazayev, A., & Ey, E. (2015a). Solutions in Under 10 Seconds for Vehicle Routing Problems with Time Windows using Commodity Computers. In *Proceedings 8th International Conference on Evolutionary Multi-Criterion Optimization (EMO 15),* pp. 418-432, March 29-April 1. Cham, Switzerland: Springer.

Cardoso, P. J. S., Guerreiro, P., Monteiro, J., & Rodrigues, J. M. F. (2018). Applying an Implicit Recommender System in the Preparation of Visits to Cultural Heritage Places. In M. Antona, & C. Stephanidis (Eds.), *Universal Access in Human-Computer Interaction 2018, LNCS 10908* (pp. 421–436)., doi:10.1007/978-3-319-92052-8_33

Cardoso, P. J., Schütz, G., Mazayev, A., Ey, E., & Corrêa, T. (2015b). A solution for a real-time stochastic capacitated vehicle routing problem with time windows. *Procedia Computer Science, 51*, 2227–2236. doi:10.1016/j.procs.2015.05.501

Cardoso, R. (2011). *Design para um mundo complexo*. São Paulo, Brazil: Cosac Naify.

Cardullo, P., & Kitchin, R. (2019). Being a 'citizen'in the smart city: Up and down the scaffold of smart citizen participation in Dublin, Ireland. *GeoJournal, 84*(1), 1–13. doi:10.100710708-018-9845-8

Carroll, A., & Heiser, G. (2010). An analysis of power consumption in a smartphone. *Proceedings of the USENIX Annual Technical Conference*. Retrieved from https://www.usenix.org/event/usenix10/tech/full_papers/Carroll.pdf

Casas, S., Morillo, P., Gimeno, J., & Fernández, M. (2009). SUED: An extensible framework for the development of low-cost DVE systems. In *Proceedings of the IEEE Virtual Reality 2009 (IEEE-VR'09). Workshop on Software Engineering and Architectures for Realtime Interactive Systems (SEARIS)*, 20–25.

Casas, S., Coma, I., Portalés, C., & Fernández, M. (2016). Towards a simulation-based tuning of motion cueing algorithms. *Simulation Modelling Practice and Theory*, *67*, 137–154. doi:10.1016/j.simpat.2016.06.002

Casas, S., Olanda, R., & Dey, N. (2017). Motion Cueing Algorithms: A Review: Algorithms, Evaluation and Tuning. [IJVAR]. *International Journal of Virtual and Augmented Reality*, *1*(1), 90–106. doi:10.4018/IJVAR.2017010107

Casas, S., Portalés, C., García-Pereira, I., & Fernández, M. (2017). On a first evaluation of romot—A robotic 3D movie theatre—For driving safety awareness. *Multimodal Technologies and Interaction*, *1*(2), 6. doi:10.3390/mti1020006

Casas, S., Portalés, C., Morillo, P., & Fernández, M. (2018). A particle swarm approach for tuning washout algorithms in vehicle simulators. *Applied Soft Computing*, *68*, 125–135. doi:10.1016/j.asoc.2018.03.044

Casas, X., Herrera, G., Coma, I., & Fernández, M. (2012). *A Kinect-based Augmented Reality System for Individuals with Autism Spectrum Disorders*. GRAPP/IVAPP.

Castelluccio, M. (2018, October 3). *TESLA'S AUTOPILOT SEEN THROUGH THE WINDSHIELD*. Retrieved from SF Magazine: https://sfmagazine.com/technotes/october-2018-teslas-autopilot-seen-through-the-windshield/

Castillo, J. C., Castro-González, Á., Fernández-Caballero, A., Latorre, J. M., Pastor, J. M., Fernández-Sotos, A., & Salichs, M. A. (2016). Software architecture for smart emotion recognition and regulation of the ageing adult. *Cognitive Computation*, *8*(2), 357–367. doi:10.100712559-016-9383-y

Ceipidor, U. B., Medaglia, C. M., Marino, A., Morena, M., Sposato, S., & Moroni, A. … Morgia, M. La. (2013). Mobile ticketing with NFC management for transport companies. Problems and solutions. In *Proceedings 2013 5th International Workshop on Near Field Communication (NFC)*, 1–6. 10.1109/NFC.2013.6482446

Cerruela García, G., Luque Ruiz, I., & Gómez-Nieto, M. (2016). State of the Art, Trends and Future of Bluetooth Low Energy, Near Field Communication and Visible Light Communication in the Development of Smart Cities. *Sensors (Basel)*, *16*(11), 1968. doi:10.339016111968 PMID:27886087

Chae, M. S., Yang, Z., Yuce, M. R., Hoang, L., & Liu, W. (2009). A 128-Channel 6 mW Wireless Neural Recording IC With Spike Feature Extraction and UWB Transmitter. *IEEE Transactions on Neural Systems and Rehabilitation Engineering*, *17*(4), 312–321. doi:10.1109/TNSRE.2009.2021607 PMID:19435684

Chahal, K., & Eldabi, T. (2010). A multi-perspective comparison between system dynamics and discrete event simulation, *Journal of Business Information Systems,* pp. 4-17.

Chan, E. S. W. (2013a). Gap analysis of green hotel marketing. *International Journal of Contemporary Hospitality Management*, *25*(7), 1017–1048. doi:10.1108/IJCHM-09-2012-0156

Chan, E. S. W. (2013b). Managing green marketing: Hong Kong hotel managers' perspective. *International Journal of Hospitality Management*, *34*, 442–461. doi:10.1016/j.ijhm.2012.12.007

Chang, T. M. (2005). The role of artificial cells in cell and organ transplantation in regenerative medicine. *Panminerva Medica*, *47*(1), 1–9. PMID:15985972

Chang, T. M. (2007). 50th anniversary of artificial cells: Their role in biotechnology, nanomedicine, regenerative medicine, blood substitutes, bioencapsulation, cell/stem cell therapy and nanorobotics. *Artificial Cells, Blood Substitutes, and Immobilization Biotechnology*, *35*(6), 545–554. doi:10.1080/10731190701730172 PMID:18097783

Cheema, M. (2019). *AWS Security releases IoT security whitepaper.* Obtido de https://aws.amazon.com/pt/blogs/security/aws-security-releases-iot-security-whitepaper/

Cheng, B., Longo, S., Cirillo, F., Bauer, M., & Kovacs, E. (2015). Building a big data platform for smart cities: Experience and lessons from santander. In *Proceedings 2015 IEEE International Congress on Big Data*, 592–599. IEEE. 10.1109/BigDataCongress.2015.91

Cheng, Y.-H., & Huang, T.-Y. (2013). High speed rail passengers' mobile ticketing adoption. *Transportation Research Part C, Emerging Technologies*, *30*, 143–160. doi:10.1016/j.trc.2013.02.001

Chen, J. J., & Adams, C. (2004). Short-range wireless technologies with mobile payments systems. *Proceedings of the 6th International Conference on Electronic Commerce - ICEC '04*, 649. 10.1145/1052220.1052302

Chen, M., Wan, J., & Li, F. (2012). Machine-to-Machine Communications: Architectures, Standards and Applications. KSII *Transactions on Internet and Information Systems*, 6(2). doi:10.3837/tiis.2012.02.002

Chen, R.-Y. (2018). A traceability chain algorithm for artificial neural networks using T–S fuzzy cognitive maps in blockchain. *Future Generation Computer Systems*, *80*, 198–210. doi:10.1016/j.future.2017.09.077

Chen, S., Ji, Y., & Tong, L. (2012). Deadline scheduling for large scale charging of electric vehicles with renewable energy. In *Proceedings of the IEEE Sensor Array and Multichannel Signal Processing Workshop*, 13–16. 10.1109/SAM.2012.6250449

Chen, Y., & Han, D. (2018). Water quality monitoring in smart city: A pilot project. *Automation in Construction*, *89*, 307–316. doi:10.1016/j.autcon.2018.02.008

Chen, Y., Krishna, T., Emer, J. S., & Sze, V. (2016). Eyeriss: An Energy-Efficient Reconfigurable Accelerator for Deep Convolutional Neural Networks. *IEEE Journal of Solid-State Circuits*, *52*(1), 127–138. doi:10.1109/JSSC.2016.2616357

Cheok, A., & Portalés Ricart, C. (2016). Welcome to MTI—A New Open Access Journal Dealing with Blue Sky Research and Future Trends in Multimodal Technologies and Interaction. *Multimodal Technologies and Interaction*, *1*(1), 1. doi:10.3390/mti1010001

Chersi, F., & Burgess, N. (2015). The Cognitive Architecture of Spatial Navigation: Hippocampal and Striatal Contributions'. *Neuron*, *88*(1), 64–77. doi:10.1016/j.neuron.2015.09.021 PMID:26447573

Cheung, F. (2006). Implementation of Nationwide Public Transport Smart Card in the Netherlands: Cost-Benefit Analysis. *Transportation Research Record: Journal of the Transportation Research Board*, 1971.

Chidamber, S. R., & Kemerer, C. F. (1991). *Towards a metrics suite for objectoriented design. Object-oriented programming systems, languages, and applications OOPSLA '91* (pp. 197–211). Phoenix, AZ: ACM; doi:10.1145/117954.117970

Chidamber, S. R., & Kemerer, C. F. (1994). A metrics suite for object-oriented design. *IEEE Transactions on Software Engineering*, *20*(6), 476–493. doi:10.1109/32.295895

Chkirbene, Z., & Hamdi, N. (2015). A survey on spectrum management in cognitive radio networks. *International Journal of Wireless and Mobile Computing*, *8*(2), 153–165. doi:10.1504/IJWMC.2015.068618

Choe, P., Kim, C., Lehto, M., Lehto, X., & Allebach, J. (2010). Evaluating and Improving a Self-Help Technical Support Web Site: Use of Focus Group Interviews. *International Journal of Human-Computer Interaction*, 333–354. doi:10.120715327590ijhc2103_4

Cho, K., Jang, H., Park, L. W., Kim, S., & Park, S. (2019). Energy Management System Based on Augmented Reality for Human-Computer Interaction in a Smart City. In *Proceedings 2019 IEEE International Conference on Consumer Electronics (ICCE)*, 1–3. IEEE. 10.1109/ICCE.2019.8662045

Christmann, A., & Steinwart, I. (2008). *Support Vector Machines*. Springer-Verlag.

Chung, N., Tyan, I., & Han, H. (2017). Enhancing the smart tourism experience through geotag. *Information Systems Frontiers*, *19*(4), 731–742. doi:10.100710796-016-9710-6

Ciesielski, K. C., Falcão, A. X., & Miranda, P. A. (2018). Path-Value Functions for Which Dijkstra's Algorithm Returns Optimal Mapping. *Journal of Mathematical Imaging and Vision*, *60*(7), 1025–1036. doi:10.100710851-018-0793-1

Civitarese, G., Bettini, C., Sztyler, T., Riboni, D., & Stuckenschmidt, H. (2019). newNECTAR: Collaborative active learning for knowledge-based probabilistic activity recognition. *Pervasive and Mobile Computing*, *56*, 88–105. doi:10.1016/j.pmcj.2019.04.006

Clark, L., Doyle, P., Garaialde, D., Gilmartin, E., Schlögl, S., Edlund, J., ... Cowan, B. R. (2019a). The State of Speech in HCI: Trends, Themes, and Challenges. *Interacting with Computers*, iwz016. doi:10.1093/iwc/iwz016

Clark, L., Pantidi, N., Cooney, O., Doyle, P., Garaialde, D., Edwards, J., ... Wade, V. (2019b). What Makes a Good Conversation?: Challenges in Designing Truly Conversational Agents. In *Proceedings of the 2019 CHI Conference on Human Factors in Computing Systems* (p. 475). ACM. 10.1145/3290605.3300705

Cocchia, A. (2014). Smart and digital city: A systematic literature review. In *Smart city* (pp. 13–43). Springer.

Colgin, L. L., Leutgeb, S., Jezek, K., Leutgeb, J. K., Moser, E. I., McNaughton, B. L., & Moser, M.-B. (2010). Attractor-map versus autoassociation basedattractor dynamics in the hippocampal network. *Journal of Neurophysiology*, *104*(1), 35–50. doi:10.1152/jn.00202.2010 PMID:20445029

Congress, T. L. (n.d.). *Who Invented the Automobile*. Retrieved from Everyday Mysteries: https://www.loc.gov/rr/scitech/mysteries/auto.html

Cordoba Tourism Office. (2015). Plan Estrategico de Córdoba 2015 2019 - Strategic Tourism Plan of Cordoba 2015-2019. EOI, Córdoba, España. Retrieved from https://www.turismodecordoba.org/archivos/2018/20180718095418000 0pdf

Corrêaa, C., & Kitanoa, C. (2015, February). Gamification in tourism: Analysis of brazil quest game. In *Proceedings of ENTER* (Vol. 2015).

Corsar, D., Edwards, P., Nelson, J., & Papangelis, K. (2014). Mobile phones, sensors & the crowd: lessons learnt from development of a real-time travel information system. *Proceedings of the First International Conference on IoT, (URB-IOT '14)* (pp. 99-101). Brussels, Belgium: ICST (Institute for Computer Sciences, Social-Informatics and Telecommunications Engineering), ICST. doi:10.4108/icst.urb-iot.2014.257328

Costa, J. P., Robb, J., & Nacke, L. E. (2014). Physiological acrophobia evaluation through in vivo exposure in a VR CAVE. *2014 IEEE Games Media Entertainment*, 1–4. doi:10.1109/GEM.2014.7047969

Costa, P. M., Fontes, T., Nunes, A. A., Ferreira, M. C., Costa, V., & Dias, T. G., ... Falcão, J. (2015). Seamless Mobility : a disruptive solution for urban public transport. In *22nd ITS World Congress*. Bordeux.

Costa, P. M., Fontes, T., Nunes, A. A., Ferreira, M. C., Costa, V., Dias, T. G., ... Cunha, J. (2016). Application of Collaborative Information Exchange in Urban Public Transport: The Seamless Mobility Solution. *Transportation Research Procedia*, *14*, 1201–1210. doi:10.1016/j.trpro.2016.05.191

Costa, V., Fontes, T., Borges, J. L., & Dias, T. G. (2019). Prediction of Journey Destination for Travelers of Urban Public Transport: A Comparison Model Study. *Lecture Notes of the Institute for Computer Sciences, Social-Informatics, and Telecommunications Engineering, LNICST*, *267*(ii), 113–132. doi:10.1007/978-3-030-14757-0_9

Costa, V., Fontes, T., Costa, P. M., & Dias, T. G. (2015a). How to Predict Journey Destination for Supporting Contextual Intelligent Information Services? In *Proceedings IEEE Conference on Intelligent Transportation Systems, ITSC*, 2959–2964. 10.1109/ITSC.2015.474

Costa, V., Fontes, T., Costa, P. M., & Dias, T. G. (2015b). Prediction of Journey Destination in Urban Public Transport. In F. Pereira, P. Machado, E. Costa, & A. Cardoso (Eds.), *Progress in Artificial Intelligence. EPIA 2015* (pp. 169–180). Lecture Notes in Computer Science, doi:10.1007/978-3-319-23485-4_18

Couch, L. W. II. (2013). *Digital and Analog Communication Systems* (8th ed.). Prentice-Hall.

Coulouris, G., Dollimore, J., Kindberg, T., & Blair, G. (2011). *Distributed Systems: Concepts and Design* (5th ed.). Addison-Wesley Publishing.

Courbariaux, M., & Bengio, Y. (2016) BinaryNet: Training Deep Neural Networks with Weights and Activations Constrained to +1 or -1. In CoRR, abs/1602.02830.

Couto, R., Leal, J., Costa, P. M., & Galvao, T. (2015). Exploring Ticketing Approaches Using Mobile Technologies: QR Codes, NFC and BLE. *IEEE Conference on Intelligent Transportation Systems, Proceedings, ITSC, 2015-October*, 7–12. 10.1109/ITSC.2015.9

Crabtree, A., & Mortier, R. (2016). *Personal Data, Privacy and the Internet of Things: The Shifting Locus of Agency and Control*. SSRN Electronic Journal; doi:10.2139srn.2874312

Crandall, J. W., Oudah, M., Ishowo-Oloko, F., Abdallah, S., Bonnefon, J. F., Cebrian, M., ... Rahwan, I. (2018). Cooperating with machines. *Nature Communications*, *9*(1), 233. doi:10.103841467-017-02597-8 PMID:29339817

Criminisi, A., Shotton, J., & Konukoglu, E. (2012). Decision Forests: A Unified Framework for Classification, Regression, Density Estimation, Manifold Learning and Semi-Supervised Learning. In *Foundations and Trends® in Computer Graphics and Vision* (Foundation, Vol. 7, pp. 81–227). Retrieved from https://www.microsoft.com/en-us/research/publication/decision-forests-a-unified-framework-for-classification-regression-density-estimation-manifold-learning-and-semi-supervised-learning/

Cruz-Neira, C., Fernández, M., & Portalés, C. (2018). Virtual Reality and Games. *Multimodal Technologies and Interaction*, *2*(1), 8. doi:10.3390/mti2010008

Csikzentmihalyi, M. (1988). The flow experience and its significance for human psy- chology. In M. Csikzentmihalyi, & I. S. Csikzentmihalyi (Eds.), *Optimal experience* (pp. 15–35). Cambridge, UK: Cambridge University Press. doi:10.1017/CBO9780511621956.002

Curran, J. (2016). The internet of history: rethinking the internet's past. In *Misunderstanding the internet* (pp. 48–84). Routledge. doi:10.4324/9781315695624-2

Dahlberg, T., & Mallat, N. (2002). Mobile Payment Service Development - Managerial Implications Of Consumer Value Perceptions. *Proceedings of the Tenth European Conference on Information Systems*, (January 2002), 649–657. Retrieved from http://is2.lse.ac.uk/asp/aspecis/20020144.pdf

Dahlberg, T., Guo, J., & Ondrus, J. (2015). A critical review of mobile payment research. *Electronic Commerce Research and Applications*, *14*(5), 265–284. doi:10.1016/j.elerap.2015.07.006

Dallinger, D. (2013). *Plug-in Electric Vehicles Integrating Fluctuating Renewable Electricity*. Kassel University Press.

Dangeti, P. (2017). *Statistics for Machine Learning*. Packt Publishing Ltd.

Daniel, S., & Doran, M.-A. (2013). geoSmartCity: Geomatics contribution to the smart city. *Proceedings of the 14th Annual International Conference on Digital Government Research*, 65–71. ACM. 10.1145/2479724.2479738

Dash Wu, D., Chen, S.-H., & Olson, D. L. (2014). Business intelligence in risk management: Some recent progresses. *Information Sciences*, *256*, 1–7. doi:10.1016/j.ins.2013.10.008

Datta, S. (2016). Learning OpenCV 3 Application Development. Chapter 6. Face Detection Using OpenCV. Retrieved from https://www.packtpub.com/mapt/book/application_development/9781784391454/6/ch06lvl1sec64/gender-classification, accessed 2019/06/05

Davachi, L. (2006, December). Item, context and relational episodic encoding in humans. *Current Opinion in Neurobiology*, *16*(6), 693–700. doi:10.1016/j.conb.2006.10.012 PMID:17097284

Davidow, J. Y., Foerde, K., Galvan, A., & Shohamy, D. (2016). An Upside to Reward Sensitivity: The Hippocampus Supports Enhanced Reinforcement Learning in Adolescence. *Neuron*, *92*(1), 93–99. doi:10.1016/j.neuron.2016.08.031 PMID:27710793

Davis, F. D., Bagozzi, R. P., & Warshaw, P. R. (1989). User acceptance of computer technology: A comparison of two theoretical models. *Management Science*, *35*(8), 982–1003. doi:10.1287/mnsc.35.8.982

Dejean, G. (2015). A methodology for classifying visitors to an amusement park. En 2015 IEEE *Conference on Visual Analytics Science and Technology (VAST)* (pp. 151-152). 10.1109/VAST.2015.7347655

Del Vecchio, P., Mele, G., Ndou, V., & Secundo, G. (2018). Creating value from social big data: Implications for smart tourism destinations. *Information Processing & Management*, *54*(5), 847–860. doi:10.1016/j.ipm.2017.10.006

DELL EMC, & Intel. (2018). *Fighting fraud the smart way — with data analytics and Artificial Intelligence: For Mastercard and other global payments processors, AI systems are a watchdog that never sleeps* [PDF file]. Retrieved from https://www.intel.ai/wp-content/uploads/sites/69/mastercard-fighting-fraud-the-smart-way-2.pdf

Dell'Era, C., Altuna, N., & Verganti, R. (2018). Designing radical innovations of meanings for society: Envisioning new scenarios for smart mobility. *Creativity and Innovation Management*, 1–14. doi:10.1111/caim.12276

Deng, W., Mayford, M., & Gage, F. H. (2013). Selection of distinct populations of dentate granule cells in response to inputs as a mechanism for pattern separation in mice. *eLife*, *2*. doi:10.7554/eLife.00312 PMID:23538967

Deterding, S., Dixon, D., Khaled, R., & Nacke, L. (2011). From game design elements to gamefulness: defining "gamification". In 15th International Academic MindTrek Conference: Envisioning Future Media Environments, 9-15, ACM, Tampere, Finland.

Deterding, S., Dixon, D., Khaled, R., & Nacke, L. (2014). Du game design au gamefulness: définir la gamification. *Sciences du jeu*, (2).

Deterding, S. (2012). Gamification: Designing for motivation. *Interaction*, *19*(4), 14–17. doi:10.1145/2212877.2212883

Dewey, J. (2010). *Arte como experiência*. São Paulo, Brazil: Martins Fontes.

Dey, S., Beerel, P. A., & Chugg, K. M. (2018). Interleaver Design for Deep Neural Networks, *arXiv:1711.06935v2*.

Di Fuccio, R., Ponticorvo, M., Ferrara, F., & Miglino, O. (2016). Digital and Multisensory Storytelling: Narration with Smell, Taste and Touch. In K. Verbert, M. Sharples, & T. Klobučar (Eds.), *Adaptive and Adaptable Learning* (pp. 509–512). Springer International Publishing. doi:10.1007/978-3-319-45153-4_51

Di Paolo, M. (2018). Analysis of harmonic impact of electric vehicle charging on the electric power grid, based on smart grid regional demonstration project - Los Angeles. In *Proceedings 2017 IEEE Green Energy and Smart Systems Conference, IGESSC 2017. Nov*ember *1–5*. doi:10.1109/IGESC.2017.8283460

Di Pierro, M. (2017). What Is the Blockchain? *Computing in Science & Engineering*, *19*(5), 92–95. doi:10.1109/MCSE.2017.3421554

Di Pietro, L., Guglielmetti Mugion, R., Mattia, G., Renzi, M. F., & Toni, M. (2015). The Integrated Model on Mobile Payment Acceptance (IMMPA): An empirical application to public transport. *Transportation Research Part C, Emerging Technologies*, *56*, 463–479. doi:10.1016/j.trc.2015.05.001

Diamler. (n.d.). *Company History*. Retrieved from Diamler: https://www.daimler.com/company/tradition/company-history/1885-1886.html

Dick, P. K. (2017). *Do androids dream of electric sheep?* New York: Del Rey.

Dinc, E., Kuscu, M., Bilgin, B. A., & Akan, O. B. (2019). Internet of Everything: A Unifying Framework Beyond Internet of Things. In Harnessing the Internet of Everything (IoE) for Accelerated Innovation Opportunities (pp. 1–30). Hershey, PA: IGI Global. doi:10.4018/978-1-5225-7332-6.ch001

Direção-Geral de Energia e Geologia. (2018). *Plano Nacional Integrado Energia e Clima 2021-2030*.

do Patrocínio, R. F., Souza, J. L. de A., Santos, C. T. O., & Martins, K. S. (2018). The vision of the Disney World: An experience marketing study at The Walt Disney Company. *Archives of Business Research*, *6*(9). doi:10.14738/abr.69.5067

Dodd, W., & Gutierrez, R. (2005). *The role of episodic memory and emotion in a cognitive robot*. In *Proceedings of the IEEE International Workshop on Robot and Human Interactive Communication (ROMAN)*. IEEE. Nashville, TN 10.1109/ROMAN.2005.1513860

Doerrfeld, B. (2016). 20+ Emotion Recognition APIs That Will Leave You Impressed, and Concerned. Retrieved from https://nordicapis.com/20-emotion-recognition-apis-that-will-leave-you-impressed-and-concerned/, accessed 2017/10/11.

Domingos, P. (2015). *The master algorithm: How the quest for the ultimate learning machine will remake our world.* Basic Books.

Dominici, G., & Guzzo, R. (2010). Customer Satisfaction in the Hotel Industry: A Case Study from Sicily. *International Journal of Marketing Studies, 2*(2). doi:10.5539/ijms.v2n2p3

Doraiswamy, H., Freire, J., Lage, M., Miranda, F., & Silva, C. (2018). Spatio-Temporal Urban Data Analysis: A Visual Analytics Perspective. *IEEE Computer Graphics and Applications, 38*(5), 26–35. doi:10.1109/MCG.2018.053491728 PMID:30273125

Doyle, M., Sharma, R., Ross, C., & Sonnad, V. (2017). *How to flourish in an uncertain future: Open banking and PSD2.* Retrieved from https://www2.deloitte.com/content/dam/Deloitte/cz/Documents/financial-services/cz-open-banking-and-psd2.pdf

Dubai Corporation of Tourism & Commerce Marketing. (2019). Visit Dubai. Retrieved January 5, 2019, from Visit Dubai website: https://www.visitdubai.com/en/business-in-dubai/why-dubai/news-and-insights/becoming-the-worlds-smartest-city

Duffy, W., Curran, K., Kelly, D., & Lunney, T. (2019). An investigation into smartphone based weakly supervised activity recognition systems. *Pervasive and Mobile Computing, 56*, 45–56. doi:10.1016/j.pmcj.2019.03.005

Dutta, J., Chowdhury, C., Roy, S., Middya, A. I., & Gazi, F. (2017). Towards smart city: Sensing air quality in city based on opportunistic crowd-sensing. *Proceedings of the 18th International Conference on Distributed Computing and Networking*, 42. ACM. 10.1145/3007748.3018286

Editors, H. (2010, April 26). *Model T.* Retrieved from HISTORY: https://www.history.com/topics/inventions/model-t

Egger, R., & Buhalis, D. (Eds.). (2008). "Mobile systems", eTourism Case Studies: Management and Marketing Issues, Butterworth-Heinemann, Oxford, UK, pp. 417-25.

Egger, R., & Bulencea, P. (2015). *Gamification in tourism: Designing memorable experiences.* BoD–Books on Demand.

Eggink, W., & Reinders, A. (2017). Design it with LSCs; an exploration of applications for Luminescent Solar Concentrator PV technologies. In *Proceedings 2017 IEEE 44th Photovoltaic Specialist Conference (PVSC)*, 2109–2113. 10.1109/PVSC.2017.8366790

Eichenbaum, H., & Cohen, N. J. (2001). *From Conditioning to Conscious Recollection: Memory Systems of the Brain.* Oxford, UK: Oxford University Press.

El Mahrsi, M. K., Come, E., Oukhellou, L., & Verleysen, M. (2017). Clustering Smart Card Data for Urban Mobility Analysis. *IEEE Transactions on Intelligent Transportation Systems, 18*(3), 712–728. doi:10.1109/TITS.2016.2600515

Elizalde-Ramírez, F., Nigenda, R. S., Martínez-Salazar, I. A., & Ríos-Solís, Y. Á. (2019). Travel Plans in Public Transit Networks Using Artificial Intelligence Planning Models. *Applied Artificial Intelligence*, 1–22.

Elman, J. L. (1990). Finding Structure in Time. *Cognitive Science, 14*(2), 179–211. doi:10.120715516709cog1402_1

Elms, C. P. (1998). Defining and measuring service availability for complex transportation networks. *Journal of Advanced Transportation, 32*(1), 75–88. doi:10.1002/atr.5670320108

Elzamly, A., Hussin, B., Abu Naser, S. S., Shibutani, T., & Doheir, M. (2017). Predicting Critical Cloud Computing Security Issues using Artificial Neural Network (ANNs) Algorithms in Banking Organizations. *International Journal of Information Technology and Electrical Engineering*, *6*(2), 40–45.

Enerdata. (2018). *Global Energy Statistical Yearbook.*

Entertainment Software Association. (2013). Essential Facts About the Computer and Video Game Industry. Entertainment Software Association. Retrieved from http://www.theesa.com/facts/pdfs/esa_ef_2013.pdf (accessed 10.10.18).

Epstein, R. A., Patai, E. Z., Julian, J. B., & Spiers, H. J. (2017). The cognitive map in humans: Spatial navigation and beyond. *Nature Neuroscience*, *20*(11), 1504–1513. doi:10.1038/nn.4656 PMID:29073650

EPTO. (2018). *Web Page*. Date recuperate 1 March, 2019, link https://disneyworld.disney.go.com/es-mx/destinations/epcot/

Erhan, D., Bengio, Y., Courville, A., & Vincent, P. (2009). Visualizing higher-layer features of a deep network. *Univ. Montr.*, *1341*, 1.

Ermilov, D., Panov, M., & Yanovich, Y. (2017). Automatic Bitcoin Address Clustering. In *2017 16th IEEE International Conference on Machine Learning and Applications (ICMLA)* (pp. 461–466). IEEE. 10.1109/ICMLA.2017.0-118

Estrada, E., Maciel, R., Zezzatti, C. A. O. O., Bernabe-Loranca, B., Oliva, D., & Larios, V. (2018). Smart City visualization tool for the Open Data georeferenced analysis utilizing machine learning. *International Journal of Combinatorial Optimization Problems and Informatics*, *9*(2), 25–40.

Ettlie, J., & Rosenthal, S. R. (2011). Service versus manufacturing innovation. *Journal of Product Innovation Management*, *28*(2), 285–299. doi:10.1111/j.1540-5885.2011.00797.x

EU Publications. (2016). *Overview of support activities and projects of the European Commission on energy efficiency and renewable energy in the heating and cooling sector.* doi:10.2826/607102

European Parliament. (2014). Mapping Smart Cities in the EU. Study, Directorate – General for Internal Policies, Brussels, PE 507.480 EN, IP/A/ITRE/ST/2013-02. Retrieved from http://www.europarl.europa.eu/studies

Ezell, S. (2009). *Explaining International IT Aplication Leadership: Contactless mobile payment. THE INFORMATION TECHNOLOGY & INNOVATION FOUNDATION.* Retrieved from https://www.itif.org/files/2009-Mobile-Payments.pdf

Face++. (2019). Face++ Cognitive Services. Retrieved from https://www.faceplusplus.com, accessed 2017/10/11

Famulari, A., De Simone, P., Verzaro, R., Iaria, G., Polisetti, F., Rascente, M., & Aureli, A. (2003). Artificial Organs as a Bridge to Transplantation. *Artificial Cells, Blood Substitutes, and Biotechnology*, *31*(2), 163–168. doi:10.1081/BIO-120020174 PMID:12751836

Fancello, G., Carta, M., & Fadda, P. (2014). A Modeling Tool for Measuring the Performance of Urban Road Networks. *Procedia: Social and Behavioral Sciences*, *111*, 559–566. doi:10.1016/j.sbspro.2014.01.089

Farhan, F., & Mushfiq, M. (2015). Biological Cell and Molecular Communication Technology: Overview and Challenges. *European Scientific Journal*, *3*, 56–60.

Farhangi, H. (2010). The path of the smart grid. *IEEE Power & Energy Magazine*, *8*(1), 18 28. doi.10.1109/MPE.2009.934876

Farooq, A., Gillani, N., Tahirkheli, R. A., & Farooq, M. U. (2018). Museum Tour Guide System. Rawalpindi: Foundation University Islamabad, Rawalpindi Campus (FURC).

Faste, H., Rachmel, N., Essary, R., & Sheehan, E. (2013). Brainstorm, Chainstorm, Cheatstorm, Tweetstorm: New Ideation Strategies for Distributed HCI Design. In *Proceedings of the SIGCHI Conference on Human Factors in Computing Systems* (pp. 1343-1352). ACM.

Fatnassi, E., Harrabi, O., & Chaouachi, J. (2017). Toward a smart mobility system: integrating electric vehicles within smart cities. *Proceedings of the Genetic and Evolutionary Computation Conference Companion (GECCO '17)* (pp. 275-276). Germany: ACM. doi:10.1145/3067695.3076075

FEELREAL. (2019). Multisensory VR Mask. Retrieved July 12, 2019, from https://feelreal.com/

Femenia-Serra, F., & Ivars-Baidal, J. A. (2018). Do smart tourism destinations really work? The case of Benidorm. *Asia Pacific Journal of Tourism Research*, 1–20. doi:10.1080/10941665.2018.1561478

Feng, L., Apers, P. M., & Jonker, W. (2004). Towards Context-Aware Data Management for Ambient Intelligence. *Lecture Notes in Computer Science, 3180*, 422-431. Retrieved from http://link.springer.com/chapter/10.1007%2F978-3-540-30075-5_41

Fernandez-Carames, T. M., & Fraga-Lamas, P. (2018). A Review on the Use of Blockchain for the Internet of Things. *IEEE Access: Practical Innovations, Open Solutions*, *6*, 32979–33001. doi:10.1109/ACCESS.2018.2842685

Ferreira, M. C., Fontes, T., Costa, V., Dias, T. G., Borges, J. L., & Cunha, J. F. E. (2017). Evaluation of an integrated mobile payment, route planner and social network solution for public transport. In Transportation Research Procedia (Vol. 24, pp. 189–196). doi:10.1016/j.trpro.2017.05.107

Ferreira, M. C., Nóvoa, H., & Dias, T. G. (2013). A Proposal for a Mobile Ticketing Solution for Metropolitan Area of Oporto Public Transport. In Lecture Notes in Business Information Processing, IESS, Vol. 143 (pp. 263–278). Springer. doi:10.1007/978-3-642-36356-6_19

Ferreira, M. C., Costa, V., Dias, T. G., Falcão, E., & Cunha, J. (2017). Understanding commercial synergies between public transport and services located around public transport stations. *Transportation Research Procedia*, *27*, 125–132. doi:10.1016/j.trpro.2017.12.052

Ferreira, M. C., Dias, T. G., & Cunha, J. F. e. (2019). Codesign of a Mobile Ticketing Service Solution Based on BLE. *Journal of Traffic and Logistics Engineering*, *7*(1), 10–17. doi:10.18178/jtle.7.1.10-17

Ferreira, M. C., Nóvoa, H., Dias, T. G., & Cunha, J. F. e. (2014). A proposal for a public transport ticketing solution based on customers' mobile devices. *Procedia: Social and Behavioral Sciences*, *111*, 232–241. doi:10.1016/j.sbspro.2014.01.056

Ferreira, P. M., Ruano, A. E., Silva, S., & Conceição, E. Z. E. (2012). Neural networks based predictive control for thermal comfort and energy savings in public buildings. *Energy and Building*, *55*, 238–251. doi:10.1016/j.enbuild.2012.08.002

Finzgar, L., & Trebar, M. (2011). Use of NFC and QR code identification in an electronic ticket system for public transport. In *Proceedings SoftCOM 2011, 19th International Conference on Software, Telecommunications and Computer Networks*, (iii), 1–6.

Fischer, G. (2001). User modeling in human–computer interaction. *User Modeling and User-Adapted Interaction*, *11*(1-2), 65–86. doi:10.1023/A:1011145532042

Fitton, D., & Read, J. C. (2016). Primed Design Activities: Scaffolding Young Designers During Ideation. In Proceedings of the 9th Nordic Conference on Human-Computer Interaction (NordiCHI '16). New York: ACM; doi:10.1145/2971485.2971529.

Flaounas, L., Omar, A., Lansdall-Welfare, T., De Bie, T., Mosdell, N., Lewis, J., & Cristianini, N. (2013). Research Methods in the Age of Digital Journalism. *Digital Journalism*, *1*(1), 102–116. doi:10.1080/21670811.2012.714928

Florida, R. (2012). *The Rise of the creative class.* New York: Basic Books.

Folsom-Kovarik, J. T., Schatz, S., Jones, R. M., Bartlett, K., & Wray, R. E. (2014). AI Challenge Problem: Scalable Models for Patterns of Life. *AI Magazine, 35*(1), 10. doi:10.1609/aimag.v35i1.2499

Fontes, T., Costa, V., Ferreira, M. C., Shengxiao, L., Zhao, P., & Dias, T. G. (2017). Mobile payments adoption in public transport. *Transportation Research Procedia, 24*, 410–417. doi:10.1016/j.trpro.2017.05.093

França, R. P., Iano, Y., Monteiro, A. C. B., Arthur, R., Estrela, V. V., Assumpção, S. L. D. L., & Razmjooy, N. (2019). Potential Proposal to Improvement of the Data Transmission in Healthcare Systems.

Fraser, N., & Gilbert, N. (1991). Simulating Speech Systems. *Computer Speech & Language, 5*(1), 81–99. doi:10.1016/0885-2308(91)90019-M

Frătean, A., & Dobra, P. (2018). The impact of control strategies upon the energy flexibility of nearly zero-energy buildings: Energy consumption minimization versus indoor thermal comfort maximization. *2018 IEEE International Conference on Automation, Quality and Testing, Robotics, AQTR 2018 - THETA 21st Edition, Proceedings*, 1–6. 10.1109/AQTR.2018.8402759

Freet, D., Agrawal, R., John, S., & Walker, J. J. (2015, October). Cloud forensics challenges from a service model standpoint: IaaS, PaaS and SaaS. In *Proceedings of the 7th International Conference on Management of computational and collective intElligence in Digital EcoSystems* (pp. 148-155). ACM. 10.1145/2857218.2857253

Friedman, B. (1997). *Human Values and the Design of Computer Technology. Nova Iorque*. Cambridge, UK: Cambridge University Press.

Friedman, B., Kahn, P. H., Borning, A., & Huldtgren, A. (2013). Value Sensitive Design and Information Systems. In *Early engagement and new technologies: Opening up the laboratory* (pp. 55–95). Springer Netherlands; doi:10.1007/978-94-007-7844-3_4

Fujii, T., Toi, T., Tanaka, T., Togawa, K., Kitaoka, T., Nishino, K., ... Motomura, M. (2018). New generation dynamically reconfigurable processor technology for accelerating embedded AI applications. In *Symposium on VLSI Circuits* (41-42). 10.1109/VLSIC.2018.8502438

Fu, L., & Yang, X. (2007). Design and Implementation of Bus–Holding Control Strategies with Real-Time Information. *Transportation Research Record: Journal of the Transportation Research Board, 1791*(1), 6–12. doi:10.3141/1791-02

Gadam, S. (2018, April 19). *Artificial Intelligence and Autonomous Vehicles*. Retrieved from Medium: https://medium.com/datadriveninvestor/artificial-intelligence-and-autonomous-vehicles-ae877feb6cd2

Gai, K., Qiu, M., Tao, L., & Zhu, Y. (2016). Intrusion detection techniques for mobile cloud computing in heterogeneous 5G. *Security and Communication Networks, 9*(16), 3049–3058. doi:10.1002ec.1224

Gallistel, C., & King, A. P. (2009). *Memory and the Computational Brain: Why Cognitive Science will Transform Neuroscience*. Wiley-Blackwell. doi:10.1002/9781444310498

Gan, L., Topcu, U., & Low, S. H. (2013). Optimal decentralized protocol for electric vehicle charging. *IEEE Transactions on Power Systems, 28*(2), 940–951. doi:10.1109/TPWRS.2012.2210288

Gan, Z., Feng, T., Wu, Y., Yang, M., Timmermans, H., Key, J., ... Technologies, T. (2019). Station-based average travel distance and its relationship with urban form and land use : An analysis of smart card data in Nanjing City, China. *Transport Policy, 79*(May), 137–154. doi:10.1016/j.tranpol.2019.05.003

Garcia, D. J. (n.d.). Metropolitan Museum of Art NYC. Retrieved from https://apps.apple.com/us/app/metropolitan-museum-of-art-nyc/id977211908

Garcia, A., Linaza, M. T., Gutierrez, A., & Garcia, E. (2019). Gamified mobile experiences: Smart technologies for tourism destinations. *Tourism Review*, *74*(1), 30–49. doi:10.1108/TR-08-2017-0131

García-Álvarez, J., González, M. A., & Vela, C. R. (2015, July). A genetic algorithm for scheduling electric vehicle charging. In *Proceedings of the 2015 Annual Conference on Genetic and Evolutionary Computation* (pp. 393-400). ACM. doi:10.1145/2739480.2754695

García-Pereira, I., Gimeno, J., Portalés, C., Vidal-González, M., & Morillo, P. (2018). *On the Design of a Mixed-Reality Annotations Tool for the Inspection of Pre-fab Buildings*. doi:10.2312/ceig.20181157

García-Pereira, I., Gimeno, J., Pérez, M., Portalés, C., & Casas, S. (2018). MIME: A Mixed-Space Collaborative System with Three Immersion Levels and Multiple Users. In *Proceedings 2018 IEEE International Symposium on Mixed and Augmented Reality Adjunct (ISMAR-Adjunct)*, 179–183. 10.1109/ISMAR-Adjunct.2018.00062

Garfinkel, S. L., & Grunspan, R. H. (2018). The Computer Book: From the Abacus to Artificial Intelligence, 250 Milestones in the History of Computer Science.

Garfinkel, H. (1967). *Studies in Ethnomethodology*. Englewood Cliffs, NJ: Prentice-Hall.

Gartner. (2011). *Gartner predicts over 70 percent of global 2000 organisations will have at least one gamified application by 2014*. Available at http://www.gartner.com/it/page.jsp?id=1844115

Gartner. (2012). *Gartner says by 2014, 80 percent of current gamified applications will fail to meet business objectives primarily due to poor design*. Available at http://www.gartner.com/it/page.jsp?id=2251015

Gartner. (2019). *IoT Glossary*. Obtido de https://www.gartner.com/it-glossary/internet-of-things/

Gaur, A., & Ondrus, J. (2012). The role of banks in the mobile payment ecosystem: a strategic asset perspective. *Proceedings of the 14th Annual International Conference on Electronic Commerce (pp. 171-177). ACM*. 10.1145/2346536.2346570

Gaur, A., Scotney, B., Parr, G., & McClean, S. (2015). Smart city architecture and its applications based on IoT. *Procedia Computer Science*, *52*, 1089–1094. doi:10.1016/j.procs.2015.05.122

Gaus, T., Olsen, K., & Deloso, M. (2018). *Synchronizing the digital supply network Using artificial intelligence for supply chain planning*. Retrieved May 10, 2019, from https://www2.deloitte.com/us/en/insights/focus/industry-4-0/artificial-intelligence-supply-chain-planning.html

Geddes, O. (2014). *A Guide to Bluetooth Beacons*. Retrieved from http://www.gsma.com/digitalcommerce/wp-content/uploads/2013/10/A-guide-to-BLE-beacons-FINAL-18-Sept-14.pdf

Geotab. (2018). *GeoTab - management by measurement*. Obtido de https://www.geotab.com/blog/what-is-smart-mobility/

Geraci, R. M. (2007). Robots and the sacred in science and science fiction: Theological implications of Artificial Intelligence. *Zygon*, *42*(4), 961–980. doi:10.1111/j.1467-9744.2007.00883.x

Gerossier, A., Girard, R., & Kariniotakis, G. (2019). Modeling and Forecasting Electric Vehicle Consumption Profiles. *Energies*, *12*(7), 1341. doi:10.3390/en12071341

Ghasemi, A., Shayeghi, H., Moradzadeh, M., & Nooshyar, M. (2016). A novel hybrid algorithm for electricity price and load forecasting in smart grids with demand-side management. *Applied Energy*, *177*, 40–59. doi:10.1016/j.apenergy.2016.05.083

Ghassemlooy, Z., Popoola, W., & Rajbhandari, S. (2019). Optical wireless communications: system and channel modelling with Matlab. Boca Raton, FL: CRC Press. doi:10.1201/9781315151724

Ghiron, S. L., Sposato, S., Medaglia, C. M., & Moroni, A. (2009). NFC Ticketing: A Prototype and Usability Test of an NFC-Based Virtual Ticketing Application. In *Proceedings 2009 First International Workshop on Near Field Communication*, 45–50. 10.1109/NFC.2009.22

Giaccardi, E., Speed, C., Nazli, C., & Caldwell, M. (2016). Things as Co-ethnographers: Implications of a Thing Perspective for Design and Anthropology. In Design Anthropological Futures, 235. London, UK: Bloomsbury Academic.

Giannopoulos, G. A. (2004). The application of information and communication technologies in transport, *152*, 302–320. doi:10.1016/S0377-2217(03)00026-2

Gimeno, J., Morillo, P., Casas, S., & Fernández, M. (2011). An augmented reality (AR) cad system at construction sites. In Augmented Reality-Some Emerging Application Areas. InTech.

Gimeno, J., Olanda, R., Martinez, B., & Sanchez, F. M. (2011). Multiuser Augmented Reality System for Indoor Exhibitions. In P. Campos, N. Graham, J. Jorge, N. Nunes, P. Palanque, & M. Winckler (Eds.), Human-Computer Interaction – INTERACT 2011 (pp. 576–579). Berlin, Germany: Springer.

Gimeno, J., Portalés, C., Coma, I., Fernández, M., & Martínez, B. (2017). Combining traditional and indirect augmented reality for indoor crowded environments. A case study on the Casa Batlló museum. *Computers & Graphics*, *69*, 92–103. doi:10.1016/j.cag.2017.09.001

Giner Martínez, F., & Portalés Ricart, C. (2005, October 3). *The Augmented User: A Wearable Augmented Reality Interface*. 417–426. Hal Thwaites.

Giner Martínez, F., & Portalés Ricart, C. (2007). Augmented teaching (L. Gómez Chova, D. Martí Belenguer, & I. Candel Torres, Eds.). International Association of Technology, Education, and Developement (IATED).

Giordano, A. A., & Levesque, A. H. (2015). *Digital Communications BER Performance in AWGN*. BPSK in Fading.

Girija, S. S. (2016). Tensorflow: Large-scale machine learning on heterogeneous distributed systems. Software available from tensorflow.org.

Global Cloud Computing Scorecard, B. S. A. (2018). Powering a Bright Future. 2018. Disponível em: https://cloudscorecard.bsa.org/2018/?sc_lang=pt-BR. Acesso em: Nov. 11, 2019.

Glorot, X., & Bengio, Y. (2010). Understanding the difficulty of training deep feedforward neural networks. In *International Conference on Artificial Intelligence and Statistics* (249–256).

Glorot, X., Bordes, A., & Bengio, Y. (2011). Deep Sparse Rectifier Neural Networks. In *Proceedings of the Fourteenth International Conference on Artificial Intelligence and Statistics* (315-323).

Göbel, J., Keeler, H. P., Krzesinski, A. E., & Taylor, P. G. (2016). Bitcoin blockchain dynamics: The selfish-mine strategy in the presence of propagation delay. *Performance Evaluation*, *104*, 23–41. doi:10.1016/j.peva.2016.07.001

Godet, M., & Durance, P. H. (2011). *La prospective stratégique - Pour les entreprises et les territoires*. Paris, France: Dunod.

Goldstein, R., Korytowski, I., & Godel, K. (2008). *Incompletude: a prova e o paradoxo de Kurt Gödel*. São Paulo, Brazil: Companhia das Letras.

Gonçalves, H. (2019). PNEC 2030 um "Admirável Mundo Novo" o que fazer?".

González-Briones, A., Valdeolmillos, D., Casado-Vara, R., Chamoso, P., Coria, J. A. G., Herrera-Viedma, E., & Corchado, J. M. (2018, June). Garbmas: Simulation of the application of gamification techniques to increase the amount of recycled waste through a multi-agent system. In *International Symposium on Distributed Computing and Artificial Intelligence* (pp. 332-343). Cham, Switzerland: Springer.

Goodfellow, I., Bengio, Y., & Courville, A. (2016). *Deep Learning*. MIT Press.

Google Cloud. (2019). *Google Maps Platform.* Obtido de https://cloud.google.com/maps-platform/?&sign=0

Google Maps Platform. (2019). Retrieved from https://cloud.google.com/maps-platform/

Google, L. L. C. (n.d.). Cloud Speech-to-Text. Retrieved from https://cloud.google.com/speech-to-text/ [Accessed 5th June 2019]

GoogleVision. (2019). Google Cloud Vision. Retrieved from https://cloud.google.com/vision/, last accessed 2019/04/09

Goon, T. (2018). Difference between random forest and Gradient boosting Algo. Retrieved May 2, 2019, from https://www.linkedin.com/pulse/difference-between-random-forest-gradient-boosting-algo-goon

Goswami, K., Ganguly, A., & Kumar Sil, D. A. (2018). Comparing Univariate and Multivariate Methods for Short Term Load Forecasting. In *Proceedings 2018 International Conference on Computing, Power, and Communication Technologies (GUCON)*, 972–976. 10.1109/GUCON.2018.8675059

Grammatico, S. (2016). Exponentially convergent decentralized charging control for large populations of plug-in electric vehicles. In *Proceedings 2016 IEEE 55th Conference on Decision and Control, CDC 2016*, 5775–5780. 10.1109/CDC.2016.7799157

Green, P. J. (2017, November). Implementation of a real-time Rayleigh, Rician and AWGN multipath channel emulator. In Proceedings TENCON 2017-2017 IEEE Region 10 Conference (pp. 35-39). IEEE.

Green, B. P. (2018). Ethical Reflections on Artificial Intelligence. *Scientia et Fides*, *6*(2), 9–31. doi:10.12775/SetF.2018.015

Gregory, J. (2003). Scandinavian Approaches to Participatory Design. *International Journal of Engineering Education*, *19*(1), 62–74.

Gretzel. (2018). Investigaciones Regionales. *Journal of Regional Research, 42,* 171-184.

Gretzel, U. (2011). Intelligent systems in tourism: A social science perspective. *Annals of Tourism Research*, *38*(3), 757–779. doi:10.1016/j.annals.2011.04.014

Gretzel, U., Ham, J., & Koo, C. (2018). Creating the City Destination of the Future: The Case of Smart Seoul. In *Managing Asian Destinations* (pp. 199–214). Singapore: Springer. doi:10.1007/978-981-10-8426-3_12

Gretzel, U., Sigala, M., Xiang, Z., & Koo, C. (2015). Smart tourism: Foundations and developments. *Electronic Markets*, *25*(3), 179–188. doi:10.100712525-015-0196-8

Gretzel, U., Zhong, L., & Koo, C. (2016). Application of smart tourism to cities. *International Journal of Tourism Cities*, *2*(2). doi:10.1108/IJTC-04-2016-0007

Grodi, R., Rawat, D. B., & Rios-Gutierrez, F. (2016). Smart parking: Parking occupancy monitoring and visualization system for smart cities. [IEEE.]. *SoutheastCon*, *2016*, 1–5.

Groenewegen, H. J., Wright, C. I., & Uylings, H. B. (1997). The anatomical relationships of the prefrontal cortex with limbic structures and the basal ganglia. *Journal of Psychopharmacology (Oxford, England)*, *11*(2), 99–106. doi:10.1177/026988119701100202 PMID:9208373

Groening, C., & Binnewies, C. (2019). “Achievement unlocked!”-The impact of digital achievements as a gamification element on motivation and performance. *Computers in Human Behavior, 97*, 151–166. doi:10.1016/j.chb.2019.02.026

Group, J. M. (n.d.). *The History of Tar Technology*. Retrieved from Jardine Motors Group: https://news.jardinemotors.co.uk/lifestyle/the-history-of-car-technology

GSMA. (2012). White Paper: Mobile NFC in Transport, (September).

Guardia, J. J., Del Olmo, J. L., Roa, I., & Berlanga, V. (2019). Innovation in the teaching-learning process: the case of Kahoot!. *On the Horizon, 27*(1), 35-45.

Guessous, Y., Aron, M., Bhouri, N., & Cohen, S. (2014). Estimating travel time distribution under different traffic conditions. *Transportation Research Procedia, 3*(July), 339–348. doi:10.1016/j.trpro.2014.10.014

Gullberg, A. T., Ohlhorst, D., & Schreurs, M. (2014). Towards a low carbon energy future – Renewable energy cooperation between Germany and Norway. *Renewable Energy, 68*, 216–222. doi:10.1016/j.renene.2014.02.001

Gulshan, V., Peng, L., Coram, M., Stumpe, M. C., Wu, D., Narayanaswamy, A., ... Webster, D. R. (2016). Development and Validation of a Deep Learning Algorithm for Detection of Diabetic Retinopathy in Retinal Fundus Photographs. *Journal of the American Medical Association, 316*(22), 2402–2410. doi:10.1001/jama.2016.17216 PMID:27898976

Guo, K., Sui, L., Qiu, J., Yu, J., Wang, J., Yao, S., ... Yang, H. (2018). Angel-Eye: A Complete Design Flow for Mapping CNN Onto Embedded FPGA. *IEEE Transactions on Computer-Aided Design of Integrated Circuits and Systems, 37*(1), 35–47. doi:10.1109/TCAD.2017.2705069

Gupta, S., Agrawal, A., Gopalakrishnan, K., & Narayanan, P. (2015) Deep Learning with Limited Numerical Precision. In *Proceedings of the 32nd International Conference on International Conference on Machine Learning:* Vol. 37. (1737–1746).

Gupta, S., Shrivastava, N. A., Khosravi, A., & Panigrahi, B. K. (2016). Wind ramp event prediction with parallelized gradient boosted regression trees. In *Proceedings 2016 International Joint Conference on Neural Networks (IJCNN)*, 5296–5301. 10.1109/IJCNN.2016.7727900

Gu, S., & Mark, L. (2014). Trip purpose inference using automated fare collection data. *Public Transport (Berlin), 6*(1-2), 1–20. doi:10.100712469-013-0077-5

Gyrfalcon Technology. (2018). Lightspeeur 2803S Neural Accelerator.

Gysel, P., Motamedi, M., & Ghiasi, S. (2016). Hardware-oriented Approximation of Convolutional Neural Networks. In *Proceedings of the 4th International Conference on Learning Representations*.

Hall, L. M. (2002). Games business experts play. Carmarthen, UK: Crown House Publishing.

Hall, C. M. (2010). Changing Paradigms and Global Change: From Sustainable to Steady-state Tourism. *Tourism Recreation Research, 35*(2), 131–143. doi:10.1080/02508281.2010.11081629

Hall, C. M. (2010). Tourism destination branding and its effects on national branding strategies: Brand New Zealand, clean and green but is it smart? *European Journal of Tourism, Hospitality, and Recreation, 1*(1), 68–89.

Halse, J. (2008). Design Anthropology: Borderland Experiments with Participation. IT University of Copenhagen. Doctoral thesis. Retrieved from www. dasts.dk/wpcontent/uploads/2008/05/joachim-halse-2008.pdf

Hamari, J., Huotari, K., & Tolvanen, J. (2015). Gamification and economics. *The gameful world: Approaches, issues, applications, 139*.

Hamari, J., Koivisto, J., & Pakkanen, T. (2014a). Do persuasive technologies persuade?-a review of empirical studies. In *International conference on persuasive technology*(pp. 118-136). Cham, Switzerland: Springer. 10.1007/978-3-319-07127-5_11

Hamari, J., Koivisto, J., & Sarsa, H. (2014b). Does gamification work?--a literature review of empirical studies on gamification. In *2014 47th Hawaii international conference on system sciences (HICSS)* (pp. 3025-3034). IEEE.

Hamari, J., & Koivisto, J. (2015). Why do people use gamification services? *International Journal of Information Management, 35*(4), 419–431. doi:10.1016/j.ijinfomgt.2015.04.006

Han, S., Mao, H., & Dally, W. J. (2015). "Deep Compression: Compressing Deep Neural Network with Pruning, Trained Quantization and Huffman Coding". *CoRR*, abs/1510.00149.

Han, D. I., Jung, T., & Gibson, A. (2013). Dublin AR: implementing augmented reality in tourism. In *Information and communication technologies in tourism 2014* (pp. 511–523). Cham, Switzerland: Springer. doi:10.1007/978-3-319-03973-2_37

Han, H., & Kim, Y. (2010). An investigation of green hotel customers' decision formation: Developing an extended model of the theory of planned behavior. *International Journal of Hospitality Management, 29*(4), 659–668. doi:10.1016/j.ijhm.2010.01.001

Harari, Y. N. (2018). *Homo deus: a brief history of tomorrow*. New York: Harper Perennial.

Hardesty, L. (2017, April 14). *Explained: Neural networks*. Retrieved from MIT News: http://news.mit.edu/2017/explained-neural-networks-deep-learning-0414

Harlev, M. A., Yin, S. H., Langenheldt, K. C., Mukkamala, R., & Vatrapu, R. (2018). Breaking Bad: De-Anonymising Entity Types on the Bitcoin Blockchain Using Supervised Machine Learning. In *Proceedings of the 51st Hawaii International Conference on System Sciences* (pp. 3497–3506).

Harris, J. (2018). Who Owns My Autonomous Vehicle? Ethics and Responsibility in Artificial and Human Intelligence. *Cambridge Quarterly of Healthcare Ethics, 27*(4), 599–609. doi:10.1017/S0963180118000038 PMID:30079847

Harris, R. A. (2005). *Voice Interaction Design*. San Francisco, CA: Elsevier.

Hartshorn, S. (2016). *Machine Learning With Random Forests And Decision Trees: A Visual Guide For Beginners*. Kindle Edition.

Hartshorn, S. (2017). *Machine Learning With Boosting: A Beginner's Guide*. Kindle Edition.

Hassabis, D., Kumaran, D., Summerfield, C., & Botvinick, M. (2017). Neuroscience-Inspired Artificial Intelligence. *Neuron, 95*(2), 245–258. doi:10.1016/j.neuron.2017.06.011 PMID:28728020

Hassani, A., Medvedev, A., Zaslavsky, A., & Jayaraman, P. P. (2018). Context-as-a-Service Platform: Exchange and Share Context in an IoT Ecosystem. In *Proceedings 9th International Workshop on Information Quality and Quality of Service for Pervasive Computing* (pp. 385-390). IEEE. 10.1109/PERCOMW.2018.8480240

Hassan, L., Dias, A., & Hamari, J. (2019). How motivational feedback increases user's benefits and continued use: A study on gamification, quantified-self and social networking. *International Journal of Information Management, 46*, 151–162. doi:10.1016/j.ijinfomgt.2018.12.004

Hastreiter, N. (2017). Huffpost: What Impact Will AI Have In The Next 20 Years? Retrieved from https://goo.gl/SdtUF3, in 2019/03/25

Hawkins, A. J. (2019, April 22). *Here are Elon Musk's wildest predictions about Tesla's self-driving cars*. Retrieved from The Verge: https://www.theverge.com/2019/4/22/18510828/tesla-elon-musk-autonomy-day-investor-comments-self-driving-cars-predictions

Haykin, S. (2008). *Neural Networks and Learning Machines* (3rd ed.). Pearson.

Heath, N. (2018, February 18). *What is AI? Everything you need to know about Artificial Intelligence*. Retrieved from ZDNet: https://www.zdnet.com/article/what-is-ai-everything-you-need-to-know-about-artificial-intelligence/

Heath, C., & Luff, P. (2000). *Technology in Action*. Cambridge University Press. doi:10.1017/CBO9780511489839

Heilig, M. L. (1962). *United States Patent No. US3050870A*. Retrieved from https://patents.google.com/patent/US3050870A/en

Heilig, M. L. (1992). EL Cine del Futuro: The Cinema of the Future. *Presence (Cambridge, Mass.)*, *1*(3), 279–294. doi:10.1162/pres.1992.1.3.279

He, K., Zhang, X., Ren, S., & Sun, J. (2015). Deep Residual Learning for Image Recognition. *Multimedia Tools and Applications*, *77*, 10437–10453.

Helkkula, A., Kowalkowski, C., & Tronvoll, B. (2018). Archetypes of Service Innovation: Implications for value cocreation. *Journal of Service Research*, *21*(3), 284–301. doi:10.1177/1094670517746776 PMID:30034213

Hidalgo, G., Cao, Z., Simon, T., Wei, S., Joo, H., Sheikh, Y., . . . Raaj, Y. (2019). OpenPose. Retrieved from https://github.com/CMU-Perceptual-Computing-Lab/openpose, accessed 2019/06/05

Hilliard, C., Wagner Cook, S., & Duff, M. C. (2016). Hippocampal declarative memory supports gesture production: Evidence from amnesia. *Cortex*, *85*, 25–36. doi:10.1016/j.cortex.2016.09.015 PMID:27810497

Hilton Waikoloa Village, Hawaii: Association of Information Systems. Retrieved from https://scholarspace.manoa.hawaii.edu/handle/10125/50331

Hinton, G. E., & Sejnowski, T. J. (1986). Learning and relearning in Boltzmann machines. In D. E. Rumelhart, J. L. McClelland, & CORPORATE PDP Research Group (Eds.), Parallel distributed processing: explorations in the microstructure of cognition. Vol. 1, MIT Press (282-317).

Hinton, G., Osindero, S., & The, Y. (2006). A fast learning algorithm for deep belief nets. *Neural Computation*, *18*(7), 1527–1554. doi:10.1162/neco.2006.18.7.1527 PMID:16764513

Hirel, J., Gaussier, P., & Quoy, M. (2011). Biologically inspired neural networks for spatio-temporal planning in robotic navigation tasks. In *Proceedings of the 2011 IEEE International Conference on Robotics and Biomimetics*, December 7-11, 2011. Phuket, Thailand IEEE. 10.1109/ROBIO.2011.6181522

Hochreiter, S. (1998). The vanishing gradient problem during learning recurrent neural nets and problem solutions. *International Journal of Uncertainty, Fuzziness and Knowledge-based Systems*, *6*(02), 107–116. doi:10.1142/S0218488598000094

Hochreiter, S., & Schmidhuber, J. (1997). Long Short-Term Memory. *Neural Computation*, *8*(8), 1735–1780. doi:10.1162/neco.1997.9.8.1735 PMID:9377276

Hodge, M. (2018). Disney 'World': The Westernization of World Music in EPCOT's "IllumiNations: Reflections of Earth". *Social Sciences*, *7*(8), 136. doi:10.3390ocsci7080136

Hölbl, M., Kompara, M., Kamišalić, A., & Nemec Zlatolas, L. (2018). A Systematic Review of the Use of Blockchain in Healthcare. *Symmetry*, *10*(10), 470. doi:10.3390ym10100470

Hollands, R. G. (2008). Will the real smart city please stand up? Intelligent, progressive or entrepreneurial? *City*, *12*(3), 303–320. doi:10.1080/13604810802479126

Hong, J.-Y., Suh, E., & Kim, S.-J. (2009). Context-aware systems: A literature review and classification. *Expert Systems with Applications*, *36*(4), 8509–8522. doi:10.1016/j.eswa.2008.10.071

Hopfield, J. (1982). Neural networks and physical systems with emergent collective computational abilities. *Proceedings of the National Academy of Sciences of the United States of America*, *79*(8), 2554–2558. doi:10.1073/pnas.79.8.2554 PMID:6953413

Hornik, K., Stinchcombe, M., & White, H. (1989). Multilayer feedforward networks are universal approximators. *Neural Networks*, *2*(5), 359–366. doi:10.1016/0893-6080(89)90020-8

Hossain, M. S., Xu, C., Li, Y., Pathan, A. S. K., Bilbao, J., Zeng, W., & El-Saddik, A. (2017). Impact of Next-Generation Mobile Technologies on IoT-Cloud Convergence. *IEEE Communications Magazine*, *55*(1), 18–19. doi:10.1109/MCOM.2017.7823332

Howard, A. G., Zhu, M., Chen, B., Kalenichenko, D., Wang, W., Weyand, T., Andreetto, M., & Adam, H. (2017). MobileNets: Efficient Convolutional Neural Networks for Mobile Vision Applications. *CoRR*, abs/1704.04861.

Hsiao, K. A., & Hsieh, P. L. (2014). Age difference in recognition of emoticons. In *Proceedings International Conference on Human Interface and the Management of Information* (pp. 394-403). Cham, Switzerland: Springer.

Hu, G. (2013) A Research Review on Theme Park. Business and Management Research, 2(4); Management School, Shanghai University of Engineering and Science, China, Management School, Shanghai University of Engineering and Science, China doi:10.5430/bmr.v2n4p83

Huang, B., Liu, Z., Chen, J., Liu, A., Liu, Q., & He, Q. (2017). Behavior pattern clustering in blockchain networks. *Multimedia Tools and Applications*, *76*(19), 20099–20110. doi:10.100711042-017-4396-4

Huang, G., Liu, Z., Maaten, L., & Weinberger, K. (2018). Densely Connected Convolutional Networks. In *IEEE Conference on Computer Vision and Pattern Recognition.*

Hubara, I., Courbariaux, M., Soudry, D., El-Yaniv, R., & Bengio, Y. (2016). Binarized Neural Networks. In D. D. Lee, M. Sugiyama, I. Guyon, & R. Garnett (Ed.), Advances in Neural Information Processing Systems: Vol. 4107–4115. *Curran Associates, Inc.*

Hu, J., Shen, L., & Sun, G. (2018). Squeeze-and-Excitation Networks, In *Proceedings IEEE Conference on Computer Vision and Pattern Recognition* (7132-7141). IEEE.

Hu, J., Wan, W., Wang, R., & Yu, X. (2012). Virtual reality platform for smart city based on sensor network and OSG engine. In *Proceedings 2012 International Conference on Audio, Language and Image Processing*, 1167–1171. IEEE. 10.1109/ICALIP.2012.6376794

Hulse, T., Daigle, M., Manzo, D., Braith, L., Harrison, A., & Ottmar, E. (2019). From here to there! Elementary: A game-based approach to developing number sense and early algebraic understanding. *Educational Technology Research and Development*, 1–19.

Hung, M. (2017). *Leading the IoT.* Obtido de https://www.gartner.com/imagesrv/books/iot/iotEbook_digital.pdf

Hunter, W. C. (2014). Virtual Reality. In J. Jafari (Ed.), *The Encyclopedia of Tourism* (2nd ed.).

Hunter, W. C., Chung, N., Gretzel, U., & Koo, C. (2015). Constructivist research in smart tourism. *Asia Pacific Journal of Information Systems*, *25*(1), 105–120. doi:10.14329/apjis.2015.25.1.105

Huotari, K., & Hamari, J. (2012). Defining gamification: a service marketing perspective. In *Proceeding of the 16th international academic MindTrek conference* (pp. 17-22). ACM. 10.1145/2393132.2393137

Hussain, A., Wenbi, R., da Silva, A. L., Nadher, M., & Mudhish, M. (2015). Health and emergency-care platform for the elderly and disabled people in the Smart City. *Journal of Systems and Software*, *110*, 253–263. doi:10.1016/j.jss.2015.08.041

Hussain, S., Wang, Z., & Toure, I. K. (2014). Performance Analysis of Web Services in Different Types of Internet Technologies. *Applied Mechanics and Materials*, 513.

Hussain, S., Zhaoshun, W., & Toure, I. K. (2013). An approach for QoS measurement and web service selectness sureness. *High Technology Letters*, *19*(3), 283–289.

Hussein, A. F., ArunKumar, N., Ramirez-Gonzalez, G., Abdulhay, E., Tavares, J. M. R. S., & de Albuquerque, V. H. C. (2018). A medical records managing and securing blockchain based system supported by a Genetic Algorithm and Discrete Wavelet Transform. *Cognitive Systems Research*, *52*, 1–11. doi:10.1016/j.cogsys.2018.05.004

Hussein, N. H., & Khalid, A. (2016). A survey of cloud computing security challenges and solutions. *International Journal of Computer Science and Information Security*, *14*(1), 52.

Iansiti, M., & Lakhani, K. (2014). Digital ubiquity: How connections, sensors, and data are revolutionizing business. *Harvard Business Review*, *92*(11), 91–99.

Ibáñez, M.-B., & Delgado-Kloos, C. (2018). Augmented reality for STEM learning: A systematic review. *Computers & Education*, *123*, 109–123. doi:10.1016/j.compedu.2018.05.002

IBM. (2017, October 15). *Watson Assistant*. Retrieved June 20, 2019, from https://www.ibm.com/cloud/watson-assistant/

IBM. (2018, June 5). *Project Debater*. Retrieved June 19, 2019, from https://www.research.ibm.com/artificial-intelligence/project-debater/

Iltaf, N., Mahmud, U., & Kamran, F. (2006). Security & trust enforcement in pervasive computing environment (STEP). In *Proceedings 2006 International Symposium on High Capacity Optical Networks and Enabling Technologies* (pp. 1-6). Charlotte, NC: IEEE. 10.1109/HONET.2006.5338409

INE. (2019). *Estadísticas turísticas*. Madrid, Spain: INE.

Intel. (2017). Movidius Myriad X VPU.

IntelDev. (2019) OpenVINO: Develop Multiplatform Computer Vision Solutions. Retrieved from https://software.intel.com/en-us/openvino-toolkit, accessed 2019/06/05

International Energy Agency. (2019). *Global Energy & CO2 Status Report 2018*.

International Transport Forum. (2012). *Policy Brief*. 10.1177/0022146513479002

Irfan, M., Abas, N., & Saleem, M. S. (2019). Net Zero Energy Buildings (NZEB): A Case Study of Net Zero Energy Home in Pakistan. In *Proceedings 4th International Conference on Power Generation Systems and Renewable Energy Technologies, PGSRET 2018*, (September), 1–6. 10.1109/PGSRET.2018.8685970

Ishii, H., Kobayashi, M., & Arita, K. (1994). Iterative Design of Seamless Collaboration Media. *Communications of the ACM*, *37*(8), 83–97. doi:10.1145/179606.179687

Islikaye, A. A., & Cetin, A. (2018). Performance of ML methods in estimating net energy produced in a combined cycle power plant. In *Proceedings 2018 6th International Istanbul Smart Grids and Cities Congress and Fair (ICSG)*, 217–220. 10.1109/SGCF.2018.8408976

Ismail, A. W., Billinghurst, M., Sunar, M. S., & Yusof, C. S. (2019). Designing an Augmented Reality Multimodal Interface for 6DOF Manipulation Techniques. In K. Arai, S. Kapoor, & R. Bhatia (Eds.), *Intelligent Systems and Applications* (pp. 309–322). Springer International Publishing. doi:10.1007/978-3-030-01054-6_22

Ivan, C., & Balag, R. (2015). An Initial Approach for a NFC M-Ticketing Urban Transport System. *Journal of Computer and Communications*, *3*(June), 42–64. doi:10.4236/jcc.2015.36006

Jackson, S. (2009). *Cult of Analytics: Driving online marketing strategies using web analytics: Driving online strategies using web analytics*. Routledge. doi:10.4324/9780080885179

Jagnow, R. (2017, July 6). Daydream Labs: Locomotion in VR. Retrieved January 7, 2019, from Daydream website: https://www.blog.google/products/daydream/daydream-labs-locomotion-vr

Jain, A., & Mahajan, N. (2017). Introduction to Cloud Computing. In The Cloud DBA-Oracle (pp. 3-10). Berkeley, CA: Apress. doi:10.1007/978-1-4842-2635-3_1

Jamshidi, P., Pahl, C., Chinenyeze, S., & Liu, X. (2015). Cloud migration patterns: a multi-cloud service architecture perspective. In *Service-Oriented Computing-ICSOC 2014 Workshops* (pp. 6–19). Cham, Switzerland: Springer. doi:10.1007/978-3-319-22885-3_2

Jánošíkova, L., Slavík, J., & Koháni, M. (2014). Estimation of a route choice model for urban public transport using smart card data. *Transportation Planning and Technology*, *37*(7), 638–648. doi:10.1080/03081060.2014.935570

Jasti, P. C., & Ram, V. V. (2016). Integrated and Sustainable Service Level Benchmarking of Urban Bus System. *Transportation Research Procedia*, *17*, 301–310. doi:10.1016/j.trpro.2016.11.096

Jeffery, K., & Casali, G. (2014). Hippocampal Neurons: Simulating the Spatial Structure of a Complex Maze. *Current Biology*, *24*(14), R643–R645. doi:10.1016/j.cub.2014.06.001 PMID:25050959

Jefsioutine, M., & Knight, J. (2004). Methods for Experience Design: The Experience Design Framework. In: J. Redmond, D. Durling, & A. de Bono (Eds.), *Proceedings of Future Ground Design Research Society International Conference 2004*, Melbourne, Australia. Nov. 17-21.

Jeihoonian, M., Ghaderi, S. F., & Piltan, M. (2010). Modeling and comparing energy consumption in basic metal industries by neural networks and ARIMA. In *Proceedings 2010 International Conference on Computer Information Systems and Industrial Management Applications, CISIM 2010*, 171–175. 10.1109/CISIM.2010.5643670

Jin, J., Gubbi, J., Marusic, S., & Palaniswami, M. (2014). An information framework for creating a smart city through internet of things. *IEEE Internet of Things Journal*, *1*(2), 112–121. doi:10.1109/JIOT.2013.2296516

Johannesen, N. J., Kolhe, M., & Goodwin, M. (2019). Relative evaluation of regression tools for urban area electrical energy demand forecasting. *Journal of Cleaner Production*, *218*, 555–564. doi:10.1016/j.jclepro.2019.01.108

Jonas, H. (2006). *O princípio responsabilidade: Ensaio de uma ética para a civilização tecnológica*. Rio de Janeiro, Brazil: Contraponto.

Jones, A., Subrahmanian, E., Hamins, A., & Grant, C. (2015). Humans' critical role in smart systems: A smart firefighting example. *IEEE Internet Computing*, *19*(3), 28–31. doi:10.1109/MIC.2015.54

Juneja, A., & Marefat, M. (2018). Leveraging blockchain for retraining deep learning architecture in patient-specific arrhythmia classification. In *2018 IEEE EMBS International Conference on Biomedical & Health Informatics (BHI)* (pp. 393–397). IEEE. 10.1109/BHI.2018.8333451

Jung, Y. (2017). Hybrid-Aware Model for Senior Wellness Service in Smart Home. *Sensors (Basel)*, *17*(5), 1–10. doi:10.339017051182 PMID:28531157

Juntunen, A., Luukkainen, S., & Tuunainen, V. K. (2010). Deploying NFC Technology for Mobile Ticketing Services – Identification of Critical Business Model Issues. In *Proceedings 2010 Ninth International Conference on Mobile Business and 2010 Ninth Global Mobility Roundtable (ICMB-GMR)*, 82–90. 10.1109/ICMB-GMR.2010.69

Kahvazadeh, A. (2016). Check In-Be Out – Innovation in Mobile Transit Payment, (April). Retrieved from https://www.apta.com/mc/revenue/presentations/Presentations/Check In-Be Out – Innovation in Mobile Transit Payment - Arash Kahvazadeh.pdf

Kang, J., Duncan, S. J., & Mavris, D. N. (2013). Real-time scheduling techniques for electric vehicle charging in support of frequency regulation. *Procedia Computer Science*, *16*, 767–775. doi:10.1016/j.procs.2013.01.080

Kapoor, K. K., Dwivedi, Y. K., & Williams, M. D. (2015). Empirical Examination of the Role of Three Sets of Innovation Attributes for Determining Adoption of IRCTC Mobile Ticketing Service. *Information Systems Management*, *32*(2), 153–173. doi:10.1080/10580530.2015.1018776

Kapsammer, E., Kimmerstorfer, E., Pröll, B., Retschitzegger, W., Schwinger, W., & Schönböck, J. ... Rets-Chitzegger, W. (2017). iVOLUNTEER-A Digital Ecosystem for Life-long Volunteering *. In iiWAS. Salzburg, Austria: ACM. doi:10.1145/3151759.3151801

Karafiloski, E., & Mishev, A. (2017). Blockchain solutions for big data challenges: A literature review. In *IEEE EUROCON 2017 -17th International Conference on Smart Technologies* (pp. 763–768). IEEE. 10.1109/EUROCON.2017.8011213

Karim, Z., & Fouad, J. (2018). Measuring urban public transport performance on route level: A literature review. *MATEC Web of Conferences, 200*. 10.1051/matecconf/201820000021

Kassambara, A. (2018). Cluster validation essentials. Retrieved from https://www.datanovia.com/en/lessons/determining-the-optimal-number-of-clusters-3-must-know-methods/

Keles, C., Karabiber, A., Akcin, M., Kaygusuz, A., Alagoz, B. B., & Gul, O. (2015). A smart building power management concept: Smart socket applications with DC distribution. *International Journal of Electrical Power & Energy Systems*, *64*, 679–688. doi:10.1016/j.ijepes.2014.07.075

Keogh, E. (2011). Instance-Based Learning. In C. Sammut, & G. I. Webb (Eds.), *Encyclopedia of Machine Learning*. Boston, MA: Springer.

Këpuska, V., & Bohouta, G. (2017). Comparing Speech Recognition Systems (Microsoft API, Google API And CMU Sphinx, In Int. Journal of Engineering Research and Application, 7(3), (Part 2), March 2017, (pp. 20–24).

Keras. (2019). Keras: The Python Deep Learning library. Retrieved from https://keras.io/, last accessed 2019/04/09

Kesavan, M., & Prabhu, J. (2018). A Survey, Design and Analysis of IoT Security and QoS Challenges. *International Journal of Information System Modeling and Design*, *9*(3), 48–66. doi:10.4018/IJISMD.2018070103

Kesner, R. P. (2007). Behavioral functions of the CA3 subregion of the hippocampus. *Learning & Memory (Cold Spring Harbor, N.Y.)*, *14*(11), 771–781. doi:10.1101/lm.688207 PMID:18007020

Kessler, E. H., & Bierly, P. E. III. (2002). Is faster really better? An empirical test of the implications of innovation speed. *IEEE Transactions on Engineering Management*, *49*(1), 2–12. doi:10.1109/17.985742

Khafash, L., Ordóñez, J. A. C., & Fraga, J. (2015). Parques Temáticos y Disneyzación. Experiencias Xcaret en la Riviera Maya. *En Turismo y ocio: Reflexiones sobre el Caribe Mexicano*, 2015, (pp. 45-84). Available at https://dialnet.unirioja.es/servlet/articulo?codigo=5386749

Khan, M. A., & Salah, K. (2018). IoT security: Review, blockchain solutions, and open challenges. *Future Generation Computer Systems*, *82*, 395–411. doi:10.1016/j.future.2017.11.022

Kimbell, L. (2009). Insights From Service Design Practice. In *Proceedings of the 8th European Academy Of Design Conference. April 1-3, 2009*. Aberdeen, Scotland: The Robert Gordon University. Available at http://www.lucykimbell.com/stuff/EAD_kimbell_final.pdf

Kimbell, L. (2011). Designing for Service as One Way of Designing Services. *International Journal of Design*, *5*(2), 41–52.

Kim, M. J., Lee, C. K., & Chung, N. (2013). Investigating the role of trust and gender in online tourism shopping in South Korea. *Journal of Hospitality & Tourism Research (Washington, D.C.)*, *37*(3), 377–401. doi:10.1177/1096348012436377

Kim, R. S., Seitz, A. R., & Shams, L. (2008). Benefits of Stimulus Congruency for Multisensory Facilitation of Visual Learning. *PLoS One*, *3*(1). doi:10.1371/journal.pone.0001532 PMID:18231612

Kim, S. A., Shin, D., Choe, Y., Seibert, T., & Walz, S. P. (2012). Integrated energy monitoring and visualization system for Smart Green City development: Designing a spatial information integrated energy monitoring model in the context of massive data management on a web-based platform. *Automation in Construction*, *22*, 51–59. doi:10.1016/j.autcon.2011.07.004

Kim, S.-J., Koh, K., Lustig, M., Boyd, S., & Gorinevsky, D. (2007). An Interior-Point Method for Large-Scale -Regularized Least Squares. *IEEE Journal of Selected Topics in Signal Processing*, *1*(4), 606–617. doi:10.1109/JSTSP.2007.910971

Kitzing, L., Mitchell, C., & Morthorst, P. E. (2012). Renewable energy policies in Europe: Converging or diverging? *Energy Policy*, *51*, 192–201. doi:10.1016/j.enpol.2012.08.064

Kiyomoto, S., & Hidano, S. (2017). Knowledge sharing framework for iKaaS platform. In *2017 2nd International Conference on Knowledge Engineering and Applications (ICKEA)* (pp. 146–150). IEEE. 10.1109/ICKEA.2017.8169919

Klopfenstein, L. C., Malatini, S., Bogliolo, A., & Carlo, U. (2017). *The Rise of Bots:A Survey of Conversational Interfaces*. Patterns, and Paradigms. doi:10.1145/3064663.3064672

Knight, J., Gibbons, C., & Ross, E. (2019). Unlocking Service Flow – Fast and Frugal Digital Healthcare Design. In Design of Assistive Technology for Ageing Population. Springer.

Kodama, R., Koge, M., Taguchi, S., & Kajimoto, H. (2017). COMS-VR: Mobile virtual reality entertainment system using electric car and head-mounted display. In *Proceedings 2017 IEEE Symposium on 3D User Interfaces (3DUI)*, 130–133. IEEE. 10.1109/3DUI.2017.7893329

Kodinariya, T. M., & Makwana, P. R. (2013). Review on determining number of Cluster in K-Means Clustering. *International Journal of Advance Research in Computer Science and Management Studies*, *1*(6), 2321–7782.

Kolumbán, G., & Kennedy, M. P. (2018). Overview of digital communications. In Chaotic Electronics in Telecommunications (pp. 131–149). Boca Raton, FL: CRC Press. doi:10.1201/9781482274516-5

Konkel, A., & Cohen, N. J. (2009). Relational memory and the hippocampus: Representations and methods. *Frontiers in Neuroscience*, *3*(2), 166–174. doi:10.3389/neuro.01.023.2009 PMID:20011138

Koo, C., Yoo, K. H., Lee, J. N., & Zanker, M. (2016). Special section on generative smart tourism systems and management: Man–machine interaction. *International Journal of Information Management*, *36*(6), 1301–1305. doi:10.1016/j.ijinfomgt.2016.05.015

Korjani, S., Facchini, A., Mureddu, M., & Damiano, A. (2017). A genetic algorithm approach for the identification of microgrids partitioning into distribution networks. In *Proceedings IECON 2017-43rd Annual Conference of the IEEE Industrial Electronics Society* (pp. 21-25). IEEE. doi:10.1109/IECON.2017.8216008

Koshizuka, N., & Sakamura, K. (2010). Ubiquitous ID: Standards for Ubiquitous Computing and the Internet of Things. *IEEE Pervasive Computing*, *9*(4), 98–101. doi:10.1109/MPRV.2010.87

Kosmopoulos, D., & Styliaras, G. (2018). A survey on developing personalized content services in museums. *Pervasive and Mobile Computing*, *47*, 54–77. doi:10.1016/j.pmcj.2018.05.002

Kowalkowski, C., Gebauer, H., & Oliva, R. (2017). Service Growth in Product Firms: Past, Present, and Future. *Industrial Marketing Management*, *60*, 82–88. doi:10.1016/j.indmarman.2016.10.015

Krajačić, G., Duić, N., & Carvalho, M. da G. (2011). How to achieve a 100% RES electricity supply for Portugal? *Applied Energy*, *88*(2), 508–517. doi:10.1016/j.apenergy.2010.09.006

Krizhevsky, A., Sutskever, I., & Hinton, G. E. (2012). ImageNet Classification with Deep Convolutional Neural Networks. In Adv. Neural Inf. Process. Syst. 1–9.

Kubrick, S. (Director). (1968). *2001: A space odyssey* [Motion picture]. United States: MGM.

Kumar, A. (2018) Demonstration of Facial Emotion Recognition on Real Time Video Using CNN: Python & Keras. Retrieved from https://appliedmachinelearning.blog/2018/11/28/demonstration-of-facial-emotion-recognition-on-real-time-video-using-cnn-python-keras/, last accessed 2019/04/09.

Kumaran, D., Hassabis, D., & McClelland, J. L. (2016). What learning systems do intelligent agents need? complementary learning systems theory updated. *Trends in Cognitive Sciences*, *20*(7), 512–534. doi:10.1016/j.tics.2016.05.004 PMID:27315762

Kusakabe, T., & Asakura, Y. (2014). Behavioural data mining of transit smart card data: A data fusion approach. *Transportation Research Part C, Emerging Technologies*, *46*, 179–191. doi:10.1016/j.trc.2014.05.012

Kusakabe, T., Iryo, T., & Asakura, Y. (2010). Estimation method for railway passengers' train choice behavior with smart card transaction data. *Transportation*, *37*(5), 731–749. doi:10.100711116-010-9290-0

Lanquar, R. (2013). Tourism in the Mediterranean: Scenarios up to 2030. MEDPRO Report No. 1/July 2011- 3rd Revision and Update May 2013. Brussels, Belgium: CEPS.

Lanquar, R. (2016). Urban coastal tourism and climate change: indicators for a Mediterranean prospective. In *Tourism in the City: Towards an Integrative Agenda on Urban Tourism, L'Aquila Conference of June 2015*. Heidelberg, Germany: Springer.

Laramee, R. S., Turkay, C., & Joshi, A. (2018). Visualization for Smart City Applications. *IEEE Computer Graphics and Applications*, *38*(5), 36–37. doi:10.1109/MCG.2018.053491729

Larios-Hernández, G. J. (2017). Blockchain entrepreneurship opportunity in the practices of the unbanked. *Business Horizons*, *60*(6), 865–874. doi:10.1016/j.bushor.2017.07.012

Larivière, B., Bowen, D., Andreassen, T. W., Kunz, W., Sirianni, N., Voss, C., ... De Keyser, A. (2017). "Service encounter 2.0": An Investigation into the Roles of Technology, Employees and Customers. *Journal of Business Research, 79*, 238–246. doi:10.1016/j.jbusres.2017.03.008

Larman, C. (2004). *Applying UML and Patterns: An Introduction to Object-Oriented Analysis and Design and Iterative Development*. Prentice Hall.

Lasi, H., Fettke, P., Kemper, H. G., Feld, T., & Hoffmann, M. (2014). Industry 4.0. *Business & Information Systems Engineering, 6*(4), 239–242. doi:10.100712599-014-0334-4

Laskowski, M., & Kim, H. M. (2016). Rapid Prototyping of a Text Mining Application for Cryptocurrency Market Intelligence. In *2016 IEEE 17th International Conference on Information Reuse and Integration (IRI)* (pp. 448–453). IEEE. 10.1109/IRI.2016.66

Lasuén, J. R., Gracia, M. I. G., & Prieto, J. L. Z. (2005). Cultura y economía. *Fundación Autor*. Recuperado de https://books.google.com.mx/books?id=EMh8PQAACAAJ

Latif, S., Afzaal, H., & Zafar, N. A. (2018). Intelligent traffic monitoring and guidance system for smart city. In *Proceedings 2018 International Conference on Computing, Mathematics and Engineering Technologies (ICoMET)*, 1–6. IEEE. 10.1109/ICOMET.2018.8346327

Leal, F. (1995). Ethics is fragile, goodness is not. *AI & Society, 9*(1), 29–42. doi:10.1007/BF01174477

LeCun, Y. (1989). Generalization and network design strategies. In Connectionism in Perspective.

LeCun, Y., Jackel, L. D., Bottou, L., Cortes, C., Denker, J. S., Drucker, H., ... & Vapnik, V. (1995). Learning algorithms for classification: A comparison on handwritten digit recognition. In Neural networks: the statistical mechanics perspective, 261-276. Mech. Perspect.

LeCun, Y., Boser, B., Denker, J. S., Henderson, D., Howard, R. E., Hubbard, W., & Jackel, L. D. (1989). Backpropagation applied to handwritten zip code recognition. *Neural Computation, 1*(4), 541–551. doi:10.1162/neco.1989.1.4.541

Lee, I. (2019). The Internet of Things for enterprises: An ecosystem, architecture, and IoT service business model. *Internet of Things, 7*, 100078. doi: 0 078 doi:10.1016/j.iot.2019.10

Lee, W. Y., & Akyildiz, I. F. (2012). Spectrum-aware mobility management in cognitive radio cellular networks. *IEEE Transactions on Mobile Computing, 11*(4), 529–542. doi:10.1109/TMC.2011.69

Lehmann, S., & Murray, M. M. (2005). The role of multisensory memories in unisensory object discrimination. *Brain Research. Cognitive Brain Research, 24*(2), 326–334. doi:10.1016/j.cogbrainres.2005.02.005 PMID:15993770

Lei, T. L., & Church, R. L. (2010). Mapping transit-based access: Integrating GIS, routes and schedules. *International Journal of Geographical Information Science, 24*(2), 283–304. doi:10.1080/13658810902835404

Le, K. N. (2019). A review of selection combining receivers over correlated Rician fading. *Digital Signal Processing, 88*, 1–22. doi:10.1016/j.dsp.2019.01.015

Leonori, S., Rizzi, A., Paschero, M., & Mascioli, F. M. F. (2018). Microgrid Energy Management by ANFIS Supported by an ESN Based Prediction Algorithm. *Proceedings of the International Joint Conference on Neural Networks, 2018-*July 1–8. 10.1109/IJCNN.2018.8489018

Le, Q. V. (2013). Building high-level features using large scale unsupervised learning. In *IEEE International Conference on Acoustics, Speech and Signal Processing* (8595–8598). 10.1109/ICASSP.2013.6639343

Leutenegger, S., Chli, M., & Siegwart, R. Y. (2011) BRISK: binary robust invariant scalable keypoints. *Proc. Int. Conf. Computer Vision (ICCV)*, 2548–2555

Levi, G., & Hassner, T. (2015). Age and gender classification using convolutional neural networks. In *Proc. of the IEEE conference on computer vision and pattern recognition workshops* (pp. 34-42). Retrieved from https://talhassner.github.io/home/publication/2015_CVPR, last accessed 2019/04/09

Li, L., Xiaoguang, H., Ke, C., & Ketai, H. (2011). The applications of WiFi-based Wireless Sensor Network in Internet of Things and Smart Grid. In *Proceedings 2011 6th IEEE Conference on Industrial Electronics and Applications* (pp. 789-793). Beijing, China: IEEE. doi:10.1109/ICIEA.2011.5975693

Lieberman, A., & Esgate, P. (2006). *La revolución del marketing del entretenimiento: acercando los magnates, los medios y la magia al mundo.* 1st. ed. Buenos Aires, Argentina: Nobuko, Universidad de Palermo.

Liedtka, J. (2015). Perspective: Linking Design Thinking with Innovation Outcomes through Cognitive Bias Reduction. *Journal of Product Innovation Management, 32*(6), 925–938. doi:10.1111/jpim.12163

Likert, R. (1932). A technique for the measurement of attitudes. Archives of Psychology, 22 - 140, 55.

Lim, A. (2008). Inter-consortia battles in mobile payments standardisation. *Electronic Commerce Research and Applications, 7*(2), 202–213. doi:10.1016/j.elerap.2007.05.003

Lin, C.-Y., Lu, Y.-L., Tsai, M.-H., & Chang, H.-L. (2018). Utilization-based parking space suggestion in smart city. In Proceedings 2018 15th IEEE Annual Consumer Communications & Networking Conference (CCNC), 1–6. IEEE.

Linaza, M. T., Gutierrez, A., & García, A. (2013). Pervasive augmented reality games to experience tourism destinations. In *Information and Communication Technologies in Tourism 2014* (pp. 497–509). Cham, Switzerland: Springer. doi:10.1007/978-3-319-03973-2_36

Lin, D. D., Talathi, S. S., & Annapureddy, V. S. (2016). Fixed Point Quantization of Deep Convolutional Networks. In *Proceedings of the 33rd International Conference on International Conference on Machine Learning*. Vol. 48. (pp. 2849–2858).

Lingamallu, M., & Garg, V. (2019). Very Short-Term HVAC Cooling Energy Forecasting for an Educational Building in Real-Time. *IOP Conference Series: Earth and Environmental Science, 238.* 10.1088/1755-1315/238/1/012069

Linley Group. (2018). Ceva NeuPro Accelerates Neural Nets.

Li, S., Xu, L., & Zhao, S. (2015). The internet of things: A survey. *Information Systems Frontiers, 17*(2), 243–259. doi:10.100710796-014-9492-7

Litchy, A. J., & Nehrir, M. H. (2014). Real-time energy management of an islanded microgrid using multi-objective Particle Swarm Optimization. In *Proceedings 2014 IEEE PES General Meeting | Conference & Exposition*, 1–5. 10.1109/PESGM.2014.6938997

Liu, C. L., & Layland, W. (1973). Scheduling Algorithms for Multiprogramming in Hard-Real-Time Environment. *Journal of the Association for Computing Machinery, 20*(1), 46–61. doi:10.1145/321738.321743

Liu, M., Phanivong, P. K., Shi, Y., & Callaway, D. S. (2019). Decentralized charging control of electric vehicles in residential distribution networks. *IEEE Transactions on Control Systems Technology, 27*(1), 266–281. doi:10.1109/TCST.2017.2771307

Liu, S., & Zhu, X. (2004). Accessibility Analyst: An integrated GIS tool for accessibility analysis in urban transportation planning. *Environment and Planning. B, Planning & Design, 31*(1), 105–124. doi:10.1068/b305

Liu, Y., Kohlberger, T., Norouzi, M., Dahl, G. E., Smith, J. L., Mohtashamian, A., ... Stumpe, M. C. (2018). Artificial Intelligence–Based Breast Cancer Nodal Metastasis Detection: Insights Into the Black Box for Pathologists. *Archives of Pathology & Laboratory Medicine*. PMID:30295070

Li, X., Jiang, P., Chen, T., Luo, X., & Wen, Q. (2017). A survey on the security of blockchain systems. *Future Generation Computer Systems*. doi:10.1016/j.future.2017.08.020

Lo, J., & Leung, P. (2015). The Preferred Theme Park. *American Journal of Economics*, (5): 472–476.

Looyestyn, J., Kernot, J., Boshoff, K., Ryan, J., Edney, S., & Maher, C. (2017). Does gamification increase engagement with online programs? A systematic review. *PLoS One*, *12*(3). doi:10.1371/journal.pone.0173403 PMID:28362821

LOPEZ RESEARCH. (2013). *An Introduction to the Internet of things (IoT)*. CISCO. Retrieved from https://www.cisco.com/c/dam/en_us/solutions/trends/iot/introduction_to_IoT_november.pdf

López, J. S. M. (2011). Sociedad del entretenimiento (2): Construcción socio-histórica, definición y caracterización de las industrias que pertenecen a este sector. *Revista Luciérnaga-Comunicación*, *3*(6), 6–16.

Lorenčík, D., Tarhaničová, M., & Sinčák, P. (2013, January). Influence of sci-fi films on Artificial Intelligence and vice-versa. In *2013 IEEE 11th international symposium on applied machine intelligence and informatics (SAMI)* (pp. 27-31). IEEE. 10.1109/SAMI.2013.6480990

Lorica, B. (2017). Practical applications of reinforcement learning in industry. Retrieved May 25, 2019, from https://www.oreilly.com/ideas/practical-applications-of-reinforcement-learning-in-industry

Lowe, D. G. (2004). Distinctive image features from scale-invariant keypoints. *International Journal of Computer Vision*, *60*(2), 91–110. doi:10.1023/B:VISI.0000029664.99615.94

Lowensteyn, I., Berberian, V., Berger, C., Da Costa, D., Joseph, L., & Grover, S. A. (2019). The Sustainability of a Workplace Wellness Program That Incorporates Gamification Principles: Participant Engagement and Health Benefits After 2 Years. *American Journal of Health Promotion*. PMID:30665309

Lüke, K., Mügge, H., Eisemann, M., & Telschow, A. (2009). Integrated Solutions and Services in Public Transport on Mobile Devices. In *Gesellschaft für Informatik (GI), I2 CS conference proceedings P-148* (pp. 109–119).

Luria, A. R. (1987). *The mind of a mnemonist: A little book about a vast memory*. Cambridge, MA, US: Harvard University Press.

Luxen, D., & Vetter, C. (2011). Real-time routing with OpenStreetMap data. In *Proc. of the 19th ACM SIGSPATIAL international conference on advances in geographic information systems* (pp. 513-516). New York, NY: ACM.

Lv, Z., Yin, T., Zhang, X., Song, H., & Chen, G. (2016). Virtual reality smart city based on WebVRGIS. *IEEE Internet of Things Journal*, *3*(6), 1015–1024. doi:10.1109/JIOT.2016.2546307

Madni, S. H. H., Latiff, M. S. A., Coulibaly, Y., & Abdulhamid, S. M. (2016). Resource scheduling for infrastructure as a service (IaaS) in cloud computing: Challenges and opportunities. *Journal of Network and Computer Applications*, *68*, 173–200. doi:10.1016/j.jnca.2016.04.016

Maggio, M., Arigliano, F., Özdemir, O., Cantera, J., Kim, E., Maestro, I., . . . Capossele, A. (2018). *SynchroniCity: Delivering an IoT enabled Digital Single Market for Europe and Beyond*. Obtido de https://synchronicity-iot.eu/wp-content/uploads/2018/09/SynchroniCity_D2.10.pdf

Mahdavian, M., Kafi, M. H., Movahedi, A., & Janghorbani, M. (2017). Improve performance in electrical power distribution system by optimal capacitor placement using genetic algorithm. In *Proceedings ECTI-CON 2017 - 2017 14th International Conference on Electrical Engineering/Electronics, Computer, Telecommunications, and Information Technology*, 749–752. 10.1109/ECTICon.2017.8096347

Mahmud, U., & Javed, M. Y. (2014). Context Inference Engine (CiE): Classifying Activity of Context using Minkowski Distance and Standard Deviation-Based Ranks. In Systems and Software Development, Modeling, and Analysis: New Perspectives and Methodologies (pp. 65-112). Hershey, PA: IGI Global.

Mahmud, U. (2015). UML based Model of a Context Aware System. In *Proceedings International Journal of Advanced Pervasive and Ubiquitous Computing*, *7*(1), 1–16. doi:10.4018/IJAPUC.2015010101

Mahmud, U. (2016). Organizing Contextual Data in Context Aware Systems: A Review. In J. Rodrigues, P. Cardoso, J. Monteiro, & M. Figueiredo (Eds.), *Handbook of Research on Human-Computer Interfaces, Developments, and Applications* (pp. 273–303). Hershey, PA: IGI Global; doi:10.4018/978-1-5225-0435-1.ch011

Mahmud, U., Hussain, S., & Yang, S. (2018). Power Profiling of Context Aware Systems: A Contemporary Analysis and Framework for Power Conservation. *Wireless Communications and Mobile Computing*, *2018*, 1–15. doi:10.1155/2018/1347967

Mahmud, U., Iltaf, N., & Kamran, F. (2007). *Context Congregator: Gathering Contextual Information in CAPP. In Proceedings 5th Frontiers of Information Technology (FIT 2007)* (pp. 134–141). Islamabad, Pakistan: COMSATS.

Mahmud, U., Iltaf, N., Rehman, A., & Kamran, F. (2007). *Context-Aware Paradigm for a Pervasive Computing Environment (CAPP). WWW\Internet 2007* (pp. 337–346). Portugal: Villa Real.

Mahmud, U., & Javed, M. Y. (2012, July). Context Inference Engine (CiE): Inferring Context. [IJAPUC]. *International Journal of Advanced Pervasive and Ubiquitous Computing*, *4*(3), 13–41. doi:10.4018/japuc.2012070102

Mahmud, U., & Malik, N. A. (2014). Flow and Threat Modelling of a Context Aware System. [IJAPUC]. *International Journal of Advanced Pervasive and Ubiquitous Computing*, *6*(2), 58–70. doi:10.4018/ijapuc.2014040105

Maj. Loren mer. (2012). *Virtual reality used to train Soldiers in new training simulator*. Retrieved from https://www.army.mil/article/84453/virtual_reality_used_to_train_soldiers_in_new_training_simulator

Malik, N. A., Mahmud, U., & Javed, M. Y. (2007). Future challenges in context aware computing. WWW/Internet 2007, (pp. 306-310).

Malik, N. A., Javed, M. Y., & Mahmud, U. (2008). *Threat Modeling in Pervasive Computing Paradigm. In Proceedings 2008 New Technologies, Mobility, and Security* (pp. 1–5). Tangier, Morocco: IEEE.

Malik, N. A., Mahmud, U., & Javed, M. Y. (2009, August). Estimating User Preferences by Managing Contextual History in Context Aware Systems. *Journal of Software*, *6*(4), 571–576.

Mallat, N. (2007). Exploring consumer adoption of mobile payments – A qualitative study. *The Journal of Strategic Information Systems*, *16*(4), 413–432. doi:10.1016/j.jsis.2007.08.001

Mallat, N., Rossi, M., Tuunainen, V. K., & Öörni, A. (2008). An empirical investigation of mobile ticketing service adoption in public transportation. *Personal and Ubiquitous Computing*, *12*(1), 57–65. doi:10.100700779-006-0126-z

Mallat, N., Rossi, M., Tuunainen, V. K., & Öörni, A. (2009). The impact of use context on mobile services acceptance: The case of mobile ticketing. *Information & Management*, *46*(3), 190–195. doi:10.1016/j.im.2008.11.008

Mangiaracina, R., Perego, A., Salvadori, G., & Tumino, A. (2017). A comprehensive view of intelligent transport systems for urban smart mobility. *International Journal of Logistics Research and Applications*, *20*(1), 39–52. doi:10.1080/13675567.2016.1241220

Marblestone, A. H., Wayne, G., & Kording, K. P. (2016). Toward an Integration of Deep Learning and Neuroscience. *Frontiers in Computational Neuroscience*, 1–41. PMID:27683554

Marín, B., Frez, J., Cruz-Lemus, J., & Genero, M. (2018). An Empirical Investigation on the Benefits of Gamification in Programming Courses. [TOCE]. *ACM Transactions on Computing Education*, *19*(1), 4. doi:10.1145/3231709

Martin, C. J., Evans, J., & Karvonen, A. (2018). Smart and sustainable? Five tensions in the visions and practices of the smart-sustainable city in Europe and North America. *Technological Forecasting and Social Change*, *133*, 269–278. doi:10.1016/j.techfore.2018.01.005

Martins, J., Gonçalves, R., Branco, F., Barbosa, L., Melo, M., & Bessa, M. (2017). A multisensory virtual experience model for thematic tourism: A Port wine tourism application proposal. *Journal of Destination Marketing & Management*, *6*(2), 103–109. doi:10.1016/j.jdmm.2017.02.002

Masabi. (n.d.). JustRide. Retrieved from http://www.masabi.com/justride-mobile-ticketing/

Master-Iesa. (2018, July 16). *Why is contemporary art important?* Retrieved June 19, 2019, from https://www.iesa.edu/paris/news-events/contemporary-art-importance

Mathworks. (n.d.). *Unsupervised Learning*. Retrieved from Mathworks: https://www.mathworks.com/discovery/unsupervised-learning.html

Matloff, N. (2017). *Statistical Regression and Classification: from Linear Models to Regression* (1st ed.). Chapman and Hall. doi:10.1201/9781315119588

Matsumoto, D., & Ekman, P. (2008). Facial expression analysis. *Scholarpedia*, *3*(5), 4237. doi:10.4249cholarpedia.4237

Ma, X., Wu, Y. J., Wang, Y., Chen, F., & Liu, J. (2013). Mining smart card data for transit riders' travel patterns. *Transportation Research Part C, Emerging Technologies*, *36*, 1–12. doi:10.1016/j.trc.2013.07.010

Ma, Y., Borrelli, F., Hencey, B., Coffey, B., Bengea, S., & Haves, P. (2012). Model predictive control for the operation of building cooling systems. *IEEE Transactions on Control Systems Technology*, *20*(3), 796–803. doi:10.1109/TCST.2011.2124461

Ma, Z. S., Callaway, D., & Hiskens, I. (2013). Decentralized Charging Control of Large Populations of Plug-in Electric Vehicles. *IEEE Transactions on Control Systems Technology*, *21*(1), 67–78. doi:10.1109/TCST.2011.2174059

Ma, Z., Zou, S., Ran, L., Shi, X., & Hiskens, I. A. (2016). Efficient decentralized coordination of large-scale plug-in electric vehicle charging. *Automatica*, *69*, 35–47. doi:10.1016/j.automatica.2016.01.035

Mazza, R., & Berre, A. (2007). Focus group methodology for evaluating information visualization techniques and tools. *11th International Conference Information Visualization (IV'07)* (pp. 74-80). IEEE. 10.1109/IV.2007.51

Mcconaghy, T., Marques, R., Müller, A., De Jonghe, D., Mcconaghy, T., & Mcmullen, G. … Granzotto, A. (2016). *BigchainDB: A Scalable Blockchain Database*. Berlin, Germany. Retrieved from http://www.noql.com

McKenzie, S., Frank, A. J., Kinsky, N. R., Porter, B., Rivière, P. D., & Eichenbaum, H. (2014). Hippocampal Representation of Relatedand Opposing Memories Develop within Distinct, Hierarchically Organized Neural Schemas. *Neuron*, *83*(1), 202–215. doi:10.1016/j.neuron.2014.05.019 PMID:24910078

Mehmood, A., Natgunanathan, I., Xiang, Y., Hua, G., & Guo, S. (2016). Protection of Big Data Privacy. *IEEE Access: Practical Innovations, Open Solutions*, *4*, 1821–1834. doi:10.1109/ACCESS.2016.2558446

Mejia, C. (2019). Influencing green technology use behavior in the hospitality industry and the role of the "green champion.". *Journal of Hospitality Marketing & Management*, *28*(5), 538–557. doi:10.1080/19368623.2019.1539935

Mell, P., & Grance, T. (2011). *The NIST definition of cloud computing*. Special Publication 800-145 - NIST.

Meng, W., Tischhauser, E. W., Wang, Q., Wang, Y., & Han, J. (2018). When Intrusion Detection Meets Blockchain Technology: A Review. *IEEE Access: Practical Innovations, Open Solutions*, *6*, 10179–10188. doi:10.1109/ACCESS.2018.2799854

Merkle, R. C. (1979). *US4309569A*. USPTO. Retrieved from https://patents.google.com/patent/US4309569

Merler, M., Ratha, N., Feris, R. S., & Smith, J. R. (2019). Diversity in Faces. *arXiv preprint arXiv:1901.10436.*

Mesáro, P., Mandičák, T., Hernandez, M. F., Sido, C., Molokáč, M., Hvizdák, L., & Delina, R. (2016). Use of Augmented Reality and Gamification techniques in tourism. *Ereview of Tourism Research*, 2.

Mesbahi, M., & Rahmani, A. M. (2016). Load balancing in cloud computing: a state-of-the-art survey. *Int. J. Mod. Educ. Comput. Sci, 8*(3), 64.4

Mezghani, M. (2008). Study on electronic ticketing in public transport Contents, (May).

Micikevicius, P., Narang, S., Alben, J., Diamos, G. F., Elsen, E., García, D., … Wu, H. (2017). Mixed Precision Training. *CoRR*, abs/1710.03740.

Milford, M. J., & Wyeth, F. G. F. (2008). Mapping a Suburb With a Single Camera Using a Biologically Inspired SLAM System. *IEEE Transactions on Robotics*, *24*(5), 1038–1053. doi:10.1109/TRO.2008.2004520

Milford, M., Jacobson, A., Chen, Z., & Wyeth, G. (2016). RatSLAM: Using Models of Rodent Hippocampus for Robot Navigation and Beyond. In M. Inaba, & P. Corke (Eds.), *Robotics Research. Springer Tracts in Advanced Robotics* (Vol. 114). Springer International Publishing. doi:10.1007/978-3-319-28872-7_27

Milgram, P., & Kishino, F. (1994). A Taxonomy of Mixed Reality Visual Displays. *IEICE Transactions on Information and Systems*, *E77-D*(12). Retrieved from http://vered.rose.utoronto.ca/people/paul_dir/IEICE94/ieice.html

Minhas, D. M., Khalid, R. R., & Frey, G. (2017). Short term load forecasting using hybrid adaptive fuzzy neural system: The performance evaluation. In *Proceedings 2017 IEEE PES-IAS PowerAfrica Conference: Harnessing Energy, Information and Communications Technology (ICT) for Affordable Electrification of Africa, PowerAfrica 2017*, 468–473. 10.1109/PowerAfrica.2017.7991270

Miorandi, D., Sicari, S., Pellegrini, F., & Chlamtac, I. (2012). Internet of things: Vision, applications and research challenges. *Ad Hoc Networks*, *10*(7), 1497–1516. doi:10.1016/j.adhoc.2012.02.016

Miraz, M. H., Ali, M., Excell, P. S., & Picking, R. (2015). *A review on Internet of Things (IoT), Internet of Everything (IoE) and Internet of Nano Things (IoNT). In Proceedings 2015 Internet Technologies and Applications (ITA)* (pp. 1–6). IEEE; doi:10.1109/ITechA.2015.7317398

Mohamed, F., & Koivo, H. (2008). *Multiobjective genetic algorithms for online management problem of microgrid. 3*, 46–54.

Moise, M. S., Gil-Saura, I., & Ruiz-Molina, M. E. (2018). Effects of green practices on guest satisfaction and loyalty. *European Journal of Tourism Research*, *20*(20), 92–104.

Momeni-K, M., Diamantas, S. C., Ruggiero, F., & Siciliano, B. (2012). Height Estimation from a Single Camera View. In *Procs Int. Conf. on Computer Vision Theory and Applications*, pp. 358-364.

Momoh, J. (2012). *Smart Grid: Fundamentals of Design and Analysis (Band 63 von IEEE Press Series on Power Engineering)*.

Monteiro, J., & Nunes, M. S. (2015). *A Renewable Source Aware Model for the Charging of Plug-in Electrical Vehicles*. (May), 51–58. doi:10.5220/0005459000510058

Moore, G. C., & Benbasat, I. (1991). Development of an Instrument to Measure the Perceptions of Adopting an Information Technology Innovation. *Information Systems Journal*, *2*(3), 192–222. doi:10.1287/isre.2.3.192

Morabit, Y. E., Mrabti, F., & Abarkan, E. H. (2019). Survey of Artificial Intelligence Approaches in Cognitive Radio Networks. *Journal of Information and Communication Convergence Engineering, 17*(1), 21-40.

Morales-Fernández, E. & Lanquar, R. (2014). El futuro turístico de una ciudad patrimonio de la humanidad: Córdoba 2031. Best Paper, Tourism, & Management Studies, No. 8.

Morelli, N. (2009). Service as value co-production: Reframing the service design process. *Journal of Manufacturing Technology Management*, *20*(5), 568–590. doi:10.1108/17410380910960993

Moro, S., Ramos, P., Esmerado, J., & Jalali, S. M. J. (2019). Can we trace back hotel online reviews' characteristics using gamification features? *International Journal of Information Management*, *44*, 88–95. doi:10.1016/j.ijinfomgt.2018.09.015

Mouli, G. R. C., Kefayati, M., Baldick, R., & Bauer, P. (2019). Integrated PV charging of EV fleet based on energy prices, V2G, and offer of reserves. *IEEE Transactions on Smart Grid*, *10*(2), 1313–1325. doi:10.1109/TSG.2017.2763683

Moustaka, V., Vakali, A., & Anthopoulos, L. G. (2018). A systematic review for smart city data analytics. [CSUR]. *ACM Computing Surveys*, *51*(5), 103. doi:10.1145/3239566

Muhamad Razali, N. M., & Hashim, A. H. (2010). Profit-based optimal generation scheduling of a microgrid. *PEOCO 2010 - 4th International Power Engineering and Optimization Conference, Program and Abstracts*, 232–237. 10.1109/PEOCO.2010.5559244

Muhammad, G., Alsulaiman, M., Amin, S. U., Ghoneim, A., & Alhamid, M. F. (2017). A Facial-Expression Monitoring System for Improved Healthcare in Smart Cities. *IEEE Access: Practical Innovations, Open Solutions*, *5*, 10871–10881. doi:10.1109/ACCESS.2017.2712788

Mukwevho, M. A., & Celik, T. (2018). Toward a smart cloud: A review of fault-tolerance methods in cloud systems. *IEEE Transactions on Services Computing*, 1. doi:10.1109/TSC.2018.2816644

Muller, M. J. (2002). Participatory Design: the Third Space in HCI. In The Human-computer Interaction Handbook (pp 1051–1068), Hillsdale, NJ: L. Erlbaum Associates.

Munizaga, M. A., & Palma, C. (2012). Estimation of a disaggregate multimodal public transport Origin-Destination matrix from passive smartcard data from Santiago, Chile. *Transportation Research Part C, Emerging Technologies*, *24*, 9–18. doi:10.1016/j.trc.2012.01.007

Munsing, E., Mather, J., & Moura, S. (2017). Blockchains for decentralized optimization of energy resources in microgrid networks. In *2017 IEEE Conference on Control Technology and Applications (CCTA)* (pp. 2164–2171). Mauna Lani, HI: IEEE. 10.1109/CCTA.2017.8062773

Murshed, S., Al-Hyari, A., Wendel, J., & Ansart, L. (2018). Design and Implementation of a 4D Web Application for Analytical Visualization of Smart City Applications. *ISPRS International Journal of Geo-Information*, *7*(7), 276. doi:10.3390/ijgi7070276

Museum Guides Ltd. (n.d.). British Museum Guide. Retrieved from https://play.google.com/store/apps/details?id=air.com.bm.london.vusiem&hl=en

Mustapa, S. I., Peng, L. Y., & Hashim, A. H. (2010). Issues and challenges of renewable energy development: A Malaysian experience. In *Proceedings International Conference on Energy and Sustainable Development: Issues and Strategies (ESD 2010)*, 1–6. 10.1109/ESD.2010.5598779

Nagasaka, K., Ando, K., Xu, Y. B., Takamori, H., Wang, J., Mitsuta, A., … Go, E. (2012). A research on operation planning of Multi Smart Micro grid. *International Journal of Advanced Mechatronic Systems*, 351–356.

Nakamoto, S. (2008). Re: Bitcoin P2P e-cash paper. Retrieved August 24, 2018, from https://satoshi.nakamotoinstitute.org/emails/cryptography/11/

Nakamoto, S. (2009). Bitcoin: A Peer-to-Peer Electronic Cash System. Retrieved August 23, 2018, from www.bitcoin.org

Nam, T., & Pardo, T. A. (2011). Conceptualizing smart city with dimensions of technology, people, and institutions. *Proceedings of the 12th Annual International Digital Government Research Conference: Digital Government Innovation in Challenging Times*, 282–291. ACM. 10.1145/2037556.2037602

Narumi, T., Nishizaka, S., Kajinami, T., Tanikawa, T., & Hirose, M. (2011). Augmented Reality Flavors: Gustatory Display Based on Edible Marker and Cross-modal Interaction. *Proceedings of the SIGCHI Conference on Human Factors in Computing Systems*, 93–102. 10.1145/1978942.1978957

Narzt, W., Mayerhofer, S., Weichselbaum, O., Haselbock, S., & Hofler, N. (2015). Be-In/Be-Out with Bluetooth Low Energy: Implicit Ticketing for Public Transportation Systems. *IEEE Conference on Intelligent Transportation Systems, Proceedings, ITSC, 2015-Octob*, 1551–1556. 10.1109/ITSC.2015.253

National League of Cities (NLC). (2017). *Trends in Smart City Development*. Retrieved from NLC website: https://www.nlc.org/sites/default/files/2017-01/Trends%20in%20Smart%20City%20Development.pdf

Naughton, J. (2019). *Don't believe the hype: the media are unwittingly selling us an AI fantasy | John Naughton*. [online] the Guardian. Available at: https://www.theguardian.com/commentisfree/2019/jan/13/dont-believe-the-hype-media-are-selling-us-an-ai-fantasy [Accessed 19 Jun. 2019].

Nehemiah, A., & Jayaraman, A. (2017, July 20). *Deep Learning for Automated Driving with MATLAB*. Retrieved from Nvidia Developer: https://devblogs.nvidia.com/deep-learning-automated-driving-matlab/

Neirotti, P., De Marco, A., Cagliano, A., Mangano, G., & Scorrano, F. (2014). Current trends in Smart City initiatives: Some stylised facts. *Cities (London, England)*, *38*, 25–36. doi:10.1016/j.cities.2013.12.010

Nemitz, P. (2018). Constitutional democracy and technology in the age of Artificial Intelligence. *Philosophical Transactions of the Royal Society A: Mathematical, Physical and Engineering Sciences*, *376*(2133). doi:10.1098/rsta.2018.0089 PMID:30323003

Nemkov, A. (2016). *The theory of generations in Russia*. Retrieved from http://sykt24.ru/novosti/teoriya-pokolenij-v-rossii-pokolenie-y/

Neuhofer, B., Buhalis, D., & Ladkin, A. (2012). Conceptualising technology enhanced destination experiences. *Journal of Destination Marketing & Management*, *1*(1-2), 36–46. doi:10.1016/j.jdmm.2012.08.001

Neuroeconomics, S. (2017). Hippocampus and Decision Making. Retrieved June 17, 2019, from http://www.shanghai-neuroeconomics.org/2017-sh-colloq/2017/5/2/hippocampus-and-decision-making

Ngai, E. W., Tao, S. S., & Moon, K. K. (2015). Social media research: Theories, constructs, and conceptual frameworks. *International Journal of Information Management*, *35*(1), 33–44. doi:10.1016/j.ijinfomgt.2014.09.004

Nielsen, J. (1993). *Usability engineering*. San Diego, CA: Academic. doi:10.1016/B978-0-08-052029-2.50007-3

Ni, J., Wu, L., Fan, X., & Yang, S. X. (2015). Bioinspired Intelligent Algorithm and Its Applications for Mobile Robot Control: A Survey. *Computational Intelligence and Neuroscience*, *2016*. PMID:26819582

Normann, R., & Ramírez, R. (1993). From Value Chain to Value Constellation: Designing Interactive Strategy. *Harvard Business Review*, *71*(4), 65–77. PMID:10127040

Nourani. (2017). Gender classification by LBP. Retrieved from https://github.com/nourani/LBP, accessed 2019/06/05.

Noy, K., & Givoni, M. (2018). Is "Smart Mobility" Sustainable? Examining the Views and Beliefs of Transport's Technological Entrepreneurs. *Sustainability, MDPI AG.*, *10*(2), 422. doi:10.3390u10020422

Nunes, F. O. (2011) *Mimo Steim*. Retrieved June 19, 2019, from http://mimosteim.me/

Nunes, F. O. (2016). *Mentira de artista: Arte (e tecnologia) que nos engana para repensarmos o mundo*. São Paulo, Brazil: Cosmogonias elétricas.

Nunes, A. A., Dias, T. G., Falcão, E., & Cunha, J. (2016). Passenger journey destination estimation from automated fare collection system data using spatial validation. *IEEE Transactions on Intelligent Transportation Systems*, *17*(1), 133–142. doi:10.1109/TITS.2015.2464335

Nunes, F. O. (2013). *O artista estah telepresente* [Performance]. Mimo Steim.

Nwankpa, C., Ijomah, W., Gachagan, A. & Marshall, S. (2018). Activation Functions: Comparison of trends in Practice and Research for Deep Learning. *Corr*. abs/1811.03378.

O'Keefe, J., & Nadel, L. (1978). *The Hippocampus as a Cognitive Map*. Oxford University Press.

Obrist, M., Comber, R., Subramanian, S., Piqueras-Fiszman, B., Velasco, C., & Spence, C. (2014). Temporal, Affective, and Embodied Characteristics of Taste Experiences: A Framework for Design. *Proceedings of the SIGCHI Conference on Human Factors in Computing Systems*, 2853–2862. 10.1145/2556288.2557007

Obrist, M., Velasco, C., Vi, C., Ranasinghe, N., Israr, A., Cheok, A. D., ... Gopalakrishnakone, P. (2016). Sensing the future of HCI: Touch, taste, and smell user interfaces. *Interaction*, *23*(5), 40–49. doi:10.1145/2973568

OECD. (March 2018). Tourism Trends and Policies 2018, Paris.

Ohwada, H., Okada, M., & Kanamori, K. (2013). Flexible route planning for amusement parks navigation. In *Proceedings 2013 IEEE 12th International Conference on Cognitive Informatics and Cognitive Computing* (pp. 421-427). 10.1109/ICCI-CC.2013.6622277

Olanda, R., Pérez, M., Morillo, P., Fernández, M., & Casas, S. (2006). Entertainment virtual reality system for simulation of spaceflights over the surface of the planet Mars. *Proceedings of the ACM Symposium on Virtual Reality Software and Technology*, 123–132. ACM. 10.1145/1180495.1180522

Oliva, R., & Kallenberg, R. (2003). Managing the transition from products to services. *International Journal of Service Industry Management*, *14*(2), 160–172. doi:10.1108/09564230310474138

Ondrus, J., Pigneur, Y., & Ecole, I. (2005). A Disruption Analysis in the Mobile Payment Market. *Proceedings of the 38th Annual Hawaii International Conference On System Sciences, 2005. HICSS'05, 00*(C), (pp. 84c-84c). IEEE. 10.1109/HICSS.2005.9

ONeil, C. (2017). Weapons of math destruction: How big data increases inequality and threatens democracy. New York: Crown.

OOMap. (2019). *BIKESHARP - Global Map of Bikeshare.* Obtido de http://bikes.oobrien.com/global.php

OpenCV. (2019). Gender Classification with OpenCV. Retrieved from https://docs.opencv.org/3.0-last-rst/modules/face/doc/tutorial/facerec_gender_classification.html, last accessed 2019/04/09

OpenCV. (2019). Open Source Computer Vision Library. Retrieved from https://opencv.org/, last accessed 2019/04/09.

Oracle. (2016). *Can Virtual Experiences Replace Reality? The future role for humans in delivering customer experience* [PDF file]. Retrieved from https://www.oracle.com/webfolder/s/delivery_production/docs/FY16h1/doc35/CXResearch-VirtualExperiences.pdf

Orth, H., Carrasco, N., Schwertner, M., & Weidmann, U. (2014). Calibration of a Public Transport Performance Measurement System for Switzerland. *Transportation Research Record: Journal of the Transportation Research Board, 2351*(1), 104–114. doi:10.3141/2351-12

Osipov, I. V., Volinsky, A. A., & Grishin, V. V. (2014). Gamification, virality and retention in educational online platform. Measurable indicators and market entry strategy. *arXiv preprint arXiv:1412.5401.*

Oskarbski, J., Birr, K., Miszewski, M., & Zarski, K. (2015). Estimating the average speed of public transport vehicles based on traffic control system data. In *Proceedings 2015 International Conference on Models and Technologies for Intelligent Transportation Systems, MT-ITS 2015*, (June), 287–293. 10.1109/MTITS.2015.7223269

Oyekan, J. (2015). A vision-based terrain morphology estimation model inspired by the avian hippocampus. *Digital Communications and Networks, 1*(2), 134–140. doi:10.1016/j.dcan.2015.04.002

Padilha, R., Iano, Y., Monteiro, A. C. B., & Loschi, H. J. (2018). Improvement of the Content Transmission in Broadcasting Systems: Potential Proposal to Rayleigh and Rician Multichannel MIMO Systems.

Padilha, R. (2018). Proposta de Um Método Complementar de Compressão de Dados Por Meio da Metodologia de Eventos Discretos Aplicada. In *Um Baixo Nível de Abstração. Dissertação (Mestrado em Engenharia Elétrica).* Campinas, Brasil: Faculdade de Engenharia Elétrica e de Computação, Universidade Estadual de Campinas.

Padilha, R., Iano, Y., Monteiro, A. C. B., Arthur, R., & Estrela, V. V. (2018, October). Betterment Proposal to Multipath Fading Channels Potential to MIMO Systems. In *Proceedings Brazilian Technology Symposium* (pp. 115-130). Cham, Switzerland: Springer.

Padovitz, A., Loke, S. W., & Zaslavsky, A. (2004). Towards a theory of context spaces. In *Proceedings Second IEEE Annual Conference on Pervasive Computing and Communications, Workshops, PerCom* (pp. 38-42). Orlando FL: IEEE. doi:10.1109/PERCOMW.2004.1276902

Pahl, C. (2015). Containerization and the paas cloud. *IEEE Cloud Computing, 2*(3), 24–31. doi:10.1109/MCC.2015.51

Paivio, A. (1991). Dual coding theory: Retrospect and current status. *Canadian Journal of Psychology/Revue. Canadian Psychology, 45*(3), 255–287. doi:10.1037/h0084295

Panetta, K. (2018) Gartner Top 10 Strategic Technology Trends for 2019. Retrieved from https://goo.gl/zqgoXP, in 2019/03/25.

Pan, J., & Yang, Z. (2018). Cybersecurity Challenges and Opportunities in the New "Edge Computing + IoT" World. In *Proceedings of the 2018 ACM International Workshop on Security in Software Defined Networks & Network Function Virtualization - SDN-NFV Sec'18* (pp. 29–32). New York, NY: ACM Press. 10.1145/3180465.3180470

Pan, S.-Y., Gao, M., Kim, H., Shah, K. J., Pei, S.-L., & Chiang, P.-C. (2018). Advances and challenges in sustainable tourism toward a green economy. *The Science of the Total Environment*, *635*, 452–469. doi:10.1016/j.scitotenv.2018.04.134 PMID:29677671

Papanek, V. (1995). *Design for the real world* (2nd ed.). London, UK: Thames & Hudson.

Papas, I., Estibals, B., Ecrepont, C., & Alonso, C. (2018). Energy Consumption Optimization through Dynamic Simulations for an Intelligent Energy Management of a BIPV Building. In *Proceedings 7th International IEEE Conference on Renewable Energy Research and Applications, ICRERA 2018*, *5*, 853–857. 10.1109/ICRERA.2018.8566915

Park, M., Han, J., Oh, H., & Lee, K. (2019). Threat Assessment for Android Environment with Connectivity to IoT Devices from the Perspective of Situational Awareness. *Wireless Communications and Mobile Computing*, *2019*, 1–14. doi:10.1155/2019/5121054

Parks, J., Ryus, P., Coffel, K., Gan, A., Perk, V., Cherrington, L., … Nakanishi, Y. (2010). *A Methodology for Performance Measurement and Peer Comparison in the Public Transportation Industry. A Methodology for Performance Measurement and Peer Comparison in the Public Transportation Industry*. doi:10.17226/14402

Park, Y., & Lee, J.-S. (2017). Artificial Synapses with Short- and Long-Term Memory for Spiking Neural Networks Based on Renewable Materials. *ACS Nano*, *11*(9), 8962–8969. doi:10.1021/acsnano.7b03347 PMID:28837313

Pasquinelli, M. (2015). Alleys of your mind: augmented intelligence and its traumas (p. 212). Meson Press.

Patel, P., Ali, M. I., & Sheth, A. (2017). On Using the Intelligent Edge for IoT Analytics. *IEEE Intelligent Systems*, *32*(5), 64–69. doi:10.1109/MIS.2017.3711653

Patrício, L., Fisk, R. P., Falcão e Cunha, J., & Constantine, L. (2011). Multilevel Service Design: From Customer Value Constellation to Service Experience Blueprinting. *Journal of Service Research*, *14*(2), 180–200. doi:10.1177/1094670511401901

Patrício, L., Gustafsson, A., & Fisk, R. (2018). Upframing Service Design and Innovation for Research Impact. *Journal of Service Research*, *21*(1), 3–16. doi:10.1177/1094670517746780

Patterson, J., & Gibson, A. (2017). Deep Learning: A Practitioner's Approach. O'Reilley Media, 1st ed.

Pawłowicz, B., Salach, M., & Trybus, B. (2018). Smart city traffic monitoring system based on 5G cellular network, RFID and machine learning. In *Proceedings KKIO Software Engineering Conference*, 151–165. Springer.

PBS. (n.d.). *Ford Installs First Moving Assembly Line 1913*. Retrieved from PBS: http://www.pbs.org/wgbh/aso/databank/entries/dt13as.html

Pearl, C. (2017). *Designing Voice User Interfaces*. California: O'Reilly Media.

Pelletier, M., Trépanier, M., & Morency, C. (2011). Smart card data use in public transit : A literature review. *Transportation Research Part C, Emerging Technologies*, *19*(4), 557–568. doi:10.1016/j.trc.2010.12.003

Peng, C., Tan, X., Gao, M., & Yao, Y. (2013). Virtual reality in smart city. In *Geo-Informatics in Resource Management and Sustainable Ecosystem* (pp. 107–118). Springer. doi:10.1007/978-3-642-45025-9_13

Perera, C., Zaslavsky, A., Christen, P., & Georgakopoulos, D. (2013). Context Aware Computing for The Internet of Things: A Survey. *IEEE Communications Surveys and Tutorials*, *16*(1), 1–41. doi:10.1109/SURV.2013.042313.00197

Petrakis, E. G., Sotiriadisb, S., Soultanopoulos, T., Rentaa, P. T., Buyya, R., & Bessis, N. (2018). Internet of Things as a Service (iTaaS): Challenges and solutions for management of sensor data on the cloud and the fog. *Internet of Things*, *3-4*, 156–174. doi:10.1016/j.iot.2018.09.009

Pieroni, A., Scarpato, N., Di Nunzio, L., Fallucchi, F., & Raso, M. (2018). Smarter city: Smart energy grid based on blockchain technology. *Int. J. Adv. Sci. Eng. Inf. Technol*, *8*(1), 298–306. doi:10.18517/ijaseit.8.1.4954

Pink, D. H. (2006). *A whole new mind: Why right-brainers will rule the future*. UK: Penguin.

Pinson, P., Mitridati, L., Ordoudis, C., & Ostergaard, J. (2017). Towards fully renewable energy systems: Experience and trends in Denmark. *CSEE Journal of Power and Energy Systems*, *3*(1), 26–35. doi:10.17775/CSEEJPES.2017.0005

Pinto, N., Cruz, D., Monteiro, J., Cabrita, C., Semião, J., & Cardoso, P. J. S. ... Rodrigues, J. M. F. (2019). IoE-Based Control and Monitoring of Electrical Grids. In Harnessing the Internet of Everything (IoE) for Accelerated Innovation Opportunities (pp. 57–82). doi:10.4018/978-1-5225-7332-6.ch003

Pirinen, A. (2016). The Barriers and Enablers of Co-design for Services. *International Journal of Design*, *10*(3), 27–42.

Pisoni, D. B. (2000). Cognitive factors and cochlear implants: Some Thoughts on Perception, Learning, and Memory in Speech Perception. *Ear and Hearing*, *21*(1), 70–78. doi:10.1097/00003446-200002000-00010 PMID:10708075

Pix 5D Cinema. (n.d.). Retrieved March 5, 2019, from https://expressavenue.in/?q=store/pix-5d-cinema

Pizam, A. (2009). Green hotels: A fad, ploy or fact of life? *International Journal of Hospitality Management*, *28*(1), 1. doi:10.1016/j.ijhm.2008.09.001

Pokric, B., Krco, S., & Pokric, M. (2014). Augmented reality based smart city services using secure iot infrastructure. In *Proceedings 2014 28th International Conference on Advanced Information Networking and Applications Workshops*, 803–808. IEEE.

Pokrovskaia, N. N. (2017). Tax, financial and social regulatory mechanisms within the knowledge-driven economy. Blockchain algorithms and fog computing for the efficient regulation. In *2017 XX IEEE International Conference on Soft Computing and Measurements (SCM)* (pp. 709–712). IEEE. 10.1109/SCM.2017.7970698

Polaine, A., Løvlie, L., & Reason, B. (2013). *Service Design*. Brooklyn, NY: Rosenfeld.

Polus, A., & Shefer, D. (1984). Evaluation of a public transportation level of service concept. *Journal of Advanced Transportation*, *18*(2), 135–144. doi:10.1002/atr.5670180204

Polydoros, A. S., & Nalpantidis, L. (2017). Survey of Model-Based Reinforcement Learning: Applications on Robotics. *Journal of Intelligent & Robotic Systems*, *86*(2), 53–173. doi:10.100710846-017-0468-y

Ponnusamy, A. (2019) cvlib: A high level easy-to-use open source Computer Vision library for Python. Retrieved from https://www.cvlib.net/, accessed 2019/06/05

Ponsignon, F., Smart, P. A., & Maull, R. S. (2011). Service Delivery System Design: Characteristics and contingencies. *International Journal of Operations & Production Management*, *31*(3), 324–349. doi:10.1108/01443571111111946

Ponzo, S., Kirsch, L. P., Fotopoulou, A., & Jenkinson, P. M. (2018). Balancing body ownership: Visual capture of proprioception and affectivity during vestibular stimulation. *Neuropsychologia*, *117*, 311–321. doi:10.1016/j.neuropsychologia.2018.06.020 PMID:29940194

Popov, A. (2006). *Marketing game. Entertain and conquer*. Moscow, Russia: Mann, Ivanov, & Ferber.

Portalés Ricart, C. (2009). *Entornos multimedia de realidad aumentada en el campo del arte*. Valencia, Spain: Universidad Politécnica de Valencia.

Portalés Ricart, C., Perales Cejudo, C. D., & Cheok, A. (2007). Exploring Social [ACM SIGCHI.]. *Cultural and Pedagogical Issues in AR-Gaming Through The Live LEGO House*, *203*, 238–239.

Portalés, C., & Perales, C. D. (2009). Sound and Movement Visualization in the AR-Jazz Scenario. 167–172. Springer-Verlag.

Portalés, C., Casas, S., & Kreuzer, K. (2019). Challenges and Trends in Home Automation: Addressing the Interoperability Problem With the Open-Source Platform OpenHAB. In Harnessing the Internet of Everything (IoE) for Accelerated Innovation Opportunities (pp. 148–174). Hershey, PA: IGI Global.

Portalés, C., Casas, S., Alonso-Monasterio, P., & Viñals, M. J. (2018). Multi-dimensional acquisition, representation, and interaction of cultural heritage tangible assets: An insight on tourism applications. In Handbook of Research on Technological Developments for Cultural Heritage and eTourism Applications (pp. 72–95). Hershey, PA: IGI Global. doi:10.4018/978-1-5225-2927-9.ch004

Portalés, C., Gimeno, J., Casas, S., Olanda, R., & Giner, F. (2016). Interacting with augmented reality mirrors. In J. Rodrigues, P. Cardoso, J. Monteiro, & M. Figueiredo (Eds.), Handbook of Research on Human-Computer Interfaces, Developments, and Applications (pp. 216–244). Hershey, PA: IGI-Global. doi:10.4018/978-1-5225-0435-1.ch009

Portalés, C., Casanova-Salas, P., Casas, S., Gimeno, J., & Fernández, M. (2019). An interactive cameraless projector calibration method. *Virtual Reality (Waltham Cross)*, 1–13.

Portalés, C., Casas, S., Coma, I., & Fernández, M. (2017). A Multi-Projector Calibration Method for Virtual Reality Simulators with Analytically Defined Screens. *Journal of Imaging*, *3*(2), 19. doi:10.3390/jimaging3020019

Portalés, C., Casas, S., Vidal-González, M., & Fernández, M. (2017). On the Use of ROMOT—A RObotized 3D-MOvie Theatre—To Enhance Romantic Movie Scenes. *Multimodal Technologies and Interaction*, *1*(2), 7. doi:10.3390/mti1020007

Portalés, C., Orduña, J. M., Morillo, P., & Gimeno, J. (2019). An efficient projector calibration method for projecting virtual reality on cylindrical surfaces. *Multimedia Tools and Applications*, *78*(2), 1457–1471. doi:10.100711042-018-6253-5

Portalés, C., Viñals, M. J., Alonso-Monasterio, P., & Morant, M. (2010). AR-Immersive Cinema at the Aula Natura Visitors Center. *IEEE MultiMedia*, *17*(4), 8–15. doi:10.1109/MMUL.2010.72

Portnoff, R. S., Huang, D. Y., Doerfler, P., Afroz, S., & McCoy, D. (2017). Backpage and Bitcoin. In *Proceedings of the 23rd ACM SIGKDD International Conference on Knowledge Discovery and Data Mining - KDD '17* (pp. 1595–1604). New York, NY: ACM Press. 10.1145/3097983.3098082

POSEUR. (2019). *Operational Programme for Sustainability and Efficient Use of Resources*. Obtido de https://poseur.portugal2020.pt/pt/candidaturas/avisos/poseur-07-2015-31-aviso-projeto-u-bike-portugal/

Prandi, F., Soave, M., Devigili, F., Andreolli, M., & De Amicis, R. (2014). Services oriented smart city platform based on 3D city model visualization. *ISPRS Annals of the Photogrammetry. Remote Sensing and Spatial Information Sciences*, *2*(4), 59.

Prendinger, H., Gajananan, K., Zaki, A. B., Fares, A., Molenaar, R., Urbano, D., ... Gomaa, W. (2013). Tokyo virtual living lab: Designing smart cities based on the 3d internet. *IEEE Internet Computing*, *17*(6), 30–38. doi:10.1109/MIC.2013.87

Puhe, M., Edelmann, M., & Reichenbach, M. (2014). *Integrated urban e-ticketing for public transport and touristic sites*. (Final Report No. IP/A/STOA/FWC/2008-096/LOT2/C1/SC12). Brussels, Belgium: European Parliament/STOA.

Punitha, S., & Mohd Rasdi, R. (2013). Corporate Social Responsibility: Adoption of Green Marketing by Hotel Industry. *Asian Social Science*, *9*(17). doi:10.5539/ass.v9n17p79

Purnima, B., & Arvind, K. (2014). EBK-Means: A Clustering Technique based on Elbow Method and K-Means in WSN. *International Journal of Computers and Applications*, *105*(9), 17–24. Retrieved from https://www.ijcaonline.org/archives/volume105/number9/18405-9674

Putintsev, N. I., Vishnevskya, O. V., & Vityaev, E. E. (2015). Development of Artificial Cognitive Systems Based on Models of the Brain of Living Organisms. *Russian Journal of Genetics: Applied Research*, *5*(6), 589–600. doi:10.1134/S207905971506012X

Putra, A. A. (2013). Transportation System Performance Analysis Urban Area Public Transport. *International Refereed Journal of Engineering and Science (IRJES)*, *2*(6), 01–15. Retrieved from www.irjes.com

Qiu, R., Badr, Y., Wang, J., & Shan, L. (2017). Developing a Smart Service System to Enrich Bike Riders' Experience. In *Proceedings 2nd International Conference on Software, Multimedia, and Communication Engineering (SMCE2017)*. Shanghai, China. 10.12783/dtcsemce2017/12468

Quadri-Felitti, D., & Fiore, A. M. (2016). Wine tourism suppliers' and visitors' experiential priorities. *International Journal of Contemporary Hospitality Management*, *28*(2), 397–417. doi:10.1108/IJCHM-05-2014-0224

Quigley, D., Georggi, N. L., Barbeau, S., Joslin, A., & Brakewood, C. (2016). *Assessment of mobile fare payment technology for future deployment in Florida*. FDOT.

Quinlan, R. (1992). *C4.5: Programs for Machine Learning* (1st ed.). Morgan Kaufmann.

Rabiner, L. R., Juang, B. H., & Lee, C. H. (1999) An Overview of Automatic Speech Recognition, In Automatic Speech and Speaker Recognition: Advanced Topics. The Netherlands: Kluwer Academic Publishers Group.

Raj, A., Gandhi, K., Nalla, B. T., & Verma, N. K. (2019). Object Detection and Recognition Using Small Labeled Datasets. In *Computational Intelligence: Theories, Applications, and Future Directions-Volume II* (pp. 407–419). Singapore: Springer.

Rama Krishna, A., Chakravarthy, A. S. N. & Sastry, A. S. C. S. (2016). Variable Modulation Schemes for AWGN Channel based Device to Device Communication. *Indian Journal of Science and Technology, 9*(20), DOI: 10.17485 / ijst / 2016 / v9i20 / 89973.

Ramesh, A., & Sadashiv, G. (2019). Essentials of Gamification in Education: A Game-Based Learning. In *Research into Design for a Connected World* (pp. 975–988). Singapore: Springer. doi:10.1007/978-981-13-5977-4_81

Ramos, C. M. Q., & Garcia, A. M. (2019). Analysis of the Contribution of ICT to Cultural and Religious Tourism: In Communicating Religious Heritage to Visitors and Tourists. In J. Álvarez-García, M.C. Río Rama, & M. Gómez-Ullate (Eds.), Handbook of Research on Socio-Economic Impacts of Religious Tourism and Pilgrimage (pp. 167-194). Hershey, PA: IGI Global.

Ramos, C. M. Q., Henriques, C., & Lanquar, R. (2016). Augmented Reality for Smart Tourism in Religious Heritage Itineraries: Tourism Experiences in the Technological Age. In J. M. F. Rodrigues, P. Cardoso, J. Monteiro, & M. Figueiredo (Eds.), Handbook of Research on Human-Computer Interfaces, Developments, and Applications (pp. 250-278). Hershey, PA: IGI Global.

Ramos, C. M. Q., Henriques, C., & Rodrigues, J. M. F. (2018) Religious Tourism Experience Model (RTEM): A Recommendation Model for Dissemination of Cultural and Religious Heritage. In J. M. F. Rodrigues, C. M. Q. Ramos, P. Cardoso, & C. Henriques (Eds.), Technological Developments for Cultural Heritage and eTourism Applications (pp. 1-29). Hershey, PA: IGI Global.

Ramos, F., Trilles, S., Torres-Sospedra, J., & Perales, F. (2018). New Trends in Using Augmented Reality Apps for Smart City Contexts. *ISPRS International Journal of Geo-Information*, *7*(12), 478. doi:10.3390/ijgi7120478

Ranasinghe, N., Karunanayaka, K., Cheok, A. D., Fernando, O. N. N., Nii, H., & Gopalakrishnakone, P. (2011). Digital Taste and Smell Communication. *Proceedings of the 6th International Conference on Body Area Networks*, 78–84. Retrieved from http://dl.acm.org/citation.cfm?id=2318776.2318795

Rao, A., & Voyles, J. R. (2017). Top 10 artificial intelligence (AI) technology trends for 2018. Retrieved June 2, 2019, from http://usblogs.pwc.com/emerging-technology/top-10-ai-tech-trends-for-2018/

Ravi, K., Khandelwal, Y., Krishna, B. S., & Ravi, V. (2018). Analytics in/for cloud-an interdependence: A review. *Journal of Network and Computer Applications*, *102*, 17–37. doi:10.1016/j.jnca.2017.11.006

Razzaque, M., & Clarke, S. (2015). Smart management of next generation bike sharing systems using Internet of Things. In *Proceedings IEEE First International Smart Cities Conference (ISC2)*, (pp. 1-8). Guadalajara, Mexico. doi:10.1109/ISC2.2015.7366219

Reguladora dos Serviços Energéticos, E. (2019). *Proposta de Regulamentação das Redes Inteligentes de eletricidade.* Retrieved from http://www.erse.pt/pt/consultaspublicas/consultas/Paginas/70_1.aspx

Regulations Regulation (EU) 2016/679 Of The European Parliament And Of The Council of 27 April 2016 on the protection of natural persons with regard to the processing of personal data and on the free movement of such data, and repealing Directive 95/46/EC (General Data Protection Regulation). (2016). Retrieved from https://eur-lex.europa.eu/legal-content/EN/TXT/PDF/?uri=CELEX:32016R0679

Reineh, A.-A., Paverd, A. J., & Martin, A. P. (2016). *Trustworthy and Secure Service-Oriented Architecture for the Internet of Things.* Retrieved June 13, 2019, from https://arxiv.org: https://arxiv.org/pdf/1606.01671.pdf

Rekimoto, J., & Nagao, K. (1995). *The world through the computer: computer augmented interaction with real world environments.* 29–36. doi:10.1145/215585.215639

Rekognition. (2019) Amazon Rekognition. Retrieved from https://docs.aws.amazon.com/rekognition/, last accessed 2019/04/09

Retyunskikh, L. T. (2003). *Socrates Academy. Philosophical games 10 years later.* Moscow, Russia: Nauka.

Reyes, A., & Peña, A. (2018). *Diversión y esparcimiento en la frontera: identidad de Ciudad Juárez. Relatos de Ciudad Juárez y otros textos.* Chihuahua, México: Ciudad Juárez.

Riazul Islam, S. M., Kwak, D., Kabir, M. H., Hossain, M., & Kwak, K.-S. (2015). The Internet of Things for Health Care: A Comprehensive Survey. *IEEE Access: Practical Innovations, Open Solutions*, *3*, 678–708. doi:10.1109/ACCESS.2015.2437951

Richard, B., & Orlowski, K. (2015). Theme Park Tourism. [Encyclopedia Entry]. In book: The SAGE International Encyclopedia of Travel and TourismChapter: Theme Park TourismPublisher. *Sage (Atlanta, Ga.).*

Riley, C. (2018, October 5). *The Top 15 Automotive Innovations of All Time.* Retrieved from AutoWise: https://autowise.com/the-top-15-car-innovations-of-all-time/

Ritchie, J., & Lewis, J. (2003). *Qualitative Research Practice: A Guide for Social Science Students and Researchers*. London, UK: Sage.

Rittinghouse, J. W., & Ransome, J. F. (2016). Cloud computing: implementation, management, and security. Boca Raton, FL: CRC Press.

Robinett, W. (1994). Interactivity and individual viewpoint in shared virtual worlds: The big screen vs. networked personal displays. *Computer Graphics*, *28*(2), 127–130. doi:10.1145/178951.178969

Robinson, L. H. C. (2017). Book Review: The Disney Musical on Stage and Screen: Critical Approaches from 'Snow White' to 'Frozen' by George Rodosthenous, Ed. London, UK: Bloomsbury Methuen Drama, pp. 97-100, 1, 257.

Robô, Ed. (2004). Retrieved June 20, 2019, from https://www.inbot.com.br/cases/roboed/

Robson, K., Plangger, K., Kietzmann, J. H., McCarthy, I., & Pitt, L. (2015). Is it all a game? Understanding the principles of gamification. *Business Horizons*, *58*(4), 411–420. doi:10.1016/j.bushor.2015.03.006

Rodrigues, J. M. F., Alves, R., Sousa, L., Negrier, A., Cardoso, P. J. S., Monteiro, J., . . . Bica, P. (2016). PRHOLO - 360° Interactive Public Relations, Chapter 7 of the Handbook of Research on Human-Computer Interfaces, Developments, and Applications (pp. 166 – 191). Hershey, PA: IGI Global. doi:10.4018/978-1-5225-0435-1.ch007

Rodrigues, J. M. F., Cardoso, P. J. S., Lessa, J., Pereira, J. A. R., Sardo, J. D. P., Freitas, M., . . . Bica, P. (2018). An Initial Framework to Develop a Mobile 5 Sense Museum System, Chapter 5 of Technological Developments for Cultural Heritage and eTourism Applications (pp. 97- 119). Hershey, PA: IGI Global. Doi:10.4018/978-1-5225-2927-9.ch005

Rodrigues, J. M. F., Martins, M., Sousa, N., & Rosa, M. (2018b) IoE Accessible Bus Stop: an initial concept. In *Proc. 8th International Conference on Software Development and Technologies for Enhancing Accessibility and Fighting Info-exclusion* (DSAI 2018), pp. 137-143. New York, NY: ACM. DOI: 10.1145/3218585.3218659

Rodrigues, H., José, R., Coelho, A., Melro, A., Ferreira, M. C., Cunha, J. F., ... Ribeiro, C. (2014). MobiPag: Integrated Mobile Payment, Ticketing and Couponing Solution Based on NFC. *Sensors (Basel)*, *14*(8), 13389–13415. doi:10.3390140813389 PMID:25061838

Rodrigues, J. M. F., Pereira, J. A. R., Sardo, J. D. P., Freitas, M. A. G., Cardoso, P. J. S., Gomes, M., & Bica, P. (2017). Adaptive Card Design UI Implementation for an Augmented Reality Museum Application. In M. Antona, & C. Stephanidis (Eds.), *Universal Access in Human-Computer Interaction 2017, Part I, LNCS 10277* (pp. 433–443). doi:10.1007/978-3-319-58706-6_35

Rodrigues, J. M. F., Veiga, R. J. M., Bajireanu, R., Lam, R., Pereira, J. A. R., Sardo, J. D. P., ... Bica, P. (2018c). Mobile Augmented Reality Framework – MIRAR. In M. Antona, & C. Stephanidis (Eds.), *Universal Access in Human-Computer Interaction 2018, LNCS 10908* (pp. 102–121). doi:10.1007/978-3-319-92052-8_9

Roeva, O., Fidanova, S., & Paprzycki, M. (2013). Influence of the population size on the genetic algorithm performance in case of cultivation process modelling. In *Proceedings 2013 Federated Conference on Computer Science and Information Systems*, (pp. 371–376). IEEE.

Rogers, E. M. (1995). *Diffusion of innovations*. New York: Free Press.

Rojas, R. (1998). *A Tutorial Introduction to the Lambda Calculus*. Berlin, Germany: Freie Universität Berlin.

Rolls, E. T. (2013). The mechanisms for pattern completion and pattern separation in the hippocampus. *Frontiers in Systems Neuroscience*, *7*, 74. doi:10.3389/fnsys.2013.00074 PMID:24198767

Roman-Belmonte, J. M., De la Corte-Rodriguez, H., & Rodriguez-Merchan, E. C. (2018). How blockchain technology can change medicine. *Postgraduate Medicine*, *130*(4), 420–427. doi:10.1080/00325481.2018.1472996 PMID:29727247

Rönneberg, M., Laakso, M., & Sarjakoski, T. (2019). Map Gretel: Social map service supporting a national mapping agency in data collection. *Journal of Geographical Systems*, *21*(1), 43–59. doi:10.100710109-018-0288-z

Rosenzweig, E. S., & Barnes, C. A. (2003). Impact of aging on hippocampal function: Plasticity, network dynamics, and cognition. *Progress in Neurobiology*, *69*(3), 143–179. doi:10.1016/S0301-0082(02)00126-0 PMID:12758108

Rousseeuw, P. J. (1987). Silhouettes: A graphical aid to the interpretation and validation of cluster analysis. *Journal of Computational and Applied Mathematics*, *20*, 53–65. doi:10.1016/0377-0427(87)90125-7

RTFE. Real Time Facial Expression Recognition with Deep Learning. Retrieved from https://github.com/a514514772/Real-Time-Facial-Expression-Recognition-with-DeepLearning, last accessed 2019/04/09

Rublee, E., Rabaud, V., Konolige, K., & Bradski, G. R. (2011). Orb: an efficient alternative to SIFT or SURF. *Proc. Int. Conf. Computer Vision (ICCV)*, 2564–2571 10.1109/ICCV.2011.6126544

Ruder, S. (2016). An overview of gradient descent optimization algorithms. In CoRR.

Rugg, G., & McGeorge, P. (2005). The Sorting Techniques: A Tutorial Paper on Card Sorts, Picture Sorts and Item Sorts. *Expert Systems: International Journal of Knowledge Engineering and Neural Networks*, *22*(3), 94–107. doi:10.1111/j.1468-0394.2005.00300.x

Rukavina, S., Gruss, S., Hoffmann, H., & Traue, H. C. (2016). Elderly People Benefit More from Positive Feedback Based on Their Reactions in the Form of Facial Expressions during Human-Computer Interaction. *Psychology (Irvine, Calif.)*, *7*(09), 1225–1230. doi:10.4236/psych.2016.79124

Russell, S. J., & Norvig, P. (2016). *Artificial intelligence: a modern approach*. Malaysia: Pearson Education Limited.

Russel, S., & Norvig, P. (2003). *Artificial Intelligence: A Modern Approach*. London, UK: Pearson Education.

Sabur, R. (2018). *Google exec: Artificial Intelligence film death scenarios 'one to two decades away'*. [online] The Telegraph. Available at: https://www.telegraph.co.uk/news/2018/03/01/google-exec-artificial-intelligence-film-death-scenarios-one/ [Accessed 18 Jun. 2019].

Saeliw, A., Hualkasin, W., Puttinaovarat, S., & Khaimook, K. (2019). Smart Car Parking Mobile Application based on RFID and IoT. [iJIM]. *International Journal of Interactive Mobile Technologies*, *13*(5). Retrieved from https://online-journals.org/index.php/i-jim/issue/view/404

Saidi, M., Lari, A. A., & Towhidkhah, F. (2013). Comparison of visual and proprioceptive training on multisensory perception using a new designed setup. In *Proceedings 21st Iranian Conference on Electrical Engineering (ICEE)*, Mashhad, pp. 1-4. 10.1109/IranianCEE.2013.6599873

Salah, K., Rehman, M. H. U., Nizamuddin, N., & Al-Fuqaha, A. (2019). Blockchain for AI: Review and Open Research Challenges. *IEEE Access: Practical Innovations, Open Solutions*, *7*, 10127–10149. doi:10.1109/ACCESS.2018.2890507

Saleiro, M., Terzic, K., Rodrigues, J. M. F., & du Buf, J. M. H. (2017). BINK: Biological Binary Keypoint Descriptor. *Bio Systems*, *162*, 147–156. doi:10.1016/j.biosystems.2017.10.007 PMID:29031966

Salonen, M., & Toivonen, T. (2013). Modelling travel time in urban networks: Comparable measures for private car and public transport. *Journal of Transport Geography*, *31*, 143–153. doi:10.1016/j.jtrangeo.2013.06.011

Sanders, E., & Stappers, P. J. (2008). Co-creation and the New Landscapes of Design. *CoDesign*, *4*(1), 5–18. doi:10.1080/15710880701875068

Sandler, M. B., Howard, A. G., Zhu, M., Zhmoginov, A., & Chen, L. (2018). MobileNetV2: Inverted Residuals and Linear Bottlenecks. In *IEEE/CVF Conference on Computer Vision and Pattern Recognition* (4510-4520). 10.1109/CVPR.2018.00474

Sangiorgi, D., Patrício, L., & Fisk, R. P. (2017). Designing for interdependence, participation and emergence in complex service systems. In Designing for Service: Key Issues and New Directions (pp. 49–64). London: Bloomsbury Academic. doi:10.5040/9781474250160.ch-004

Sardo, J. D. P., Semião, J., Monteiro, J. M., Esteves, E., Pereira, J., Freitas, M., Rodrigues, J. M. F. (2017). Portable Device for Touch, Taste and Smell Sensations in Augmented Reality Experiences, In Proceedings INCREaSE, SPRINGER LNCS XXX, pp. 307–322. doi:10.1007/978-3-319-70272-8_26

Sarma, S. (2014). Bus Tracking & Ticketing using USSD Real-time application of USSD Protocol in Traffic Monitoring. *Journal of Emerging Technologies and Innovative Research*, *1*(7), 628–634.

Saxena, G., & Ilbery, B. (2008). Integrated rural tourism: A border case study. *Annals of Tourism Research*, *35*(1), 233–254. doi:10.1016/j.annals.2007.07.010

Sayre, S., & King, C. (2003). *Entertainment and Society: Influences, Impacts, and Innovations* (1st ed.). Kindle Edition.

Schiermeier, Q. (2013). Renewable power: Germany's energy gamble. *Nature*, *496*(7444), 156–158. doi:10.1038/496156a PMID:23579661

Scoville, W. B., & Milner, B. (2000). Loss of recent memory after bilateral hippocampal lesions, 1957. *The Journal of Neuropsychiatry and Clinical Neurosciences*, *12*(1), 103–113. doi:10.1176/jnp.12.1.103-a PMID:10678523

Seaborn, K., & Fels, D. I. (2015). Gamificación en la teoría y la acción: una encuesta. *Revista internacional de estudios humano-computador, 74*, 14-31.

Secall, R. E. (2001). Nuevo segmento emergente de turismo: Los parques temáticos. *Cuadernos de Turismo*, (7): 35–54.

SEGITTUR. (2017). *Informe destinos turísticos inteligentes: construyendo el futuro*. Madrid, Spain: Ministerio de Industria, Energía y Turismo, Secretaria de Estado de Comunicaciones y para la Sociedad de la Información.

SEGITTUR. (2018). *INTELITUR, Guía Aplicaciones turística ("236 aplicaciones para facilitar la vida al viajero")*. Madrid, Spain: Ministerio de Energía, Turismo y Agenda Digital, Secretaria de Estado de Turismo.

Segura, M. (2018). INTELIGÊNCIA ARTIFICIAL APLICADA A NEGÓCIOS. *Revista Inteligência Competitiva*, *8*(3), 101–110.

Seibel, P. (2005). *Practical Common Lisp*. New York: Apress. doi:10.1007/978-1-4302-0017-8

Senix. (n.d.). *Ultrasonic Sensors FAQs*. Retrieved from Senix: https://www.senix.com/ultrasonic-sensor-faqs/

Sentient Play, L. L. C. (2016, September 13). *KOMRAD*. Retrieved September 01, 2018, from https://apps.apple.com/us/app/komrad/id1020876671

Serpanos, D., & Wolf, M. (2017). The IoT Landscape. In Proceedings *Internet-of-Things (IoT) Systems: Architectures, Algorithms, Methodologies* (pp. 1–6). Cham, Switzerland: Springer; doi:10.1007/978-3-319-69715-4_1

Sever, N. S., Sever, G. N., & Kuhzady, S. (2015). The evaluation of potentials of gamification in tourism marketing communication. *International Journal of Academic Research in Business and Social Sciences*, *5*(10), 188–202. doi:10.6007/IJARBSS/v5-i10/1867

Shae, Z., & Tsai, J. J. P. (2017). On the Design of a Blockchain Platform for Clinical Trial and Precision Medicine. In *2017 IEEE 37th International Conference on Distributed Computing Systems (ICDCS)* (pp. 1972–1980). IEEE. 10.1109/ICDCS.2017.61

Shaikh, I. (2016). Emotion Recognition Using Scikit-learn & OpenCV. Retrieved from https://github.com/its-izhar/Emotion-Recognition-Using-SVMs, accessed 2017/10/11

Shams, L., & Seitz, A. R. (2008). Benefits of multisensory learning. *Trends in Cognitive Sciences*, *12*(11), 411–417. doi:10.1016/j.tics.2008.07.006 PMID:18805039

Shawahna, A., Sait, S. M., & El-Maleh, A. H. (2018). FPGA-Based Accelerators of Deep Learning Networks for Learning and Classification: A Review. *IEEE Access: Practical Innovations, Open Solutions*, *7*, 7823–7859. doi:10.1109/ACCESS.2018.2890150

Sherman, W. R., & Craig, A. B. (2018). *Understanding virtual reality: Interface, application, and design*. Morgan Kaufmann.

Shih, D.-H., Shih, P.-Y., & Wu, T.-W. (2018). An infrastructure of multi-pollutant air quality deterioration early warning system in spark platform. In *2018 IEEE 3rd International Conference on Cloud Computing and Big Data Analysis (ICCCBDA)* (pp. 648–652). IEEE. 10.1109/ICCCBDA.2018.8386595

Shin, D., Lee, J., Lee, J., & Yoo, H. (2017). 14.2 DNPU: An 8.1TOPS/W reconfigurable CNN-RNN processor for general-purpose deep neural networks. In *IEEE International Solid-State Circuits Conference* (240-241). 10.1109/ISSCC.2017.7870350

Sigala, M., Toni, M., Renzi, M. F., Di Pietro, L., & Mugion, R. G. (2019). Gamification in Airbnb: Benefits and Risks. *e-Review of Tourism Research, 16*(2/3).

Sigala, M. (2015a). Gamification for crowdsourcing marketing practices: Applications and benefits in tourism. In *Advances in crowdsourcing* (pp. 129–145). Cham, Switzerland: Springer. doi:10.1007/978-3-319-18341-1_11

Sigala, M. (2015b). The application and impact of gamification funware on trip planning and experiences: The case of TripAdvisor's funware. *Electronic Markets*, *25*(3), 189–209. doi:10.100712525-014-0179-1

Silva, A. (2016). Facial Expression Recognition with OpenCV 3+. Retrieved from http://andrew-silva.com/2016/11/09/facial-expression-recognition-with.html, accessed 2017/10/11

Silva, B. N., Khan, M., & Han, K. (2018). Towards sustainable smart cities: A review of trends, architectures, components, and open challenges in smart cities. *Sustainable Cities and Society*, *38*, 697–713. doi:10.1016/j.scs.2018.01.053

Simonyan, K., & Zisserman, A. (2014). Very deep convolutional networks for large-scale image recognition. In *arXiv preprint arXiv:1409.1556.*

Skålén, P., Gummerus, J., von Koskull, C., & Magnusson, P. R. (2015). Exploring value propositions and service innovation: A service-dominant logic study. *Journal of the Academy of Marketing Science*, *43*(2), 137–158. doi:10.100711747-013-0365-2

Skinner, H., Sarpong, D., & White, G. R. (2018). Meeting the needs of the Millennials and Generation Z: Gamification in tourism through geocaching. *Journal of Tourism Futures*, *4*(1), 93–104. doi:10.1108/JTF-12-2017-0060

Smith, R. C., Vangkilde, K. T., Kjærsgaard, M. G., Otto, T., Halse, J., & Binder, T. (Eds.). (2016). Design Anthropological Futures. London, UK: Bloomsbury Academic.

Sniedovich, M. (2006). Dijkstra's algorithm revisited: The dynamic programming connection. *Control and Cybernetics*, *35*(3).

Sommerville, I. (2016). Software Engineering (10th ed.). Pearson.

Song, W., & Giszter, S. F. (2011). Adaptation to a cortex-controlled robot attached at the pelvis and engaged during locomotion in rats. *The Journal of Neuroscience*, *31*(8), 3110–3128. doi:10.1523/JNEUROSCI.2335-10.2011 PMID:21414932

Sortomme, E., Hindi, M. M., MacPherson, S. D. J., & Venkata, S. S. (2011). Coordinated charging of plug-in hybrid electric vehicles to minimize distribution system losses. *IEEE Transactions on Smart Grid*, *2*(1), 186–193. doi:10.1109/TSG.2010.2090913

Sotres, P., de la Torre, C. L., Sánchez, L., Jeong, S., & Kim, J. (2018). Smart City Services Over a Global Interoperable Internet-of-Things System: The Smart Parking Case. 2018 Global Internet of Things Summit (GIoTS), 1–6. IEEE.

Sousa, L., Alves, A., & Rodrigues, J. M. F. (2016). Augmented reality system to assist inexperienced pool players. *Computacional Visual Media*, *2*(2), 183–193. doi:10.100741095-016-0047-3

Spielberg, S. (Director). (2001). A.I. Artificial Intelligence [Motion picture]. The United States. A.I. Artificial Intelligence Movie.

Spiess, P., Karnouskos, S., Guinard, D., Savio, D., Baecker, O., de Souza, L. M., & Trifa, V. (2009). SOA-Based Integration of the Internet of Things in Enterprise Services. In *Proceedings 2009 IEEE International Conference on Web Services*. Los Angeles, CA: IEEE. 10.1109/ICWS.2009.98

Squire, L. R., Stark, C. E., & Clark, R. E. (2004). The medial temporal lobe. *Annual Review of Neuroscience*, *27*(1), 279–306. doi:10.1146/annurev.neuro.27.070203.144130 PMID:15217334

Srinivasa, K. G., Sowmya, B. J., Shikhar, A., Utkarsha, R., & Singh, A. (2018). Data Analytics Assisted Internet of Things Towards Building Intelligent Healthcare Monitoring Systems: IoT for Healthcare. *Journal of Organizational and End User Computing*, *30*(4), 83–103. doi:10.4018/JOEUC.2018100106

Stachenfeld, K. L., Botvinick, M. M., & Gershman, S. J. (2017). The hippocampus as a predictive map. *Nature Neuroscience*, *20*(11), 1643–1653. doi:10.1038/nn.4650 PMID:28967910

Stickdorn, M., & Schneider, J. (Eds.). (2010). *This Is Service Design Thinking*. Amsterdam, The Netherlands: BIS Publishers.

Strang, T., & Linnhoff-Popien, C. (2004). A Context Modelling Survey. *First International Workshop on Advanced Context Modelling, Reasoning and Management (Ubicomp 2004)*. Nottingham, UK: ACM.

Struckmeier, O., Tiwari, K., Salman, M., Pearson, M. J., & Kyrki, V. (2019). ViTa-SLAM: A Bio-inspired Visuo-Tactile SLAM for Navigation while Interacting with Aliased Environments, arXiv:1906.06422v4. Retrieved August, 01, 2019, from https://arxiv.org/pdf/1906.06422.pdf

Studios, B. L. (n.d.). The Night Cafe. Retrieved July 12, 2019, from https://www.borrowedlightvr.com/the-night-cafe/

Suchman, L. (1987). *Plans and Situated Actions: The Problem of Human-Machine Communication*. Cambridge, UK: Cambridge University Press.

Suzuki, J., Balasubramaniam, S., Pautor, S., Meza, V. D. P., & Koucheryavy, Y. (2014). A Service-Oriented Architecture for Body Area NanoNetworks with Neuron-based Molecular Communication. *Mobile Networks and Applications*, *19*(6), 707–717. doi:10.100711036-014-0549-0

Swan, M. (2015). *Blockchain : blueprint for a new economy*. Sebastopol, CA: O'Reilly; Retrieved from https://books.google.co.in/books?hl=en&lr=&id=RHJmBgAAQBAJ&oi=fnd&pg=PR3&ots=XQtFD1ZSe0&sig=Dkx_avuMX-VeCtYv5LD-6N_TG7uQ&redir_esc=y#v=onepage&q&f=false

Swan, M. (2018). Blockchain for Business: Next-Generation Enterprise Artificial Intelligence Systems. *Advances in Computers*, *111*, 121–162. doi:10.1016/bs.adcom.2018.03.013

Swathipriya, P. (2019). *Performance Evaluation of Intermediate Public Transport by Benchmarking and Numerical Rating Approach*, *6*(4), 1–11.

SynchroniCity. (2019). *Technical Framework*. Obtido de https://synchronicity-iot.eu/tech/

Synopsys DesignWare. (2017). EV6x Vision Processors.

Szegedy, C., Liu, W., Jia, Y., Sermanet, P., Reed, S., Anguelov, D., ... & Rabinovich, A. (2014). Going Deeper with Convolutions. *arXiv:1409.4842*.

Szegedy, C., Vanhoucke, V., Ioffe, S., Shlens, J., & Wojna, Z. (2016). Rethinking the Inception Architecture for Computer Vision. In *IEEE Conference on Computer Vision and Pattern Recognition*, (2818-2826). 10.1109/CVPR.2016.308

Sze, V., Chen, Y., Yang, T., & Emer, J. S. (2017). Efficient processing of deep neural networks: A tutorial and survey. *Proceedings of the IEEE*, *105*(12), 2295–2329. doi:10.1109/JPROC.2017.2761740

Tavilla, E. (2015). *Transit Mobile Payments: Driving Consumer Experience and Adoption*. Federal Reserve Bank of Boston Research Report.

Taylor, M. (2016, October 8). Self-Driving Mercedes-Benzes Will Prioritize Occupant Safety over Pedestrians. Retrieved June 13, 2019, from https://www.caranddriver.com/news/a15344706/self-driving-mercedes-will-prioritize-occupant-safety-over-pedestrians/

TEA. (2019). *The Themed Entertainment Association (TEA)*. Retrieved from http://www.teaconnect.org

Technology, O. (2016). Olorama. Retrieved from http://www.olorama.com/en/

Tehrani, N. H., Khan, U. T., & Crawford, C. (2016). Baseline load forecasting using a Bayesian approach. In *Proceedings 2016 IEEE Canadian Conference on Electrical and Computer Engineering (CCECE)*, 1–4. 10.1109/CCECE.2016.7726749

Templates. (2010.). Retrieved June 22, 2019, from http://flask.pocoo.org/docs/1.0/tutorial/templates/

Teng, C.-C., Horng, J.-S., Hu, M.-L., Chien, L.-H., & Shen, Y.-C. (2012). Developing energy conservation and carbon reduction indicators for the hotel industry in Taiwan. *International Journal of Hospitality Management*, *31*(1), 199–208. doi:10.1016/j.ijhm.2011.06.006

TensorFlow. (2019). TensorFlow: An end-to-end open source machine learning platform. Retrieved from https://www.tensorflow.org/, last accessed 2019/04/09

Terzic, K., Rodrigues, J. M. F., & du Buf, J. M. H. (2015). BIMP: A real-time biological model of multi-scale keypoint detection in V1. *Neurocomputing*, *150*, 227–237. doi:10.1016/j.neucom.2014.09.054

Tesla. (2018). *Autopilot*. Retrieved from Tesla: https://www.tesla.com/autopilot

TfL. (n.d.). Contactless and mobile pay as you go. Retrieved from https://tfl.gov.uk/fares/how-to-pay-and-where-to-buy-tickets-and-oyster/pay-as-you-go/contactless-and-mobile-pay-as-you-go

The most extreme examples of Virtual Reality: Immersive Experience. (2017, November 6). Retrieved July 10, 2019, from Premo website: https://3dcoil.grupopremo.com/blog/extreme-examples-virtual-reality-immersive-experience/

The University of Hong Kong. (2017). New functions of hippocampus unveiled: Scientists achieve major breakthrough in untangling mysteries of the brain. Retrieved July 12, 2019, from www.sciencedaily.com/releases/2017/09/170929093215.htm

The Visual Workspace. (2019). Retrieved July 07, 2018, from https://whimsical.com/

Thompson, C. (2016, June 10). *The 3 biggest ways self-driving cars will improve our lives*. Retrieved from Business Insider: https://www.businessinsider.com/advantages-of-driverless-cars-2016-6

Tian, J., Li, H., & Chen, R. (2018). Individual Behavior Change under Smart City Environment—A Proposal of Smart Citizen Concept with Four Dimensions. *IConference 2018 Proceedings.*

Tian, Y., Li, X., Wang, K., & Wang, F. Y. (2018). Training and testing object detectors with virtual images. *IEEE/CAA Journal of Automatica Sinica, 5*(2), 539-546.

Tibshirani, R., Walther, G., & Hastie, T. (2001). Estimating the number of clusters in a data set via the gap statistic. *Journal of the Royal Statistical Society. Series B, Statistical Methodology, 63*(2), 411–423. doi:10.1111/1467-9868.00293

Tie, S. F., & Tan, C. W. (2013). A review of energy sources and energy management system in electric vehicles. *Renewable & Sustainable Energy Reviews, 20*, 82–102. doi:10.1016/j.rser.2012.11.077

Tomaszewska, E., & Florea, A. (2018). Urban smart mobility in the scientific literature — bibliometric doi:10.2478/emj-2018-0010

Tomaszewska, E., & Florea, A. (2018). Urban smart mobility in the scientific literature — bibliometric analysis. Engineering Management in Production and Services, 10(2).

Toscano, L. (2017) Emotions Detection Via Facial Expressions with Python & OpenCV. Retrieved from https://www.apprendimentoautomatico.it/articles/17/emotions-detection-via-facial-expressions-with-python-opencv, accessed 2017/10/11

Tourism4.0. (2018). What is Tourism 4.0? Available at https://www.tourism4-0.org/ last accessed 2019/07/08.

Touzani, S., Granderson, J., & Fernandes, S. (2018). Gradient boosting machine for modeling the energy consumption of commercial buildings. *Energy and Building, 158*, 1533–1543. doi:10.1016/j.enbuild.2017.11.039

Tozer, E. P. (2012). *Broadcast Engineer's Reference Book* (1st ed.). FOCAL PRESS. doi:10.4324/9780080490564

Trischler, J., Pervan, S. J., Kelly, S. J., & Scott, D. R. (2018). The value of codesign: The effect of customer involvement in service design teams. *Journal of Service Research, 21*(1), 75–100. doi:10.1177/1094670517714060

Tryon, C. (2013). Reboot cinema. *Convergence, 19*(4), 432–437. doi:10.1177/1354856513494179

Trzcinski, T., Christoudias, C. M., Fua, P., & Lepetit, V. (2013) Boosting binary keypoint descriptors. *Proc. Int. Conf. Computer Vision and Pattern Recognition (CVPR)*, 2874–2881.

Tsai, P. (2018, April 02). *Data snapshot: AI Chatbots and Intelligent Assistants in the Workplace.* Retrieved June 19, 2019, from https://community.spiceworks.com/blog/2964-data-snapshot-ai-chatbots-and-intelligent-assistants-in-the-workplace

Tsai, W.-L., Hsu, Y.-L., Lin, C.-P., Zhu, C.-Y., Chen, Y.-C., & Hu, M.-C. (2016). Immersive Virtual Reality with Multimodal Interaction and Streaming Technology. *Proceedings of the 18th ACM International Conference on Multimodal Interaction*, 416–416. 10.1145/2993148.2998526

Tuballa, M. L., & Abundo, M. L. (2016). A review of the development of Smart Grid technologies. *Renewable & Sustainable Energy Reviews, 59*, 710–725. doi:10.1016/j.rser.2016.01.011

Tulving, E. (1985). How many memory systems are there? *The American Psychologist, 40*(4), 385–398. doi:10.1037/0003-066X.40.4.385

Tulving, E. (2002). Episodic memory: From mind to brain. *Annual Review of Psychology, 53*(1), 1–25. doi:10.1146/annurev.psych.53.100901.135114 PMID:11752477

Turing, A. M. (1950). Computing machinery and intelligence. *Mind, 59*(236), 433–460. doi:10.1093/mind/LIX.236.433

Turner, D., Rodrigues, J. M. F., & Rosa, M. (2019). Describing People: An Integrated Framework for Human Attributes Classification, accepted INternational CongRess on Engineering and Sustainability in the XXI cEntury – INCREaSE 2019.

Turner, M., & Wilson, R. (2010). Smart and integrated ticketing in the UK: Piecing together the jigsaw. *Computer Law & Security Review, 26*(2), 170–177. doi:10.1016/j.clsr.2010.01.015

Turnquist, M. A., & Blume, S. (1980). Evaluating Potential Effectiveness of Headway Control Strategies for Transit Systems. *Transportation Research Record: Journal of the Transportation Research Board*, (746): 25–29.

Udoh, I. S., & Kotonya, G. (2017). Developing IoT applications: Challenges and frameworks. *IET Cyber-Physical Systems: Theory & Applications, 3*(2), 65–72. doi:10.1049/iet-cps.2017.0068

Umuroglu, Y., Fraser, N. J., Gambardella, G., Blott, M., Leong, P. H. W., Jahre, M., & Vissers, K. A. (2016). FINN: A Framework for Fast, Scalable Binarized Neural Network Inference. In CoRR, abs/1612.07119.

UNWTO. (2012). *Global Report on City Tourism. AM Reports: Volume six.* Madrid, Spain: UNWTO.

Upadhyay, N. K., Joshi, S., & Yan, J. J. (2016). Synaptic electronics and neuromorphic computing. *Science China. Information Sciences, 59*(6). doi:10.100711432-016-5565-1

Vakali, A., Anthopoulos, L., & Krco, S. (2014). Smart Cities Data Streams Integration: Experimenting with Internet of Things and social data flows. *Proceedings of the 4th International Conference on Web Intelligence, Mining and Semantics (WIMS14)*, 60. ACM. 10.1145/2611040.2611094

Van Der Meer, D., Mouli, G. R. C., Mouli, G. M. E., Elizondo, L. R., & Bauer, P. (2018). Energy Management System with PV Power Forecast to Optimally Charge EVs at the Workplace. *IEEE Transactions on Industrial Informatics, 14*(1), 311–320. doi:10.1109/TII.2016.2634624

van Gent, P. (2016). Emotion Recognition With Python, OpenCV and a Face Dataset. A tech blog about fun things with Python and embedded electronics. Retrieved from http://www.paulvangent.com/2016/04/01/emotion-recognition-with-python-opencv-and-a-face-dataset/, accessed 2017/10/11

van Maanen, P. P., Lindenberg, J., & Neerincx, M. A. (2005). Integrating human factors and artificial intelligence in the development of human-machine cooperation. In *Proc. of the 2005 international conference on artificial intelligence (ICAI'05)*

Van, A., Guck, J. C., Machuca, M., & Kellerer, W. (2019). Bounded Dijkstra (BD): Search Space Reduction for Expediting Shortest Path Subroutines. Computer Science: Networking and Internet Architecture. Retrieved from https://arxiv.org/abs/1903.00436

Vargheese, R., & Dahir, H. (2014). An IoT/IoE enabled architecture framework for precision on shelf availability: Enhancing proactive shopper experience. In *Proceedings 2014 IEEE International Conference on Big Data (Big Data)* (pp. 21-26). Washington, DC: IEEE. 10.1109/BigData.2014.7004418

Vargo, S. L., & Lusch, R. F. (2017). Service-dominant logic 2025. *International Journal of Research Marketing, 34*(1), *46-67.*

Vargo, S. L., & Lusch, R. F. (2004). Evolving to a new dominant logic for marketing. *Journal of Marketing*, *68*(1), 1–17. doi:10.1509/jmkg.68.1.1.24036

Vasantha Kumar, S., & Vanajakshi, L. (2013). Modewise Travel Time Estimation on Urban Arterial Travel Time Estimation Using Buses as Probes. In *Proceedings of the 92nd Annual Meeting of the Transportation Research Board*, Washington, D.C. 10.100713369-014-1332-z

Veldman, E., & Verzijlbergh, R. A. (2015). Distribution grid impacts of smart electric vehicle charging from different perspectives. *IEEE Transactions on Smart Grid*, *6*(1), 333–342. doi:10.1109/TSG.2014.2355494

Vendrúscolo, J. D. B. G., & Moré, R. P. O. (2018). Contribuições Da Inteligência Artificial Nos Sistemas De Informação De Apoio A Gestão Universitária. Available at https://repositorio.ufsc.br/handle/123456789/190471

Venieris, S. I., & Bouganis, C. (2018). fpgaConvNet: Mapping Regular and Irregular Convolutional Neural Networks on FPGAs. *IEEE Transactions on Neural Networks and Learning Systems*, *30*(2), 326–342. doi:10.1109/TNNLS.2018.2844093 PMID:29994725

Venkatesh, V., Morris, M. G., Davis, G. B., & Davis, F. D. (2003). User Acceptance of Information Technology : Toward a Unified View Author (s): Davis Published by : Management Information Systems Research Center, University of Minnesota. *Management Information Systems Quarterly*, *27*(3), 425–478. doi:10.2307/30036540

Vera, L., Gimeno, J., Casas, S., García-Pereira, I., & Portalés, C. (2017). A hybrid virtual-augmented serious game to improve driving safety awareness. *International Conference on Advances in Computer Entertainment*, 293–310. Springer.

Vera, L., Gimeno, J., Coma, I., & Fernández, M. (2011). Augmented Mirror: Interactive Augmented Reality System Based on Kinect. In P. Campos, N. Graham, J. Jorge, N. Nunes, P. Palanque, & M. Winckler (Eds.), In *Human-Computer Interaction – INTERACT 2011: 13th IFIP TC 13 International Conference Proceedings, Part IV* (pp. 483–486), Lisbon, Portugal, September 5-9, 2011. 10.1007/978-3-642-23768-3_63

Véstias, M. P., Duarte, R. P., Sousa, J. T., & Neto, H. C. (2018). Lite-CNN: A High-Performance Architecture to Execute CNNs in Low Density FPGAs. In *28th International Conference on Field Programmable Logic and Applications* (pp. 393-399). 10.1109/FPL.2018.00075

Viechnicki, P., Khuperkar, A., Fishman, T., & Eggers, W. (2015). *Smart mobility: Reducing congestion and fostering faster, greener, and cheaper transportation options.* Obtido de https://www2.deloitte.com/insights/us/en/industry/public-sector/smart-mobility-trends.html

Vlahakis, V., Ioannidis, N., Tsotros, M., Stricker, D., Gleue, T., Daehne, P., & Almeida, L. (2002). Archeoguide: An Augmented Reality Guide for Archaeological Sites. *IEEE Computer Graphics and Applications*, 9.

Vosmeer, M., & Schouten, B. (2014). Interactive Cinema: Engagement and Interaction. In A. Mitchell, C. Fernández-Vara, & D. Thue (Eds.), *Interactive Storytelling* (pp. 140–147). Springer International Publishing.

Wallace, R. B., Liu, H., Goubran, R., Bilodeau, M., & Knoefel, F. (2019). Cloud based Artificial Intelligence Processing of Ambient Home Sensors. In *ISPIM Conference Proceedings* (pp. 1-12). The International Society for Professional Innovation Management (ISPIM).

Wang, N., Choi, J., Brand, D., Chen, C., & Gopalakrishnan, K. (2018). Training Deep Neural Networks with 8-bit Floating Point Numbers. *CoRR* abs/1812.08011.

Wang, R., Wu, S., Wang, C., An, S., Sun, Z., Li, W., … Fu, M. (2018). *Optimized Operation and Control of Microgrid based on Multi-objective Genetic Algorithm.* 1539–1544. doi:10.1109/POWERCON.2018.8601845

Wang, Y.-C., & Tsai, P.-H. (2014). A Study on the Applications of the Cultural History Promotion via Gamification Based. Digital Curation: Taking "History Hero" as an Example. Available at http://pnclink.org/document/2014%20 Poster_Third%20Place_Poster%20Image.pdf

Wang, D., Ohnishi, K., & Xu, W. (2019). Multimodal Haptic Display for Virtual Reality: A Survey. *IEEE Transactions on Industrial Electronics*, 1–1. doi:10.1109/TIE.2019.2920602

Wang, J., Gong, B., Liu, H., & Li, S. (2015). Multidisciplinary approaches to artificial swarm intelligence for heterogeneous computing and cloud scheduling. *Applied Intelligence*, *43*(3), 662–675. doi:10.100710489-015-0676-8

Wang, R., Jin, L., Xiao, R., Guo, S., & Li, S. (2012). 3D Reconstruction and interaction for smart city based on world wind. *2012 International Conference on Audio, Language and Image Processing*, 953–956. IEEE. 10.1109/ICALIP.2012.6376751

Wang, X. V., Wang, L., Mohammed, A., & Givehchi, M. (2017). Ubiquitous manufacturing system based on Cloud: A robotics application. *Robotics and Computer-integrated Manufacturing*, *45*, 116–125. doi:10.1016/j.rcim.2016.01.007

Wardman, M. (2004). Public transport values of time. *Transport Policy*, *11*(4), 363–377. doi:10.1016/j.tranpol.2004.05.001

Wasim, M. U., Ibrahim, A. A. Z. A., Bouvry, P., & Limba, T. (2017). Law as a service (LaaS): Enabling legal protection over a blockchain network. In *2017 14th International Conference on Smart Cities: Improving Quality of Life Using ICT & IoT (HONET-ICT)* (pp. 110–114). IEEE. 10.1109/HONET.2017.8102214

Weichert, F., Bachmann, D., Rudak, B., & Fisseler, D. (2013). Analysis of the Accuracy and Robustness of the Leap Motion Controller. *Sensors (Basel)*, *13*(5), 6380–6393. doi:10.3390130506380 PMID:23673678

Welcome to Jinja2. (2007). Retrieved June 22, 2019, from http://jinja.pocoo.org/docs/2.10/

Weng, X., Liu, Y., Song, H., Yao, S., & Zhang, P. (2018). Mining urban passengers' travel patterns from incomplete data with use cases. *Computer Networks*, *134*, 116–126. doi:10.1016/j.comnet.2018.01.048

Werbach, K., & Hunter, D. (2015). *For the Win: How Game Thinking Can Revolutionize Your Business*. Moscow, Russia: Mann, Ivanov, & Ferber.

Werthner, H., Alzua-Sorzabal, A., Cantoni, L., Dickinger, A., Gretzel, U., Jannach, D., & Stangl, B. (2015). Future research issues in IT and tourism. *Information Technology & Tourism*, *15*(1), 1–15. doi:10.100740558-014-0021-9

Wilson, N. (2013). Making Use of Automated Data Collection to Improve Transit Effectiveness. Retrieved from https://pt.slideshare.net/BRTCoE/making-use-of-automated-data-collection-to-improve-transit-effectiveness

Wimmer, G. E., & Shohamy, D. (2012). Preference by association: How memory mechanisms in the hippocampus bias decisions. *Science*, *338*(6104), 270–273. doi:10.1126cience.1223252 PMID:23066083

Wingfield, C. (2012). All the world's a game, and business is a player. *The New York Times*, A1.

Witten, I. H., Frank, E., Hall, M. A., & Pal, C. J. (2016). *Data Mining: Practical Machine Learning Tools and Techniques*. Morgan Kaufmann.

Wongso, R., Cin, C., & Suhartono, J. (2018). TransTrip: A Shortest Path Finding Application for Jakarta Public Transportation using Dijkstra Algorithm. *Journal of Computational Science*, *14*(7), 939–944. doi:10.3844/jcssp.2018.939.944

World Tourism Organization. (2011). *World travel market global trends report.*

World Tourism Organization. (2019a). Tourism and the SDGs. Retrieved May 2, 2019, from https://icr.unwto.org/content/tourism-and-sdgs

World Tourism Organization. (2019b). Tourism for SDGs – Welcome To The Tourism For SDGs Platform! Retrieved September 20, 2005, from http://tourism4sdgs.org/

Wu, Q.-T., Wu, C., & Wang, Y. M. (2016). *Application of Reinforcement Learning on High-Speed Rail Cognitive Radio*. Paper presented at the 2016 International Conference on Artificial Intelligence: Techniques and Applications (AITA 2016). Shanghai, China

Xilinx, V. (2018). The first adaptive compute acceleration platform (acap).

Xu, C., Wang, K., & Guo, M. (2017). Intelligent Resource Management in Blockchain-Based Cloud Datacenters. *IEEE Cloud Computing*, *4*(6), 50–59. doi:10.1109/MCC.2018.1081060

Xu, F., Buhalis, D., & Weber, J. (2017). Serious games and the gamification of tourism. *Tourism Management*, *60*, 244–256. doi:10.1016/j.tourman.2016.11.020

Xu, F., Tian, F., Buhalis, D., Weber, J., & Zhang, H. (2016). Tourists as mobile gamers: Gamification for tourism marketing. *Journal of Travel & Tourism Marketing*, *33*(8), 1124–1142. doi:10.1080/10548408.2015.1093999

Xu, F., Weber, J., & Buhalis, D. (2013). Gamification in tourism. In *Information and communication technologies in tourism 2014* (pp. 525–537). Cham, Switzerland: Springer. doi:10.1007/978-3-319-03973-2_38

Xu, L., Yu, X., Wang, H., Dong, X., Liu, Y., Lin, W., ... Wang, J. (2019). Physical layer security performance of mobile vehicular networks. *Mobile Networks and Applications*, 1–7.

Xu, W., Cho, H., Kim, Y. H., Kim, Y. T., Wolf, C., Park, C. G., & Lee, T. W. (2016). Organometal Halide Perovskite Artificial Synapses. *Advanced Materials*, *28*(28), 5916–5922. doi:10.1002/adma.201506363 PMID:27167384

Yadav, R., Kumar Dokania, A., & Swaroop Pathak, G. (2016). The influence of green marketing functions in building corporate image. *International Journal of Contemporary Hospitality Management*, *28*(10), 2178–2196. doi:10.1108/IJCHM-05-2015-0233

Yaddaden, Y., Bouzouane, A., Adda, M., & Bouchard, B. (2016). A New Approach of Facial Expression Recognition for Ambient Assisted Living. In *Proceedings of the 9th ACM International Conference on PErvasive Technologies Related to Assistive Environments* (p. 14). New York, NY: ACM. 10.1145/2910674.2910703

Yang, D., Kou, L., & Liu, A. (2014). *US9672499B2*. USPTO. Retrieved from https://patents.google.com/patent/US9672499B2/en

Yankelovich, N. (2008). Using Natural Dialogs as the Basis for Speech Interface Design. In *Human Factors and Voice Interactive Systems. Signals and Communication Technology*. Boston, MA: Springer. doi:10.1007/978-0-387-68439-0_9

Yasaweerasinghelage, R., Staples, M., & Weber, I. (2017). Predicting Latency of Blockchain-Based Systems Using Architectural Modelling and Simulation. In *2017 IEEE International Conference on Software Architecture (ICSA)* (pp. 253–256). Gothenburg, Sweden: IEEE. 10.1109/ICSA.2017.22

Yecies, B. (2016). Transnational collaboration of the multisensory kind: Exploiting Korean 4D cinema in China. *Media International Australia*, *159*(1), 22–31. doi:10.1177/1329878X16640104

Yeom, S. (2011). Augmented reality for learning anatomy. *Changing Demands, Changing Directions. Proceedings Ascilite Hobart*, 1377–1383.

Yılmaz, H., & Coşkun, İ. O. (2016). New toy of marketing communication in tourism: Gamification. In e-Consumers in the era of new tourism (pp. 53-71). Springer, Singapore.

Yin, S. H., & Vatrapu, R. (2017). A first estimation of the proportion of cybercriminal entities in the bitcoin ecosystem using supervised machine learning. In *2017 IEEE International Conference on Big Data (Big Data)* (pp. 3690–3699). IEEE. 10.1109/BigData.2017.8258365

Yin, Z., Yu, F. R., Bu, S., & Han, Z. (2015). Joint cloud and wireless networks operations in mobile cloud computing environments with telecom operator cloud. *IEEE Transactions on Wireless Communications, 14*(7), 4020-4033.

Yin, C., Xiong, Z., Chen, H., Wang, J., Cooper, D., & David, B. (2015). A literature survey on smart cities. *Science China. Information Sciences*, *58*(10), 1–18. doi:10.100711432-015-5397-4

Yli-Huumo, J., Ko, D., Choi, S., Park, S., & Smolander, K. (2016). Where Is Current Research on Blockchain Technology?— A Systematic Review. *PLoS One*, *11*(10). doi:10.1371/journal.pone.0163477 PMID:27695049

Yoh, A. C., Iseki, H., Taylor, B. D., & King, D. A. (2006). Institutional Issues and Arrangements in Interoperable Transit Smart Card Systems: A Review of the Literature on California, United States, and International Systems UCB-ITS-PWP-2006-2, (March).

Yu, X., Salmon, C. T., & Leung, C. (2015). Emotional interactions between artificial companion agents and the elderly. In *Proceedings of the 2015 International Conference on Autonomous Agents and Multiagent Systems* (pp. 1991-1992). International Foundation for Autonomous Agents and Multiagent Systems.

Yuan, X.-C., Sun, X., Zhao, W., Mi, Z., Wang, B., & Wei, Y.-M. (2017). Forecasting China's regional energy demand by 2030: A Bayesian approach. *Resources, Conservation, and Recycling*, *127*, 85–95. doi:10.1016/j.resconrec.2017.08.016

Yue, L., Junqin, H., Shengzhi, Q., & Ruijin, W. (2017). Big Data Model of Security Sharing Based on Blockchain. In *2017 3rd International Conference on Big Data Computing and Communications (BIGCOM)* (pp. 117–121). IEEE. 10.1109/BIGCOM.2017.31

Yu, E., & Sangiorgi, D. (2018). Service Design as an Approach to Implement the Value Cocreation Perspective in New Service Development. *Journal of Service Research*, *21*(1), 40–58. doi:10.1177/1094670517709356

Yu, Z., Chen, S., & Tong, L. (2016). An intelligent energy management system for large-scale charging of electric vehicles. *CSEE Journal of Power and Energy Systems*, *2*(1), 47–53. doi:10.17775/CSEEJPES.2016.00008

Zahavy, T., Zrihem, N. B., & Mannor, S. (2016). Graying the black box: understanding DQNs, *arXiv:160202658*

Zanella, A., Bui, N., Castellani, A. P., Vangelista, L., & Zorzi, M. (2014). Internet of Things for Smart Cities. *IEEE Internet of Things Journal*, *1*(1), 22–32. doi:10.1109/JIOT.2014.2306328

Zeiler, M. D., & Fergus, R. (2013). Visualizing and Understanding Convolutional Networks. arXiv. vol. 30 (pp. 225–231).

Zeng, W., & Ye, Y. (2018). VitalVizor: A Visual Analytics System for Studying Urban Vitality. *IEEE Computer Graphics and Applications*, *38*(5), 38–53. doi:10.1109/MCG.2018.053491730 PMID:30273126

Zhang, C. (2018). Easy Real time gender age prediction from webcam video with Keras. Retrieved from https://www.dlology.com/blog/easy-real-time-gender-age-prediction-from-webcam-video-with-keras/, last accessed 2019/04/09

Zhang, C., & Ma, Y. (2012). *Ensemble Machine Learning*. New York: Springer-Verlag. doi:10.1007/978-1-4419-9326-7

Zhang, C., Romagnoli, A., Zhou, L., & Kraft, M. (2017). From Numerical Model to Computational Intelligence: The Digital Transition of Urban Energy System. *Energy Procedia*, *143*, 884–890. doi:10.1016/j.egypro.2017.12.778

Zhang, C., Wu, D., Sun, J., Sun, G., Luo, G., & Cong, J. (2016). Energy-Efficient CNN Implementation on a Deeply Pipelined FPGA Cluster. In *Proceedings of the International Symposium on Low Power Electronics and Design* (pp. 326–331). 10.1145/2934583.2934644

Zhang, C., & Zhang, S. (2002). Association Rule Mining. In *Lecture Notes in Artificial Intelligence*. Springer-Verlag.

Zhang, J., Dai, L., He, Z., Jin, S., & Li, X. (2017). Performance analysis of mixed-ADC massive MIMO systems over Rician fading channels. *IEEE Journal on Selected Areas in Communications*, *35*(6), 1327–1338. doi:10.1109/JSAC.2017.2687278

Zhang, L., Chen, S., Dong, H., & El Saddik, A. (2018). Visualizing Toronto city data with Hololens: Using augmented reality for a city model. *IEEE Consumer Electronics Magazine*, *7*(3), 73–80. doi:10.1109/MCE.2018.2797658

Zhang, N., Yang, P., Ren, J., Chen, D., Yu, L., & Shen, X. (2018). Synergy of Big Data and 5G Wireless Networks: Opportunities, Approaches, and Challenges. *IEEE Wireless Communications*, *25*(1), 12–18. doi:10.1109/MWC.2018.1700193

Zhang, Q., Guo, Y., Laffont, P., Martin, T., & Gross, M. (2017). A Virtual Try-On System for Prescription Eyeglasses. *IEEE Computer Graphics and Applications*, *37*(4), 84–93. doi:10.1109/MCG.2017.3271458 PMID:28829296

Zhang, T., Pota, H., Chu, C.-C., & Gadh, R. (2018). Real-time renewable energy incentive system for electric vehicles using prioritization and cryptocurrency. *Applied Energy*, *226*, 582–594. doi:10.1016/j.apenergy.2018.06.025

Zhang, X., Zhou, X., Lin, M., & Sun, J. (2018) ShuffleNet: An Extremely Efficient Convolutional Neural Network for Mobile Devices. In *IEEE/CVF Conference on Computer Vision and Pattern Recognition* (pp. 6848–6856). 10.1109/CVPR.2018.00716

Zhong, R. Y., Lan, S., Xu, C., Dai, Q., & Huang, G. Q. (2016). Visualization of RFID-enabled shopfloor logistics Big Data in Cloud Manufacturing. *International Journal of Advanced Manufacturing Technology*, *84*(1-4), 5–16. doi:10.100700170-015-7702-1

Zhong, R. Y., Xu, X., Klotz, E., & Newman, S. T. (2017). Intelligent manufacturing in the context of industry 4.0: A review. *Engineering*, *3*(5), 616–630. doi:10.1016/J.ENG.2017.05.015

Zhou, B., Dastjerdi, A. V., Calheiros, R. N., Srirama, S. N., & Buyya, R. (2015). mCloud: A context-aware offloading framework for heterogeneous mobile cloud. *IEEE Transactions on Services Computing*, *10*(5), 797–810. doi:10.1109/TSC.2015.2511002

Zhou, J., Murphy, E., & Long, Y. (2014). Commuting efficiency in the Beijing metropolitan area: An exploration combining smartcard and travel survey data. *Journal of Transport Geography*, *41*, 175–183. doi:10.1016/j.jtrangeo.2014.09.006

Zichermann, G., & Cunningham, C. (2011). *Gamification by design: Implementing game mechanics in web and mobile apps*. O'Reilly Media.

Zickermann, G., & Lynder, J. (2014). *Gamification in business: How to break through the noise and capture the attention of co-workers and clients: Mann*. Moscow, Russia: Ivanov and Ferber.

Zomerdijk, L., & Voss, C. (2010). Service design for experience-centric services. *Journal of Service Research*, *13*(1), 67–82. doi:10.1177/1094670509351960

Zomerdijk, L., & Voss, C. (2011). NSD processes and practices in experiential services. *Journal of Product Innovation Management*, *28*(1), 63–80. doi:10.1111/j.1540-5885.2010.00781.x

About the Contributors

João Rodrigues graduated in Electrical Engineering in 1993, he got his M.Sc. in Computer Systems Engineering in 1998 and Ph.D. Electronics and Computer Engineering in 2008 from University of the Algarve (UAlg), Portugal. He is Adjunct Professor at Instituto Superior de Engenharia, also in the UAlg, where he lectures Computer Science and Computer Vision since 1994. He is member of associative laboratory LARSyS (ISR-Lisbon), CIAC (UAlg) and the Associations APRP, IAPR, SUPERA, and ARTECH. He participated in more than 15 financed scientific projects, 4 of them as PI. He is co-author more than 190 scientific publications, including several books. His major research interests lie on computer vision, augmented reality, human-machine cooperation, human-computer interaction and augmented intelligence [https://bit.ly/369wlf2].

Pedro Cardoso holds a PhD in the field of Operational Research from the University of Seville (Spain), a Master in Computational Mathematics from the University of Minho (Portugal) and a Degree in Mathematics - Computer Science from the University of Coimbra (Portugal). He teaches Computer Science and Mathematics at University the Algarve (UAlg) and is member of LARSys/UAlg. He has high knowledge in the fields of databases, algorithms and data structures, machine learning, data science, and Operational Research. Over the past few years he edited 3 books with IGI, has been involved in more than 10 national and international scientific and development projects and is the co-author of more than 70 scientific publications.

Jânio Monteiro graduated in Electrical and Computers Engineering in 1995 from the University of Porto, and later obtained a Master and PhD degrees respectively in 2003 and 2010, also in Electrical and Computer Engineering from Instituto Superior Técnico, Technical University of Lisbon. Since 2003 is a member INESC-ID in Lisbon, where he participated in several European R&TD projects funded by the Information Society Technologies Programme (IST), of the European Commission, namely: Olympic, My-e-Director and Saracen. As part of the Algarve University he has also participated in an international project called ENERGEIA and several national projects with regional companies. He is co-author of more than three dozen publications, including journal articles, papers in scientific conferences, several book chapters, deliverables of projects and national patents. His main areas of interest involve Communication Networks, Smart Grids, Wireless Sensor Networks, Human Machine Interfaces and Internet of Things.

Célia M. Q. Ramos graduated in Computer Engineering from the University of Coimbra, obtained her Master in Electrical and Computers Engineering from the Higher Technical Institute, Lisbon University, and the PhD in Econometrics in the University of the Algarve (UALG), Faculty of Economics,

Portugal. She is Adjunct Professor at School for Management, Hospitality and Tourism, also in the UALG, where she lectures computer science. Areas of research and special interest include conception and development of information systems, tourism information systems, big data, etourism, econometric modeling and panel-data models. Célia Ramos has published in the fields of information systems and tourism, namely, she has authored a book, book chapters, conference papers and journal articles. At the level of applied research, she has participated in several funded projects.

* * *

Manolo Pérez Aixendri studied Computer Engineering (1998) and Telematic Engineering (2007) at the University of Valencia (UV). He started developing his work in the Artec group of the UV in 1995. He obtained the Diploma of Advanced Studies (DEA) in the doctoral program "Information Technology, Computing and Communication" of the UV. He is Associate Professor in the Department of Computer Science of the UV from 2001 until today. His research areas are Advanced Multimodal Interfaces, Virtual Reality and Augmented Reality applications, Serious Games as well as Physical Simulation and Animation in Realtime and Projection Systems.

Pedro Bento is a Professor at the Polytechnic Institute of Beja. PhD in Physical Education and Sport. Coordinator of the Short-Cycle Degree Outro Sports and Leisure, Teacher of the Master in Physical Activity and Health. Researcher in physical activity and active tourism area. Member of the U-Bike Project - Programa Operacional Sustentabilidade e Eficiência no Uso de Recursos, POSEUR-01-1407-FC-000003, 2016 – 2019.

Jose Borges has a PhD in Computer Science from the University College of London, U.K., MSc in Electronic Engineering and Computers from the Faculty of Engineering, University of Porto and graduation in Mechanical Engineering from the Faculty of Engineering, University of Porto. Associate Professor Department of Industrial Engineering and Management at the Faculty of Engineering, University of Porto, and Researcher at the INESC TEC. Teaches courses in Statistics, Data Mining, Information Systems and Human Computer Interaction. 30+ papers in international journals, ISI proceedings and book chapters. Supervision or co-supervision of 3 completed PhD thesis and of 3 ongoing thesis. Supervision of +45 master thesis.

Isabel Sofia Brito is a professor at the Polytechnic Institute of Beja. Her research interests include Requirements Engineering, System Modeling, Software Engineering for Sustainability, Big Data and IoT. She teaches graduate and undergraduate courses on computer science. She has participated in several research and teaching projects, such as, DECIdE - Sistema de Apoio à Investigação Científica e Tecnológica, SAICT-POL/24135/2016 | LISBOA-01-0145-FEDER-024135, 2018 –2019; Naming the Pain in Requirements Engineering (NaPiRE) 2017 –2019 and; U-Bike, POSEUR – Programa Operacional Sustentabilidade e Eficiência no Uso de Recursos, POSEUR-01-1407-FC-000003, 2016 – 2019. She has occupied several conference positions in software engineering conferences, mainly as program chair. She has organized international conferences and workshops on software and requirements engineering-related topics.

Lucília Cardoso has a PhD in Business Sciences and Tourism since 2010. Integrate Member of CITUR – Centre for Tourism Research, Development and Innovation, Portugal. Is postdoctoral researcher of Aveiro University, in Switzerland she's studding the Swiss Higher Education in Tourism & Hospitality. Is International researcher of Favourite Destinations project of the Euro-Asian Tourism Studies Association. Has publications in destination brand-choice, destination image and imagery, food and mountains tourism and smart destinations. She is Jury member of Art&Tur International Tourism Film Festival since 2016.

Pablo Casanova-Salas received the multimedia engineering degree from the University of Valencia, Valencia,Spain, in 2016. He is currently a researcher at the ARTEC (Advanced Research and Technological Expansion in Computer Graphics) group at the University of Valencia. His research interests include augmented reality,virtual reality, advanced user interfaces, human-computer interaction and mobile computing.

Sergio Casas-Yrurzum has a master's degree in Computer Engineering and also a bachelor's degree in Telecommunications Engineering - Telematics Specialty. He received the Spanish National Award on University Studies in 2008. He received his PhD in Computational Mathematics at the University of Valencia in 2014. He works as a senior researcher in the Robotics Institute (IRTIC) of the University of Valencia, where he is also a part-time professor at the School of Engineering (ETSE). His expertise is in the simulation field with special focus on Virtual Reality, Augmented Reality and motion cueing.

Damian Copeland is an experienced academic and practitioner specialising in the areas of interaction design and speech automation. He completed his PhD in interaction design at Leeds Beckett University in 2009 and is currently a director at ICR Speech Solutions, a UK-based independent specialist in the design and delivery of speech technologies and automation. His main academic interests are in speech AI and the design of assistive technologies for people with disabilities.

Vera Costa has a degree in Mathematics at the Faculty of Sciences of the University of Porto and a Master Degree in Data Analysis and Decision Support Systems at the Faculty of Economics of the same university. She started to teach mathematics to high school and, after that, information systems to college students. She participated in several research projects, and she contributes with her knowledge of statistical analysis and software programming, in different application fields, such as health, politics, and transportation. Currently, she is a Ph.D. candidate of Transportation Systems of the MIT Portugal program.

Dario Cruz obtained a Bachelor Degree in Computer Engineering and Automation in 2013 at the University of Cape Verde and later, in 2018, a Master degree in Electrical and Electronic Engineering at the University of Algarve. Since 2017, he has been working in several research projects at the University of Algarve. His knowledge fields include algorithms for the scheduling of electric vehicles charging, load of electric vehicles protocols and machine learning algorithms.

Konstantinos Domdouzis is a computer scientist with a BSc (Hons) in Computer Science from the University of Luton (currently renamed to University of Bedfordshire) and an MSc in Computer Networks & Communications from the University of Westminster. Konstantinos has realized his PhD at the

Department of Civil & Building Engineering of Loughborough University focusing on the applications of Wireless Sensor Technologies in the Construction Industry. He undertook postdoctoral research at the Department of Agricultural & Biological Engineering at the University of Illinois at Urbana-Champaign, United States, focusing on Systems Informatics for Biomass Feedstock Production. Since then, Kostas joined the Greek Army in order to realize his compulsory service and also worked as a KTP Associate/ Software & Systems Developer at Whole Systems Partnership, a healthcare modelling consultancy, in collaboration with Brunel University. Konstantinos was also a researcher within the Centre of Excellence in Terrorism, Resilience, Intelligence and Organised Crime Research (CENTRIC), and his work was focused on the EU-FP7-funded project ATHENA; his main role in the project was the development of computational tools for crisis management. In 2015, he was hired as a Lecturer in Database Systems at the Department of Computing, Sheffield Hallam University and at 2018, he was promoted to Senior Lecturer.

João Falcão e Cunha is a lecturer and researcher at the School of Engineering of the University of Porto. He holds a PhD in Computing Science from Imperial College London (1989), a MSc in Operations Research from Cranfield University (1984) and a first degree in Electrical Engineering from the School of Engineering of the University of Porto (1983). He is a member of IEEE - Computer Society. He has been involved with theoretical and experimental work in Software Engineering and Information Systems for the past 25 years. His research interests include decision support systems, graphical user interfaces, object-oriented modeling and service engineering and management. He was the coordinator at FEUP the integrated master programme in Industrial Engineering and Management, and the master programme in Service Engineering and Management. He was a member of the Executive Committee of ERCIM, the European Research Consortium for Informatics and Mathematics, and the Academic Director of the IBM Center for Advanced Studies in Portugal (IBM CAS Portugal). He is currently the Dean of FEUP.

Sherjeel Farooqui is currently working as an Assistant Professor at the Department of Software Engineering, Foundation University, Islamabad, Pakistan. He did his M.S. in Computer System Engineering from National University of Sciences &Technology Islamabad, Pakistan, in 2006. His research interests include IoT, Self Organizing Networks, Network Security and Lora Wifi. He has publications in prestigious international conferences.

Marta Campos Ferreira is a Postdoc and Invited Assistant Professor in the Faculty of Engineering of University of Porto (FEUP). She holds a PhD in Transportation Systems from FEUP (MIT Portugal Program), a M.Sc. in Service Engineering and Management from FEUP and a Lic. in Economics from the Faculty of Economics of University of Porto (FEP). She has been involved in R&D projects in areas such as technology enabled services, transport and mobility. Her main research interests include mobile services design, human computer interaction, and intelligent transportation systems. She is the author of several publications in conferences and international scientific journals.

Guilherme Fidélio is a User Interface and Interaction Designer currently based in São Paulo, Brazil. He studied Digital Design at Senac University Center, wherein 2018 he worked as an academic researcher in the area of interactive environments focused on Game Design. His subsequent works based on the Edgard Project have been published at the VI International Symposium on Innovation in Interactive Media, on the Human-Computer Interaction International 2019 and Medium. As a member of the Apple

Developer Academy program, he developed iOS applications with a focus on user experience and interfaces and was involved in projects for the Malala Foundation. He especially enjoys creating interfaces that delight and successfully guide its users.

Dan Fitton is a Reader in User Experience Design and a member of the Child-Computer Interaction Research Group (www.ChiCI.org). He is particularly interested in user-centred design techniques for novel interactive technologies, particularly in the context of child and teenage users.

Fernando Fogliano has a postdoctoral degree in the Graduate Studies Program in Arts at UNESP, a Ph.D., and a Master's degree in Communication and Semiotics at PUC-SP. He is the co-coordinator of the cAt, Science Art and Technology research group at UNESP, and a member of the International Research Group on convergences between Art, Science and Technology of UNESP, GIIP. Fernando studies the application of technologies for natural interaction with conversational robots at Senac University Center. Amongst his research interests are subjects as visual and interactive narratives; aesthetic, cognitive and technological implications of the relationship between image and interactivity; and mediation and interaction strategies in artistic production in the field of neuroaesthetics and language. He is a member of the SCIArts art group and teaches at undergraduate courses in Design and Photography at Senac University Center.

Reinaldo França has a B.Sc. in Computer Engineering in 2014. Currently, he is an Ph.D. degree candidate by Department of Semiconductors, Instruments and Photonics, Faculty of Electrical and Computer Engineering at the LCV-UNICAMP working with technological and scientific research as well as in programming and development in C/C ++, Java and .NET languages. His main topics of interest are simulation, operating systems, software engineering, wireless networks, internet of things, broadcasting and telecommunications systems

Teresa Galvão has a background in Mathematics from the University of Coimbra, a Master in Electrical and Computers Engineering, and a PhD in Sciences of Engineering from the University of Porto. She is Assistant Professor in the Faculty of Engineering of University of Porto and a senior researcher at INESC TEC in Porto. She has participated in several national and European R&D projects in areas related to transportation systems and mobility. She collaborates regularly with the largest public transport companies in Portugal as researcher and consultant and was responsible for the development and implementation of several innovative systems for the operational planning, mobile ticketing, and passenger information in several of those companies. The academic and professional background led her to have a broad and multidisciplinary perspective of the current transportation and mobility challenges. Her main research interests are operational research, data science, human-computer interaction, and transportation systems. She has more than 70 scientific publications, supervised 9 PhD students and more than 100 MSc students. She is co-founder and CEO of OPT-Otimização e Planeamento de Transportes, SA, a company that develops innovative solutions for the optimization of public transports operation, the provision of passenger information and mobility management.

Inma García-Pereira was born in Valencia, Spain in 1985. She received a M.S. degree in Audiovisual Communication (2008) and a M.S. degree in Computer Engineering (2016), both from the University of Valencia, Spain. She received the FPU grant (2016-present) by the Spanish Ministry of Science, Innovation and Universities to complete her Ph.D. at the Institute of Robotics and Information and Communication Technologies (IRTIC) of the University of Valencia, Spain. She currently works as a Junior Researcher in the IRTIC at the University of Valencia, where she is also a part-time professor at the School of Engineering (ETSE). She belongs to the ARTEC group, dedicated to 3D interactive graphics, virtual reality, augmented reality and civil simulation. Currently, she is developing her doctorate in the field of indoor real-time locating systems and its application to augmented reality. Her research interests include augmented and virtual reality, RTLS, collaborative shared environments, multimedia applications and multimodal interaction systems.

Anthony Gephardt has recently graduated from Waynesburg University. An interest in technology and cars produced a spark for autonomous vehicles. This realization heavily influenced his first professional writings. Mainly focusing on providing complex information in an understandable package.

André Gomes de Souza is an interaction designer based in São Paulo, Brazil. He is a member of the Research and Application of Technologies for Natural Interaction with Conversational Robots Group at Senac University Center, and his research led to papers in the VI International Symposium on Innovation in Interactive Media and on the Human-Computer Interaction International 2019. Also, he has done projects for global companies in the last three years for clients such as Amazon, Google, Samsung, Uber, Paypal, Itaú, Vivo. His work can be seen at andregsouza.com.br.

Shariq Hussain received his Master's degree in Computer Science from PMAS Arid Agriculture University, Rawalpindi, Pakistan, in 2007, and Ph.D. degree in Applied Computer Technology from University of Science and Technology Beijing, China, in 2014. Since 2014, he has been with the Department of Software Engineering, Foundation University Islamabad, Rawalpindi Campus, where he is currently an Assistant Professor. His main research interests include web services, QoS in web services, web service testing, IoT and e-learning. He served as Workshop/Special Session Co-chair for IEEE International Conference on Internet of People 2018 (IoP 2018) and publicity chair for IEEE International Conference on Internet of People (IoP 2015). He is currently an editorial board member of Journal of Next Generation Information Technology (JNIT) - AICIT, Rep. of S. Korea

Rachel Jones is a highly experienced innovation and strategy leader who helps organizations to shape new strategies, products, and services for people. Rachel has spent 20 years at the forefront of digital innovation, framing and scoping opportunities, forging partnerships, and developing products and services across different industry sectors. Her clients include Vodafone, Mircosoft, Skype, Virgin and Reuters.

John Knight is a Design Strategy Director at Avanade's London Digital Studio and a Doctoral Candidate at Aalto University of Arts, Design and Architecture. Having studied Fine Art with telematic's pioneer Roy Ascott, he subsequently worked in the creative industries for over thirty years. In the last twenty years, he has published over 100 articles on all aspects of digital design and has recently focused on adapting Service Dominant Logic to agile digital health services.

Gutha Jaya Krishna received his B.Tech. and M.Tech. degrees in Information Technology and Artificial Intelligence from Jawaharlal Nehru Technology University and University of Hyderabad, Hyderabad, India, in 2007 and 2010, respectively. He is currently pursuing his Ph.D. in Computer Science form University of Hyderabad, India and working as Research Fellow at Center of Excellence in Analytics, Institute for Development and Research in Banking Technology, Hyderabad. His research interests include Analytics, Machine Learning and Evolutionary Computing. He is the holder of GATE scholarship for pursuing his M.Tech degree from 2008-2010. He achieved an All India Rank of 38th in Council of Scientific and Industrial Research - National Eligibility Test for Junior Research Fellowship in Engineering Sciences, June 2014. He also achieved an All India Rank of 91 in Joint Entrance Screening Test, Theoretical Computer Science, 2016. He has published his works as two journal articles (Applied Soft Computing; Engineering Applications of Artificial Intelligence), one book chapter and attended five conferences.

Robert Lanquar, Ph.D. (Université Aix-Marseille III & Texas A&M University), after a career of journalist, was in charge of the marketing and products unit in the research department of UNWTO, the UN agency dedicated to tourism. He was a long time in charge of tourism coordinator of the Blue Plan for the Mediterranean and taught in different universities (Paris La Sorbonne Pantheon, Université Libre de Bruxelles, Glion Hospitality and Tourism School). He is still a World Bank expert and professor at the LR Tourism and Hospitality School - Group Excelia.

Nuno Loureiro is full professor at the Polytechnic Institute of Beja - Portugal. He is member of the technical, scientific and pedagogical committee of the master degree of Physical Activity and Health. His research interests include Physical Activity, Health and Environmental Health. Member of the U-Bike Project - Programa Operacional Sustentabilidade e Eficiência no Uso de Recursos, POSEUR-01-1407-FC-000003, 2016 – 2019.

Juliana Machado is a Brazilian student of Digital Design at Senac University Center. She is part of the Research and Application of Technologies for Natural Interaction with Conversational Robots Group at Senac University Center, developing her research on chatbots and affective computing. Together with the Research Group, she contributed to the production of the Edgard Project and published papers about Art and Design in Artificial Intelligence at the International Symposium on Innovation in Interactive Media 2019 and Human-Computer Interaction International 2019.

Umar Mahmud is a Software Engineer having 13+ years of teaching experience in the fields of Software Engineering and Computer Sciences. He has completed BE (Computer Software) in 2003 and MS (Computer Software Engineering) in 2007 from Military College of Signals (MCS), National University of Sciences and Technology (NUST), Islamabad, Pakistan. His research interests are in Context Awareness, IoE, Power Awareness, Software Engineering, Machine Learning and Military History. He has published 15+ papers in journals and as chapters. He is currently engaged as a faculty member in Department of Software Engineering (DSE), Foundation University Islamabad, Rawalpindi Campus (FURC).

Arif Jamal Malik is currently working as an Assistant Professor at the Department of Software Engineering, Foundation University, Islamabad, Pakistan. He did his M.S. and Ph.D. in Computer Science from National University of Computer and Emerging Sciences, Islamabad, Pakistan, in 2008 and 2014 respectively. His research interests include Evolutionary Computation, Swarm Intelligence, Network

Security, and Machine Learning. He has several publications in prestigious international journals and conferences.

Nazir Ahmed Malik did his B. Sc (Hons.) in Naval Sciences from Karachi University and Masters in Computer Science from Uni of Agri, Faisalabad, Pakistan. After his Masters, he has worked as System Programmer for Command, Control and Communication System (C3) at Maritime Headquarters, Pakistan Navy. Thereafter, he did his Masters Leading to PhD (Information Security) from National University of Sciences and Technology (NUST), Pakistan. During his PhD he has been research fellow at Information Security Group (ISG) at Royal Holloway, University of London (RHUL), UK and University of Texas at Arlington, USA. He has worked as Assistant Professor and Head of Distance Learning at Pakistan Navy Engineering College (PNEC), NUST from 2010 to 2014 at Department of Management Information Systems (MIS), Karachi. His research interests are Information Security Auditing, Information Security Management and Policy Making. He has worked on several research projects involving Information Security Audits of various Information Systems and Networks including Data Centers, Security and Privacy in pervasive Computing Environment, Threat Modeling of Context aware systems, Privacy and Consent in pervasive networks. He has published several research papers in reputed international journals and conferences and has also appeared on National TV programs as guest.

Luís Murta is a Professor at the Polytechnic Institute of Beja. PhD in Physical Education and Sport. Coordinator of the Degree Course in Sport, Teacher of the Master in Physical Activity and Health. Research in the areas of sports science, Pedagogy of sport, sports coach training, teacher training, physical activity and active tourism, cycling and bike mobility. Coordinator of the U-Bike Project - Programa Operacional Sustentabilidade e Eficiência no Uso de Recursos, POSEUR-01-1407-FC-000003, 2016–2019.

Alberto Ochoa has a Bs'94–Eng.Master'00; PhD'04-Postdoctoral Researcher'06, Industrial Postdoctoral Research'09 & Energy Studies Postdoctoral Research'18). He joined the Autonomous University of Ciudad Juarez in June 2008. He has written scientific articles in 17 languages. It has 14 books, 87 chapters of books related to AI and more than 227 papers related mainly to logistics for Smart Cities and Social Modeling using different techniques of artificial intelligence. He has directed 47 doctoral theses, 49 master's theses and 57 Bachelor's degrees. Participate in the organization of several international conferences. His research interests include ubiquitous computing, evolutionary computation, natural language processing, social models for a Smart City, anthropometric characterization and social data mining. In 2018 he ended a sabbatical stay of 67 weeks in the INEEL (National Institute of Electricity and Clean Energies) -With a research project focused on optimizing Intelligent Logistics processes in a Smart City. During the first semester of 2018, together with students of the Videogame Programming career, is doing Serious Games for the improvement of Environmental Education in a Smart City. Currently has a social service and inquiry project related to intelligent applications to help children with some type of disability using a Humanoid Nao Robot. During June of 2018 he is attending for the XVII occasion students of the Scientific Summer. During August 2018, it is developing specialized software for Multicriteria Analysis. Since September 2016, he has received the distinction of SNI Level 2.

Luís M. R. Oliveira received the Electrical Engineering diploma, the MSc degree in Electrical Engineering and the PhD degree in Electrical Engineering from the University of Coimbra, Coimbra, Portugal, in 1995, 2001 and 2014, respectively. In 1996 he joined the University of Algarve, Portugal, where he is currently an Adjunct Professor. He is a researcher at CISE - Electromechatronic Systems Research Centre. His research interests include modelling and simulation, fault diagnostics, and protection of power transformers.

Pedro Pacheco is a CEO and founder at Easy Cicle and Urban Cicle Café and Fábrica de Bicicletas do Porto – Enterpreneur in development market strategy and product development in e-mobility, smart city, system integration and bicycle manufacturing. Experience in internacional markets, trading and joint-venture product development.

Alexandra Matos Pereira holds a Ph.D. on Tourism, Leisure and Culture from the University of Coimbra and a Diploma on Advanced Studies in the same scientific area. Her Master in Translation Studies is from the University of Porto as well as her Licentiate Degree in French and English Modern Languages and Literatures. She is Assistant Professor at Universidade Lusófona do Porto and ISLA – Instituto Politécnico de Gestão e Tecnologia and Invited Assistant Professor at University of Coimbra where she lectures Tourism Studies, focusing mainly on the social and cultural aspects of tourism. She served as academic director and executive manager on a private polytechnic institute, in Porto, where she lectured Tourism Studies and Applied English to undergraduate and graduate students. She also tutored internships in tourism, translation and interpretation, and took part in dissertations' defense juries. She has been a full member of organizing and scientific committees of international conferences on Tourism and on Translation, where she has participated as keynote speaker, moderator and reviewer. She is Vice-President of the Scientific Committee of CEPESE - Centre for the Study of Population, Economics and Society, University of Porto, being also Research Associate in the Heritage, Culture and Tourism Research Group. Her research interests are mainly focused on Touring tourism (Ph.D. thesis), Roots tourism, Tour Guiding, Tourism Destination Image, having already published several papers on those areas. She is an editorial board member of several international journals of tourism and hospitality.

Nelson Pinto graduated in Electronic and Telecommunications Engineering Sciences in 2016 and received a Master's degree in Electrical Engineering during 2018, from the University of Algarve. He is an researcher in European funded project Agerar, focusing in "Renewable Energies", "Energy Management and Efficiency" and "Micro and Smart Grids". His main areas of interest are telecommunications, wireless sensor networks and microelectronics.

Vadlamani Ravi is a Professor at the Institute for Development and Research in Banking Technology, Hyderabad since June 2014. He obtained his Ph.D. in the area of Soft Computing from Osmania University, Hyderabad and RWTH Aachen, Germany (2001); MS (Science and Technology) from BITS, Pilani (1991) and M.Sc. (Statistics & Operations Research) from IIT, Bombay (1987). At IDRBT, he spearheads the Center of Excellence in Analytics, first-of-its-kind in India. He has 207 papers to his credit with the break-up of 85 papers in refereed International Journals, 6 papers in refereed National Journals, 94 papers in refereed International Conferences and 3 papers in refereed National Conferences and 19 invited book chapters. He also edited a Book entitled "Advances in Banking Technology and Management: Impacts of ICT and CRM" (http://www.igi-global.com/reference/details.asp?id=6995), published by IGI Global, USA, 2007. Some of his research papers are listed in Top 25 Hottest Articles by Elsevier and World Scientific. He has an H-index of 34 and 5136 citations for his papers (http://scholar.google.co.in/). His profile was among the Top 10% Most Viewed Profiles in LiknkedIn in 2012. He is recognized as a Ph.D. supervisor at Department of Computer and Information Sciences, University of Hyderabad and Department of Computer Sciences, Berhampur University, Orissa. So far, 6 Ph.D. students graduated under his supervision and 2 more are currently working towards Ph.D. So far, he advised more than 70 M.Tech./MCA/M.Sc projects and 50 Summer Interns from various IITs. He currently supervises 2 M.Tech students. He is on the IT Advisory Committee of several commercial Indian for their DWH, CRM and analytics projects He is a referee for 40 International Journals of repute. Moreover, he is a member of the Editorial Review Board for the "SWARM and Evolutionary Computation Journal from ELSEVIER"; Managing Editor, Journal of Banking and Financial Technology, IDRBT and Springer; and 6 other International Journal from IGI Global, USA and Inderscience, Switzerland; Editorial Board Member for Book Series in Banking, Inderscience Switzerland, He is on the PC for some International Conferences and chaired many sessions in International Conferences in India and abroad. His research interests include Fuzzy Computing, Neuro Computing, Soft Computing, Evolutionary Computing, Data Mining, Text Mining, Web Mining, Big Data Analytics, Deep Learning, Social Media Analytics, Privacy Preserving Data Mining, Global/Multi-Criteria/Combinatorial Optimization, Bankruptcy Prediction, Risk Measurement, Customer Relationship Management (CRM), Fraud Analytics, Sentiment Analysis, Churn Prediction in Banks and firms and Asset Liability Management through Optimization. He has 30.5 years of research and 17 years of teaching experience. As part of academic outreach, he is a Guest Speaker in IIM Kolkata's PGP programme and an invited Resource Person in various National Workshops and Faculty development programmes on Soft Computing, Data Mining, Big Data funded by AICTE in various colleges. He is an External examiner for Ph.D. in Auckland University of Technology, New Zealand and some Indian Universities. Further, he is an External Expert to review Research Project Proposals on Analytics and ACRM submitted by the Belgian Academics to Belgian Government for funding. During 2002-2005, he worked as a Faculty at the Institute of Systems Science (ISS), National University of Singapore (April 2002 - March 2005). Earlier, he worked as Assistant Director (Scientist E1) from 1996 -2002 and Scientist C from 1993-1996 respectively at the Indian Institute of Chemical Technology (IICT), Hyderabad. He was deputed to RWTH Aachen (Aachen University of Technology) Germany under the DAAD Long Term Fellowship to carry out advanced research during 1997-1999.

Bheemidi Vikram Reddy received his B.Tech. degree in Electronics and Communication Engineering from Jawaharlal Nehru Technology University, Hyderabad, India, in 2016. He is currently working as Research Associate in Center of Excellence in Analytics, Institute for Development and Research in Banking Technology, Hyderabad. His research interests include Big Data Analytics, Machine Learning and Deep Learning.

Aida Reyes has a PhD in Administration Sciences from the National Autonomous University of Mexico. Master in Administration Sciences and Industrial Engineer in Production, by the Technological Institute of Ciudad Juárez. Full-time research Professor at the Institute of Social Sciences and Administration of the UACJ since the year 2008. It has a 15-year experience in the industry. Member of the Mexican Network of Organizational Studies. Member of the Tourism Studies Network (SECTUR). Member of the National System of Investigators (SNI). Founder of the Association of Academics of Chihuahua. Founder of the Tourism and Landscape Network. The author of books written in co-authorship on the topics of methodology research and environmentalism. A dozen published scientific articles and book chapters. Director of Bachelor's degree in Master's and doctoral studies. Currently, is the Coordinator of the Academy of Research Seminars. Her research interests include sustainability, tourism and leisure, cultural organization and organization studies.

Cristina Portalés Ricart is MSc. in Geodesy and Cartography from the Universidad Politécnica de Valencia and MSc. in Surveying and Geoinformation from Technische Universität Wien (2003). Research fellow at the Mixed Reality Laboratory of the University of Nottingham (2005) and at the Interaction and Entertainment Research Centre of the Nanyang University of Singapore (2006), both in the area of computer vision. In 2008 she received the Ph.D. degree by the UPV. From 2011-2012 she worked at AIDO (Industrial Association of Optics, Colour and Imaging), where she was primarily involved in technical management of the FP7 funded project SYDDARTA. In 2012 she received the Juan de la Cierva post-doctoral grant by the Spanish Ministry to join IRTIC (Research Institute of Robotics and Information and Communication Technologies), where she continued researching in accurate 3D reconstruction. Her current research interests are focused on geometric calibration, image processing, 3D reconstruction, multispectral imaging, augmented reality and human-computer-interaction.

José Vicente Riera was born in Valencia in 1984. He received the Degree in Computer Science Engineering and Technical Telecommunication Engineering by the University of Valencia both in 2009. He has also completed the Master in Advanced Computing and Intelligent System in 2011. In 2008 he received a Collaboration Fellowship from the Ministry of Education and Science, and in 2009 a Research Fellowship from the University of Valencia. Since 2008 he has been involved in a number of scientific and technical projects on virtual reality and different kind of simulators (immersive systems, visual systems, motion platforms, etc). He worked in these projects as researcher at the Robotics Institute of the University of Valencia, and more concretely, at the Artec Research Group of that Institute, where he is actually working. His current research interests include the VR simulation hardware for all kind of vehicles (cars, trucks, heavy machinery, cranes, planes, helicopters, etc), motion cueing algorithms and especially motion platform systems. Also is very interested in Augmented Reality applications and systems.

Jesús Gimeno Sancho received the master degree (2008) and PhD in Computer Science from the University of Valencia (2016), with a dissertation about "Contributions to the Authoring of Augmented Reality Contents for Education, Industry and Construction Sectors". Currently he is researcher at the IRTIC Institute and parttime professor at the University of Valencia. His research interests include augmented reality, virtual reality, motion capture, real time simulation, advanced user interfaces and mobile computing. In the last years he has served also as visiting scientist at the Augmented Reality Group at Bauhaus Universitat (Weimar, Germany, 2008) and at VRAC Center at Iowa State University (Ames IO, USA, 2010).

Diego Sandoval has a Ph.D. Candidate in Management of Sustainable Systems, El Colegio de Chihuahua, Ciudad Juárez, México2018 – present. Master of Science in Manufacturing Engineering, The University of Texas at El Paso. El Paso TX. Master of Science in Industrial Engineering, Instituto Tecnológico de Ciudad Juárez. Bachelor of Science in Industrial Engineering, Instituto Tecnológico de Ciudad Juárez. Relevant experience of Professor of Engineering Management, Environmental Management, Economic Analysis of Environmental Systems, Environmental-sensible Plant Layout and Material Handling. Research activities: Director of 30+ Master thesis, Authored and coauthored 100+ peer-reviewed papers in the areas of Engineering Management, Greenspaces Planning, and Urban Environmental Systems.

Rachel Lerner Sarra is a User Interface and Interaction Designer based in São Paulo, Brazil. She participated in projects for companies such as Itaú, Avon, Visa, Livelo, Via Varejo, ConectCar, Plenum, Heineken, WGSN, Saint Paul, and Veloe. As a member of the Research and Application of Technologies for Natural Interaction with Conversational Robots Group at Senac University Center, she developed the Edgard Project and published papers regarding applied technology in interactive environments for the VI International Symposium on Innovation in Interactive Media and Human-Computer Interaction International 2019. Her recent researches encompass interactive narratives, entertainment, the impact of surprise in experiences, game design, and visual strategies for experience design.

Deniz Sayar received her BID from the Middle East Technical University (METU), and her PhD from Istanbul Technical University (ITU). She is currently an Assistant Professor at Izmir University of Economics (IUE), Department of Industrial Design. Her research interests cover service design, design management, and digitally enabled business model innovation.

Jorge Semião graduated in Electrical and Computer Engineering from Instituto Superior Técnico - Technical University of Lisbon in 1996, and obtained the Master, in 2001, and the PhD, in 2010, also in Electrical and Computer Engineering from the same university, with specialization in Microelectronics and Electronic and Computer Systems. He started working in research in 1995 at INESC, and now he is a senior researcher at INESC-ID in Lisbon. His main research interests are the design and optimization of digital integrated circuits, the design for testability and the test of electronic circuits, the project of fault-tolerant circuits and the aging of electronic circuits. He participated in several research projects (12), and is the author and co-author of more than 90 scientific articles published, four national patents and four patents still under patent pending. Also obtained awards for "Best INESC-ID PhD student in 2010," for "Best Internship in Electrical Engineering" (awarded by the Ordem dos Engenheiros in 2001), and "Best Teacher in the Bachelor of Electrical and Electronics Engineering " (EST - University of Algarve, in 1999). (http://w3.ualg.pt/~jsemiao/).

Evelyn Teran is a student of the Universidad Autónoma de Ciudad Juárez (UACJ) - ICSA Career: Tourism. Work Experience: Texas Ribs Restaurant, Ciudad Juarez, Chihuahua (June 2015– March 2017). Food and Beverage Quick Service "La Cantina de San Angel Inn" Walt Disney World-Epcot (May to October 2017). Food and Beverage "San Angel Inn Restaurant" Walt Disney World- Epcot (October 2017 to April 2018). "La Diana" Restaurant, Ciudad Juarez, Chihuahua (June – November 2018). Internship in the Tourism Department of Chihuahua state "Ah Chihuahua", Ciudad Juarez, Chihuahua (January – current date). Speak to Languages: Spanish and English

Diego R. Toubes is an Associate professor in the Faculty of Administration and Tourism at the University de Vigo (Spain). His research interests include tourism risk management and sustainable tourism. He has published academic papers and collaborations in the fields of flood risk, risk perception and crisis communication.

Lucia Vera studied Computer Engineering at the University of Valencia (UV) (1999), being Extraordinary Prize of her class. She received a Predoctoral Scholarship from the Generalitat Valenciana for 4 years, developing her work in the Artec group of the UV. Shee obtained the Diploma of Advanced Studies (DEA) in the doctoral program "Information Technology, Computing and Communication" of the UV (2001). Later she was Associate Professor in the Department of Computer Science of the UV. She studied the Master in Administration and Project Management of UV and Polibienestar (2011). She is currently a Researcher in the Artec group of the Robotics Institute (UV), an expert in Advanced Multimodal Interfaces, Virtual Reality and Augmented Reality applications, as well as in the simulation and animation of virtual characters for interactive environments.

Mário Véstias received the electrical engineering degree in 1993, and the PhD in electrical and computer engineering in 2002, both from the Technical University of Lisbon, Portugal. He is a Coordinate Professor at the Polytechnic Institute of Lisbon, School of Engineering (ISEL), Department of Electronic, Telecommunications and Computer Engineering (DEETC), where he is responsible for undergraduate and graduate courses on computer architecture and digital systems design. He is also a senior researcher at the ESDA (Electronic Systems Design and Automation) group at the research institute INESC-ID in Lisbon. His current research interests are about architectures for on-chip many-core systems for high-performance reconfigurable computing and for high-performance embedded computing and computer arithmetic.

Noelia Araújo Vila is PhD in Tourism Management and Planning (2013) and Bachelor in Business Administration and Management (2006). Since 2007 she collaborates in the coordination tasks of the University Master in Tourism Management and Planning and teaches as professor in the Departments of Business Organization and Marketing and Financial Economics and Accounting (University of Vigo). She participates in various research projects linked to the tourism and business sector. She is a member of the research group of the University of Vigo Emitur (Marketing and Tourism). She holds a Master in Management and Management of SMEs with a prize for the best business project.

Elizabeth Baoying Wang is a professor in Waynesburg University. She received her PhD degree in computer science from North Dakota State University, Master's degree from Minnesota State University of St. Cloud, and Bachelor's degree from Beijing University of Science and Technology. Her research interests include data mining, data warehouse, bioinformatics, parallel computing. She is a member of ACM, ISCA and SIGMOD. As professional activities, she serves as a reviewer and/or a committee member of many international conferences and journals.

Mohammad Zaheeruddin received his B.Tech. degree in Computer Science and Engineering from Jawaharlal Nehru Technology University, Hyderabad, India, in 2014. He is currently working as Research Associate in Center of Excellence in Analytics, Institute for Development and Research in Banking Technology, Hyderabad. His research interests include Analytics, Machine Learning and Deep Learning.

Index

E

F

G

H

I

L

M

N

O

P

Q

R

S

T

U

V

W

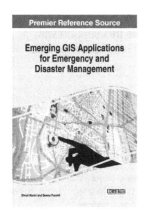

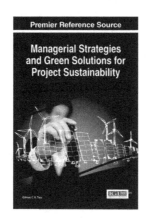

Printed in the United States
By Bookmasters